Play and Exploration in Children and Animals

Play and Exploration in Children and Animals

Thomas G. Power
University of Houston

LAWRENCE ERLBAUM ASSOCIATES, PUBLISHERS
2000 Mahwah, New Jersey London

Lawrence Erlbaum Associates, Inc., Publishers
10 Industrial Avenue
Mahwah, NJ 07430

Cover design by Kathryn Houghtaling Lacey

Library of Congress Cataloging-in-Publication Data

Power, Thomas George, 1954–
Play and exploration in children and animals / Thomas G. Power.

p. cm.

Includes bibliographical references and index.
ISBN 0-8058-2241-0 (cloth : alk. paper) — ISBN 0-8058-2242-9 (pbk: alk. paper)
1. Play—Psychological aspects. 2. Parent and child. 3. Play behavior in animals. 4. Psychology, Comparative. I. Title.
BF717.P69 1999
155.4′18—dc21 99-30955
CIP

Books published by Lawrence Erlbaum Associates are printed on acid-free paper, and their bindings are chosen for strength and durability.

Printed in the United States of America
10 9 8 7 6 5 4 3 2 1

To Brenda, Eric, and Gregory

Contents

Preface

Since the 1930s, children's play has been the focus of intensive empirical research. Although the specific topics examined have varied with the theories and perspectives of the day, the underlying premise has been that play is a "special" context in which unique opportunities for learning occur. In contrast to the learning in more structured situations (e.g., the classroom), learning during play results from children's active attempts to manipulate, replicate, and experiment with their social and nonsocial environments.

Empirical studies of children's play peaked in the 1970s. As described in Rubin, Fein, and Vandenberg's (1983) frequently cited review, researchers during that period had significant optimism about the importance of play for child development. By the 1980s, however, researchers had become more skeptical about the importance of play, and raised many questions about the meaning of previous findings. This skepticism has continued into the 1990s, and research activity has waned.

The study of animal play has a shorter history. Although biologists have long acknowledged that many young animals play (e.g., Groos, 1898), it wasn't until the 1960s that much empirical research was devoted to this topic. Since then a small but growing empirical literature on animal play has emerged. In 1981, Fagen wrote a book entitled *Animal Play Behavior,* which summarized the empirical research in the area and provided a detailed natural history of animal play based on naturalists' observations of young animals in various settings. He also set the agenda for the next two decades of research. Much of the research published since the Fagen volume is reviewed in an excellent book edited by Marc Bekoff and John Byers (1998), entitled *Animal Play: Evolutionary, Comparative, and Ecological Approaches.*

An issue of considerable debate among play researchers has been the relevance of the animal research for understanding the play of humans. Does it help answer important questions about the nature and functions of human play, or is play in animals simply an area of interest in its own right with only tangential relevance? Smith (1982) published one of the few major attempts to integrate the two literatures. Rather than stimulating further attempts to integrate the areas, however, his review seems to have further fueled the debate, with most child development researchers finding the research on animal play "interesting" but not all that relevant.

This book has two purposes. The first is to provide a state-of-the-art review of empirical research on play, in order to summarize what we know, what we don't know, and what we still need to learn. The second is to explore in some detail the relevance of the literature on animal play for understanding the play of humans. The nature and amount of children's play varies significantly across cultures, and an effort to consider cross-cultural work has been made.

The book should be of interest to several audiences: researchers in child development; graduate and upper-level undergraduate students in psychology, human development, and early childhood education; and professionals who work in early child care and clinical settings. The presentation is nontechnical and includes numerous references for further reading and study. Those working with young children should find the information valuable in understanding and developing methods to enrich their play, and those who conduct (or plan to conduct) research should gain a good "starting point" for future research and study.

ACKNOWLEDGMENTS

I would like to acknowledge the support of several individuals. First, I would like to thank my editor, Susan Milmoe, for all of her support, assistance, and encouragement throughout this project. If all authors were blessed with such an editor, there's no question the world would be a much better place. I am also deeply indebted to the production staff at Lawrence Erlbaum Associates, who have made the publication process a positive experience.

I would like to thank Gerald Gratch for our ongoing dialogue about young children and their development. His observations and insights have contributed much to my thinking in this (and many) areas, and his unconditional support has been greatly valued and appreciated. I would also like to thank Coleen Carlson and Marc Bekoff for their comments on parts of the manuscript, and to thank the anonymous reviewer provided by Lawrence Erlbaum Associates for the extensive, thoughtful, and extremely helpful comments on the entire book. I also acknowledge the support of the College of Social Sciences at the University of Houston for granting me a one-semester leave to initiate this project.

Finally, I would like to thank my family for their love and support throughout this seemingly endless process. This book is the result of three long years of immersing myself in a literature far larger than I ever imagined. I am thankful to my wife Brenda for each and every time she convinced me (at least for an hour or two) that I would and could finish the book; for her tolerance for the books and boxes of publications tucked into every nook and cranny of our home; and for her never-ending support in every form imaginable. I also would like to acknowledge the contributions of my sons Eric and Gregory, who, from the ages of 4 to 6, provided me with valuable inspiration and recreation throughout the entire time that Daddy was "working" on play.

—Thomas G. Power

1

Introduction

Probably no activity is as closely associated with childhood as play. Children spend a great deal of their time playing—at home, at school, in parks, in stores, at the dinner table—wherever and whenever the opportunity arises. Although children from different cultural backgrounds may differ in the amount and nature of their play, the major types of play are found across cultures. Play is not limited to humans however; many young mammals and birds also play, often in ways similar to the play of children.

Given the ubiquity of play, it is not surprising that the major developmental theorists have devoted considerable attention to it (e.g., Bruner, 1972; Erikson, 1977; Freud, 1959; Piaget, 1962; Vygotsky, 1967), as have students of animal behavior (Bekoff & Byers, 1981; Burghardt, in press; Fagen, 1981; Martin & Caro, 1985; Panksepp, Siviy, & Normansell, 1984). The study of play, however, is still in its infancy. There is little agreement about many basic issues, and, partially as a result of this confusion, interest in children's play appears to have declined in the last 2 decades. Although a core group of play researchers continues to address important questions, much of the major research in children's play was conducted prior to 1980. Several play-related topics have generated considerable interest since this time (e.g., theory of mind, emotional regulation, attachment), but play has shifted from a topic of interest to a context in which to study other developmental phenomena.

The decline in interest in play as a topic is illustrated by comparing the last two editions of the *Handbook of Child Psychology* (i.e., Damon, 1998; Mussen, 1983). Rubin, Fein, and Vandenberg (1983) wrote a chapter for the 1983 edition that reviewed work through the 1970s—the decade of greatest attention to children's play. In the much-expanded 1998 edition, play had been dropped as a chapter topic. A review of the index of the 4,000-plus-page 1998 edition showed approximately 15 pages referring to play or pretense. Of these, most were short references to the topic, with the only reviews being brief, 2-page sections on the topics of play with child caregivers and the development of pretense. Related items not in the index included exploration, curiosity, and object manipulation, although a short 3-page review of the literature on the manual exploration of objects was found in the text.

This is not to say that play does not continue to be of interest to some child development researchers. An emerging focus is on children's beliefs about play, particularly their understanding of the "fantasy–reality distinction" and their "theory of mind" (e.g., Harris & Kavanaugh, 1993; Joseph, 1998; Leslie, 1987; Lillard, 1993, 1998; Woolley, 1997). The emphasis, however, is more on understanding the development of children's thinking, than on the study of play per se. That is, play is viewed as one of many contexts in which to explore the nature of young children's thinking. Also, researchers outside the traditional child development community—including some anthropologists, folklorists, and sociologists—continue to study play.

There is a natural ebb and flow of interest in topics in child development, and the decline in research on children's play may have been inevitable. However, additional factors may have contributed, including: a growing recognition of the problems involved in defining, identifying, and studying play (e.g., Smith, Takhvar, Gore, & Vollstedt, 1985); research demonstrating that earlier arguments about the importance of play for development may have been overstated (Martin & Caro, 1985); and the belief—usually unstated—that play is not a serious topic for scientific inquiry (Rubin et al., 1983). My goal in this book is to attempt to rekindle an interest in children's play among child development researchers by offering a comprehensive and current review of the literature, and by demonstrating the value of an evolutionary approach to its understanding. By identifying what is known and still needs to be known about children's play, I hope to point out future directions for research.

UNANSWERED QUESTIONS ABOUT CHILDREN'S PLAY

Despite the numerous books and articles that have been written on play, many important questions remain unanswered. Peter Smith identified a number of these in his 1982 review. These and several others are briefly considered here.

Is Play a Distinct Motivational/Behavioral Category?

One problem with research on play is that some researchers and theorists do not even agree whether play exists! Although this controversy more often arises in research on animals than on humans, the issues are relevant equally to animal's and children's play. The two major arguments against treating play as a separate behavioral category are: There is nothing unique about behaviors labeled as "play" that cannot be described and accounted for by other, more basic principles of animal and human behavior (e.g., Lazar & Beckhorn, 1974; Schlosberg, 1947); and the category "play" is not useful because it refers to a wide range of behavior patterns that have little in common in either their structure or the psychological and physiological processes involved in their control (e.g., Berlyne, 1969). Moreover, even researchers who find the term *play* useful for describing certain patterns of behavior sometimes argue that play may not represent a distinct motivational system (e.g., Pellis, 1991).

The problem in demonstrating that play exists lies at least partially in its definition. Because play is often defined as behavior that is intrinsically motivated and not the consequence of external pressures or appetitive drives (see Rubin et al., 1983, for a review), the skeptic can always identify possible external motivators or immediate biological needs that account for a behavior pattern thought by others to be the result of intrinsic motivation.

Are Comparisons Between Animal and Human Play Meaningful?

Nothing distresses some writers more than examining the behavior of other animals for further insights into the behavior of humans. Although they may find evolution a useful concept when it comes to understanding other animals, they often hold the belief that the evolution of higher level cognitive functioning in humans (particularly language and the development of reason) makes the evolutionary analysis of human behavior inappropriate. They argue that because humans can think abstractly, communicate verbally, and engage in complex forms of self-regulation, we are so different from our animal relatives that cross-species comparisons of behavior patterns are meaningless. In fact, some theorists have gone so far as to argue that any comparative analysis of human behavior is of questionable utility and even irresponsible (e.g., Hirsch, 1976; Lerner & von Eye, 1992).

These arguments have been made regarding play. Even though many forms of children's play are similar to the play-fighting, play-chasing, and object play of other animals—especially other primates—some feel that these similarities are superficial and can bring about misleading conclusions. Even some researchers who began their careers studying children's play from an evolutionary perspective report an increasing awareness of the limitations of the comparative approach (e.g., Vandenberg, 1985).

What Exactly Is Play?

The problems in defining play and differentiating it from other behavior patterns have been well documented (e.g., Fagen, 1981; Martin & Caro, 1985; Smith et al., 1985). The main problems are that: Although numerous criteria for defining play have been identified, given the wide range of play behaviors none of the criteria alone, or in specific combination, reliably differentiate between play and nonplay; and although play is usually defined as behavior that lacks an immediate purpose or benefit, it is difficult to make this determination without fully understanding the behavior under study—something troublesome to accomplish without a satisfactory definition with which to start.

Rubin et al. (1983), in their comprehensive review of the literature on children's play, argued that there are three kinds of play definitions: play as disposition, play as observable behavior, and play as context. Under the dispositional category, six features of play were identified: intrinsic motivation, attention to means rather than ends, active engagement of the organism, freedom from externally imposed rules, nonliterality, and behavior dominated more by the organism than by the environment. Under the topic of play as observable behavior, Rubin et al. described various play taxonomies that have been used in defining children's play, including Piaget's (1962) distinction among practice play, symbolic play, and games with rules. And finally, under the topic of play as context, Rubin et al. described the characteristics of settings most likely to elicit children's play. These include: availability of engaging materials or play partners, freedom of choice granted regarding the child's behavior, minimal adult direction, a secure atmosphere, and scheduling at a time when appetitive biological needs (e.g., hunger, sleep) have been met.

Bekoff and Byers (1981) and Fagen (1981) discussed many of the features that have been used to define play in animals. A comparison of these features reveals many similarities with the discussion of Rubin et al.

(1983). Play, according to most researchers who study animals, consists of behaviors similar to those in other nonplay contexts (e.g., foraging, fighting, or prey capture) but that occur out of context, are triggered by different stimuli, may be directed toward substitute objects, and/or are modified in their structure. Compared to behaviors in functional contexts, behaviors during play are often exaggerated, reordered, incomplete, brief, repeated, varied in sequence, inhibited, and/or otherwise transformed. Moreover, play is apparently not motivated by appetitive drives or extrinsic goals, and as a consequence lacks the consummatory acts and biological consequences of functional behavior. Fagen noted that animals will search for a play partner and that play appears pleasurable to the performers. Finally, Fagen wrote that play occurs when general environmental conditions are supportive or benign, and is rarely seen is situations involving subjective uncertainty or fear. Young animals in populations experiencing food shortages, for example, often show very low levels of play (e.g., Baldwin & Baldwin, 1976; Barrett, Dunbar, & Dunbar, 1992; Lee, 1984; Sommer & Mendoza-Granados, 1995). Other environmental stressors—such as death of a group member, relocation to a new environment, or introduction of a new animal to a group—can also decrease the frequency of play (Merrick, 1977).

Because few examples of play include all (or even most) of these features, and one or more of these characteristics are often found during nonplay, Martin and Caro (1985) argued that play is "polythetically" defined, and therefore difficult to define by its individual attributes alone. Fagen (1981) pointed out that these definitional issues are not unique to play. Many well-studied areas of animal behavior—including aggression, sexual behavior, and cooperation—also lack satisfactory definitions.

What Are the Effects of Play on Development?

A factor that apparently drove much of the research on children's play in the 1960s and 1970s was the widely held (and by no means new) view that play was important for child development. Influenced largely by Freudian theory and Piaget's observations of the behavior of infants and children, researchers believed that play contributes greatly to children's emotional and intellectual growth (e.g., Singer, 1973; Smilansky, 1968; Sylva, 1977). This belief was reinforced by observations in the late 1960s that children from impoverished backgrounds showed deficits in their play behavior (particularly pretend play) and other areas of social and intellectual functioning (e.g., Rosen, 1974; Smilansky, 1968).

Although there was (and continues to be) more skepticism about the importance of play among researchers who study animals, early researchers studying animal play often showed equal levels of optimism (e.g., Aldis, 1975; Birch, 1945; Poirier & Smith, 1974). This belief was based on the observation that despite many risks of play (e.g., injury, alerting predators, energy expenditure), the young of so many species devote a great deal of time and energy to this activity. Therefore, play could have evolved only if it served some important biological function(s) in influencing the rate of survival and, ultimately, success in later reproduction.

Biologists who write about play make the distinction between its functions and its effects (e.g., Allen & Bekoff, 1995; Symons, 1978). Put simply, the term *function* is reserved only for those consequences of a behavior pattern that led to its selection during the course of evolution; the term *effects* refers to any additional consequences that may have resulted. For example, play-chasing may have evolved because it provides practice in the skills involved in predator avoidance (its *function*), but it may also provide a good deal of aerobic exercise. Its contribution to the general physical fitness of the individual, in this case, is labeled a beneficial *effect*. Because it is virtually impossible to prove that a given behavior pattern serves a given evolutionary function, in practice it is difficult to separate functions and effects (see Buss, Haselton, Shackelford, Bleske, & Wakefield, 1998, for a recent discussion of these and related concepts).

Martin and Caro (1985) pointed out that there have been three major empirical approaches for examining the functions and effects of play: experimental, correlational, and optimal design studies. More specifically, these involve:

> (1) Experimental manipulations of early play experience, for example, to give fewer or more opportunities for particular types of play; (2) searching for correlations between play and other forms of behavior, exploiting natural variations within populations; and (3) using the indirect "argument from design" approach, in which the observed structure of play is matched against the requirements of its hypothesized function. (p. 66)

Although the experimental approach provides the strongest evidence for the effects of play (experimental documentation does not necessarily demonstrate function), it has been underutilized, and most studies take the second or third approach. This makes it difficult to draw causal inferences.

In spite of the early enthusiasm about play in both children and animals, research in the late 1970s and early 1980s showed that many of the presumed functions and effects of play were hard to demonstrate empirically. Several researchers identified methodological problems in the earlier studies of play, and conducted carefully designed experiments that showed little or no benefits of play experience (e.g., Caro, 1980; Simon & Smith, 1983; Smith, Simon, & Emberton, 1985; Smith & Witney, 1987; Vandenberg, 1990). Caro (1980), for example, found that providing cats with extensive opportunities for object play did not increase their success in prey capture, and Smith and associates (e.g., Simon & Smith, 1983) found that the often-documented association between play and problem solving in children may have been the result of subtle experimenter effects. Moreover, more recent studies of energy expenditure during play raised the possibility that play was not as costly as is commonly assumed, although this issue is far from being resolved (Bekoff & Byers, 1992; Martin, 1982, 1984b; Miller & Byers, 1991; Siviy & Atrens, 1992). Although most of these researchers called for further studies of the effects and functions of play, their critical methodological reviews may have had the unintended effect of discouraging further research on this topic.

PURPOSE OF THIS BOOK

For each of five major play categories—solitary object exploration/play, play-fighting, locomotor play, social object/social pretend play, and parent–child play—relevant evidence from the animal literature is reviewed from an evolutionary perspective to identify any possible functions or effects. Because the ultimate purpose of this book is to better understand the play of human children, the play of primates—especially the apes—is considered in the most detail. When deemed appropriate, however, the play of other species is also considered, in order to develop hypotheses about how a given behavior pattern evolved.

In each section, the review of the animal literature is followed by a review of the empirical literature on children's play in Western, industrialized cultures (in which the vast majority of play studies have been conducted), and finally by a review of the research on children's play in non-Western cultures. Cross-cultural comparisons, to the extent they can be made, are critical, because if the nature of children's play reflects the operation of evolutionary factors, then commonalties across diverse human cultures should be identified. Data on children's play in hunting-and-gathering cultures are particularly important in this regard because these are the settings in which human behavior evolved (Eibl-Eibesfeldt, 1989).

ASSUMPTIONS OF THE PRESENT APPROACH

The major assumptions guiding the discussion in this book are that play is a distinct motivational/behavioral system that has evolved in humans and other animals, and that patterns of children's play in a given culture reflect the influence of both biological evolution and cultural factors. Although some behavior sequences are easier to identify as play than are others, the term *play,* as used in this book, refers to the occurrence of behaviors similar to those in seen functional contexts (e.g., foraging, fighting) but occurring out of context and having some of the structural modifications listed earlier. Because it covers a diverse assortment of behaviors, more specific definitions of play are offered in the chapters that follow.

There are further assumptions underlying this text. As has been suggested by others (e.g., Baldwin & Baldwin, 1973; Martin & Caro, 1985), play is not *essential* for normative development, but instead has evolved as one pathway through which individuals can learn about their environment, establish relationships with others, and practice and refine skills that facilitate their survival or otherwise increase their reproductive success. Moreover, as suggested by Fagen (1981) and others, play serves as a source of behavioral innovation within a given population. Additionally, play may have evolved to serve short-term and/or long-term functions (e.g., Barber, 1991; Martin & Caro, 1985), and its effects on development may be general and/or specific (depending on the species and the type of play under consideration).

A second set of assumptions concerns the relationship between play and development. From infancy to middle childhood, play both reflects and contributes to the development of various motor, cognitive, emotional, and social systems. For example, 16-month-olds who repeatedly drop food to the floor during dinner (e.g., Piaget, 1952) do so in a way that both reflects and contributes to their developing understanding of cause and effect. Similarly, fantasy play with peers both reflects and contributes to the child's developing social cognitive skills. Although there has been considerable debate about whether play contributes to or reflects development, the path of influence is usually bidirectional.

Finally, because play (even solitary play) develops in a social context, cultural practices and individual life experiences have a significant impact on the amount and kind of play that children exhibit. Parents, siblings, friends, neighbors, relatives, teachers, and others all may serve as play partners and contribute greatly to the development of children's play. This does not mean that their contributions are purely cultural—evolution may have led to the development of behavior patterns that influence the initiation and maintenance of social play—both in children and in their playmates. This point illustrates the futility of trying to separate "environmental" and "biological" influences—even though patterns of social play may have resulted from the operation of evolutionary processes, the degree to which their functions are realized is a function of the responsiveness of the social environment.

DIFFICULTIES WITH THE PRESENT APPROACH

Any time one takes an evolutionary approach to human behavior, one encounters a number of problems. First, the data available to investigate and assess evolutionary hypotheses are limited, in that no record of the behavior of our extinct biological ancestors exists. Unlike paleontologists or archaeologists who have access to fossils and physical artifacts to derive and test their hypotheses, play researchers have few physical traces at their disposal to give hints about the behavior of young prehistoric humans. Additionally, our biological relatives that would be the most important to study (both primate and nonprimate) are extinct. Moreover, the environments in which play behavior evolved have changed, and even the few hunting-and-gathering cultures that currently exist likely differ considerably from those in which play evolved.

The best the play researcher can do is to observe the behavior of living animals that vary in their biological relatedness to each other and to humans, and then, based on knowledge of their biological ancestry, draw tentative hypotheses about evolutionary processes. In addition, because human beings and their closest ancestors have spent over 99% of their time on the planet in a hunting-and-gathering existence (Eibl-Eibesfeldt, 1989), the behavior of current hunter-gatherers can be studied (albeit taking into consideration whenever possible the impact of recent contacts with industrial cultures). Finally, similarities and differences in human behavior across diverse contemporary cultures can be studied in order to draw hypotheses about biological and cultural evolution.

Although this enterprise may seem highly speculative (see Latour & Strum, 1986), whole theories of the nature of prehistoric civilizations, or even the origin of the universe, are based on piecing together similar bits of empirical evidence. Archaeologists, for example, often must construct theories of the social structure, subsistence practices, and religious beliefs of prehistoric cultures from a few shards of pottery; the place in which they were found; and the tools, bones, and ashes that surround them.

A major problem with comparative analyses is the difficulty of distinguishing between behavioral homologies and analogies. A homology is a similarity in the behavior of members of two related species that is the result of evolution from a common ancestor (Atz, 1970). A behavioral analogy is a similarity that is the result of parallel, independent evolution. Thus, for example, the fact that both lion and human infants suckle from their mothers to obtain nourishment is probably a homology, whereas the fact that both ants and humans build multi-room dwellings is likely an analogy. Because it is difficult (if not impossible) to prove that a given similarity in behavior patterns is a homology, some critics (Lerner & von Eye, 1992) have argued that the comparative approach is of limited utility for studying humans and should be abandoned. According to Lerner and von Eye (1992), "The mere portrayal of behaviors in two species as appearing comparable is no proof at all of their common evolutionary heritage. Nor is it any proof at all regarding the extent to which such behaviors are genetically constrained or produced," and "to claim that such descriptive analogies are indicative of common evolutionary histories is, at best, naïve, and at worst, poor scholarship" (p. 21).

Although these authors are unquestionably correct in their first two assertions (proving homology is one of the most difficult tasks in behavioral biology), to argue that one should not take the comparative approach to the study of humans because of this difficulty is short sighted. Accepting this argument would result in ignoring a wealth of potentially relevant information for understanding the biological bases of human behavior. By comparing similarities and differences in the behavior of various species as a function of their biological relatedness, one can develop, test, and reject hypotheses about the evolution of individual behavior patterns. Although we may never be able to prove how a given behavior pattern evolved, psychologists historically have been limited in their ability to prove much of anything about human development. Ethical constraints greatly limit the use of experimental methods with humans, making it impossible to prove, for example, that parental childrearing practices influence children's emotional adjustment, or that environmental stimulation in infancy influences cognitive growth. The comparative approach is simply another correlational method that (when used in conjunction with other methods) allows researchers to develop and refine theories of human behavior. Although comparative analyses that consider only a single pair of species are limited, analyses that include a wide range of related species are much less so. This is the approach I use in this current book.

A final problem with both comparative and cross-cultural analyses is the nature of the available data. Much of what we know about the behavior of many species comes primarily from observations of animals in captivity. These observations may reveal little about animal behavior in natural environments (e.g., Stevenson, 1983), and reliance on such reports can lead to faulty conclusions about evolutionary processes. Likewise, children's behavior has not been extensively studied and documented outside of Western industrialized cultures, and the data on children in other cultures range in both quantity and quality. They are often anecdotal in nature, and derive from single observers spending a short period of time in single places to study something besides children's play.

OVERVIEW

To reiterate, the purpose of this book is to review the literature on animal and children's play from an evolutionary perspective. Given the voluminous literature on children's play, the scope had to be limited. Although researchers have documented the ways in which the nature of children's play varies as a function of their physical or mental health status, only studies of healthy, "normally developing" children are reviewed here; studies of play in children with chronic illnesses, sensory or cognitive deficits, or psychopathology are omitted. The scope is also limited to young children—primarily infants, toddlers, preschoolers, and children in the early grade-school years. This age range was chosen because it is the period during which children spend the greatest amount of time playing. Finally, again due to space limitations, studies of environmental effects on play (e.g., neighborhoods, play materials, playground design) are excluded. Types of animal play that receive little attention are precocious sexual behaviors (e.g., Kramer & Burghardt, 1998) and allomothering or play-mothering behavior (e.g., Lancaster, 1971)—the former because empirical research on the normative sexual play in young children for comparison is virtually nonexistent, and the latter because it is not clear whether such behavior is play or simply early caregiving behavior (Fagen, 1981).

Chapters 2 through 9 are devoted to the five types of play listed earlier: solitary object exploration/play, play-fighting, locomotor play, social object/social pretend play, and parent–child play. For each type of play, structure is considered before functions and effects. Without a description of what animals or children actually do during play, it is difficult to consider hypotheses about functions or effects. Because of the focus on the play of infants, toddlers, and young children, the literatures on organized sports (e.g., soccer, baseball) and rule-governed children's games (e.g., hide-and-seek, marbles) are not reviewed. Although children usually begin to engage in such formal, rule-governed activities between the ages of 4 and 6, these activities do not dominate play until middle childhood (Piaget, 1962). A review of the voluminous literature on sports and games (much of which comes from the fields of anthropology and leisure studies) is beyond the scope of this book.

What is presented is empirical research on children's exploration and play. Beyond attempting to illustrate the value of an evolutionary approach, this book does not set out to evaluate the numerous play theories that have been offered over the years. (An excellent review of these can be found in Rubin et al., 1983.) Children's exploration and play is much too diverse to be adequately accounted for by any single theory, so research conducted from a wide variety of theoretical perspectives is considered here. As is typical in the discipline of psychology today, "mini-theories" devoted to specific content areas characterize most current thinking about children's play, and no single theory dominates the field.

Three issues that cross-cut the review of structure, function, and effects receive considerable attention throughout this book: the developmental course of children's play, gender differences in play behavior, and the impact of children's play partners.

In Part I (chaps. 2 and 3), the literature on solitary object play is reviewed. Because of the difficulties inherent in differentiating among various forms of object use—especially in the young (Weisler & McCall, 1976)— four areas of research are considered: studies of undifferentiated object manipulation (including research on curiosity and Piagetian sensorimotor development), object exploration, object play, and tool-use. A new scheme for understanding the relation among these forms of object manipulation is then presented.

The discussions of play-fighting and locomotor play in Part II (chaps. 4–7) focus on species similarities and differences in play, including play signals, play behaviors, the initiation and termination of play, and differences in behavior during play and more "serious" contexts (e.g., aggression, prey capture, and prey avoidance). The developmental course of play, play partners, and gender differences in play are emphasized. Part II highlights many similarities in the structure and possible functions of these play types in humans and in other animals.

In contrast, chapter 8, on social object and social pretend play, highlights many more differences than similarities between children's and animals' play. Specifically, the advanced symbolic and language abilities of humans provide for the development of elaborate forms of sociodramatic play and games with rules that are not found in even our closest animal relatives. Given significant changes in the nature of children's play during the third year of life, the literatures on preschool children and infants and toddlers are reviewed separately. Because many studies of children's prosocial behavior, social skills, sociometric popularity, conflict resolution, dominance hierarchies, aggression, and sibling relationships involve observations of children in "free-play" contexts, much of this research is reviewed as well, along with more traditional "play" topics such as sociodramatic play and gender differences in play activities.

Because play with parents differs significantly from play with peers or siblings, a separate chapter, chapter 9, is devoted to it. The review of parent–child play during infancy is presented developmentally, using Sander's (1962) periods of mother–infant interaction as an organizing scheme. Because it is often difficult to distinguish between play and other affiliative interactions between parents and infants (especially during the first few months of life), the entire literature on parent–infant interaction in noncaretaking contexts is reviewed, including a selective assessment of studies of face-to-face interactions, object mediated play, parental language, and so on. In the section on gender differences, differences between the play of mothers and fathers, and between their play with girls and with boys, are considered. Finally, studies of direct and indirect parental influences on children's play are reviewed, considering separately each of the types of play considered in chapters 2 through 8. Chapter 10 offers some conclusions, suggests directions for future research, and lays out the implications of play research for professional service and practice.

I

Solitary Object Exploration and Play

From the time that they can grasp objects and bring them to their mouths to suck, infants in many cultures spend a great deal of their time exploring, playing with, and otherwise manipulating objects. This is particularly true during the second half year of life, when infants seem bent on putting their hands (or mouths) on everything that they possibly can. Although this is the beginning of an ongoing battle for parents (who quickly learn the limits of their babyproofing skills), it reflects the active attempts of infants to learn about the physical world around them through their actions and their senses. Numerous theorists have written about the intense interest of infants and young children in everyday objects (e.g., Hunt, 1965; Piaget, 1952; Wenar, 1976; White, 1959), and toy manufacturers have benefited greatly from parents' (and especially grandparents'!) awareness of this newfound interest.

Although human children show the greatest levels of object manipulation and play, other animals (especially some birds, carnivores, and primates) also play with objects. Biologists have offered several evolutionary explanations for the existence of object play in animals, many of which may apply equally well to humans.

A problem with the research on object exploration and play is that it is often difficult to differentiate between the playful and nonplayful manipulation of objects (Weisler & McCall, 1976). At least three forms of object manipulation have been discussed in humans and other animals: object exploration, object play, and tool-use. Numerous studies of undifferentiated object manipulation have been conducted as well, making a review of this literature difficult.

Because of the sometimes subtle differences among object exploration, play, and tool-use, all three behavior patterns are considered in this section.

Given the purpose of this volume, the review of research on tool-use is limited to a consideration of its relation to object exploration and play. The literature on nonhuman animals is considered first, followed by the literature on humans. Each chapter has two major parts: the first devoted to the structure of object-directed behavior, and the second devoted to empirical research on its functions and effects.

2

Solitary Object Exploration and Play in Animals

According to biologists, selection pressures have led to the evolution of object exploration, play, and use in many animal species. By investigating and acting on objects, animals can increase their likelihood of survival by finding or constructing shelters, procuring food, identifying or protecting themselves from danger, marking their territories, and so on. Behavior with objects is found across the phylogenetic spectrum, from the construction activities of ants to the shell exploration of hermit crabs to the nest building of birds to the nut cracking of chimpanzees.

Despite the apparent similarities in object use across species (e.g., shelter construction or food retrieval), a closer look reveals that species vary in the specificity of their adaptations (Beck, 1980; Morris, 1964; C. Parker, 1978). Members of specialist species that exploit objects, such as some invertebrates, show patterns of object-directed behavior that are well adapted to the particular environment in which they evolved, but are of limited usefulness outside these settings. In contrast, members of generalist species, especially the primates, show flexible patterns of object use that enable them to adapt to a wider range of environmental conditions (Napier & Napier, 1985; Poirier & Smith, 1974).

Species also differ in the sensory modalities devoted to object use. Nocturnal mammals—including the tarsiers and many of the prosimians (the suborder of primates, including the lemurs and lorises; see Fig. 2.1)—rely primarily on their senses of smell and hearing when exploring the environment, whereas the diurnal anthropoids (the suborder of primates that includes the New and Old World monkeys, apes, and humans; see Fig. 2.1) rely primarily on visual and manual exploration (Napier & Napier, 1985).

Differences between the physical structures of these groups of primates correspond to patterns of behavioral adaptation. For example, apparently to facilitate survival in a low-light environment, most prosimians have large mobile ears that enhance their auditory sensitivity, and moist snouts on long muzzles that increase their sense of smell. Their visual abilities are limited in that they see in black and white, and their slightly side-facing eyes limit the stereoscopic abilities needed for accurate depth and distance perception.

Most anthropoids, who have adapted to a diurnal (i.e., daytime) existence, possess color vision, and have front-facing, mobile eyes that give them fully developed stereoscopic vision. They exhibit numerous adaptations for the manual exploration of objects. For instance, Old World monkeys, apes, and human beings have truly oppositional thumbs and are able to employ both precision (index finger with thumb) and power (whole-hand) grips. Apes and humans have complex fingerprints with multiple ridges that provide tactile sensitivity. However, as diurnal organisms, the senses of smell and hearing are less well developed in the anthropoids. Compared to prosimians, most anthropoids have shortened muzzles, dry noses, and small, virtually immobile ears (Napier & Napier, 1985).

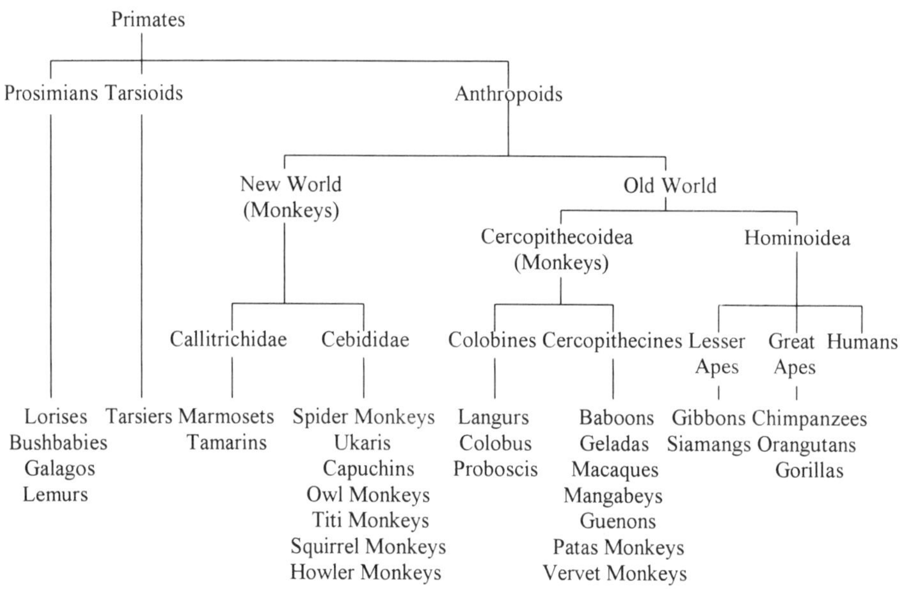

FIG. 2.1. Classification of primates.

The sections that follow review the empirical research on object exploration, manipulation, and use in nonhuman animals. Because the ultimate purpose of this review is to increase our understanding of object-directed behavior in humans, research on olfactory investigation and research on the exploration of large-scale environments are not reviewed. We focus primarily on nonhuman anthropoids (see Fig. 2.1), because members of this suborder most closely resemble humans in both the structural adaptations that have evolved for object use and the functions of these behaviors for the organism.

MODES OF OBJECT MANIPULATION

Biologists, psychologists, and others commonly use three terms to describe animals' behavior with objects: *exploration*, *play*, and *tool-use*. Although a particular instance of object-directed behavior may fit into more than one of these categories (Weisler & McCall, 1976), these distinctions are useful in that they describe behavior patterns with apparently very different functions.

Exploration

Researchers usually define *exploration* as behavior that reduces uncertainty about the external environment through information gathering (Cowan, 1983; Renner, 1990; Voss & Keller, 1983; Weisler & McCall, 1976). Exploration typically occurs upon an animal's initial exposure to an object or environment, or when some change in the environment has occurred (e.g., Berlyne, 1950; Inglis, 1983; Montgomery, 1953). Exploration, however, is not the only common response to novelty—fear and avoidance can also occur, especially if the novelty is too great (e.g., Barnett, 1958a; Menzel, 1962; Montgomery, 1955; Murphy, 1978). The degree of neophobia varies with the ecological plasticity of the species as well (Greenberg, 1990).

Exploration could not be easily explained by the early learning theorists (e.g., Dashiell, 1925; Nisson, 1930) because it is not associated with an obvious, primary biological need such as eating, drinking, or procreating. In fact, animals often delay eating when hungry or risk bodily harm in order to explore (Berlyne, 1966). Generally, the opportunity to explore can serve as a powerful reinforcer (e.g., Girdner, 1953; Montgomery & Segall, 1955). The fact that animals explore for no obvious reason also was inconsistent with early homeostatic theories of motivation, which postulated that behavior in the absence of pressure from primary drives moves toward a state of minimal stimulation and excitement (e.g., Miller & Dollard, 1941).

In the 1950s, theorists addressed the lack of a clear motivational basis for exploratory behavior by proposing that animals have an innate exploratory or manipulatory drive (e.g., Harlow, Harlow, & Meyer, 1950; Montgomery, 1954). In direct contrast to earlier views, such theorists argued that exploratory behavior results from a "stimulus hunger" (Barnett, 1958b), and that organisms seek out stimuli "for their own sake" (Berlyne, 1966) or to reduce a "boredom drive" (Myers & Miller, 1954). Although more recent theories have moved away from drive notions and now focus on cognitive processes such as discrepancy reduction and mental representations (e.g., Inglis, 1983; Lowenstein, 1994; Toates, 1983), they still share the assumption that exploratory behavior has a strong motivational basis independent of the satisfaction of primary biological needs. Exploration is now viewed in evolutionary terms, because of its likely survival functions such as locating food, avoiding predators, and stimulating learning (Barnett, 1958b; Cowan, 1983; Inglis, 1983; Renner, 1990). Such benefits, however, must be weighed against the risks of energy expenditure and detection by predators (Birke & Archer, 1983; Renner, 1990).

Object Play

Hutt (1970b) argued that play with objects differs both structurally and functionally from object exploration. Based on Berlyne's (1960, 1966) distinction between specific and diverse exploration, Hutt described motivational, stimulus, and response factors that differentiate object exploration and play. Exploration is behavior directed toward answering the "implicit query" of "What does this object do?" It is elicited by novel stimuli, and consists of prolonged, stereotyped sequences of behavior involving synchrony of the visual and tactile receptors. It is accompanied by an intent facial expression and declines over time.

Object play answers the implicit query of "What can *I* do with this object?" Play usually follows investigation, and is made up of brief, variable, and often idiosyncratic sequences that involve desynchrony or only transient synchrony of the receptors. It is accompanied by a relaxed facial expression and is a quadratic function of time.

Other definitions of object play (e.g., Fagen, 1981; Lancy, 1980; Loizos, 1969; Weisler & McCall, 1976) are similar to Hutt's (1970b). Although theorists provide different lists of criteria, most agree that object play consists of behaviors similar to those used in other contexts (such as foraging or prey capture), but that the behaviors occur out of context and are modified in structure. As described in chapter 1, behaviors during object play are often exaggerated, reordered, incomplete, brief, repeated, varied in sequence, and/or otherwise transformed, and they lack the biological consequences of functional behaviors.

Although Hutt's (1970b) distinction between exploration and play is based on differences between behaviors that are information seeking versus stimulus seeking (Berlyne, 1960, 1966), this may be an oversimplification. Beck (1980), for example, wrote that "in both [exploration and play], the animal appears to acquire information about the object that would promote its potential use as a tool" (p. 151). Moreover, Weisler and McCall (1976) argued that it is often difficult to separate exploration and play in practice. When manipulating objects, animals often shift rapidly between exploratory and play behaviors in the ongoing stream of behavior, making it hard to determine when one stops and the other starts. Fagen (1981) found concerns about dividing the behavior stream unconvincing: "The validity of this criticism is questionable. Would we include feeding and predator avoidance in the same causal category if a foraging baboon interrupted its search for edible roots every five seconds in order to scan for predators in the vicinity?" (p. 59).

As described in the first chapter, the motivation behind object play has long puzzled theorists, because it is behavior that, by definition, appears to have no apparent short-term functional goal. In fact, a number of costs of object play have been identified, including energy expenditure, injury, exposure to predators, and loss of time that might otherwise be devoted to more immediate survival-related activities (Fagen, 1981). Evolutionary analyses of object play generally conclude that its long-term benefits outweigh its short-term costs. The major evolutionary functions (or effects) of object play that have been hypothesized include: (a) providing general information about the properties of environmental objects, such as hardness, flexibility, ease of handling, and durability; (b) providing practice in the motor skills involved in various predatory and food acquisition behaviors, including the use of tools; (c) providing for the occurrence of novel behavioral sequences that serve as the raw materials for developing innovative forms of tool-use; and (d) promoting sensorimotor maturity, cognitive development, and general behavioral plasticity (Beck, 1980; Bruner, 1972; Fagen, 1981; Vandenberg, 1978).

Tool-Use

Although many definitions of tool-use have been offered (e.g., Alcock, 1972; S. Parker & Gibson, 1977; van Lawick-Goodall, 1970), one of the most complete, although "grammatically and phonetically tortuous" (Beck, 1980, p. 10), was offered by Beck (1980): "...tool-use is the external employment of an unattached environmental object to alter more efficiently the form, position, or condition of another object, another organism, or the user itself when the user holds or carries the tool during or just prior to use and is responsible for the proper and effective orientation of the tool" (p. 10).

Tool-use, therefore, involves using objects in the environment to more easily achieve specific behavioral goals. Like object exploration—and unlike object play—tool-use is goal directed behavior that has a clear immediate function: The tool is used as a means to an end. Because it is impossible to unequivocally deduce an animal's motives from observation of its behavior, it is sometimes difficult to differentiate tool-use from other forms of object manipulation (especially when an attempt at tool-use is unsuccessful and the assumed goal is not achieved). However, upon repeated observation of different members of a species, common patterns of tool-use usually can be identified and differentiated from nontool forms.

Vandenberg (1978) offered a perspective on object-directed behavior that combines all three specific forms of object manipulation discussed previously. In it, he argued for the existence of an "exploration-play-application [i.e., tool-use]" (p. 724) sequence: "The exploration-play aspect of the sequence is more likely to occur when the animal is confronted with a significantly foreign object or situation that it must first explore and determine if it is safe. Having found it safe, the animal is then likely to play" (p. 735). Thus, play follows exploration. Additionally, play "develops an array of potential skills and strategies, some of which may later be utilized by the organism in a more 'serious' [i.e., tool-use] context" (p. 735).

Therefore, play serves an important role in the development of tool-use. This is a commonly held assumption about the importance of play (e.g.,

Bruner, 1972). Although it was once believed that the use and manufacture of tools is what separates humans from other animals, research has identified patterns of tool-use in a wide range of species, from insects to primates (see Beck, 1980, for a comprehensive review). Among the nonhuman primates, the great apes—especially the chimpanzees—are the most prolific tool-users.

STUDIES OF UNDIFFERENTIATED OBJECT MANIPULATION

Unfortunately, most studies of object use in animals have not differentiated between the three modes of object manipulation. In these studies, the researchers code and describe animals' behavior toward objects with little consideration of their function. Instead, they simply provide information on the total amount of time that animals spend with objects, or on how much time is devoted to specific, object-directed behaviors. Beyond some anecdotal, qualitative data, the timing or sequencing of behaviors is not addressed. Therefore, a touch is a touch, whether it occurs during the initial investigation of the object or after the object has become familiar. Although most of these studies involve animals in captivity (e.g., Glickman & Sroges, 1966; Parker, 1974a, 1974b; Torigoe, 1985), occasionally studies involve naturalistic observations of interactions with objects in the wild (e.g., Baldwin & Baldwin, 1978a; McGrew, 1977; Menzel, 1966). Because the behaviors that are observed and reported on in these studies often include examples of exploration, play, and tool-use, we review them first before moving on to research in these more specific areas.

Several studies have compared the object manipulation of animals from multiple species under controlled experimental conditions (e.g, Glickman & Sroges, 1966; Parker, 1974a, 1974b; Torigoe, 1985). In these studies, members of different species are presented with the same stimuli under the same conditions. Their responses are observed, coded, and quantitatively compared. Because these are studies of animal "curiosity," the quantitative measures do not differentiate among object exploration, play, or tool-use. Moreover, because all concern animals in captivity, the results may or may not generalize to the behavior of animals in natural settings (Glickman & Sroges, 1966; O'Neill & Price, 1991; Stevenson, 1983).

Despite these limitations, this research reveals some consistent phylogenetic trends in the nature of object manipulation that provide a context for some of the more detailed descriptive data in the sections that follow. Two major findings of this research are: (a) object manipulation is more common in juveniles than in infants or adults (e.g., Fedigan, 1972; Fragaszy & Adams-Curtis, 1991; Glickman & Sroges, 1966; Natale, 1989; Starin, 1990; Torigoe, 1987; Visalberghi, 1988), and (b) the behaviors that animals direct toward novel objects tend to reflect the primary survival functions of object manipulation for that species. For example, when Glickman and Sroges (1966) gave animals a set of novel objects, they found that primates showed mostly food-consumption behaviors (i.e., approaching, grasping, sniffing, and chewing), whereas carnivores showed behaviors used in subduing and consuming prey (i.e., batting, steadying, sniffing, and chewing).

These studies also help to identify the specific forms of object manipulation. Several studies of primates are summarized in Table 2.1. As depicted in the table, object-directed behavior can be classified into one of five general categories: procuring, investigating, transforming, relating, and other large motor behaviors. One category (investigating) corresponds to exploration (i.e., What does *this object* do?), and three (transforming, relating, and other large motor behaviors) correspond to types of object play (i.e., What can *I* do with this object?). The first category in the table (procuring) facilitates both.

With the notable exception of the capuchin monkeys, species differences in the complexity of object manipulation vary as a function of phylogenetic proximity to humans. The most common definitions of complexity include: the number or diversity of manipulatory actions, and/or the degree to which maniplation involves one versus multiple objects. Across studies, the great apes tend to score the highest on complexity, followed by the lesser apes, and then by the capuchin and the cercopithecine monkeys. Colobine monkeys, New World monkeys (with the exception of capuchins), and prosimians show the lowest levels of complexity (see Fig. 2.1). These conclusions are supported by the results of studies including both multiple and single species (Adams-Curtis & Fragaszy, 1994; Baldwin & Baldwin, 1978a; Deriagina, 1982; Fedigan, 1972; Fragaszy & Adams-Curtis, 1991; Fragaszy, Baer, & Adams-Curtis, 1991; Natale, 1989; Parker, 1974a, 1974b; Renner, Bennett, Ford, & Pierre, 1992; Starin, 1990; Toriogoe, 1985, 1986, 1987; Visalberghi, 1988; Westergaard, 1992, 1993).

Torigoe (1985) conducted one of the largest multiple-species studies of primates to date. As noted in Table 2.1 (see the footnotes to the table), behaviors fell into three general levels of complexity: low-complexity manipulatory behaviors found in most primate species (primate-common

TABLE 2.1

Types of Object Manipulation Identified in Primates in Multi-Species Studies

	Glickman and Sroges (1966)	*C. Parker (1974b)*	*Torigoe (1985)*	*Natale (1989)*
Procuring				
Buccal prehension				+
Picking up[a]		+	+	+
Holding[a]		+	+	+
Carrying[a]	+		+	
Transferring[a]			+	
Investigating				
Sniffing				+
Poking/touching	+		+	+
Mouthing/licking/biting[a]		+	+	+
Scratching				+
Pressing		+		+
Rotating[a]			+	+
Bringing to eyes				+
Transforming				
Tearing/breaking	+			+
Twisting/untwisting		+	+	
Wadding/forming		+	+	
Relating				
Rubbing[b]	+	+	+	+
Rolling[b]	+		+	+
Draping/Wrapping[c]		+	+	
Hitting/Striking		+	+	+
Dropping[c]			+	+
Lining up				+
Other large-motor behaviors				
Waving/shaking[c]	+	+	+	+
Pushing/pulling[a]		+	+	+
Throwing[c]		+	+	+

[a]Primate-common actions (Torigoe, 1985). [b]Simple secondary actions (Torigoe, 1985). [c]Complex secondary actions (Torigoe, 1985).

actions); intermediate-complexity behaviors that involve relating an object to the substrate (simple secondary actions); and high-complexity behaviors that involve relating two separate objects (complex secondary actions). Species differences in object manipulation are summarized in Table 2.2. Results showed that, with the exception of the capuchins, secondary manipulation was rare among the New World monkeys, the leaf-eaters, and the lemurs, and was highest among the great apes. Only the apes and the capuchins showed the more complex patterns of secondary manipulation (i.e., drop, drape, swing, and throw), whereas the great apes, the capuchins, and the cercopithecines showed the simpler pattern (i.e., roll, slide, and rub). Marmosets and lemurs were the most likely to manipulate with their mouths, the lesser apes with their feet, and the apes and the cercopithecines were the most likely to use the fingers opposably or independently of one another. Only chimpanzees and baboons explored with the index finger alone, and only great apes untwisted the rope stimulus (two findings not summarized in Table 2.2).

Other findings not reported in the table further demonstrated the limited manipulatory patterns of the lemurs, spider monkeys, marmosets, and leaf-eaters. Of the primate-common actions listed in Table 2.1, lemurs did not rotate or shift objects, marmosets did not pull or engage in passive holding, and leaf-eaters did not shift objects. Finally, spider monkeys did not show many of the primate-common behaviors, including rotate, transfer, pull, passive holding, and shift. In fact, the only actions exhibited by this group were mouth, pick up, flip, and transport.

Longitudinal studies of capuchin (Adams-Curtis & Fragaszy, 1994) and Japanese macaque (Torigoe, 1986) infants show that there is an association between the ages at which manipulatory behaviors first emerge and their complexity. In both studies, primate-common behaviors emerged earlier than did simple secondary behaviors, which emerged before complex secondary behaviors.

Similar developmental patterns have been found in longitudinal studies of sensorimotor development in nonhuman primates during the first 2 years of life. In these studies, sensorimotor stages have been assessed using standardized Piagetian assessments (most commonly the Uzgiris-Hunt Ordinal Scales of Sensorimotor Development; Uzgiris & Hunt, 1975) and/or observations of object manipulation. Species studied include capuchins (Natale, 1989; Spinozzi, 1989), macaques (Natale, 1989; S. Parker, 1977; Poti, 1989), orangutans (Chevalier-Skolnikoff, 1983), chimpanzees (Chevalier-Skolnikoff, 1977; Hallock & Worobey, 1984; Mathieu & Bergeron, 1981; Mathieu, Daudelin, Dagennis, & Decarie, 1980; Mignault,

TABLE 2.2

Object Manipulation Scores for 11 Groups of Primates (Summary of Findings of Torigoe, 1985)

Primates	*Number of Kinds*	*Roll/Slide/Rub*	*Drop/Drape Swing/Throw*	*Secondary Manipulation*	*Mouth*	*Feet*	*Finger*
Apes							
Great apes	High	Yes	Yes	High	Medium	Low	Yes
Lesser apes	Medium	No	Yes	Medium	Medium	High	Yes
Old World monkeys							
Cercopithecines							
Macaques	High	Yes	No	Medium	Medium	Low	Yes
Baboons	Medium	Yes	No	Medium	Medium	Low	Yes
Mangabeys	Medium	Yes	No	Medium	Medium	Low	Yes
Guenons	Medium	Yes	No	Medium	Medium	Low	Yes
Colobines							
Leaf-eaters (Colobus, langurs, proboscis)	Low	No	No	Low	Medium	Low	No
New World monkeys							
Capuchins	High	Yes	Yes	Medium	Medium	Low	No
Spider monkeys	Low	No	No	Low	Medium	Low	No
Marmosets	Low	No	No	Low	High	Low	No
Prosimians							
Lemurs	Low	No	No	Low	High	Low	No

1985; Poti & Spinozzi, 1994; Vauclair, 1984; Vauclair & Bard, 1983), and gorillas (Chevalier-Skolnikoff, 1977; Natale, 1989; Redshaw, 1978; Spinozzi & Natale, 1989). Although most studies have involved comparing nonhuman primate behavior to human norms, in some studies human infants were tested as well (Redshaw, 1978; Vauclair, 1984; Vauclair & Bard, 1983).

In general, the results of these studies have demonstrated remarkable similarities between primates species (both human and nonhuman) in the course of sensorimotor development during the first year of life. With just a few exceptions (see discussion later in this chap.) human and nonhuman primates show the same sensorimotor stages in the same order, and demonstrate similar rates of developmental change. Thus, virtually all of the primates studied went through Piaget's first four stages: coordination of reflexes, primary circular reactions and the first acquired adaptations to the environment, secondary circular reactions, and the coordination of secondary schemata and their application to new situations. The major difference is that due to the earlier development of motor skills in nonhuman primates, they usually reach developmental milestones at ages earlier than do humans.

Although human and nonhuman primates show very similar patterns of sensorimotor behavior during the first year of life, there are some minor differences. Specifically, nonhuman primates often show less interest in mechanical toys than do humans, especially in toys with mobile parts (Mathieu et al., 1980; Mathieu & Bergeron, 1981); macaques and gorillas (in contrast to capuchins and humans) frequently engage in buccal prehension (pick up objects with their mouths alone; Antinucci, 1989b); and macaques apparently do not show secondary circular reactions (S. Parker, 1977; Poti, 1989). Whether gorillas engage in secondary circular reactions is a matter of controversy (Poti & Spinozzi, 1994). Finally, in a detailed study of two chimpanzees, one bonobo (pygmy chimpanzee), and one human infant all between the ages of 7 and 12 months, Vauclair and Bard (1983) found that the human infant was more likely to pick objects off the floor for the purpose of manipulation, explore objects by holding and rotating them, extract one object from another, transfer objects from one hand to another, manipulate objects with both hands at once, and wave and shake objects. The chimpanzees were more likely to mouth, to sniff, and to push or pull objects on the floor.

More significant differences in sensorimotor development begin to emerge after 12 months (Hallock & Worobey, 1984; Poti & Spinozzi, 1994; Redshaw, 1978). The main difference at this age is that nonhuman primates engage in fewer relational behaviors with multiple objects, such as tower building or placing a cube into a cup (Knobloch & Pasamanick, 1959; Mignault, 1985; Poti & Spinozzi, 1994; Redshaw, 1978). Such behaviors are frequently shown by human infants; sometimes shown by chimpanzees; and rarely or never shown by capuchins, gorillas, or macaques. Into the second year other differences occur, particularly in tasks involving complex cause-and-effect relationships such as tertiary circular reactions (Mathieu et al., 1980; Mathieu & Bergeron, 1981; Natale, 1989; S. Parker, 1977). These behaviors are reported frequently for humans; occasionally for chimpanzees, capuchins, and gorillas; and never for macaques. A striking difference is that symbolic (i.e., pretense) behavior is rare in nonhumans (Mignault, 1985), although isolated anecdotal examples of such behaviors in home-reared orangutans, chimpanzees and gorillas have been reported in the literature (for reviews, see Call & Tomasello, 1996; Mitchell, 1994a).

Despite the differences in the nature of sensorimotor behavior in the second year, at least one study of orangutans (Chevalier-Skolnikoff, 1983), two studies of gorillas (Chevalier-Skolnikoff, 1977; Redshaw, 1978), and four studies of chimpanzees (Chevalier-Skolnikoff, 1977; Hallock & Worobey, 1894; Mathieu et al., 1980; Mathieu & Bergeron, 1981) have provided evidence of achievement of Stage 5 (tertiary circular reactions and the discovery of new means through active experimentation) and, in most cases, Stage 6 (representation and the invention of new means through mental combinations). Chevalier-Skolnikoff (1983) pointed out that Stage 5 and 6 behaviors emerge much more slowly in apes than in humans, often not being demonstrated until the third or fourth year, and often are not complete until ages 5 to 8. Although there have been some suggestions that Stages 5 and 6 are attained in some monkey species as well (e.g., S. Parker, 1977; S. Parker & Gibson, 1977), a closer examination of the evidence raises serious doubts about this conclusion (see Dore & Duman, 1987; Fragaszy & Adams-Curtis, 1991).

The results of these studies suggest remarkable similarities across primate species in the course of early sensorimotor development and patterns of object manipulation. Although rates of development may vary, species show very similar sequences in the development of the primate-common manipulatory behaviors. Species differences tend to emerge at later ages, with the development of simple and complex secondary behaviors. Although this discussion has focused more on differences than on similarities, the fact that orangutans, gorillas, and chimpanzees go through the same six sensorimotor stages as do humans shows a great deal of consistency across closely related species. Therefore, patterns of sensorimotor development in prelinguistic human infants are likely to reflect evolutionary adaptations that are shared with the other primate groups.

The nature of cross-species similarities and differences in the nature of object manipulation has led to much speculation about the evolution of object manipulation in primates. The most common view (e.g., Fragaszy & Adams-Curtis, 1991; Glickman & Sroges, 1966; Parker & Gibson, 1977; Torigoe, 1985; Westergaard, 1992) is that object manipulation evolved as a feeding adaptation, and the most complex forms of manipulation evolved in species that engaged in "omnivorous extractive foraging", that is, obtaining plant and insect foods by such means as using stones to open nuts, using probes to access hidden foods, using leaf sponges to drink liquids, foraging through dead leaves, peeling bark and breaking hollow twigs and branches to get to insects, and pounding nuts on surfaces (examples from Visalberghi, 1988). This form of foraging (which requires a high degree of digit coordination and often the use of tools) is thought to have evolved among primates who frequently left the arboreal environment during daylight hours to explore for hidden food on the forest floor, and was most likely to have evolved in populations with few natural predators (i.e., where prolonged terrestrial excursions were possible).

Differences in the object manipulation of current primate species are consistent with this interpretation. Glickman and Sroges (1966) argued, for example, that among Old World monkeys, the colobines (who show a limited range of manipulatory behaviors) have adapted primarily to an arboreal existence where food is readily available in the form of fruits, leaves, and twigs. Food for these herbivores is abundant and can be easily procured by plucking it from trees. In contrast, the cercopithecines (who show much higher levels of manipulatory skills), have adapted to both an arboreal and a terrestrial existence, and their more varied diet of fruits, berries, insects, lizards, and small mammals requires more object reactivity, because "the search for food is more active and involves a wider range of manipulatory behavior" (Glickman & Sroges, 1966, p. 177). Glickman and Sroges maintained that differences in susceptibility to predators may account for the differences in the object reactivity they found between baboons (who have few predators) and the much smaller and relatively more susceptible guenons. The nocturnal, arboreal prosimians explore mostly through sniffing, chewing, and licking objects, and show limited manipulatory skills.

Several authors (e.g., Fragaszy & Adams-Curtis, 1991; Torigoe, 1985; Visalberghi, 1988; Westergaard, 1992, 1993) argued that the evolution of secondary forms of manipulation supported the development of tool-use in extractive foraging contexts (e.g., cracking nuts with stones, using probes to access food). Numerous studies have shown that the most proficient tool-using primates (i.e., chimpanzees, orangutans, macaques, baboons, and capuchin monkeys) engage in the highest levels of secondary manipulation. The manipulatory abilities of capuchin monkeys have received considerable attention (e.g., Fragaszy & Adams-Curtis, 1991; Visalberghi, 1988), because this is the only genus of New World monkeys that shows extensive tool-use. Given the evolutionary history of Old and New World monkeys, the case of capuchins suggests that the co-occurrence of secondary manipulation and tool-use probably evolved independently in Africa and South America.

Westergaard (1992, 1993), in his research on baboons, provided some of the most detailed data on the relation between secondary manipulation and tool-use in primates. When infant baboons were given objects that allowed for combinatory actions (i.e., cups, balls, paper towels, and a container of water), they engaged in a high number of relational actions. In a study of 6-month-old baboons: "The baboons placed several types of objects into plastic cups, including rubber balls, paper towels, and food biscuits. The baboons used cups to concurrently contain as many as two objects and repeated the action of inserting and removing an object as many as three times in one uninterrupted behavioral sequence" (p. 399).

Moreover, longitudinal analyses revealed that the development of relational actions preceded and appeared to support the development of tool-use (Westergaard, 1993). The developmental sequence was as follows: detaching, chewing, and shredding paper towels and rubbing them on the floor of the cage (prior to 3 months); rubbing towels along the rim of the juice container and dropping towels into the container (3 to 5 months); and dipping a towel into the container and then putting it in the mouth to chew (6 months).

The development of complex manipulatory skills in species showing omnivorous, extractive foraging is supported by anatomical adaptations that apparently have evolved for that purpose. Besides the adaptations supporting a diurnal existence (e.g., color and stereoscopic vision), these adaptations primarily concern hand anatomy and those that support bipedal locomotion. Theorists have long argued that the evolution of hand functions and bipedal locomotion played a role in the development of manipulatory abilities (e.g., Connolly & Elliott, 1972; Costello & Fragaszy, 1988; Napier, 1962). Napier (1962), for example, argued that whereas prehensile abilities evolved so that large primates who could not be tree runners could successfully navigate the arboreal environment, the evolution of bipedal locomotion in the early terrestrial hominoids was "the ultimate mechanism for emancipation of the hands" (p. 61). Bipedal locomotion frees up the forepaws for object manipulation, as well as increases the likelihood of bimanual manipulation (Marshack, 1984)—a skill important for many forms of tool-use. Although much has been made of the opposable thumb and its role in human evolution, comparison of the hand and foot anatomies of various primates (see Schultz, 1956) could lead to the improbable hypothesis that, with the exception of the capuchins, gross anatomy of the foot is more correlated with manipulatory abilities of the hands than is the gross anatomy of the hands. Inspection of Fig. 2.2, for example, shows that with the exception of the capuchin monkeys (*Cebus*), the primates showing the greatest manipulatory skills (i.e., macaques—*Macaca*, baboons—*Papio*, orangutans—*Pongo*, gorillas—*Gorilla*, chimpanzees—*Pan*, and humans—*Homo*) are those with feet that are wide and/or have the shortest external digits. Thus, primates with feet reflecting adap-

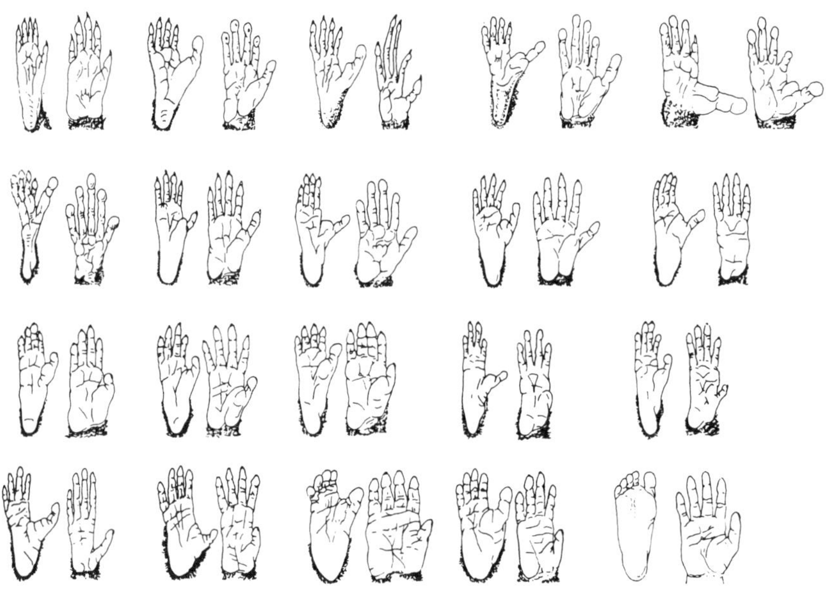

FIG. 2.2. Ventral views of the right hands and feet of a tree-shrew and of selected primate species. For each pair of extremities, the foot is shown on the left and the hand on the right. Key (from left to right): Top row—*Tupaia, Lemur, Daubentonia, Galago,* and *Nycticebus*. Second row—*Tarsius, Leontopithecus, Aotus, Cebus,* and *Ateles*. Third row—*Cersopithecus, Macaca, Papio, Colobus,* and *Presbytis*. Bottom row—*Hylobates, Pongo, Gorilla, Pan,* and *Homo*. From Schultz (1956). Reproduced with permission of S. Karger AG, Basel.

tations to a terrestrial or semiterrestrial lifestyle show the greatest manipulatory abilities with their hands. Therefore, freeing up the hands was apparently at least as important as the opposable thumb in the evolution of manipulatory skills.

Gross hand anatomy apparently is more important for accounting for species that show the most limited manipulatory abilities. Lemurs and marmosets, for example, cannot move the digits of their hands independently, and spider monkeys (*Ateles*) and leaf-eaters (*Colobus* and *Presbytis*) have a missing or a shortened thumb (see Fig. 2.2). It is likely that given the relative unimportance of manipulatory skills for food gathering in these vegetarian, arboreal species, structural changes that supported arboreal locomotion rather than object manipulation led to the loss or reduction of the thumb.

A more complex way in which locomotor development may influence the development of manipulatory skills was offered by scholars at the Instituto di Psicologia in Rome (see volume edited by Antinucci, 1989a). Vauclair (1984), for example, argued that species differences in the ontogenetic timing of locomotion versus prehension impacts the development of manipulatory skills. Although it has long been recognized that prolonged immaturity is associated with greater levels of learning and later flexibility in behavior (e.g., Wilson, 1975), Antinucci (1989b) pointed out that asynchronies in the rates of development in different areas have predictable effects. Regarding object manipulation, he argued that because humans (and, to a lesser extent, capuchins) develop the ability to independently locomote long after they develop the ability to grasp, they have considerable "time of exposure" for the manual manipulation of objects. In contrast, macaques (and, to a lesser extent, gorillas) develop the ability to independently locomote *before* they develop the ability to grasp. Thus, their time of exposure is limited, because once they can independently locomote they can retrieve objects using their mouths alone (buccal prehension). Buccal prehension, Antinucci argued, is a more direct means of acting on the environment than is manual prehension. In manual prehension, the nature of interactions with the environment is more indirect in that the animal acts on objects "almost entirely through the use of intermediaries: the hand at the more primitive levels, (the hand plus) other objects at more advanced ones" (p. 82). Therefore, "The temporal interaction of prehension and locomotion development...channels early cognitive interaction with the external world into different modes, which in turn 'snowball,' for the very effect of development, into larger and larger differences in later structures" (p. 84).

Although most manipulatory behaviors in primates can be seen as feeding adaptations, Torigoe (1985) argued that certain behaviors, especially some shown only by the apes and capuchins (e.g., drop, swing, throw), may be related to agonistic behaviors that evolved in a defensive context. These are discussed further in the section on tool-use.

Although the studies reviewed earlier help identify commonalities and differences in object manipulation across the phylogenetic spectrum, they have their limitations—the animals usually were studied in captivity (often in small, relatively barren cages with little else to do), the sample sizes were small, a limited range of species was studied, the number and variety of stimuli were limited, and the coding systems and analyses provided little information about the timing or sequencing of behavior. The relative lack of field data on object manipulation is cause for concern because, with the exception of the findings of research on chimpanzees (e.g., McGrew, 1977; van Lawick-Goodall, 1968), the field research that has been done shows that the manipulation of nonfood objects is rare in natural settings (Baldwin & Baldwin, 1978a; Menzel, 1966; Starin, 1990; Symons, 1978). Clearly, additional fieldwork that focuses on the nature of object manipulation in natural settings is necessary. In order to make a case that manipulatory skills evolved in response to ecological pressures, it is important to demonstrate the existence of complex manipulatory patterns in naturalistic settings.

In summary, whatever the specific evolutionary developments that led to the sophisticated object manipulatory skills of humans, the results reviewed here are consistent with the conclusion that the primate species showing the greatest levels of manipulatory skills and the greatest flexibility in environmental adaptation are those who are diurnal; terrestrial; bipedal; have few or no natural predators; have prolonged periods of immaturity in development where prehensile abilities precede locomotor ones; and possess anatomical characteristics that support visual and motor exploration, including color vision, stereoscopy, opposability, digit independence, tactile sensitivity, and large brains with well-developed visual and motor cortexes. Compared to the other primates, human beings clearly possess these characteristics.

We now turn to an analysis of the data on the more specific modes of object manipulation. Research on exploration, object play, and tool-use are considered in that order. Each section is made up of two parts; the first reviewing empirical data on the structure of the behavior, and the second reviewing empirical evidence on its possible functions and developmental effects. Following these sections is a review of the research on gender differences in solitary object-directed behavior, covering all the modes of object manipulation considered here.

STUDIES OF OBJECT EXPLORATION

Although isolated accounts of exploratory behavior have appeared in the literature since the late 1800s (e.g., Darwin, 1874; Kinnaman, 1902; Romanes, 1892; Slonaker, 1912; Small, 1899), it was not until the early 1950s that the systematic study of animal exploration began. The early studies primarily involved the locomotor search behavior of laboratory rats in mazes (e.g., Berlyne, 1950; Montgomery, 1953; Myers & Miller, 1954). Exceptions involved several pioneering studies of rodents in the field (e.g., Barnett, 1958a; Southern, 1954), and early studies of exploration in captive monkeys (Harlow et al., 1950; Menzel, 1962) and chimpanzees (Menzel, Davenport, & Rogers, 1961; Welker, 1956a). More recently, exploratory behavior has been documented in a range of nonmammalian species, including insects (e.g., Bakchine, Pham-Delegue, Kaiser, & Masson, 1990; Heidecker & Leuthold, 1984), snakes (Chiszar, Scudder, Knight, & Smith, 1978; Scudder & Chiszar, 1977), lizards (Gove & Burghardt, 1983), and birds (Haemig, 1989; Verbeek, Drent, & Wiepkema, 1994).

Early research on exploration in laboratory rats was designed primarily to test theories of motivation (for a review, see Inglis, 1983), and to identify the stimulus characteristics that influenced exploration of the environment (see Berlyne, 1960). The early work on primates focused on motivational issues (e.g., Harlow, Blazek, & McClearn, 1956; Mason, Harlow, & Rueping, 1959) and stimulus characteristics (e.g., Carr & Brown, 1959; Mason, 1961; Menzel et al., 1961; Menzel & Menzel, 1979; Welker, 1956b). Weisler and McCall (1976) summarized the results of these and other studies in their description of the stimulus attributes that elicit exploratory behavior: "(a) physical stimulus properties (e.g., number of visual elements or brightness contours, intricacy of configuration, responsiveness to manipulation, potential to change shape or move); and/or (b) novelty or discrepancy relative to the organism's remembered experience" (p. 493).

Structure

Despite the research devoted to trying to understand both the motivation for exploratory behavior and the conditions that influence its occurrence, descriptive research on the structure of exploratory behavior and its correlates is rare. As recently as 1990, Renner stated in a review of the literature that "precious little is known of the nature of object investigation as part of exploration" (p. 19). Similarly, Renner and Seltzer (1994) noted that: "No published work has attempted to identify the sequential patterns of acts that underlie object investigation or to ascertain whether these sequences contain indivisible behavioral units" (p. 335)

Although few systematic, quantitative analyses of the nature of animal behavior during object exploration have been conducted, numerous qualitative descriptions of these behaviors have been offered. Despite the finding that the exploration of invertebrates and other nonmammals sometimes differs considerably from the exploration of mammals (e.g., tongue flicking in snakes and lizards; the laying of chemical trails in certain insects), exploratory behavior, at least on the surface, is remarkably similar across mammalian species. Consider, for example, the following descriptions:

> When first introduced into its cage, each rat showed a marked degree of curiosity. It examined every corner and object within the inclosure, using mainly the sense of smell assisted by the sense of touch and to a still less degree that of sight. After about an hour of such investigation most of those in the revolving cages had found their way to the nest boxes in which they remained for a time. (Slonaker, 1912, p. 23)

> On encountering the novel object the piglets oriented towards it, approached it, sniffed it and then generally started to bite it. (Wood-Gush & Westergaard, 1991, p. 603)

> Exploration: visual regard, licking, sniffing and/or touching without manipulation of an object with paws or mouth while slowly moving toward or around an object. (West, 1977, p. 54—a study of domestic kittens)

> The typical primate within our sample approached an object within the first minute, after an initial orienting response, picked it up in the forepaws, and chewed upon it, stopping at various points to sniff it or to inspect it visually while turning the object about in its forepaws. (Glickman & Sroges, 1966, p. 164)

These descriptions are consistent with Weisler and McCall's (1976) conclusion that exploratory behavior is relatively stereotyped, and generally follows a fixed sequence:

> (a) A*lerting* or the recruiting of attention to the new stimulus situation, (b) *distance-receptor scanning or examination* of the new situation, (c) *motor-aided* perceptual examination of the situation in which the organism may

> perambulate the environment or manipulate the object to explore it more completely, and (d) *active physical interaction* apparently for the purpose of discerning what will happen as a consequence of the organism's interactions with the object or situation. (p. 493)

On the surface, then, it appears that over and above species differences in the tendency to utilize olfactory versus manual exploration, the nature of exploratory behavior in mammals varies little from species to species. That animals are motivated to explore, even in situations that may be dangerous, is well illustrated in a study of three Old World monkeys (all species of guenons) by Macdonald and Pinel (1991). When their subjects were startled by the unexpected triggering of a jack-in-the-box, "within seconds ($M = 8$ s), each monkey returned to the jack-in-the-box and spent much of the 20-min test session inspecting it ($M = 905$ s) while largely ignoring the stuffed-bear control object ($M = 203$ s)" (p. 207). MacDonald and Pinel (1991) went on to note that:

> Behavior during the test sessions was similar for each species. When the jack-in-the-box was triggered, each monkey vocalized loudly, quickly climbed the metal "tree," and withdrew to the top corner of the enclosure farthest away from the jack-in-the-box. However, within a few seconds, ($M = 8$ s), each monkey cautiously approached the jack-in-the-box. The L'Hoest and Diana monkeys both left the metal tree, moved toward the jack-in-the-box on the ground, and touched the box. In contrast, the Yellow-Nosed monkey swung out over the jack-in-the-box and inspected it while hanging from the metal tree.... Each monkey withdrew from the jack-in-the-box after the initial approach and spent the remainder of the test sessions in a series of approach sequences directed to the jack-in-the-box. (pp. 211–213)

These observations illustrate the simultaneous operation of the fear and exploration systems in the presence of a threatening stimulus (see Murphy, 1978). A similar tendency to explore in the face of threatening stimuli was documented in a laboratory study by Inglis and Shepard (1994). These researchers found that in spite of the continuous availability of wholesome food, wild rats (*Rattus norvegicus*) would repeatedly push a lever to obtain poisonous food that they later rejected. Inglis and Shepard interpreted this finding as evidence that animals will work "to gather information about a rare, but very important event, namely the presence of dangerous food at a previously preferred and normally safe feeding site" (p. 154).

In one of the few molecular, quantitative descriptions of the nature of exploratory behavior, Renner and Seltzer (1994) conducted sequential analyses of the object exploration of 12 male Long-Evans rats in a laboratory setting. Individual behaviors were coded into 12 different categories, and sequences were examined by quasi-grammatical analysis. Results showed that exploration of objects was "orderly but not stereotyped" (p. 335), in that the nature of the sequences varied with characteristics of the objects, and that individual animals varied considerably in the sequences they exhibited. Clearly, the conclusion that exploration represents stereotyped behavior is premature, and additional molecular analyses like that of Renner and Seltzer (1994) need to be conducted.

Functions/Effects

Although there is considerable consensus that object exploration serves an information-gathering function, very little empirical work has been conducted with animals to confirm this hypothesis. Of what little evidence there is, probably the best involves several studies of cross-modal perception in monkeys and apes. In these studies, animals were given an opportunity to manually explore an object without being able to see it in the process. Then, through either a conditioning or a visual preference task, researchers examined whether the animal matched the object they had explored manually with one they could see. The fact that such cross-modal transfer has been demonstrated in macaques (Cowey & Weikrantz, 1975; Gunderson, Rose, & Grant-Webster, 1990; Tolan, Rogers, & Malone, 1981), orangutans (Davenport & Rogers, 1970), and chimpanzees (Davenport & Rogers, 1970; Davenport, Rogers, & Russel, 1973, 1975) demonstrates that in these species information is acquired through the process of manual exploration.

Finally, in an experimental study of male great titmice (Verbeek et al., 1994), it was demonstrated that individual differences in birds' exploratory style predicted their behavioral responsiveness to changes in the environment. When the experimenters changed the location of the birds' feeding sites, animals who had earlier adopted a slower and more thorough approach to exploration of a novel environment were less likely to repeatedly return to the initial location in search of food than were birds who adopted a quicker, more superficial style. Birds evidenced stable, individual differences in their exploratory style, showing similarities in the way they explored a large-scale environment and novel objects. Verbeek and colleagues argued that the quick, superficial exploratory style might be most adaptive in a stable environment, whereas the slow, thorough style may be more adaptive in an unstable one.

STUDIES OF OBJECT PLAY

Research has shown that object play is found in many birds and mammals; is not found in invertebrates, fishes, and amphibians; and is rare in reptiles (Burghardt, 1984, 1988; Fagen, 1981). However, many species of mammals have not been observed to play (Bekoff & Byers, 1981), and both Fagen (1981) and Burghardt (in press) stated that it is possible that play in some reptiles and other animals may have not yet been identified.

Robert Fagen devoted a large section of his often-cited 1981 book on animal play to a section entitled "Natural History of Play Behavior." In this section, he presented a phylogenetically organized review of the literature on animal play—a literature that varies significantly in source and detail: "The accounts range in length from a few words to entire volumes. Their quality also varies. Many existing descriptions are verbal and anecdotal, some are brief or ambiguous, and only a few include quantitative data suitable for testing theory" (p. 77). However, a review of this history provides a good first pass at identifying the range and variety of animal play behaviors. In the book, Fagen (1981) stated that: "Mustelids (i.e., weasels and ferrets), like humans, chimpanzees, ravens, large parrots, bottle-nosed dolphins, goats, red deer, and felids, seem to have adopted play as a way of life" (p. 148). Besides identifying the most playful species, he also argued that both birds and mammals show high levels of playfulness: "Given the evolutionary opportunity, avian phylogeny can produce play like that of the kea, hyacinthine macaw, and common raven that appears to surpass most mammalian play in elaborateness and complexity" (p. 218).

His natural history shows that object play varies by order. In general, object play was most commonly mentioned in primate, carnivore, and bird species. Examples of object play mentioned within the remaining land-living mammalian orders—marsupiala (e.g., kangaroos), insectivora (e.g., hedgehogs), edentata (e.g., anteaters), rodentia (e.g., porcupines), and artiodactyla (e.g., pigs)—were less frequently mentioned and usually were described as less complex, involving carrying, shaking, and touching. Although Fagen (1981) acknowledged that the natural history is incomplete and preliminary (much more descriptive research needs to be done), he reported no instances of object play in members of the following orders: monotremata (e.g., platypuses), lagomorpha (e.g., rabbits), proboscidea (e.g., elephants), and perissodactyla (e.g., horses).

Studies of object play published since Fagen (1981) are consistent with these conclusions. For example, in studies of the play of equids (e.g., horses) and artiodactyls (e.g., pigs), it was noted that object play was rare (Crowell-Davis, Houpt, & Kane, 1987; Newberry, Wood-Gush, & Hall, 1988). This is not surprising given their limited manipulatory abilities.

Burghardt (1984) pointed out differences between the physiology and life histories of mammals and reptiles that may explain why play is so common in the former and so rare in the latter. In making this contrast, he specified characteristics of mammals that may have allowed for the evolution of play in their young:

> The frequent observation that play occurs only when animals are in a "relaxed field," sated, warm, content, and so forth supports the view that play of various kinds originated in animals with (1) sufficient metabolic resources; (2) an efficient aerobic system; (3) a relatively familiar and often risk-free juvenile environment; (4) the ability to accumulate more energy than can be shunted to growth; (5) escape from the need to behaviourally thermoregulate to bring the body to the optimum for vigorous behavior; (6) innate precocial exploratory and arousal mechanisms; (7) specific motivational and behavioural systems; and (8) relatively high survival rate of offspring. (p. 31)

Burghardt (1984) offered the intriguing hypothesis that play evolved not directly as a "means of incorporating flexibility, modifiability, plasticity, or innovation within a behavioral repertoire" but instead as a "serendipitous bonus for having to find other ways to do what reptiles did 'naturally'" (p. 35). Specifically, he stated that the nonstereotyped sequences of abbreviated behaviors found in play may have been a consequence of the fact that young mammals receiving parental care are nutritionally well-off and have "a surplus of metabolic resources, that, unlike in reptiles, could be activated quickly with aerobic metabolism. The remnants of behavior patterns, physiological mechanisms, and motivations, coupled with exploratory tendencies, were readily triggered as the enforced inactivity associated with parental care led to 'boredom' and lowered thresholds" (p. 34). Play, therefore, "originated from boredom and deteriorated behavior, an outrageous speculation that may, after all, also be true of the writer and his thesis" (Burghardt, 1984, p. 36). Burghardt (1988) elaborated on these ideas and offered a number of specific predictions about within- and between-species differences in play that are predicted by the theory. Moreover, in a recent study (Burghardt, Ward, & Rosscoe, 1996), he and his colleagues described high levels of object play in a zoo turtle receiving environmental enrichment.

Consistent with Burghardt's (1984) theory is the widely held view that play is most common in "generalist" species that can adapt to a range of environments (e.g., Wilson, 1975). Members of generalist species often need a prolonged period of immaturity and parental protection to learn to survive in a particular environmental setting. This has led theorists to hypothesize that, for many species with prolonged infancy or juvenile periods, object play is a context in which young animals learn and refine survival skills, such as various predatory and tool-use behaviors. For example, if one considers the primates showing the highest levels of tool-use (e.g., capuchins, baboons, macaques, chimpanzees, and humans), one can see that they tend to have the longest infancy and juvenile periods, and spend a great deal of their time engaged in object manipulation and play. Ortega and Bekoff (1987) provided further support for the relationship between play and precocity in their brief review of the literature on avian play. Of 26 avian orders they considered, 12 included species in which play had been observed. Object play was the most common form of play observed. As shown in Table 2.3, species with altricial young (where the young are born helpless and require extended parental care) were more likely to exhibit play than were species with precocial young (where the young show a great deal of independence from birth). The association between status of the young and play was statistically significant.

Structure

Presented in Table 2.4 is a list of the object play behaviors shown by birds and/or land-living, nonprimate mammals mentioned in Fagen's natural history of play. Although various aquatic mammals play as well, their behaviors are

TABLE 2.3

Frequency of Avian Orders Showing Play as a Function of Maturity of Young (Data From Ortega & Bekoff, 1987)

	Was Play Observed?	
Maturity of Young	*Yes*	*No*
Altricial young	10	6
Precocial young	2	8

TABLE 2.4

Types of Object Play by Birds and/or Land-Living Nonprimate Mammals in Fagen's (1981) Natural History of Play Behavior

Procuring
Picking up and carrying
Investigating
Mouthing/biting
Transforming
Tearing/breaking
Relating
Rolling
Wrapping around/hugging
Hitting/striking/pecking
Dropping/catching
Other large-motor behaviors
Shaking
Pushing/pulling
Throwing/kicking

not listed in Table 2.4, due to differences in object manipulation in an aquatic medium. Note also that primate play behaviors are not listed in the table, because they corresponded almost exactly to the transforming, relating, and other large motor behaviors already listed in Table 2.1.

Comparison of Tables 2.1 and 2.4 show that, at least superficially, primates and nonprimates differ little in the specific play behaviors they exhibit. Although the first two behaviors in Table 2.4 (i.e., picking up/carrying and mouthing/biting) refer to behaviors that might be considered investigation, it is likely that the specific way in which these behaviors were exhibited made them playful in nature. Examination of the rest of the behaviors in Table 2.4 shows direct correspondence with Table 2.1, with only a few exceptions. The only play behaviors in Table 2.1 that are not in Table 2.4 are twisting/untwisting, wadding/forming, rubbing, and lining up. Some of these behaviors (i.e., twisting/untwisting, wadding/forming) may only be in Table 2.1 because of the rope stimuli that were used in the object manipulation studies (C. Parker, 1974a, 1974b; Torigoe, 1985). The others (i.e., rubbing and lining up) are relational or secondary behaviors that might be more common in primates than in other mammals. Given these minor differences, and the sketchy and incomparable nature of many of the reports making up the natural history, the evidence for similarities across species is much stronger than the evidence for differences.

Examination of individual species differences in object play in the Fagen (1981) natural history shows that, consistent with a training hypothesis, the nature of object play is related to the nature of adult behavior with objects. For example, object play in carnivores and birds of prey appears to be tied to prey-subduing and prey-consumption behaviors, object play in other birds is tied to the use of tools in food attainment, and object play in primates appears to be tied to foraging practices and tool-use.

For carnivores, behavior during object play has structural similarities with behavior during play capturing, and these animals have a preference for playing with objects that resemble their natural prey (Negro, Bustamante, Milward, & Bird, 1996). Jane Egan (1976) made both of these points in her often-cited study of play in domestic cats:

> Typically an object is first sniffed and/or patted; the stimulation received determines whether or not it is bitten (fur is the stimulus most effective in eliciting biting). If it is bitten it might then be picked up in the mouth, then shaken or thrown (with live prey this would serve to stun it) or carried (with prey to a corner perhaps, where it might eventually be eaten). Alternatively, the initial patting might cause the object to roll, in which case it might elicit crouching and pouncing (movement being the stimulus for these patterns) which, as with prey, would serve to immobilize it. A fur-less object is seldom bitten, carried, shaken, or thrown; playing with it therefore entails the repetition of other than mouth patterns. (pp. 161–162)

Negro et al. (1996) found that juvenile kestrels spent equal amounts of time involved in object play whether preylike objects or other objects were available. Thus, "raptors devote a certain amount of time and energy to manipulation of objects during the post-fledgling period. These will be directed toward prey, or objects resembling prey, if they are available but to other objects if not" (p. 712). Negro et al. also argued that play may serve to facilitate the development of hunting skills, in that kestrels played with objects only at the age during which hunting skills were developing—adult kestrels did not play with the objects provided.

The relationship between object play and predatory behavior is also well illustrated in Biben's (1982a) study of object play in two canids—bush dogs and crab-eating foxes. In this longitudinal research, the young from four captive litters were followed from the first to the sixth month of life. Object play (i.e., nonexploratory manipulation) emerged between 5 and 7 weeks for both species, but the nature of the play differed significantly. Specifically, crab-eating foxes—who are omnivorous and capture and eat prey nonsocially—showed a greater variety of play behaviors than did bush dogs—who are carnivores who hunt and eat in packs. Biben argued that these differences are consistent with the fact that crab-eating foxes use a greater variety of techniques for capturing prey than do bush dogs.

Rasa (1984) also found an association between prey-capturing behaviors and object play in her study of object play in juvenile dwarf mongooses. Mongooses are a particularly informative species for play research because they have a unique vocalization that only occurs in play contexts (whether the play is social or solitary). Moreover, the vocalization is continuous, so it is easy for an observer to determine when play is occurring and when it is not. Detailed sequential and cluster analyses of data from molecular coding of behavior during play with a novel object (a folded piece of typing paper) helped identify two clusters of mongoose behavior—one corresponding to investigation and one corresponding to play. The investigation cluster was made up of cheek marking, sniffing, placing the nose under the object, and scratching the object with one forefoot, whereas the play cluster was made up of a wide range of 32 animal behaviors. Spectrographic analyses of the play vocalizations showed three levels of vocal intensity that demonstrated clear associations with animal behavior. Specifically, investigatory and first object contact behaviors were associated with the lowest intensity vocalizations (nonplay); manipulation of objects with the paws and enemy-directed behavior patterns, such as standing on the hind legs with body vertical, were associated with the second level of intensity (low-intensity play); and the highest level of intensity was associated with egg-smashing behaviors (high-intensity play). Therefore, not only did the study provide evidence for the association between object play and prey-catching behavior, it also provided some empirical support for the distinction between object exploration and play.

In a detailed study of the motor movements of Australian magpies, Pellis (1983) demonstrated the association between observed object play and predation in a bird of prey. In four family groups of magpies observed weekly from the fledgling period to the 16th week of life, Pellis documented the gradual development of the motor skills required in leg flicks, where the foot comes up and grasps an object being held in the bird's beak. Young birds would frequently engage in object play where the apparent goal was to grasp with the foot an object held in the beak. The frequency of these play behaviors peaked at 4 to 6 weeks postfledgling. Interestingly, the first successful leg flicks did not occur until 4 weeks, at which approximately 50% were successful. The success rate increased to 93% by 8 weeks. Because magpies do not catch prey large enough for foot use until 8 to 10 weeks postfledgling, Pellis argued that object play serves an important practice function in this species for the acquisition of complex motor skills.

A similar study is Beck's (1980) research on predatory shell dropping in Herring gulls. Beck observed almost 3,000 occurrences of shell dropping over a 44-day period. Adult gulls clearly preferred hard surfaces over soft ones for their drops; flew directly to dropping sites once a shell was obtained; headed into the wind when dropping shells; and severed the prey immediately at extraction apparently, in order to avoid a painful bite. Accuracy in shell dropping increased from the yearling to the immature to the adult groups. Yearling gulls rarely dropped shells, but instead spent considerable time dropping other objects, such as bits of vegetation, shell fragments, and flotsam. Beck noted that, "Unlike predatory droppers, play droppers showed no preferences for hard substrates, rarely pecked at or repositioned the objects once dropped, often caught objects in the air and redropped them, dropped in close proximity to other gulls without competition, and preferentially dropped at high wind velocities" (pp. 208–209). Beck argued, as do others (e.g., Negro et al., 1996; Pellis, 1983), that object play in the young facilitates the development of important predatory skills.

Research on the importance of object play in primates primarily comes from research on chimpanzees in the Gombe Stream Reserve (McGrew, 1977; van Lawick-Goodall, 1968). In their naturalistic observations of termite fishing (extracting termites from their nests using a stick for a tool; van Lawick-Goodall, 1968) and leaf sponging (obtaining water using a leaf as a receptacle; McGrew, 1977), these researchers noticed that infants watched their mothers engage in the behaviors, played with the objects that their mothers used as tools, and developed the skills gradually over time, with practice. These researchers suggested that the early play experiences facilitated the development of these complex skills. Interestingly, due to the likelihood of painful bites, infants did not play by their mothers when they were ant dipping (a procedure similar to termite fishing, but with ants as the prey). Subsequently, the development of ant dipping was much delayed compared to that of termite fishing (McGrew, 1977). Nishida and Hiraiwa (1982), however, did observe infant play during maternal ant fishing in a natural setting, and offered a five-stage sequence for the development of tool-use in this setting: "(i) manipulatory play, (ii) tool-manufacture, (iii) motor-skill of tool-use, (iv) knowledge of quality of the tool and efficiency of its use, (v) motor-skill involved in response to the anti-predatory behavior of the insect-prey" (p. 89).

Functions/Effects

Conclusive research on the effects of object play is limited. In two studies of the "manipulation drive" in rhesus monkeys in captivity, Harlow and associates (Harlow, 1950; Harlow et al., 1950) found that, when given the opportunity to manipulate complex mechanical puzzles involving locks and latches, adult rhesus monkeys showed a great deal of interest in the puzzles and effected an increase in correct responses to the puzzles over days. Moreover, in an experimental study in which one group was given the puzzles to play with for 12 days, these monkeys outperformed a control group at a test session at the end of this period (Harlow et al., 1950).

Second, in an experimental study of the predation hypothesis, Caro (1980) randomly assigned domestic cats to a condition where they either had a chance between the ages of 28 and 84 days to manipulate objects (thirty-four 40-minute exposure sessions) or they did not. Cats were later tested on their predatory abilities with several prey objects, including a mouse, a rat, a canary, and a goldfish. Thirteen predatory behaviors were coded, and these were used to calculate 25 behavior scores for each of the four prey animals. The results of 100 statistical comparisons (4 prey × 25 behavior measures) yielded only five statistically significant comparisons (no more than what one would expect by chance alone). Moreover, the significant findings were in the unpredicted direction. Although Caro (1980) listed many possible reasons for the lack of significant findings (e.g., control animals had access in their cages to objects for play that were not part of the experimental manipulation, such as wood shavings; the objects given to the experimental subjects were not appropriate; the wrong behaviors were observed), the study provides no support for the commonly held belief that object play facilitates the development of predatory skills in cats.

Finally, in three classic studies of tool-use (Birch, 1945; Kohler, 1925; Schiller, 1952), chimpanzees appeared to learn behaviors during object play that were later applied in a tool-use context. Unfortunately, these were not experimental studies where tool-use among animals previously allowed to play with the objects was compared to tool-use in animals who were not. Therefore, it is difficult to differentiate between the possibility that play experience was important versus the simpler explanation that nonplay behaviors during object manipulation eventually led to the discovery of tool-use behaviors (Vandenberg, 1978).

In conclusion, there is currently very little empirical support for the hypothesis that object play has effects on the development of functional behaviors, such as prey capture or tool-use. Martin and Caro (1985), in a discussion of the functions of play, concluded that "at present, there is no direct evidence that play has any important benefits" (p. 97), and called for future research. They argued that one likely possibility is that the benefits of play are minor, not major, and that previous researchers have overestimated the costs of play. If the costs are not that large, "then play need only have minor biological benefits in order to be maintained by natural selection" (p. 81). They also argued that play may have short- as well as long-term benefits, that its effects may be general versus specific, and that it may be one of several alternative pathways through which skill development can be achieved. Given the diversity in play behavior between closely related species and even between individuals or populations of a given species that play, Martin and Caro maintained that play may facilitate, but not be critical for, development, and therefore its effects are difficult to demonstrate empirically.

STUDIES OF TOOL-USE

In an often-cited book on animal tool-use, Beck (1980) provided a detailed "catalogue of animal tool behavior." Similar to the natural history on play presented in Fagen (1981), Beck's catalogue is based on data on animal behavior from a wide variety of sources. In this catalogue, Beck presented evidence of tool-use for animals in the following groups: invertebrates (i.e., insects, crustaceans, cephalopods, and gastropods); fish (several species that shoot water); birds (multiple species); nonprimate mammals, such as rodents (three cases), ungulates (primarily elephants), and carnivores (primarily bears and sea otters); monkeys (primarily capuchins, macaques, and baboons); and apes (primarily orangutans and chimpanzees, although also seen in gibbons and gorillas). Beck (1980) could find no reported cases of tool-use in either reptiles or amphibians. Furthermore, he stated that: "chimpanzees show the greatest frequency and diversity of tool behavior of any nonhuman animal and, were it not for them, this book would be a slim volume" (p. 79).

More recent studies are consistent with Beck's (1980) catalogue, with most studies involving either chimpanzees (e.g., Fay & Carroll, 1994;

McGrew & Marchant, 1992; Suzuki, Kuroda, & Nishihara, 1995) or capuchin monkeys (e.g., Fernandes, 1991; Fragaszy, Vitale, & Richie, 1994; Westergaard & Suomi, 1993, 1994b). There are also occasional studies of nonprimate species (e.g., Barber, Ellgaard, Thien, & Stack, 1989; Chevalier-Skolnikoff & Liska, 1993; Hart & Hart, 1994; Henschel, 1995). For a review of some of the more recent research with primates, see Chevalier-Skolnikoff (1989), Visalberghi (1990), and Gibson and Ingold (1993).

A comparison of Beck's (1980) tool-use catalogue with Fagen's (1981) natural history of play behavior reveals some similarities and differences. Concerning similarities, two groups of animals showing high levels of object play—the birds and the primates—also show multiple instances of tool-use. Moreover, within these groups, many of the species (especially the primates) showing high levels of object play also exhibit considerable tool-use (e.g., ravens, capuchins, baboons, macaques, and chimpanzees). However, for other species where object play is common (e.g., cats, canids), the animals do not commonly use tools. Finally, the invertebrates and fish that show tool-use apparently do not play with objects.

Beck (1980) cautioned that, given the diversity of species showing tool-use as well as the diversity of behaviors falling into this category, one should not assume that all forms of tool-use have the same ontogenetic determinants. Specifically, based on the work of Bateson (1976), Beck argued that tool-use among invertebrates and fish may largely reflect behavior patterns that have "specific, inherited determinants, with only general, environmental determinants" (p. 137), whereas most instances of tool-use in birds and primates have "specific environmental determinants, and only general inherited determinants" (p. 139). Therefore, given the importance of specific environmental experiences for the development of tool-use in birds and primates, it seems likely that, for many of these species, a significant portion of the refinement of these skills occurs in the object play context. Practice during play apparently is not necessary for tool-use in invertebrates and fish, given the specific genetic determinants of these behavior patterns. Finally, for species that show object play and little tool-use, it is likely that object play serves other functions for them (e.g., practice of behaviors for prey capture and consumption in cats).

Structure

In his review of the literature, Beck (1980) identified 21 general modes of tool-use (see Table 2.5). Six modes apparently function to extend the

TABLE 2.5

Modes and Functions of Tool-Use Identified by Beck (1980)

Extend Reach	*Amplify Mechanical force*	*Augment Agonistic Displays*	*Control Flow of Liquids*
Reach	Pound, hammer	Drop, throw down	Wipe, apply
Prop and climb	Aimed throw	Unaimed throw	Contain
Balance and climb	Club, hit	Brandish, wave	Sponge, absorb
Stack	Pry	Drag, kick, roll, slap	
Bait	Dig		
Insert and probe	Prod, jab		

Note: Two modes did not fit into these functional categories: drape/affix, and hang and swing.

animals' reach, six to amplify mechanical force, four to augment agonistic displays, and three to control the flow of fluids. Only two of the modes he identified (i.e., drape/affix, and hang and swing) do not fit into these four general functional categories.

In a more recent review of tool-use in apes and monkeys, Chevalier-Skolnikoff (1989) identified no new forms of tool-use that fall outside of Beck's (1980) 21 categories. Most recent papers on tool-use in primates have either identified an otherwise unobserved form among members of a particular species (e.g., Fernandes, 1991; Machida, 1990; Rogers & Kaplan, 1994; Starin, 1990; Tokida, Tanaka, Takefushi, & Hagiwara, 1994; Westergaard & Suomi, 1993), or have explored further the nature of nut cracking and termite fishing in chimpanzees (e.g., Boesch & Boesch, 1990; Fay & Carroll, 1994; Gonzalez-Kirchner & Sainz-de-la-Maza, 1992; McGrew & Marchant, 1992; Suzuki et al., 1995). Moreover, in a series of studies by Westergaard and colleagues (e.g., Westergaard, 1995; Westergaard, Greene, Menuhin-Hauser, & Suomi, 1996; Westergaard & Suomi, 1995a, 1995b), the process of tool manufacture and use in captive capuchin monkeys has been explored as a primate model for understanding the emergence of tool-use in hominids.

Although most of the discussion of object play and tool-use to this point has focused on behaviors involved in food acquisition and consumption, one form of tool-use that is common in primates and may have evolved in an agonistic context is aimed throwing. Hamilton, Buskirk, and Buskirk (1975), for example, reported that throwing objects, a behavior usually not attributed to baboons, was frequently observed in agonistic settings during their fieldwork in Southwest Africa. In her classic monograph on wild chimpanzees in the Gombe Stream Reserve in Tanzania, Jane van Lawick-Goodall (1968) reported that aimed throwing of objects, although rare, did occur in the wild, and that it was usually used by adults in an agonistic context (with humans and baboons being the favorite targets). She also observed aggressive threats that involved the whipping of branches and sticks. Similarly, Kortland and Kooij (1963) summarized the results of a survey of zookeepers, game wardens, and other observers of primate behavior who reported that throwing objects was a common problem in zoos, and that it occurred most frequently in gorillas and chimpanzees, although it occasionally was seen in orangutans and capuchins. Kortland and Kooij (1963) argued that throwing behaviors evolved as an adaptation to terrestrial life in open or semi-open habitats, and that they served an agonistic function. According to Kortland and Kooij (1963), throwing is instinctive, but aim develops with experience. Consistent with this hypothesis is Westergaard and Suomi's (1994a) finding that captive capuchin monkeys showed aimed throwing and that they could be trained to reliably hit both stationary and moving targets. Others have gone further to argue that the evolution of throwing contributed to the evolution of handedness, hunting, and even language in early hominids (e.g., Calvin, 1983; Fifer, 1987; Knusel, 1992). Kortland and Kooij (1963) reported that clubbing was sometimes seen in chimpanzees as well—at least in zoo settings—and that it may also serve an agonistic function.

That throwing is well-developed in chimpanzees is evidenced in Hopkins, Bard, Jones, and Bales' (1993) study of hand preferences in chimpanzee throwing and infant cradling. In this study of 18 males and 18 females in captivity, Hopkins et al. (1993) witnessed an average of 10.6 throwing acts per animal during a 2-month period of daily observations. Chimps threw small objects (often food), usually at unfamiliar humans either walking or running by their cages. What is striking is the structure of this behavior pattern (described for right-handed chimps): "The modal pattern of throwing can best be described as resembling that of a baseball pitcher. The chimpanzee stood bipedally, stepped forward placing its body weight on the left leg, and hurled the material. In nearly every throwing event, the material hit the target" (p. 787).

Therefore, in contrast to the possibility that fine-motor manipulatory skills evolved as an adaptation for food procurement and consumption, the large-motor skills involved in throwing may have evolved in an agonistic context. Examination of the development of throwing skills in humans shows that large-motor skills predominate in throwing behavior. For example, advanced throwing skills involve shifting the weight from one foot to another, differential rotation of different areas of the trunk, pelvic rotation, swinging of the arm, stepping, trunk bending, arm bending, coordinated upper and forearm movements, and object release (Haywood, 1986). It would be informative to speculate on how arm, shoulder, and hand anatomies that are apparent adaptations to arboreal arm swinging (e.g., in gibbons and chimpanzees; see Martin, 1990) may have provided the anatomical structures that allowed for the evolution of throwing behavior.

Function/Effects

Studies on the functions and effects of tool-use would have to demonstrate that animals who use tools in a particular situation have greater success in achieving their goals than animals who do not. Although the majority of descriptive studies of tool-use seem to suggest that this would be true, no study was found that directly addressed this issue.

GENDER DIFFERENCES IN SOLITARY OBJECT-DIRECTED BEHAVIOR

Few animal studies have examined gender differences in solitary object-directed behavior. Usually, the sample size is too small for the issue to be examined; when the sample size is sufficient, investigators apparently have little interest in addressing the issue. Glickman and Sroges (1966) found no statistically significant gender differences in object manipulation among four broad taxonomic categories: marsupials, carnivores, rodents, and primates. Of the few studies of nonprimates that could be found, two found no gender differences (exploratory behavior in Norway rats: Renner, Bennet, & White, 1992; object play in hyenas: Pedersen, Glickman, Frank, & Beach, 1990), and one found that males engaged in more object contact than did females (domestic cats: Barrett & Bateson, 1978).

Studies of gender differences in primates show considerable variation among closely related species. Although most studies do not analyze for gender differences, of those that do, studies of rhesus and colobus monkeys find that females engage in more object manipulation than do males (rhesus: Novak et al., 1993; O'Neill & Price, 1991; Wilcoxon, Meier, Orlando, & Paulson, 1968; colobus: Starin, 1990); studies of capuchin monkeys and chimpanzees find that males engage in more object manipulation than do females (capuchin monkeys: Visalberghi, 1988; chimpanzees: Parquette & Prescott, 1989); and a study of gorillas (Gold, 1992) found no gender differences. Studies of tool-use are no less consistent. Females have been found to show more tool-use than males in both rhesus monkeys (Parks & Novak, 1993; Tsumori, 1966) and chimpanzees (Boesch & Boesch, 1984), whereas males have been found to show more tool-use than females in baboons (Beck, 1972).

It is therefore difficult to draw any firm conclusions about gender differences in object manipulation at this point. Gender differences are likely to be a function of many factors, including: the type of behavior being studied (e.g., exploration, play, tool-use); the nature, type, and novelty of the objects available; the setting in which the observations occur (e.g., captivity, wild); the animal's age; the other animals present (gender composition and size of the group, dominance status of members); and gender differences in the function of object manipulation for a particular species. Therefore, considerable additional research needs to be conducted before hypotheses about the nature and functions/effects of gender differences in object-directed behavior can be entertained.

SUMMARY AND CONCLUSIONS: ANIMAL RESEARCH

The evidence reviewed in this chapter suggests that, as a result of selection pressures, patterns of behavior toward objects by the young have evolved in many animal species. Such behaviors take on many forms and appear to be closely related to the functions of object-directed behavior for older members of these species.

Behavior with objects falls into three general categories: exploration, object play, and tool-use. Exploratory behavior (i.e., behavior that serves to reduce uncertainty about the environment) appears in the widest range of species and is similar in its gross structure in most mammals. Object play occurs in fewer species than does exploration and apparently serves a very different set of functions: Whereas exploration is somewhat stereotyped and functions to reduce uncertainty, play is more varied and complex, and may function to facilitate the acquisition of motor skills and to increase behavioral variability. Play appears to be most common in generalist species, especially those where the young are protected by and dependent on their parents for long periods before maturity. Because these are the species receiving the most specific environmental inputs during their development, play may provide sufficient time for the acquisition and refinement of various skills before they are required for survival.

Like play, tool-use is not as common as is exploration. Although tool-use is found in invertebrates and fish, these behaviors appear to be largely under the influence of specific genetic determinants for these species, whereas specific environmental influences are more important for tool-use in birds and mammals (especially primates). Thus, for some (but definitely not all) tool-using species, the exploration-play-application sequence offered by Vandenberg (1978) may apply—especially for certain birds and primates (i.e., macaques, baboons, capuchins, and the apes).

Research on the structure and function of object manipulation in nonhuman primates suggests that it may have evolved to facilitate the effectiveness of food acquisition and specific agonistic behaviors. Fine motor manipulatory skills appear more tied to the feeding context, whereas certain large motor behaviors are tied to the agonistic context. When comparing across various primate species, those showing the most complex manipulatory behaviors and tool-use patterns are those that are diurnal, terrestrial, bipedal, have few or no natural predators, have prolonged periods of immaturity in development where prehensile abilities precede locomotor ones, and possess anatomical characteristics that support visual and motor exploration (including color vision, stereoscopy, opposability, digit independence, tactile sensitivity, and large brains with well-developed visual and motor cortexes).

By integrating the findings of several studies, we have identified common forms of object manipulation in primates, including exploratory behaviors (investigating), play behaviors (transforming, relating, and other large motor behaviors), and patterns of tool-use (extending reach, amplifying mechanical force, augmenting agonistic displays, and controlling the flow of liquids; see Tables 2.1, 2.4, and 2.5). At least on the surface, the exploratory and play behaviors of primates are similar to those of many other mammals. However, when examining differences in primates across the phylogenetic spectrum, the primates closest to humans (i.e., the great apes and especially the chimpanzees) are most likely to engage in diverse and relational actions with objects. Humans are more likely than other primates to lift objects from their backgrounds for the purpose of manipulation, manipulate objects bimanually, manipulate objects with mobile parts, and engage in pretense. They are less likely to engage in buccal prehension. The few longitudinal studies that have been conducted on early object manipulation show that the more complex forms of manipulation emerge developmentally in a similar manner and order across several monkey and ape species.

The sophisticated tool-using and manipulative behaviors of the great apes (especially the chimpanzees) appears to be the consequence of an extended period of immaturity where opportunities for extended play and practice contribute to the acquisition and refinement of these skills. These flexible patterns of behavior make it possible for chimpanzees to adapt to a wide range of environments by perfecting those skills that function well within a particular environment. Finally, because object play is an effective means for generating behavioral diversity, it may, in fact, also facilitate adaptation to a wide range of environments in this way.

Although the picture presented in this chapter may appear coherent and internally consistent, the research on which it is based is rather limited. The cross-species comparisons are based mostly on gross descriptions of animal behavior, with little data on variations in the sequencing, timing, intensity, or quality of behavior. Much of the data were collected on animals in captivity, responding to limited or unnatural stimuli. Moreover, descriptions of animal behavior in natural settings are limited to only a few species. Finally, experimental evidence for the effects of exploration and object play are virtually nonexistent, and the hypothesized functions or effects of object-directed behavior can only be inferred from an examination of its structure and its relation to the environments in which it has presumably evolved. Clearly, several directions for future theory and inquiry have been identified.

3

Solitary Object Exploration and Play in Children

Child development researchers have long been interested in children's object-directed behavior. Major theories that have focused on the role of object-related experiences in child development include Piaget's (1952) theory of sensorimotor development, White's (1959) theory of mastery motivation, Hunt's (1965) theory of intrinsic motivation, and Gibson's (1988) theory of perception. These theories, along with many others, have stimulated a great deal of research on children's object-directed behavior.

The current review yields several conclusions. First, regardless of culture, the object-directed behaviors of humans during the first year of life are very similar to those of other primates during that same age period. There is, however, considerable divergence between human and other primates in the second year, with the development of complex relational and pretend play. Second, despite numerous cross-cultural similarities in the development of object-directed behaviors in young children, the total amount and quality of such behavior varies considerably by culture—possibly a result of cultural differences in opportunities for and the encouragement of object manipulation. Third, during the preschool and early childhood years, the nature of object manipulation changes dramatically—apparently as a function of the child's developing cognitive and motor skills. And finally, despite differences in the functions of exploratory, play, and tool-use behaviors, the nature of developmental differences in these contexts are remarkably similar.

The review is organized as follows. First, studies of undifferentiated object manipulation during the first 2 years of life are reviewed, to identify similarities and differences in the object manipulation of human infants and other primates. Also considered are studies employing higher level measures of manipulatory style and studies of developmental differences. Following this review are separate sections devoted to object exploration, play, and tool-use. Within each of these sections, the literature on the structure of object-directed behavior is reviewed, followed by a consideration of possible functions and effects. Following this is a presentation of studies of object-directed behavior in non-Western cultures, and then a consideration of the literature on gender differences in object-directed behavior. The chapter concludes with a new scheme for differentiating various types of object-directed behavior as a guide for future research.

STUDIES OF UNDIFFERENTIATED OBJECT MANIPULATION DURING THE FIRST 2 YEARS

As was the case in research with other animals, most studies of object-directed behavior during the first 2 years of life do not differentiate among object exploration, play, and tool-use, but instead provide general descriptive information on infants' manipulation of objects. In the typical study, infants and/or toddlers are given a standard object or series of objects to manipulate for a short period of time, usually in a laboratory setting. These are reviewed here, starting with the individual behaviors identified in the object manipulation of human infants.

Individual Behaviors

In general, the varieties of object-directed behaviors shown by infants differ little from those identified in other primates (especially the great apes). This can be seen by comparing the list of the object-directed behaviors in Table 3.1 with the behaviors for nonhuman primates in Table 2.1. The behaviors in Table 3.1 come from five descriptive studies of human children under the age of 2, and include all behaviors identified for children in this age range in the literature except pretend play, functional play (a category described later), and several categories of constructive play identified by Largo and Howard (i.e., container play, stacking, and grouping; 1979). If the Largo and Howard (1979) study had been included in the table, these behaviors would fall into the "relating" category and would be elaborations of "touching together."

Comparison of Tables 2.1 and 3.1 shows that the nonpretend, object-directed behaviors of both human and nonhuman primates fall into the same five categories: procuring, investigating, transforming, relating, and other large motor behaviors. Moreover, the specific behaviors within each cate-

TABLE 3.1

Types of Nonpretend Object Manipulation Identified in Studies of Human Infants and Toddlers

	Uzgiris (1967)	*Collard (1971)*	*Karniol (1989)*	*Palmer (1989)*	*Caruso (1993)*
Procuring					
Picking up[a,]			+	+	
Holding[a]	+		+		
Carrying[a]				+	
Transferring[a]		+	+	+	
Bilateral hold			+		
Two-handed hold			+		
Investigating					
Poking/touching	+	+	+	+	+
Mouthing/biting[a]	+	+	+	+	+
Scratching		+			
Pressing				+	
Rotating[a]	+	+	+	+	+
Bringing to eyes			+		
Instrumental sequential action			+		
Transforming					
Tearing/breaking	+		+		
Twisting/untwisting/bending			+		+
Wadding/squeezing	+		+	+	+
Relating					
Rubbing[b]	+	+			+
Hitting/striking	+	+	+	+	+
Dropping[c]	+	+		+	+
Touching together		+	+		+
Other large-motor behaviors					
Waving/shaking[c]	+	+	+	+	+
Pushing/pulling[a]	+	+	+	+	+
Throwing[c]	+	+			+

[a]Primate-common actions (Torigoe, 1985), [b]Simple secondary actions (Torigoe, 1985), [c]Complex secondary actions (Torigoe, 1985).

gory are, for the most part, the same. Examination of the differences show that humans were not described as engaging in buccal prehension, sniffing, licking, forming, rolling, draping/wrapping, and lining up objects; and nonhuman primates were not described as showing bilateral holding (holding a different object in each hand), two-handed holding (holding one object with two hands), instrumental sequential actions, bending, squeezing, and touching objects together. Examination of these differences in the context of the larger literature, however, suggests that it is unlikely that most reflect true differences between human and nonhuman primate behavior. Many undoubtedly reflect differences in the stimuli and coding systems employed in the studies used to make up the tables. For example, for human infants, the lack of forming and wrapping/draping is likely a consequence of the fact that none of the human studies used a rope for a stimulus, and the lack of rolling likely reflects the absence of stimulus materials that could be easily rolled. Moreover, human infants frequently line up objects during play (e.g., Gesell, 1940). Studies of tool-use in monkeys and apes have shown that nonhuman primates engage in bilateral and two-handed holding (Vauclair, 1993), instrumental sequential actions (Chevalier-Skolnikoff, 1989), and relational behaviors (i.e., touching objects together; Westergaard, 1993). Therefore, it appears that the only likely differences between human infants and nonhuman primates in the types of manipulative behaviors they exhibit are the absence of buccal prehension, object sniffing, and object licking in humans (probably a function of the more limited role of olfactory exploration in humans vs. other primates); and the limited pretend and high-level constructive capacitites in nonhuman primates (likely a function of humans' greater symbolic abilities).

The similarities in the major types of object manipulation observed in human and nonhuman primates is striking, and suggests that these behaviors have an evolutionary basis. At least in terms of the gross structure of manipulatory behaviors, human infants do not approach objects much differently than do other primates, especially the great apes. This is not to say that differences do not exist. Possible differences could involve differences in the timing, sequencing, or quality of object-directed behaviors that are not captured by these gross behavioral labels; or differences in functions of these behaviors or in the degree to which object-directed behaviors vary with specific object properties or situational variables. Moreover, although object-directed behaviors may be similar in their observed structure, they may have very different *meanings* for the participants. Reynolds (1993) made this point well in a discussion of object play in preschool children:

> Indeed, when the differences in locomotion type and hand morphology are taken into account, the video images of low-level action patterns of humans and nonhuman primates are close to identical. Nor are there profound differences in the complexity of the objects produced by preschool children.... Thus, a thorough-going behaviorist who compares human action patterns in a preschool to action patterns in a chimpanzee colony is right to insist that there are no profound discrepancies between the kinds of objects produced by apes and by children—for none are apparent on the video. However, anthropologists learned a long time ago to attend to the audio as well as the video, and a pretend cake can be discriminated from a lump of mud because the children *tell* you it is a cake. (pp. 408–409)

Higher Order Measures

As in the animal studies (e.g., Parker, 1974a, 1974b; Torigoe, 1985), research on children's manipulation of objects has gone beyond simply describing the individual behaviors that children engage in to deriving higher order measures of the type, style, and complexity of object-directed behavior. These are summarized in Table 3.2. Given the large number of constructs and measures in this area, several factor analytic studies have been conducted to identify the underlying dimensions of object manipulation. These studies show that object manipulation is a multidimensional phenomenon, and its structure cannot be summarized with just a few measures. Presented in Table 3.3 is a summary of three factor analytic studies of object manipulation in 9- to 12-month-old infants.

Although the results of these analyses are difficult to compare—given differences in sample sizes, ages of the participants, coding systems, measures, stimuli, and settings in which the observations occurred—enough similarities in the results exist to suggest some directions for future research. Despite the fact that each researcher presented a different multifactor solution, four factors were similar in at least two of the three studies: (a) a *sophisticated behavior* factor (Caruso, 1993; van den Boom, 1994); (b) an *attention span* factor, with perseverance as a major component (McCall, 1974; van den Boom, 1994); (c) an *examining* factor that combined fine-motor manipulation and total object time (McCall, 1974; van den Boom, 1994); and (d) a *mouthing* factor that contrasted mouthing with a more developmentally advanced behavior (McCall, 1974; van den Boom, 1994). Two additional factors not listed in the table are suggested by the results of Caruso's (1993) analysis. He found separate factors assessing the breadth and the depth of manipulation. *Breadth*

TABLE 3.2

Higher Order Measures of Type, Style, and Complexity of Object Manipulation

Type of manipulation

- "Appropriate" or functional behaviors—that is behaviors tied to the functional characteristics of objects (Belsky & Most, 1981; Largo & Howard, 1979; McCall, 1974; Rosenblatt, 1977; van den Boon, 1994; Zelazo & Kearsley, 1980).
- Relational manipulation (Belsky & Most, 1981; Largo & Howard, 1979; Power, Chapieski, & McGrath, 1985; Rosenblatt, 1977; van den Boom, 1994; Zelazo & Kearsley, 1980).
- Circular reactions or producing auditory/visual effects (McCall, 1974; Power et al., 1985; Yarrow et al., 1983).
- Vigorous manipulation (Caruso, 1993; Kopp, 1974; Largo & Howard, 1979; McCall, 1974; van den Boom, 1994).

Style of manipulation

- Physical intensity (Keller, Scholmerich, Miranda, & Gauda, 1987).
- Rate of object change or play tempo (e.g., Caruso, 1993; Fein & Apfel, 1979; Fenson, Sapper, & Minner, 1974; Johnson & Brody, 1977; Kagan, Lapidus, & Moore, 1978; Keller et al., 1987; McCall, 1974; Wenckstern, Weizmann, & Leenaars, 1984; van den Boom, 1994).
- Focused or sustained attention on individual objects (Czerniecki, Deitz, Crowe, & Booth, 1993; Fein & Apfel, 1979; Kopp & Vaughn, 1982; Power et al., 1985; Reppucci, 1970; Ruff & Lawson, 1990; Ruff, Lawson, Parrinello, & Weissberg, 1990; van den Boom, 1994).
- Persistence in goal-directed behavior (Power et al., 1985; Yarrow et al., 1983).
- Uniqueness (McCall, 1974; Power et al., 1985).
- Diversity: Behaviors performed (Caruso, 1993; Keller et al., 1987; Power et al., 1985); objects manipulated (Caruso, 1993; Fein & Apfel, 1979; Keller et al., 1987).

Complexity of manipulation

- Ordered play scales (Belsky & Most, 1981; Karniol, 1989; Uzgiris, 1967).

referred to the diversity of object manipulation, and included measures of the mean number of different schemes employed, as well as measures of behavioral tempo (the ratio of total behaviors to total time spent in object manipulation). *Depth* referred to the mean number of uses per manipulatory scheme. Whether these two factors are unique to Caruso's sample is impossible to determine because he was the only one of the three researchers to include diversity measures in his factor analyses.

The possible existence of these six factors is consistent with Belsky's (1980) notion that measures of the developmental level of object manipulation are conceptually distinct from measures of behavioral style. Sophisticated behavior—the developmental-level factor in these analyses—was

TABLE 3.3

Common Factors Identified in Studies of Infants' Object Manipulation

Study (n)	*Infant Age in Months*	*Sophisticated Behavior*	*Attention Span*	*Examining*	*Mouthing*
McCall, 1974 (44)	9–10		Perseveration vs. manipulate with visual regard	Manipulate with visual regard Length of session Secondary & tertiary circular reactions	Appropriate vs. mouthing
Caruso, 1993 (40)	11–12	Fine motor Combining			
van den Boom, 1994 (100)	9	Relational Developmental level Functional	Perseverance Attention span	Simple manipulation Total time exploring vs. vigorous play	Mouthing vs. undifferentiated exploration

separate from stylistic measures in two of the three studies, and many aspects of manipulatory style were distinct as well. The existence of separate examining and mouthing factors in two of the studies is consistent with the work of Ruff (1984), who suggested that mouthing can serve exploratory or nonexploratory functions, depending on the situation and the age of the child. Finally, the relative independence of the various dimensions is consistent with the findings of Power, Chapieski, and McGrath (1985), who found few significant correlations between their measures of developmental level, attention span, and exploratory diversity.

Developmental Course

Numerous studies have identified age differences in object manipulation during the infancy and toddler years. Together, these studies show that during the first 2 years of life, infants develop a wider range of schemes for the manipulation of objects, and advance from predominantly oral and/or large-motor manipulation of single objects to manipulative patterns that involve the coordination of the eyes and both hands, and/or the coordination of multiple objects. During this time period, children's manipulation becomes both more focused and sustained and more likely to exploit the functional and symbolic possibilities of the stimuli. These changes likely reflect the rapid motor and cognitive developments of this period, particularly the child's developing symbolic abilities, as well as the increasing planfulness of child behavior and the developing ability to decenter and focus on multiple aspects of an object (e.g., Case, 1985; Piaget, 1952).

A review of specific studies shows changes in manipulatory behaviors during the first 2 years of life for each of the major categories in Table 3.1. Object manipulation in the first few months is primarily oral (e.g., Rochat, 1989), with most developments in other manipulatory behaviors beginning at about 4 months. For procuring—the first category in the table—with increasing age infants are more likely to pick objects up for manipulation, as well as to transfer objects from one hand to another (Eppler, 1995; Palmer, 1989). They are also more likely to pick up more than one object at a time and to engage in two-handed holding (Eppler, 1995; Karniol, 1989).

Regarding investigating, with increasing age infants are less likely to mouth objects (Belsky & Most, 1981; Largo & Howard, 1979; McCall, 1974; Palmer, 1989; Ruff, 1984; Ruff & Dubiner, 1987), and they are more likely to examine objects through visually guided rotation (Eppler, 1995; McCall, 1974; Ruff, 1986; Ruff & Dubiner, 1987). Ruff (1984), however, noted that although two-handed rotation increased with age, one-handed rotation decreased. She also found that the nature of mouthing changed with age, with 6-month-olds showing more active, exploratory mouthing; 9-month-olds showing more gnawing; and 12-month-olds showing more passive mouthing (e.g., the mouth as a placeholder). Also with age infants are more likely to engage in instrumental sequential actions (Karniol, 1989; Piaget, 1952) and in role-differentiated manipulation with their two hands (Kimmerle, Mick, & Michel, 1995). Research on the fingering of objects is inconsistent: Three studies showed increases during this period (Eppler, 1995; Ruff, 1984 [Experiment 1], 1986 [Experiment 2]), two studies showed decreases (Oakes & Tellinghuisen, 1994; Palmer, 1989), and two showed no change (Ruff, 1984 [Experiment 2], 1986 [Experiment 1]).

Only one study could be found that documented a change in transforming behaviors (although these behaviors have been studied much less extensively than behaviors in the other categories). Palmer (1989) found that object squeezing increased from 9 to 12 months.

Relational behaviors all show increases with age during this time period, although many peak before 18 months. Nonfunctional relational manipulation, for example, increases between 9 and 12 or 13 months and then declines thereafter (Belsky & Most, 1981; Largo & Howard, 1979). Functional relational manipulation increases after 9 months as well, but peaks later, usually between 15 and 20 months (Belsky & Most, 1981; Fenson, Kagan, Kearsley, & Zelazo, 1976; Largo & Howard, 1979). Largo and Howard (1979) found that container play peaked at 15 months, and that stacking and grouping peaked between 15 and 21 months.

Finally, large-motor behaviors generally decrease during this time period. Largo and Howard (1979) reported decreases in vigorous play, and Fenson et al. (1976) reported decreases in banging. Eppler (1995), however, did document increases in sliding, rolling, and banging between 3½ and 5½ months, as did Palmer (1989) between 6 and 12 months.

In general, there is a decline in the kinds of object manipulation presented in Table 3.1 after 18 months of age (e.g., Wenar, 1976), with an increasing focus on pretend play during the latter half of the second year (see section on object play). However, research using higher order variables shows that both diversity (Fein & Apfel, 1979; Fenson et al., 1976) and attention span (Czcrniccki Deitz, Crowe, & Booth, 1993; Ruff & Lawson, 1990; Sarid & Breznitz, 1997) increase during the second, third, and fourth years.

Several researchers have offered ordinal scales of object manipulation during the first 2 years of life (e.g., Belsky & Most, 1981; Karniol, 1989; Uzgiris, 1967). Two that provide detailed information on nonpretend behavior are presented in Table 3.4. Both researchers argue that, with increasing age, infants develop a wider range of schemas for manipulating objects—not the notion that new schemas necessarily replace old ones. Note that although both scales in the table are made up of behaviors already discussed (see Tables 2.1 and 3.1), there is very little overlap in the two scales. Both, however, show that during the first 9 months of life, infants demonstrate patterns of manipulation that reflect their growing cognitive and motor skills. Future studies of this type could further identify the specific skills that develop during this time period.

STUDIES OF OBJECT EXPLORATION

Structure

Research on infants' responses to novelty show that although infants often display little hesitation in approaching novel stimuli before the age of 6 months at some point during the second half of the first year most infants begin to demonstrate signs of hesitation or wariness to the unfamiliar (Emde, Gaensbauer, & Harmon, 1976). Although wariness is greater for

TABLE 3.4
Two Ordinal Schemes for Describing Infant Object Manipulation During the First 9 Months of Life

Age	*Uzgiris (1967)*	*Karniol (1989)*
1½ months		Rotation
2 months	Mouthing	
2½ months		Translation
3 months	Visual inspection	
3½ months		Vibration
4 months	Hitting on surface	Bilateral hold
		Two-handed hold
4½ months		Transfer
5 months	Shaking	
5½ months		Coordinated action (single object)
6 months	Examining	
7 months	Differentiated schemes (tearing, pulling, rubbing, sliding, etc.)	Coordinated action (two objects)
7½ months		Deformations
8-9 months	Dropping/throwing	Instrumental sequential actions

social than for nonsocial stimuli (Eckerman & Rheingold, 1974) and declines over the second year of life (Sroufe, 1977), children continue to show wariness in response to objects or events that represent too great a violation of their expectations (Sroufe, Waters, & Matas, 1974). Research on attachment (see chap. 9) has shown that the presence of an attachment figure usually reduces the degree of wariness (Morgan & Ricciuti, 1969), and that infants often look to attachment figures to help them interpret unfamiliar objects or events (Feinman, 1992).

In an extensive case study of a 7-month-old infant yet to show much wariness, Hutt (1967) described the structure of infant exploration as follows:

> In general, when an object was offered it was grasped immediately, i.e., within 3 sec. It was then held in both hands for a few moments. With a novel object there was often accompanying apnoea lasting several seconds. At the beginning of the 7th month a "clutching" pattern usually with one, but sometimes both hands, was predominant. This consisted of fairly rapidly opening and closing the fist.... This would often be followed by half-turns of the wrist which resulted in different aspects of the object being viewed.

> Both the clutching and turning movements of the hand would then be performed with arm extended to the side.... Another motor pattern which appeared in this period was the transfer of the object from hand-to-hand.... A pattern which appeared later in the month was turning of the object over and over again while held in both hands.... All these manipulatory patterns were accompanied by visual inspection and an intent facial expression—the mouth would often be half-open, but sometimes the upper gum would bite the lower lip. (pp. 243–244)

Rochat (1989), in a study of 2- to 5-month-old infants, emphasized the predominance of oral exploration early in this time period, and documented a shift from oral to visual examination. In this study, "At 2 and 3 months, over two thirds of the analyzed infants start their multimodal exploration of the object with oral contact. Conversely, at 4 and 5 months, the majority of infants start a transport of the object into the field of view for visual inspection" (p. 876).

Piaget (1952) argued that, before 8 to 9 months, children's behavior with objects is focused on the practice of various sensorimotor schemata, and that the child shows limited interest in the object per se: "Interest is consequently not centered on the object as such but on its utilization" (p. 258). By 8 or 9 months, however, there is a shift of attention to the object itself:

> When the 8-month-old child examines a cigarette case or a hanging necktie everything transpires as though such objects presented a problem to his mind, as though he were trying to "understand." Not only does he look at such objects for a much longer time than the 4- to 5-month-old child before proceeding to acts, but furthermore, he engages in an ensemble of exploratory movements relating to the object and not to himself. He feels, explores the surface, the edges, turns over and slowly displaces, etc., and the last behavior patterns are very significant of a new attitude...it is as though the child said to himself when confronted by the new object: "What is this thing? I see it, hear it, grasp it, feel it, turn it over, without recognizing it: what more can I do with it?" (pp. 258–259)

Hutt (1966) presented a much less detailed description of exploration in 3- to 5-year-olds (the subjects explored a large box with numerous lights, buttons, and a lever): "In general, when the children entered the room they looked at the novel object immediately, or approached it, often asking the observer what it was. They would then examine the object manually or inspect it visually while holding the lever, and finally engage in active manipulation of the lever" (p. 67).

Comparison of these descriptions with descriptions of animal exploration reveals virtually the same patterns of behavior. Moreover, comparison of the specific behaviors Hutt (1967) and Piaget (1952) described in infants with those in Tables 2.1 and 3.1 show that all of the behaviors fell into the *procuring* and *investigating* categories—all behaviors observed in both human and nonhuman primates.

Most descriptive research on object exploration beyond the infancy period has been conducted to examine the differences in child behavior during exploration versus play. Across studies, the behavior codes for exploratory behaviors tend to fall into the *object procuring*, *manual investigation*, and *question-asking* categories (e.g., Hughes, 1978; Hutt, 1966; Schneider, Moch, Sandfort, Averswald, & Walther-Weckman, 1983; Vlietstra, 1978). Results of these studies confirm Hutt's (1970b) hypotheses about the differences between object exploration and play, in showing that exploratory behavior (as compared to behavior during play) becomes less likely as the object manipulation session progresses (Hutt, 1966; Schneider et al., 1983; Vlietstra, 1978; Wohwill, 1984), has sequences of longer duration (Hughes, 1978), is accompanied by an intent facial expression (Hutt, 1966), is accompanied by a lower heart rate (Hughes & Hutt, 1979), and is made up of more stereotyped (or less varied) sequences of behavior (Hughes, 1978). In support of this final characteristic, Hughes (1978) found that although nearly two thirds of all possible two-behavior sequences occurred during play, less than one quarter of all possible sequences were seen during exploration. Finally, in a cluster analysis, Hughes (1979) found that the behavioral clusters during exploration were more consistent and well defined than were those identified in play.

The other source of information on exploratory behavior postinfancy comes from several factor analytic studies of curiosity (Henderson, 1988; Henderson & Moore, 1979; Kreitler, Zigler, & Kreitler, 1975). In the typical study, children are given a series of curiosity assessments, including at least one task involving the manipulation of novel objects. In general, the results of these studies suggest that object manipulation, visual preference, task choice, and verbal measures of curiosity usually tap independent dimensions and, as such, do not assess a single, underlying construct. Such an interpretation is consistent with the results of Vandenberg's (1984) study of 4- to 12-year-olds that yielded no significant correlations among measures of preference for complexity, preference for the unknown, depth and breadth of object search, and question asking in a sample of over 100 children. Coie (1974) also found little evidence for convergence across tasks. Therefore, just as was the case for object manipulation in infants, curiosity is a multidimensional construct that requires a variety of measures for its assessment.

Developmental Course. Studies of developmental differences in the amount of time that children spend exploring are inconsistent. Some find that the amount of object exploration increases from the preschool to the early school years (Kleinman, 1979; Schneider et al., 1983; Switzky, Haywood, & Isett, 1974), some find that it decreases (Henderson, 1988; Henderson & Moore, 1979), and some find no differences. Given that the amount of exploration is at least partially a function of stimulus complexity (see later discussion), it is possible that the inconsistencies between studies could be due to differences in the stimuli employed. For example, if complex stimuli are used, one might find that exploration increases with age, because older children are more likely to appreciate stimulus features, whereas with simple stimuli, exploration might decrease with age, because older children process the rather limited amount of information quickly, and then turn to other activities.

Although there are no clear age differences in the *amount* of exploration, several studies suggest that the *nature* of exploration does show developmental changes. In general, with increasing age, children's exploratory behavior becomes more systematic and planful in nature. This has been documented in a number of studies.

Piaget and Inhelder (1956) were among the first to document such changes in a haptic exploration task, where children manipulated objects without seeing them. Three stages of exploration were identified. Stage I (3½ to 4½ years) is when the child engages in mostly passive and/or random exploration, such as holding the object or tactually examining parts of the object discovered by chance. Stage II (4½ to 6 years) is when exploration becomes more active, but was not always systematic. Stage III (6 to 7 years) is when exploration follows a systematic, general plan.

Abravanel (1968) and Zaporozhets (1969) further elaborated on Piaget and Inhelder's (1956) observations. Both researchers observed that although 3-year-olds spent most of their time indiscriminately "playing" with the figures, the behavior of 6- to 7-year-olds was more systematic. With increasing child age, exploration shifted from the use of the palms to the fingertips, was increasingly bimanual in nature, involved greater mobility of the hands, and was more systematic in that the entire outline or length of the stimulus was examined in an orderly fashion (vs. random explorations of various parts or segments). The final characteristic—orderly examination—suggests that, with age, children's behavior increasingly comes under the influence of their goals and plans to systematically familiarize themselves with every aspect of the stimulus (Vandenberg, 1984). Similar changes have been noted in more recent studies as well (e.g., Kleinman, 1979; Vandenberg, 1984; Wohwill, 1975).

Object Characteristics and Exploration. A significant amount of research has been conducted to examine the properties of stimuli that affect the amount and nature of exploratory behavior. Objects that elicit the greatest amount of exploration tend to be those that are *novel* (e.g., Fenson, Sapper, & Minner, 1974; Hutt, 1967; Oakes & Tellinghuisen, 1994; Ross, Rheingold, & Eckerman, 1972; Rubenstein, 1974; Ruff, 1981; Schaffer & Parry, 1970; Steele & Pederson, 1977; Willatts, 1983), *physically responsive* (McCall, 1974), and *make sounds when touched* (McCall, 1974; Rosenberg & Butler, 1982). Studies have also shown that the amount of *configural complexity* positively predicts the amount of exploration (e.g., McCall, 1974; Oakes & Tellinghuisen, 1994; Switsky et al., 1974; Wohwill, 1975). Although configural complexity effects are not always significant (McCall, 1974) and may vary with child age, younger children tend to prefer stimuli of intermediate complexity (Switzky et al., 1974). Although researchers initially thought that novelty effects on manipulation also varied by child age (e.g., Schaffer & Parry, 1970), more recent research has shown that this is not the case (e.g., Ruff, 1981). The effects of object responsiveness and sound potential appear to operate similarly across age, at least for the age ranges studied.

Functions/Effects

A variety of theorists have argued that considerable information is obtained through the oral and manual exploration of objects (e.g., Gibson, 1988; Piaget, 1952). Through object exploration, the child can obtain information about an object's: (a) size, shape, hardness, surface texture, weight, volume, consistency, rigidity, and material composition; (b) construction and components; and (c) functional/tool properties, sound production potential, and relationship with other objects—both actual (e.g., distance, parallelism) and possible (e.g., stackability, combinability; Derevensky, 1979; Gibson, 1962; Lederman & Klatsky, 1987; Pierce, 1991). According to Gibson (1988): "The active exploration of objects, leading to observable consequences and more specialized exploratory activities, has important results for learning about what an object affords, what can be done with it, its functional possibilities and uses" (p. 24). She also argued that object exploration is a basis for the classification of objects, as well as for learning about the causal relationships between events.

Researchers in the areas of perception and motor development have distinguished between various kinds of object manipulation in an attempt to describe how information is derived through manual exploration. One of the earliest distinctions was between *active* and *passive* touch (Gibson, 1962). More recently, a much wider range of manipulatory movements has been identified. Lederman and Klatsky (1987), for example, identified a number of specific "exploratory procedures" (p. 344) that they argued are used in deriving particular kinds of knowledge about object characteristics and properties (see Table 3.5). Turvey (1996) extended these further by examining how information about object length, weight, size, and orientation can be estimated by the manner in which subjects wield objects (i.e., the nature of *dynamic touch*).

Compared to the research on nonhuman primates, a substantial amount of evidence has been collected that demonstrates the information-gathering function of object manipulation in humans. The research methods include studies of: cross-modal perception, habituation, distractibility, expectancy violation, discriminative behavior, performance correlates of strategy use,

TABLE 3.5

Postulated Links Between Knowledge About Objects and Exploratory Procedures

Knowledge About Object	*Exploratory Procedure*
Substance-related properties	
Texture	Lateral motion
Hardness	Pressure
Temperature	Static contact
Weight	Unsupported holding
Structure-related properties	
Weight	Unsupported holding
Volume	Enclosure, contour following
Global shape	Enclosure
Exact shape	Contour following
Functional properties	
Part motion	Part motion test
Specific function	Function test

Note. From Lederman and Klatsky (1987). Copyright © 1987 by Academic Press. Adapted with permission.

experimental studies of exploratory mode, and correlates of individual differences in exploratory behavior. Together, these studies demonstrate, using a variety of methods, that through manual exploration children derive a significant amount of information about the objects that they explore, and that individual differences in exploratory behavior are associated with measures of general intelligence and problem-solving abilities.

Cross-Modal Perception. As was the case for nonhuman primates, studies of cross-modal (touch–vision) perception have demonstrated that infants can visually recognize objects that previously they had only been given the opportunity to touch (for review, see Rose & Ruff, 1987). This ability is well documented after the age of 6 months; the extent of younger infants' abilities in this domain is a matter of some disagreement (Brown & Gottfried, 1986; Gibson & Walker, 1984; Meltzoff & Borton, 1979).

Although the cross-modal paradigm confirms that information is gained through manipulation when objects cannot be seen, this does not demonstrate that nonredundant information is obtained through manual exploration when the object being explored can be seen (Ruff, 1989). Research, in fact, has demonstrated that manual behavior during visually guided exploration differs from behavior when the exploration cannot be visually monitored. Abravanel (1972), for example, found that although 4- and 5-year-olds in the haptic-only condition spent a great deal of time tracing surfaces, grasping parts of the objects, and locating features by hand, children allowed to explore both manually and visually used the hands primarily as "pedestals" for visual examination. This contrasts with the case for *infants*, where the opportunity to explore both manually and visually results in *more* object rotation and/or fingering than does manual exploration alone (Gibson & Walker, 1984; Rochat, 1989).

Habituation. Ruff (1984, 1989) argued that one way to study information gathering during visually guided manual manipulation is to examine the effects of repeated object exposure. If information is being gained through a particular manipulatory behavior, then the behavior should become less frequent with repeated exposure as information about the object is acquired, and become more frequent once a new object is presented. In a summary of the results of several studies using this approach, Ruff (1989) identified four behaviors (i.e., visual inspection paired with handling, fingering, transferring, and object rotation) that can be labeled as *examining behaviors* (after Uzgiris, 1967), because they each decrease with repeated object exposure and recover after exposure to novelty. Other behaviors such as throwing or banging either increase or show no changes in occurrence over time (see the section on object play later in this chapter), and therefore cannot be considered examining by these criteria. Ruff and her colleagues (Ruff, Saltarelli, Capozzoli, & Dubiner, 1992) used the same methods to differentiate between several kinds of object mouthing: active mouthing (mouthing followed by a look at the object), gnawing (probably related to teething), sucking, and place holding. Of these, only active mouthing appears to have a clear information-gathering function. Moreover, in a study of 6-, 9-, and 12-month-olds, she found that active mouthing was most common at 6 months and declined thereafter: "One method—mouthing—drops out as the infant becomes a more efficient explorer of objects and the surviving method—manual activity associated with vision—provides the infant with a system in which there is useful redundancy from two modalities and a great deal of flexibility" (Ruff et al., 1992, p. 860).

Oakes, Madole, and Cohen (1991) demonstrated another use of the habituation paradigm. They found that when 6- and 10-month-old infants were allowed to examine different exemplars of an object category, over trials their examining behavior decreased and then increased again when they were given an object from a different category.

Distractibility. Oakes and Tellinghuisen (1994) devised a creative way of demonstrating that active cognitive processing goes on when infants are engaged in the kinds of examining behaviors observed by Ruff (1989) and Oakes et al. (1991). They found that 7- and 12-month-old infants were less distractible when engaged in examining than in nonexamining behaviors (see also Tellinghuisen & Oakes, 1997). Both Ruff, Capozzoli, and Saltarelli (1996) and Lansink and Richards (1997) found this same tendency, with the latter study using both behavioral and heart-rate measures of focused attention.

Expectancy Violation. Expectancy violation studies also demonstrate the information gathering that occurs during visually guided object manipulation. Pieraut-Le-Bonniec (1985) found that, by 9 months (but not before), infants spent more time examining "deceptive" objects that "seemed to be open but resisted the introduction of the hand" (p. 413) than their solid, opaque counterparts. The examining behaviors observed were tied to the nature of the deception. Nine-month-olds scratched and pushed on the transparent surface with their index finger, thumb, or palm in apparent attempts to test the object's solidity.

Bushnell (1982), in a study of 8-, 9-½, and 11-month-olds, used (a) a "virtual object task" that employed mirrors, in which infants saw an image of an object that was not there, and (b) two cross-modal tasks, in which the object that the infant felt was different in shape and texture than what the infant could see. Compared to their responses in the control trials, 9½- and 11-month-old infants showed greater touching and looking in the "trick" trials, whereas 8-month-olds did not respond differentially across conditions.

Finally, Baldwin, Markman, and Melartin (1993) demonstrated that 9-month-olds showed increases in specific object-related behaviors (e.g., squeezing a horn) when expectations about the object developed during earlier exploration had been violated.

Discriminative Behavior. A further way to demonstrate the information-gathering function of exploratory behaviors is to examine the relationship between specific behaviors and characteristics of the objects manipulated:

> After the infant picks up some initial information about the object...exploratory behaviour might be expected to become more specific to the properties of the object being explored.... Such discriminative behavior...though...may not be sufficient to show that the behaviour is used for the purpose of gathering information about the objects. To probe further, we need to isolate the property being attended to and determine whether different manipulative behaviours are employed to explore different novel properties. (Ruff, 1989, p. 305)

Several studies have done just that. As summarized in Table 3.6, familiarizing infants with an object or objects and then presenting a new object that varies along a single dimension generally leads to an increase in the kinds of exploratory behaviors that would best yield information about that dimension. As shown in the table, this has been demonstrated during the first year of life for changes in hardness, shape, texture, weight, size, rigidity, and sound potential. Moreover, the behaviors tied to the specific dimensions are similar to those identified by Lederman and Klatsky (1987) in adults (cf. Table 3.5).

That discriminative behavior toward objects develops gradually during the first year was demonstrated by Kimmerle et al. (1995). They found that the degree to which infants showed role differentiation of their two hands in the manipulation of objects did not vary by object characteristics at 7 months, but the type of toy did influence bimanual role differentiation at 9, 11, and 13 months.

TABLE 3.6

Discriminative Exploratory Behavior as a Function of Stimulus Dimensions

Study	*Age in Months*	*Stimulus Dimension(s)*	*Stimulus-Behavior Connections*
Gibson & Walker (1984)	12	Hard-soft	Hard—banging Soft—presses (squeezes)
Ruff (1984)			
Study 1	6, 9, 12	Shape	Shape—active mouthing, transferring
		Texture	Texture—fingering
Study 2	9, 12	Shape	Shape—looking, fingering, rotating, transferring
		Texture	Texture—looking, fingering
		Weight	Weight—less examining
Rochat (1989)	3, 4	Shape, size, color, texture (all at once)	Scratching, mouth to eyes, eyes to mouth
Palmer (1989) Experiment 2	6, 9, 12	Weight, size, texture, rigidity, sound potential	Weight—waving, mouthing, switching, squeezing, scooting
			Size—switching, picking up, releasing, holding
			Texture—picking up, scooting, releasing, squeezing, mouthing
			Rigidity—mouthing, squeezing
			Sound—waving, mouthing

Performance Correlates of Manual Strategy Use. Studies with older children have shown that the use of specific exploration strategies is associated with the acquisition of specific object information. Wolff (1972), in a study of 4- to 6-year-olds found that the amount of "stimulus correlated activity" during visually guided manual exploration (i.e., "motor activity produced by S which was in some way correlated with the shape of the stimulus form," p. 432) correlated negatively with trials to criterion in a later visual recognition task. Interestingly, the association between trials to criterion- and stimulus-correlated activity decreased with age: 4 years, $r = -.72$; 5 years, $r = -.42$; 6 years, $r = -.18$. This appeared to be due to the fact that older children did well on the task, regardless of the amount of manual activity they exhibited. Kleinman (1979) identified specific exploratory behaviors that correlated with later stimulus-matching accuracy in a study of haptic exploration among kindergartners, second graders, fourth graders, and college students. Regardless of age, matching accuracy was positively correlated with thoroughness of exploration ($r = .73$), tracing the stimulus object ($r = .54$), and systematically comparing the stimulus features ($r = .71$).

Experimental Studies of Exploratory Mode. Further evidence for the role of manual exploration in information gathering comes from experimental studies showing that subjects assigned to active manual exploration display better later visual recognition performance than do subjects assigned to either passive touch, experimenter-guided touch, or visual inspection alone (e.g., Jones, 1972; Wolff, 1972). In an extension of this paradigm, Goodnow (1969) found that kindergartners who were given the chance to actively handle objects (i.e., a tissue, a paper clip, and a screwdriver) later generated on a task of divergent thinking more nonstandard uses for the objects than did students in a visual-inspection-only condition. The groups did not differ on the number of standard uses. Fu, Kelso, and Moran (1984) replicated this result in a more recent study.

Individual Differences in Exploratory Behavior. Finally, researchers interested in the function or effects of exploratory behavior have tried to identify individual differences in children's exploratory behavior and to examine their relationships with other aspects of child functioning. These relationships are summarized in Table 3.7. Yarrow and associates (Jennings, Harmon, Morgan, Gaiter, & Yarrow, 1979; Messer et al., 1986; Messer, Rachford, McCarthy, & Yarrow, 1987; Yarrow et al., 1983), in a multiple-method, longitudinal study of mastery motivation in infants, reported several relationships between infant exploratory behavior and measures of intellectual functioning. Although the results of this study were complex, and often at variance with their hypotheses, several findings revealed positive associations between exploratory behavior and infant performance on standardized assessments of cognitive functioning. For example, the amount of exploratory behavior at 6 months predicted Bayley scores at 12 months (Yarrow et al., 1983), and the practice of "sensorimotor skills" at 6 months predicted McCarthy scores at 30 months (Messer et al.,

TABLE 3.7

Correlates of Individual Differences in Exploratory Behavior in Infants and Toddlers

Exploratory Behavior Measure	*General Cognitive Functioning*	*Problem Solving*
Amount of exploratory behavior	Yarrow et al. (1983)	Yarrow et al. (1983)
	Messer et al. (1987)	Caruso (1993)
Sensorimotor Practice	Messer et al. (1986)	
Cognitive maturity	Jennings et al. (1979)	
Active object engagement	Messer et al. (1987)	
Examining	Ruff & Dubiner (1987)	
Sustained attention/continuity	Pedersen & Wender (1968)	Jennings et al. (1979)
Play tempo		Kagan et al. (1978)
		Caruso (1993)
Breadth		Messer et al. (1987)
		Caruso (1993)

1986). At 12 months, the cognitive maturity of object manipulation was positively correlated with 12-month Bayley scores (Jennings et al., 1979), and at 30 months, active object engagement and the extent of investigation were both positively associated with McCarthy scores, whereas the amount of "no engagement" showed a negative correlation (Messer et al., 1987). In a different study, Ruff and Dubiner (1987) found that the amount of time that 12-month-olds spent examining during novelty trials (i.e., looking, rotating, and fingering) was positively correlated with Bayley performance, whereas the amount of time they spent not exploring (i.e., mouthing, banging, and dropping) showed a negative correlation.

Two studies examined the relationship between measures of sustained attention during object manipulation and developmental assessments. Although these investigators did not differentiate between exploration and play, they are reviewed here because much of what they were assessing was exploratory behavior. In the first study, Pedersen and Wender (1968) found that teacher ratings of "sustained directed activity" at age 2½ were positively correlated with performance IQ on the WISC at age 6½. In the second, Kagan, Lapidus, and Moore (1978) examined the relationship between play tempo at 8 and 27 months, and later developmental outcomes. They found that slow play tempo in infancy predicted greater response times and fewer errors in the Embedded Figures Test at age 10.

Yarrow and associates also examined the relationship between exploratory behavior and infant problem solving. Again, although the results were mixed, exploratory behavior at 6 months was positively associated with exploratory behavior during problem solving at 12 months (Yarrow et al., 1983), and the continuity of exploration at 6 months predicted persistence in 12-month problem solving (Jennings et al., 1979). At 30 months, the extensiveness of exploration (i.e., breadth) was positively associated with positive problem engagement during problem solving, whereas the distribution of investigation (the degree to which the child focused for long periods of time on individual objects) was negatively related (Messer et al., 1987).

Caruso (1993) further examined the relationship between exploratory behavior and problem solving in a study of 11- to 12-month-olds. He found that the breadth of exploration was positively correlated with success in problem solving, and with ratings of the cognitive sophistication of problem solving. Other positive relationships were found between behavioral tempo during exploration and persistence in problem solving, and between the total amount of time exploring and the number of strategies used during problem-solving.

Although individually confusing and often inconsistent, the results of these studies suggest possible relationships between individual differences in exploratory behavior in infancy and later measures of problem solving and general cognitive functioning. By considering the results of these studies in the context of the individual difference factors presented in Table 3.3, it is interesting to note that the measures of sophisticated behavior (practicing sensorimotor skills, cognitive maturity) and examining (examining, active object engagement) tended to be associated with measures of general cognitive functioning, whereas other measures of exploratory style (e.g., continuity, breadth, tempo) tended to be associated with behavior during problem solving (see Table 3.7). Total amount of exploratory behavior was associated with both.

Observational assessments of object exploration during the preschool years are rarely associated with standardized measures of intelligence (e.g., Henderson & Wilson, 1991), although teacher ratings of curiosity often are (e.g., Langevin, 1971). The observed amount of object exploration by children (ages 4 to 7) in object manipulation tasks, however, does correlate positively with success in tasks involving discrimination learning (Kreitler, Zigler, & Kreitler, 1984), exploratory learning (Inagaki, 1978), and social problem solving (Minuchin, 1971). Although the number of studies is limited, the range of skills involved varies considerably (from identifying schedules of reinforcement to understanding the operation of an electromagnet to generating resolutions to social dilemmas). Reppucci (1970), in a study of 2-year-olds, also found that object exploration predicted performance in a discrimination learning task, but in this case the best predictor was the amount of sustained activity with objects during free play.

Behavior during exploration also has been found to be associated with psychological adjustment. McReynolds, Acker, and Pietila (1961) found that the amount and diversity of exploratory behavior during an object curiosity task in 11-year-olds was positively correlated with teacher ratings of psychological adjustment. Additionally, Ruff, Lawson, Parrinello, and Weissberg (1990) found that both quantitative and qualitative measures of sustained attention during exploration at 2 years were negatively associated with mother ratings of behavior problems at 3.5 years.

Conclusions. In summary, exploratory behavior in infancy and childhood appears to serve an information-gathering function. Using a variety of methods, researchers have demonstrated that during exploration infants and young children extract at least short-term information about the characteristics of objects, including information about texture, hardness, weight, shape, size, and sound potential. Researchers have also shown the existence of specific exploratory procedures to derive such information. These exploratory behaviors become more planful and systematic, and are less driven by stimulus characteristics, with increasing child age. Moreover, the use of systematic exploratory strategies is associated with greater information yield. There appear to be reliable individual differences in exploratory style that are related to child development outcomes. Although insufficient data have been collected to draw any firm conclusions about the pattern of correlations, the data suggest that promising outcome areas for future research include psychological adjustment and problem-solving skills. One might expect that exploratory style might be particularly relevant for tasks that require the active seeking, generation, and integration of information. Because correlation does not imply causation, whether exploratory style plays a causal role in the development of problem-solving skills is still an open question.

STUDIES OF OBJECT PLAY

As discussed earlier, Hutt (1970b) and others have argued that object play usually follows object exploration, beginning after the child has answered the implicit query "What does this *object* do?" and has moved on to the question of "What can *I* do with this object?"

Structure

In her classic study of exploration and play in 3- to 5-year-olds mentioned earlier, Hutt (1966) described the transition from exploration to play as follows:

> As investigation of the object decreased other activities involving it increased. When analyzed these consisted of repetitive motor movements, manipulations of long duration accompanied by visual inspection of other objects, and a sequence of activities incorporating both the novel object and other toys—in other words a "game." Examples of these were respectively: patting the lever repeatedly, leaning on the lever making the bell ring continuously while looking around the room, and running round with the truck ringing the bell each time the object was passed. There is another group of responses that can be termed "transposition-of-function"—those responses that resulted in the object explicitly fulfilling another function, e.g., something to climb, a bridge, or a seat. All these activities (i.e., repetitive movements, "games", and 'transposition-of-function' responses) are those which an observer would recognize and label as *play*. (p. 70)

In studies of infants and toddlers, researchers usually reserve the phrase *object play* for instances of pretend behavior, or for behaviors in the "relating" and "other large-motor behaviors" categories in Table 3.1. In their extensive review chapter on children's play, Rubin et al. (1983) stated that the first pretend gestures usually occur around 12 to 13 months of age, although they may occur somewhat earlier. Early forms of pretend are generally "decontextualized" behaviors: "The behavior is detached from the situational context in which it ordinarily occurs (e.g., bedtime, mealtime), from the outcome with which it is ordinarily associated (e.g., sleep, food consumption), and from the behavioral signs that ordinarily accompany a real need (e.g., crankiness for sleep, finger sucking for nourishment)" (p. 717).

A review of the literature shows that the early pretend behaviors identified by researchers fall into the following categories: eating and food preparation, grooming, caretaking, bedtime routines, talking on the telephone, and "animating" inanimate objects (e.g., having dolls or stuffed animals perform actions, having a toy car make engine sounds; Belsky & Most, 1981; Fein & Apfel, 1979; Largo & Howard, 1979; Lowe, 1975; Nicolich, 1977; Rosenblatt, 1977).

Nonpretend behaviors explicitly labeled "play" by infancy researchers include rubbing, hitting, dropping, waving, shaking, pushing, pulling, throwing, and other relational behaviors (Collard, 1971; Largo & Howard, 1979; Ruff, 1984, 1986). In research on older children, the term *play* is also used to describe pretend, relational, and large-motor behavior with objects. Other behaviors labeled *play* include repetitive motor movements, rule-bound "games," and "unconventional" manipulation (Hughes, 1978; Hutt, 1966; Schneider et al., 1983; Switzky et al., 1974). Although a review of the infant and child literature revealed no studies mentioning behaviors in Table 3.1's "transforming" category as examples of play, these behaviors seem to be likely candidates as well.

Developmental Course. Studies of age differences in play behaviors reveal a decrease in large-motor behaviors with age (especially vigorous object play and banging—see the earlier section on undifferentiated object manipulation), and numerous transformations in the nature of pretend behavior as well. The remaining studies concern age differences in the nature of relational play—especially play involving construction activities with blocks and similar objects.

Haight and Miller (1993), in a longitudinal home observational study from 1 to 4 years, found that the mean time pretending (solitary and social combined) increase significantly after 12 months of age. In this study, the rates of pretending at 12, 24, 36, and 48 months were .06, 3.3, 7.8, and 12.4 minutes/hour, respectively. The nature of pretend play changes as well. By observing age differences in the pretend play of children in solitary free-play settings, as well as observing the kinds of pretend behaviors that children of different ages will imitate in structured experimental settings (for a methodological review, see McCune-Nicolich & Fenson, 1984), three major changes have been identified (see reviews by Fenson, 1984; Rubin et al., 1983). The changes, first noted by Piaget (1962), reflect the increasing tendency with age for children to: (a) direct pretend behaviors toward others versus the self ("decentration"—e.g., feeding a doll with a spoon); (b) engage in pretend behaviors where one object is substituted for another ("decontextualization"—e.g., using a block to represent a dish); and (c) engage in extended sequences of planful, pretend behavior ("integration"—e.g., the child "pours" pretend liquid into a cup and then gives a stuffed animal a "drink"). Rubin et al. (1983) also noted, as did Piaget (1962), that with increasing age, pretend play is more likely to occur in social versus solitary contexts.

These developmental changes in the nature of pretend behavior are so reliable that they serve as the basis for several pretend play scales that have been validated by various scaling procedures (e.g., Guttman scalograms).

Four of these scales are summarized in Table 3.8. As can be seen in the table, with the exception of the Corrigan (1987) scale (which focuses on object-agent relationships—see the discussion of this topic later in this chapter), the scales are very similar and incorporate the developmental changes described earlier.

TABLE 3.8

Pretend Play Scales

Nicolich (1977)	*Belsky & Most (1981)*	*Corrigan (1987)*	*Tamis-Lemonda & Bornstein (1991)*
	Mouthing		
	Simple manipulation		
	Functional		Unitary functional
	Relational		Inappropriate combinatory activity
	Functional relational		Appropriate combinatory activity
Presymbolic schemes (recognitory gestures)	Enactive naming (approximate pretend activity)		Tangential pretend
		Agent Object	
Auto-symbolic (self-directed)	Pretend self	Self-realistic	Self-directed pretense
Single-scheme symbolic games	Pretend other	Other-realistic (passive)	Other-directed pretense
	Substitution	Self-substitute	
		Other-Substitute (passive)	
Combinatorial symbolic games	Sequence pretend		Sequential pretense
Planned symbolic games	Sequence pretend substitution	Substitute-realistic (passive)	Substitution pretense
	Double-substitution	Substitute-substitute (passive)	
		Other-realistic (active)	
		Other-substitute (active)	

Studies since the Rubin et al. (1983) and Fenson (1984) reviews have elaborated on the developmental changes described earlier, some of which have been incorporated into the scales presented in Table 3.8. For example, as pretend play becomes more decentered, children first employ the "other" as a recipient of the pretend act (e.g., the doll *gets* a drink), and only later begin to employ the other as the agent of the act (e.g., the doll *gives* someone or something else a drink; e.g., Corrigan, 1982, 1987; Fenson & Ramsay, 1980; Lowe, 1975; Lyytinen, 1991a; Watson & Fisher, 1977; Watson & Jackowitz, 1984).

Regarding decontextualization, or object substitutions, children show age-related changes as well. Although children first engage in these actions with objects similar to the object to be symbolized, the ability to substitute a dissimilar object or to pretend with no object is not demonstrated until later. Double substitutions are seen at the latest age of all (e.g., Casby & Corte, 1987; Copple, Cocking, & Matthews, 1984; Crum, Thornburg, Benninga, & Bridge, 1983; Elder & Pederson, 1978; Fein, 1975; Jackowitz & Watson, 1980; Overton & Jackson, 1973; Pederson, Rook-Green, & Elder, 1981).

Finally, researchers have developed new ways to demonstrate the increasing sequential integration of solitary pretend play. Bretherton (1984) interpreted age-related changes in pretend play in terms of Nelson's (1981) script framework, and both Le Normand (1986) and Eckler and Weininger (1989) described age-related changes in pretend play that demonstrate its evolving sequential nature. Le Normand (1986, p. 125) identified four developmental levels of pretend play between 2 and 4 years: representational ("representation of something by something else by acting out"), conceptual ("classification of actions with language as anticipating"), programmatic ("organization of sequences of actions into scenes with or without language"), and scripts ("organization of play with themes").

In one of the few studies of age differences in the nature of solitary pretend play from early to middle childhood, Eckler and Weininger (1989) demonstrated the utility of applying story grammar methods to the study of pretend play. Children were given a set of toys, told to pretend, and then asked to explain to the experimenter what they were doing during their play. Play sessions were coded using a five-level system for assessing the highest level of structural complexity of children's play demonstrated during the sessions (see Table 3.9). Coding of play complexity was accomplished by determining whether the play narrative could be divided into a series of problem-solving "episodes" taking the form presented at the bottom of Table 3.9. Play sessions not made up of episodes were classified as "iden-

TABLE 3.9

Levels of Complexity of Children's Play Using Story Grammar Analyses (Eckler & Weininger, 1989)

Pre-episode play	
Identification	Play focusing on identifying the objects and understanding their functions
Setting up	Play focusing on what the objects *could* be or do, and what child wants the objects to be or do
Episode play	
Level 0	Play involving one episode[a]
Level 1	Play involving two or more episodes linked by *then* or *cause* (or both); for example, episode *then* episode
Level 2	Play involving three or more episodes following the form:
	episode *then* episode
	then
	episode

[a]Episodes took the form:
Initiating event causes protagonist to desire goal
Protagonist tries to get goal
Protagonist does method toward goal
Consequence = Goal
Outcome = Goal

tification" and "setting up" (see Table 3.9); play sessions including episodes were coded at Levels 0, 1, or 2, depending on the highest level of play observed. Results showed that play complexity assessed with this five-point scale showed a correlation of $r = .70$ with child age.

Consideration of these age-related differences in the nature of children's solitary pretend play reveal many of the same developmental differences identified in the studies of exploration reviewed earlier, as well as studies of nonpretend play and tool-use to be reviewed subsequently. As mentioned previously, with increasing age, the child's pretend play involved a focus on and coordination of multiple objects; a consideration of multiple aspects and dimensions of the situation; a move from individual behaviors to complex, hierarchically organized behavior patterns; and sustained attention and longer sequences of behavior. Therefore, some of the same cognitive processes that underlie developmental changes in the nature of nonpretend behavior likely underlie the development of pretend behavior as well.

The developmental changes unique to pretend play involved the child's developing symbolic abilities: the ability to shift one's perspective from the self to another (i.e., the shifts from self-directed to other-directed pretend, as well as from object to agent pretend); the ability to represent one object with another, even when the substitute object differs considerably in its physical structure and function; the ability to pretend that an object is present when it is not; and the ability to engage in "double substitutions" where both agent and object are substituted objects.

Piaget (1962) argued that these age-related differences in pretend reflected the development of the "symbolic function," which also makes possible the development of both language and cognitive operations. His hypotheses have been adopted by others, and have been the impetus for numerous studies investigating the relationship between language development and the nature of pretend play. It is now well documented that there is a strong relationship between the level of language development and the level of solitary pretend play behavior (e.g., Casby & Corte, 1987; Corrigan, 1987; Kelly & Dale, 1989; Lyytinen, Poikkeus, & Laakso, 1997; McCune, 1995; Shore, 1986; Ungerer & Sigman, 1984). Most investigators believe that there is a reciprocal relationship between the development of language and the development of pretend play, in which both are "a function of the underlying capacity for material representation, emerging in the context of a system of related skills" (McCune, 1995, p. 204).

Besides documenting developmental changes in the nature of pretend behavior, researchers have also examined changes in children's understanding of pretense versus reality (for reviews, see Harris & Kavanaugh, 1993; Joseph, 1998; Lillard, 1998; Woolley, 1997). Several theorists have argued that this ability and the ability to pretend in general relate to the development of children's "theory of mind"—a topic that has been the focus of considerable recent research and theoretical attention (for reviews, see Hobson, 1990; Leslie, 1987; Lillard, 1993). However, review of these topics is beyond the scope of the current volume.

Studies of development in nonpretend forms of play show similar developmental progressions. Gesell (1940), for example, identified five stages in children's block construction. Between 12 and 18 months, children show little interest in construction, but instead usually play with the blocks as single units. Between 18 and 24 months, most children build a tower or put the blocks in a row. At 3 years, children usually arrange the blocks in a row, with the vertical arrangement "superseded by horizontal alignment" (p. 111). At 4 years, elaborate two-dimensional structures are constructed, and at 5 years complicated three-dimensional structures are common. At both 4 and 5 years, children usually name the structures.

Other researchers have confirmed and elaborated on Gesell's observations. They have noted the increasing complexity of block constructions with age (e.g., Johnson, 1974; Reifel, 1984; Stiles-Davis, 1988; Zervigon-Hakes, 1984), as well as confirmed the shift from vertical to horizontal constructions during the third year (Stiles-Davis, 1988). Reifel (1984) elaborated on the changes around age 4, noting that enclosures replace towers, rows, and other simple spatial relations at this time. Several authors observed that, with increasing age, children's constructions become representational, that is, models of real-world objects (Johnson, 1974; Zervigon-Hakes, 1984). When this is the case, children become more concerned about ends (i.e., what they are trying to build—the end product) than means (the building process itself; Christie & Johnsen, 1987). Christie and Johnsen (1987) maintained that this trend continues into the elementary school years, and serves as the transition into craft activities.

In one of the most thorough examinations of age-related changes in construction play to date, Stiles-Davis (1988) conducted molecular analyses of the activities and constructions of children between the ages of 18 and 42 months. She observed increases as a function of child age in the number of spatial loci in the structures, the number of spatial directions the constructions extended, and the number of spatial relationships demonstrated (e.g., on, next to, between, lean against). Examination of construction processes showed that simple reiterative processes became less common with age, whereas multiple sequential and multiple alternation processes increased. Stiles-Davis argued that these developments: "reflect changes in [the] ability both to construct spatial relationships between objects and to coordinate sets of relations in order to form more complex spatial groupings" (p. 530). She noted that the shift from vertical to horizontal structures during the third year reflects differences in the spatial requirements of these two construction forms.

Other changes in children's constructive play during the preschool years have been documented in two detailed studies of children's manipulation of nesting cups (DeLoache, Sugarman, & Brown, 1985; Greenfield, Nelson, & Saltzman, 1972). Both studies showed that children's constructive behavior becomes more hierarchically organized with age, as children replace the "pot" strategy for construction (i.e., putting single cups into other single cups or stacks of cups) with the "subassembly" strategy (i.e., putting one cup into another and then putting this assembled unit into a third cup or stacked unit). Children's error correction strategies in such tasks change with age as well (DeLoache et al., 1985). When two cups do not fit, younger children are more likely than older children to use the "decompose" strategy (i.e., take part or all of the stack apart and then either leave the pieces disassembled or reassemble them as they originally were), whereas older children are more likely to either take the stack apart and reassemble it in a "different and improved way" (the "insert" strategy) or to put the nonfitting top cup on the bottom (the "reverse" strategy). These differences reflect "important developmental trends of increasing flexibility and expanding scope of thought and action: The children became increasingly able to undo an unsuccessful action and institute an alternative approach, and they became increasingly able to coordinate multiple relationships rather than focusing on a single object or relation" (DeLoache et al., 1985, p. 938).

Greenfield (1991), in a broadranging, interdisciplinary theory, argued that the increasing hierarchically organized nature of children's sequential behavior during construction play parallels developments in the area of language production, and that both reflect the operation of a common neural substrate that evolved to support language and tool-use in ancestors of humans and the great apes. In an extended description of this theory, she presented behavioral, language, neuropsychological, and comparative data to support her hypotheses. A number of commentaries follow.

Functions/Effects

Despite the wide range of functions that have been offered for object play, research has focused mostly on its role in generating behavioral diversity and facilitating problem solving. Hutt and Bhavnani (1972), in a follow-up of Hutt's original preschool subjects, found that children who demonstrated the most uniqueness in their patterns of object play at ages 3 to 5 scored higher on creativity assessments at ages 7 to 10 than did preschoolers who simply investigated the novel object but did little else. Based on parent-, teacher-, and self-reports at ages 7 to 10, children who earlier had demonstrated the most unique object play were described as being more independent, curious, and socially skilled than were children from the other two groups. The relationship among diversity, play, and problem solving was illustrated by Cheyne and Rubin (1983), who found that configural richness during free play with objects was negatively associated with solution time in a later problem-solving task.

Although these results are consistent with the hypothesis that play has a causal role in the development of creativity and problem solving, they tell us little about the direction of effects, nor do they rule out the possibility that a third variable, such as intelligence, might account for the shared variance between the variables of interest. For example, Johnson, Ershler, and Lawton (1982) found that the observed amount of construction play was positively associated with children's performance on standardized assessments of intelligence. Similarly, Cheyne and Rubin (1983) found that solution time on their problem-solving task was negatively associated with children's mental age.

Such problems in the interpretation of correlational data led Sylva, Bruner, and Genova (1976) to develop an experimental method for demonstrating the effects of object play on children's problem solving. Basing their study on earlier work with nonhuman primates, Sylva et al. (1976) employed a variation of Kohler's (1925) lure retrieval task. They found that children who were given a chance to play with the task materials before problem solving required fewer hints to complete the task once problem solving began than did a group who only observed the experimenter demonstrate the principles required in solving the problem. Given the positive nature of these results, a number of investigations were conducted in the mid- to late 1970s that experimentally investigated the role of play in problem solving (e.g., Pepler & Ross, 1981; Smith & Dutton, 1979; Vandenberg, 1981) and in divergent thinking (i.e., studies of associative fluency employing alternative uses tasks; e.g., Dansky, 1980a; Dansky & Silverman, 1973, 1975).

Although the results of these early studies were promising, suggesting that play facilitates problem solving and divergent thinking, several critiques of the literature were published in the early 1980s that identified a number of methodological problems with this research (e.g., Cheyne, 1982; Pellegrini, 1984–1985; Rubin et al., 1983; Smith & Simon, 1984). The major problems identified were: The effects of play were usually not differentiated from the effects of exploration—in fact, there was concern that play did not even occur in some of the studies; some designs confounded experimental condition with total exposure time to the task materials; some studies had inappropriate or no control groups; and very few studies used double-blind procedures, opening up the possibility that experimenter effects might have accounted for the results.

In a detailed review of 11 early experimental studies, Smith and Simon (1984) found that only two used experimenters blind to the hypotheses of the study, one of which found no effects of play on problem solving. The issue of experimenter effects was particularly worrisome for these early studies, because usually the same experimenter who was present during the play/nonplay sessions later administered the problem-solving tasks. Furthermore, the design of many studies involved having the experimenter give subjects "standardized" hints during the problem-solving sessions, a procedure that could allow for subtle differential patterns of treatment by condition. Even in the alternative-uses tasks—in which hints were not given—the behavior of an experimenter not blind to the study hypotheses might elicit differential patterns of response. For example, experimenters interviewing children in the play conditions might persist longer in trying to elicit alternative uses than would experimenters interviewing children in the other conditions (Smith & Simon, 1984).

When Smith and colleagues conducted a series of more carefully controlled studies using double-blind procedures, they found very little evidence for the causal role of play. Three studies showed no effects of play on success in lure retrieval (for a review, see Simon & Smith, 1986), and one study (Smith & Whitney, 1987) showed no differences in associative frequency between children randomly assigned to conditions of fantasy play, free play, imitation, or a control group. Even Dansky (1985), in a reply to a critique of his work by Simon and Smith, acknowledged that the "short term facilitative effects of play on associative fluency are rather fragile set effects" (p. 282)—hardly a convincing testimonial for the causal role of play in the development of divergent thinking.

More recent studies have confirmed Smith and colleagues' rather discouraging conclusions. In lure retrieval tasks, L. A. Barnett (1985) found no differences in performance among several demonstration, practice, and play conditions, and Hughes (1987) found no differences between children randomly assigned to exploration versus play. Moreover, Pellegrini, in a series of studies (see Pellegrini, 1984–1985, for a review), found that free play was inferior to a procedure where the experimenter interacted with the child asking a set of standardized questions in generating alternative uses, and Vandenberg (1990) found that children who engaged in "broad-focus" goal-directed activity performed better on a subsequent memory task than did children in his "narrow-focused" or play conditions.

In summary, there is currently very little experimental evidence demonstrating the facilitative role of object play on tasks of either convergent or divergent thinking. As stated by Smith and Simon (1984), "On the most pessimistic assessment, the studies reviewed in this chapter may be showing little more than that children need time to feel secure in a new environment (hence the poor performance in no treatment controls), and that experimenters can unwittingly affect the performance of their subjects (hence the patchy superiority of the play children in some studies)" (p. 211).

Despite this rather discouraging assessment, most researchers believe that this question is far from being answered. Researchers argue that, given the methodological problems in the current research, more controlled studies need to be conducted employing double-blind procedures, in which the play sessions are videotaped and later coded, and a variety of relevant control groups are included (Rubin et al., 1983; Smith & Simon, 1984). Moreover, it has been argued that research has to go beyond the single-session study to designs in which play and nonplay experiences are experienced over a longer period of time in order for their effects to be observed (Pepler, 1982). Smith and Simon (1984) pointed out that it is unlikely that the typical 8- to 10-minute play session has a noticeable effect on child behavior:

> It may well be the case that the one-session play experiment has poor ecological validity. Some important things may indeed be learnt from a single experience; for example, avoidance of a frightening or harmful event (such as the appearance of a predator, or a nasty food-taste). But what we know of naturally occurring play in animals and humans does not suggest that learning from play is of this kind. Typically, play bouts are repeated, fairly short and often incorporate a fair amount of repetition or overlap of experience with previous bouts. The probable inference is that any learning from play bouts is piecemeal, and cumulative over bouts or episodes. (pp. 211–212)

Researchers also need to use a wider range of outcome measures and tasks. Often, the dependent variable is dichotomous—did the child successfully solve the problem or not? Tasks with graded success levels would be preferable, as would tasks that allow for comparisons of the child's approach to or strategies for problem solving (Smith & Simon, 1984). In choosing play materials, situations, and problem-solving tasks, researchers need to be sensitive to the study of psychological processes. By carefully selecting play and problem-solving materials and procedures, it may be possible to document more specifically what is transferred from play to problem solving when a relationship between these two variables is found (Pepler, 1982).

The relationship between play and problem solving also needs to be investigated developmentally, in that the nature of this relationship might change with the child's developing motor and cognitive abilities (Rubin et al., 1983). Finally, it could be argued that the designs of many of the empirical studies testing the effects of play on problem solving do not derive directly from the theories. If one wants to demonstrate that play functions to increase behavioral diversity leading to the emergence of novel, useful behavior forms that can be applied to subsequent problem-solving, then it seems that the problems to be solved should *not* be of the kind that have a single solution already known by the investigator, and sufficient time and freedom should be provided to allow for novel forms of behavior to emerge. Furthermore, the novelty and utility of the behaviors generated during play should be documented, and then the subjects should be allowed to not only turn their attention to solving problems, but also try to find problems to which their innovations can be applied. Finally, as suggested by Rubin et al. (1983), researchers also should look at how play occurring *after* the child is given a problem aids in problem solving, as well as the more typical sequence of play followed by problem solving. The typical experimental study of play and problem solving has few, if any, of these characteristics.

Future research on the functions and effects of object play should focus on areas other than response diversity and problem solving. Although much of what children learn through the manipulation of objects is probably acquired through exploratory behavior, object play undoubtedly provides information as well. When banging, pushing, pulling, disassembling, or relating one object to another, children likely acquire information about the object's hardness, flexibility, consistency, weight, ease of handling, construction, durability, and so on. Moreover, during the course of pretend, children may learn more about an object's possible functions. In fact, Ruff (1989), who did not include banging in her "examining" category, acknowledged that banging probably serves multiple functions—information gathering being one. It is unlikely that the acquisition of information ends when play begins. Further research on the information-gathering functions of large motor, constructive, and pretend play behaviors would broaden our understanding of the ways in which information is acquired through object manipulation.

Researchers have also argued that object play may be important in promoting general cognitive, social, and motor development in childhood. More correlational and experimental studies on the relationship between specific object play experiences and general indicators of child development need to be conducted.

Finally, an effect of play that has been considered mostly in the literature on humans concerns the role of play in helping children cope with stress and/or anxiety (e.g., Erikson, 1950; Waelder, 1933). A largely untested assumption in this area is that when children incorporate real-life, stressful material into their play, it helps them work through the anxiety that these events create, by putting threatening, otherwise uncontrollable events under the child's control. This assumption serves as the basis for many forms of play therapy, especially for children who have experienced psychological trauma (e.g., Axline, 1947; Klein, 1955). This effect of play has largely been limited to discussions of humans, because such anxiety-reducing play usually involves pretend or fantasy. Empirical research on the anxiety-reducing functions of children's play is rare. Clinicians working with survivors of natural disasters have observed that elements of the traumatic event are often incorporated into children's play (e.g., Bloch, Siber, & Perry, 1956; Lacey, 1972; McFarlane, 1987; Saylor, Swenson, & Powell, 1992; Terr, 1981). Other studies have noted an increase in pretend play after the birth of a sibling (for reviews, see Kramer, 1996; Kramer & Schaefer-Hernan, 1994) or prior to surgery (Gilmore, 1966). It is not clear, however, if these increases in pretend play are any higher than are found for equally salient, nonanxiety-provoking events, nor are there much data showing that children who engage in such play are more or less likely to suffer long-term adjustment problems in reaction to the event. Related to this research are numerous studies conducted during the 1940s and 1950s of preschoolers' behavior during doll play, which attempted to use doll play as a way of assessing children's reactions to frustrating or stressful events in the home (for review, see Levin & Wardwell, 1962). However, in these studies, doll play was used as a method for assessing the child's unconscious motivations, rather than studied as a way to alleviate anxiety.

Based on clinical work with a group of kidnap victims and 12 other children who had experienced some other form of psychic trauma (e.g., dog attack, sexual abuse, witnessing a murder), Terr (1981) identified 11 characteristics of post-traumatic play: "compulsive repetitiveness, unconscious link to the traumatic event, literalness, failure to relieve anxiety, wide age range of players, varying lag time prior to its development, carrying power to involve nontraumatized children, contagion to new generations, danger, art and talk as alternative modes of playing, and usefulness of tracing post-traumatic play to an earlier trauma" (p. 741).

Sugar (1989), in a review of the literature on children and disaster, listed the following as possible functions of reenactment of traumatic events during play: "efforts at mastery, identification with the aggressor..., efforts to communicate with adults and others to get help or revenge, and to validate the experience" (p. 169).

Several experimental investigations were located that directly addressed the anxiety-reduction hypothesis. Two types of studies have been conducted: those showing that children's play in anxiety-inducing situations incorporates content relevant to the nature of the stressful experience (Gilmore, 1966; Watson, 1994), and those showing that play relieves child anxiety due to separation (Barnett, 1984), a fear-inducing film or story (Barnett

& Storm, 1981; Watson, 1994), or threats of noxious auditory or visual stimulation (Gilmore, 1966). Studies of the first type show that there appears to be no simple linear relationship between the level of child anxiety and children's play. Although the results of the Gilmore (1966) and Watson (1994) studies were complex (each investigator reported the results of three separate studies with conflicting findings), probably the most general conclusion is that play appears to be most likely to occur in conditions of moderate rather than high or low anxiety. Regarding studies of the second type, in which children were randomly assigned to play or control conditions after exposure to a potentially stressful event, Barnett and colleagues (Barnett, 1984; Barnett & Storm, 1981) found that children's physiological arousal decreases after they are given the opportunity to play. In the study employing the most elaborate design, Barnett (1984) found that play only reduced arousal for children showing high levels of initial behavioral and/or physiological arousal to the stressor, and only if given the opportunity to engage in solitary play. High-anxiety children assigned to a peer play condition or to one of two story-listening control conditions (peers present or absent) did not show reductions in physiological arousal. Moreover, high-anxiety children in the solitary play condition were most likely to engage in fantasy play.

Given the widely held beliefs about the anxiety-reducing effects of play, there is a great need for future research in this area. The effects of the Barnett (1984) study are particularly intriguing, in that they suggest that solitary object play may be more important for anxiety reduction than is social play. Because other studies (e.g., Kramer, 1996; Kramer & Schaefer-Hernan, 1994) have found that children's social play also incorporates elements of stressful events, future studies should focus on both kinds. An interesting finding not reviewed here concerns the temporal relationship between the occurrence of the potentially stressful event and children's play activity. Watson (1994) reported that the mother of one of his subjects who did not engage in anxiety-relevant play in the laboratory after exposure to a fear-inducing filmstrip (the Hansel and Gretel story) reported that her child spent the next 2 days engaged in Hansel and Gretel play. Watson argued, "A cooling-off period may be necessary for children to be able to use play to process information or to rework the anxiety, because when children do not have control over their immediate anxiety, they may not be able to deal with an additional anxiety, even if it is only in fantasy" (p. 44).

A related finding is Kramer and Schafer-Hernan's (1994) finding that "high-sibling acceptance children" (children who later engage in high amounts of positive interaction with their siblings) actually showed a suppression in fantasy during social play immediately following the sibling's birth—a suppression not evidenced by low-sibling acceptance children. Thus, they argued, "temporary disruption in fantasy play may indicate adaptive coping" (p. 749).

Future research on play and anxiety would benefit by investigating both solitary and social play across a range of potentially stressful situations that vary in their severity. If possible, studies should examine child behavior and physiological functioning before, during, and after exposure, taking into account the possibility that the effects of exposure on play may be delayed. Multiple assessments of child behavior and anxiety should be obtained, including reports from teachers, parents, and children themselves, and (whenever possible) multiple measures of physiological arousal. Experimental studies should compare the effects of various play conditions, including those promoting play that is solitary or social, and fantasy or nonfantasy, with anxiety-relevant or anxiety-irrelevant materials. Moreover, more detailed analyses of the content of children's fantasy would be useful in order to explore the specific ways that anxiety-related themes are incorporated into play sequences.

In summary, much work needs to be done to study the effects and functions of object play on child development. There is currently very little evidence that object play serves any more functions than could be accounted for by exploratory behavior alone. Future research will require that researchers think in a much broader way about possible play functions, as well as about their methods for demonstrating relationships. It is important that such studies distinguish between the effects of exploration and play, as well as demonstrate the specific processes through which play facilitates problem solving, divergent thinking, anxiety reduction, and other outcomes. Finally, studies on play and associative fluency need to get beyond simply examining the number of associations, to examine their originality and utility as well. "Play has been described as an elusive phenomenon. Pinning down the effects of play is proving just as elusive. If indeed the effects of play are slowly cumulative, and perhaps in a sense incidental to the play experience itself, this creates special difficulties, but also special challenges; in this case, challenges to the problem-solving skills and creativity of the investigator" (Smith & Simon, 1984, p. 212).

STUDIES OF TOOL-USE

Compared to the four categories of tool-use identified in nonhumans described earlier in this text, (i.e., extend the animals' reach, amplify mechanical force, augment agonistic displays, and control the flow of fluids; Beck, 1980), the range and variety of tool-use in humans is staggering. Humans use tools to achieve many survival-related goals (e.g., eating, drinking, grooming, sleeping, communication, shelter construction, health maintenance, reproduction), as well as in activities less critical to survival (e.g., entertainment, cognitive stimulation). Given the range and variety of tool-use behaviors in humans, a broad-based discussion of this issue is beyond the scope of the present volume. Studies of tool-use in human infants and toddlers, however, are useful in furthering our understanding of the nature of early object manipulation, so some selected examples are considered here. As was the case in the studies of age-related changes in children's exploration and play, these studies demonstrate the increasing planfulness and hierarchical organization of children's behaviors that accrue with age.

Probably one of the best general descriptions of the development of early skilled action is found in Bruner (1972). Although this description focuses mostly on the development of skilled motor actions, such as reaching and grasping in the first year of life, it can be applied to the development of tool-use as well (Koslowski & Bruner, 1972). Bruner (1972) described the nature of skilled behaviors and their development from intention to outcome in a way worth representing in detail here. He started his analysis with the issue of intention:

> Intention, viewed behaviorally, has several measurable features: anticipation of the outcome of an act, selection among appropriate means for achievement of an end state, sustained direction of behavior during deployment of means, a stop order defined by an end state, and finally some form of substitution rule whereby alternative means can be deployed for correction of deviation or to fit idiosyncratic conditions. (p. 2)

According to this analysis, once a child has the intention to act:

> Initial arousal is often followed by a loosely ordered sequence of constituent acts that will later occur in an appropriate serial order to achieve the end state toward which the intention appears to be steering. Meanwhile, during this clumsy athetoid phase, the consummatory response will continue, and even the wrongly orchestrated constituent acts can be shown to have an appropriate adjustment with respect to the goal. (p. 2)

> In time, and probably by virtue of sheer practice of the act in the presence of the releasing stimulus, which permits the coordination of internal and peripheral feedback, the act is successfully executed. (p. 3)

Bruner (1972) argued that once the child develops the requisite skills to successfully achieve his or her goal, three subsequent developments occur:

The child begins to demonstrate "anticipatory patterning" in his or her behavior, the behavior becomes "modularized" and therefore more efficient in its execution, and the newly mastered act is "supplanted by a 'higher-order action' that usually encompasses [the original act] as a subroutine" (p. 4).

Bruner (1972) attributed the "puzzling" replacement of newly acquired skills with more complex ones to information-processing capacity: "Modularization frees available information-processing capacity for further use in task analysis, just by virtue of constituent subroutines requiring less attention.... [The] act can then be incorporated into a higher-order, longer-sequence act without requiring so much attention as to disrupt regulation of the higher-order act" (p. 4). He went on to note that, "Goal-directed, skilled action, then, may be conceived of as the *construction* of serially ordered constituent acts whose performance is modified toward less variability, more anticipation, and greater economy by benefit of feedforward, feedback, and knowledge of results" (p. 5).

Despite this detailed theoretical analysis, few researchers have undertaken the kinds of molecular descriptions of child behavior necessary to document changes in the tool-use context. However, there have been some exceptions (e.g., Koslowski & Bruner, 1972). Another notable exception is Connolly and Dalgleish's (1989) study of the development of spoon use during the first 2 years of life. Using a longitudinal, cross-sectional design, these researchers followed groups of 11- and 17-month-old infants for 6 months, videotaping the subjects on a monthly basis. Detailed coding and analysis of the videotapes yielded evidence to support Bruner's (1972) view of the development of skilled action. Results showed that the toddlers' use of the spoon became more efficient and less varied with age. Specifically, with increasing age, children used a smaller number of grasping patterns, the trajectory of the spoon from the bowl to the mouth followed a smoother path, and less time was required to complete each behavioral unit. Older children also spent more time watching the spoon, and demonstrated increased consistency in the timing of anticipatory mouth opening.

Several qualitative shifts in behavior occurred in the way that children approached the task with age, probably representing the kinds of "quantum jumps in performance" to which Bruner (1972, p. 4) referred when describing the development of higher order sequences. Older infants, for example, were less likely to simply dip the spoon vertically into the bowl to get food, and more likely to use wrist rotation for this purpose. The behavior of multiple systems appeared to get more integrated with age, because older infants were more likely to use the contralateral hand for task-related activities—especially holding the bowl, flex their elbow when lifting the spoon, and move their head and flex their shoulders in order to facilitate the spoon's entry into the mouth.

Connolly and Dalgleish (1989) also observed that infants commonly demonstrated parts of the "spoon-to-dish-to-mouth sequence," such as repeatedly putting the spoon into the dish and taking it out, or repeatedly taking the spoon in and out of the mouth without getting more food. Finally, Connolly and Dalgleish identified four general stages in the development of spoon use (see Fig. 3.1): "(a) the appearance of simple repetitive actions with the spoon; (b) the outline construction of the action sequence, spoon-to-dish-to-mouth; (c) the incorporation of function, the transfer of food by manipulation of the tool; and (d) the incorporation of correction routines, which are deployed when necessary" (p. 911).

Thus (as predicted by Bruner, 1972), with increasing development and practice, infants showed evidence of developing anticipatory behavior (i.e., mouth opening), increasing efficiency (less time per unit, smoother trajectories, and less variation in grasping patterns), and the development of higher order sequences (i.e., qualitative changes in food attainment, increasing integration of behavioral systems, and the development of complex hierarchically ordered subroutines derived from earlier behavior patterns—see Fig. 3.1).

Several other studies have examined age-related differences in the kinds of problem-solving tasks that infants and toddlers can perform. Brown (1990), in reviewing work on lure retrieval tasks conducted in her own laboratory and by Bates, Carlson-Luden, and Bretherton (1980), concluded that:

> Although children in habituation paradigms seem to understand the need for point of contact early (5–7 months), they cannot apply that knowledge to tool-use tasks at 10 months unless the contact between the tool and the goal is provided in the physical layout of the task itself, that is, the solution is physically situated, if you will, in the environment itself. Several months later, the younger children in the Brown and Slattery studies (13–18 months) can learn, with a demonstration, to envision the point of contact that is not specified in the visual array but is invited by the pulling features afforded by the tools. By 24 months children readily note the pulling potential of unattached tools. Although young children "have" the requisite knowledge very early on, they need help in the form of a demonstration, or pulling specific cues provided by the environment, to prompt the application of their knowledge. As they mature, they need less environmental support and can imagine pulling functions not demonstrated or physically specified in the environment. (p. 123)

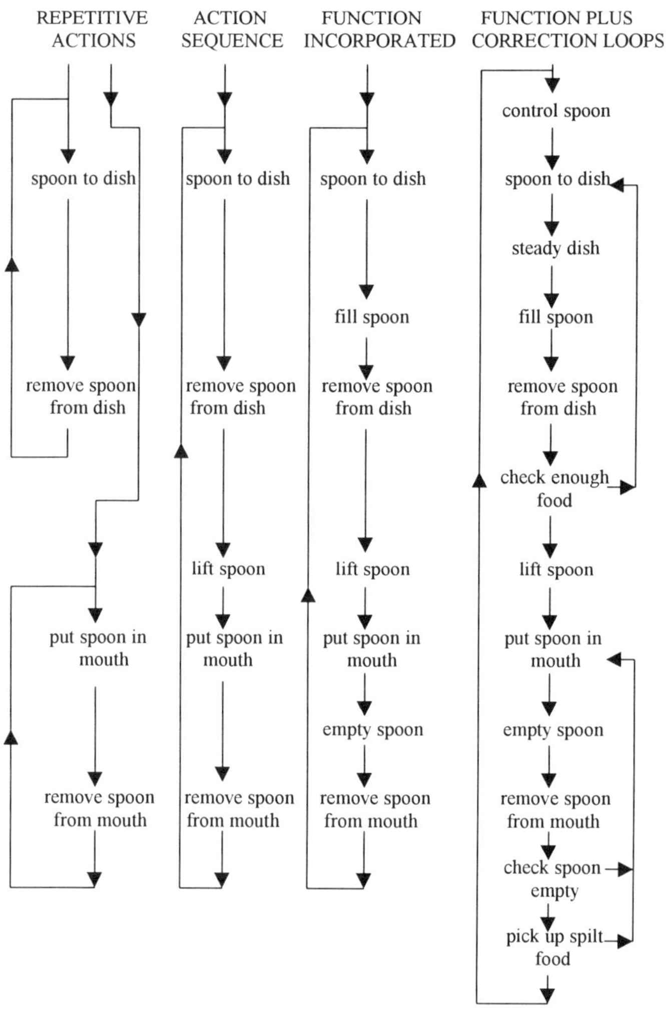

FIG. 3.1. Stages in the development of spoon use skill. From Connolly and Dalgleish (1989). Copyright © 1989 by the American Psychological Association. Adapted with permission.

van Leeuwen, Smitsman, and van Leeuwen (1994) further examined the task characteristics underlying developmental progressions in lure-retrieval tasks. Specifically, they demonstrated that the difficulty of these tasks is a function of the "complexity of the affordance structures involved" in the problem (p. 189). Figure 3.2 presents five configurations of a hook tool and

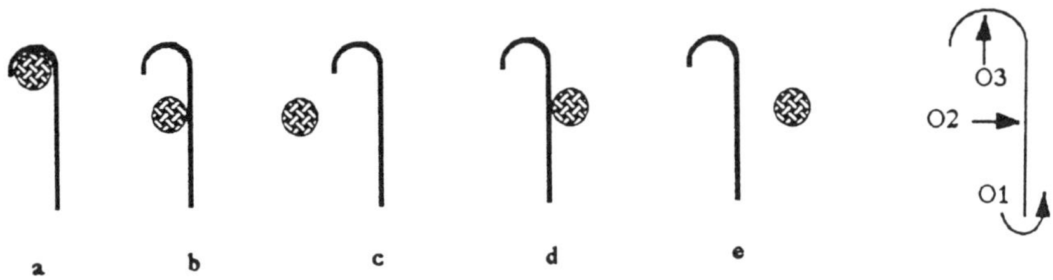

FIG. 3.2. Configurations of tool–target relationships in van Leeuwen, Smitman, and van Leeuwen (1994). Copyright © 1994 by the American Psychological Association. Reprinted by permission.

goal object that vary in terms of theoretical complexity. Notice that three relationships between the tool and the goal determine a configuration's complexity:

> O1 = crook orientation of hook with respect to the target;
> O2 = spatial contact between hook and target; and
> O3 = height of target with respect to hook. (p. 178)

Notice also that their hypothesized complexity is partially a function of the number of relationships that require operations for successful task completion. Thus, for example, Configuration e, which requires three operations for successful task completion (i.e., changing orientation, establishing contact, and adjusting height), is rated as more complex than are Configurations c and d, which require two operations. As well, Configurations c and d are rated as more complex than Configuration b, which requires only one operation. Notice additionally that orientation operations yield higher complexity ratings than do contact operations (i.e., of the two configurations requiring two operations, d is rated as more complex than is c).

A series of studies of 9-month-old to 4-year-old children showed that the configurations presented in Fig. 3.2 are indeed presented in order of increasing complexity. The percentage of subjects successfully solving the problems declined with increasing complexity, although the difference between Configurations a and b was not statistically significant (probably as a result of a ceiling effect for these two problems).

In summary, the age-related changes in tool-use are similar to those observed in the exploration and play contexts. Before discussing this further, however, we briefly consider the research on object manipulation in non-Western cultures. If, indeed, patterns of object exploration, play, and tool-use in humans are a product of evolution and natural selection, then the diversity of object-directed behaviors documented in Western cultures should be found in non-Western cultures as well.

STUDIES IN NON-WESTERN CULTURES

Cross-cultural comparisons show that the amount of object manipulation demonstrated by young children varies widely across cultural settings. Although object manipulation appears to be most common in Western societies (e.g., Blurton-Jones & Konner, 1973; Finley & Layne, 1971), the amount of manipulation shown by children in hunter–gatherer and traditional subsistence agriculture cultures varies greatly. In studies of traditional cultures from South America, Central America, Africa, and Asia (see discussion to follow), the amount of object manipulation observed in young children ranges from no reported instances (Dansen's 1973 study of young Baoule children in Africa) to between 30% and 35% of all time intervals observed (Bakeman, Adamson, Konner, & Barr, 1990—!Kung infants and toddlers in Botswana; Sigman et al., 1988—Embu toddlers in Kenya).

One possible reason that object manipulation is less frequent in traditional, non-Western cultures is that usually few or no toys are available to children in these settings (e.g., Brazelton, Robey, & Collier, 1969; Kopp, Khoka, & Sigman, 1977; Landau, 1981; Whiting, 1963). Children, however, have been observed to manipulate a range of natural and nontoy materials such as sticks, grass, stones, sand, food remains (e.g., bones, nutshells), beads, cans, bottles, boxes, and pieces of cloth (e.g., Bakeman et al., 1990; Eibl-Eibesfeldt, 1989; Konner, 1972; Leacock, 1971; Prosser, Hutt, Hutt, Mahindadasa, & Goonetilleke, 1986; Whiting, 1963).

Observers who have spent time in cultures where the rate of object manipulation is low note that this appears to be a consequence of infant and child care practices adapted to the demands of the ecological setting. In most of these cultures, infants and toddlers have very few opportunities to independently explore the physical environment, because they spend the majority of their waking hours in contact with a caregiver—either in a sling, on someone's back, or in someone's lap or arms (Brazelton et al., 1969; Hogbin, 1943; Kaplan & Dove, 1987; Kopp et al., 1977; Whiting, 1963). Usually the mother has this caregiving role, although in some cultures other women, elderly men, and/or siblings assist in the activity (Whiting, 1963). Young children in these cultures spend so much time in contact with a caregiver for two major reasons: so that the caregiver can go about his or her daily chores with the infant in tow, or so that the infant is protected from a physical environment that is perceived as either too dangerous or too unclean for infant locomotion and exploration. Kaplan and Dove (1987), for example, noted that when the hunter–gatherer Ache of Eastern Paraguay make foraging trips into the forest, children under the age of 3 spend 80% to 100% of their time in physical contact with a caregiver, and when not in physical contact are not allowed to venture more than a meter away, out of fear of harm to the child.

There are exceptions, however, to this general trend. !Kung infants, who spend a great deal of time in a sling, show high levels of object manipulation, mostly because they can manipulate their mother's braids or the beads around her neck (Bakeman et al., 1990). Moreover, low levels of object experience do not necessarily mean low levels of stimulation—many observers point out that the lack of stimulation from the physical environment is more than compensated for by the level of social, tactual, and kinesthetic stimulation afforded by such high levels of close physical contact (e.g., Konner, 1977).

Studies of specific object-directed behaviors in traditional, non-Western cultures (although few in number) suggest that young children, regardless of the cultural setting, show the same behaviors and the same stages of sensorimotor development (e.g., Bakeman et al., 1990; Brazelton et al., 1969; Dansen, 1973; Goldberg, 1977; Kagan, Kearsley, & Zelazo, 1978; Kopp et al., 1977). However, in many non-Western cultures, infants often lag behind their Western counterparts, especially after the ninth month of life (Brazelton et al., 1969; Goldberg, 1972; Kaplan & Dove, 1987; Kopp et al., 1977). Although such differences are not always found (e.g., Dansen, 1973; Konner, 1976), the fact that they have been documented in so many cultures is in direct contrast with a number of studies showing that African infants achieve some large-motor developmental milestones (such as sitting and walking) at an earlier age than do their western counterparts (for a review, see Super, 1981).

Consistent with the slower rates of development among infants and toddlers in some non-Western cultures are the reports of many observers who have noted that children in these cultures often show very little interest in objects, and/or engage in listless, aimless manipulation. Gaskins (1994), in her naturalistic observations of "exploratory play" in Mayan infants, noted that:

> Most of the playtime was spent in picking up and dropping objects or manipulating objects in simple ways, such as mouthing and banging, actions that do not take into account the characteristics of the particular object being

> used. Unlike American infants, Mayan infants spent almost no time in play that related two objects or accommodated to the particular characteristics of objects. (1994, p. 329)

Similarly, in a study of infants, toddlers, and older children in rural Nepal, Anderson and Mitchell (1984) noted that, in over 7 hours of naturalistic observations, "toddlers were found to engage in aimless or passive behavior 96.3% of the time" (p. 738). Regarding older children:

> We did observe play activities consistent with developmental theories of play.... Sensorimotor play, including play on an improvised swing and see-saw, imitative dancing, aimless manipulation of natural materials, rolling hoops, and pulling objects was observed. Much of this play, even among older children, was unvaried and repetitive. Little experimentation, exploration, or diversification of the activity toward higher levels of mastery were observed. (p. 737)

Finally, several investigators have noted that, when given standardized sensorimotor tests, infants in some non-Western cultures often show little interest in the novel objects offered. In their study of Mayan infants in southern Mexico, Brazelton et al. (1969) observed, "Novelty was met with impassive faces in tested children and parents alike. Repeatedly, infants watched us carefully as we demonstrated the use of test objects, imitated each movement we had made to score a success on the test, and then dropped the object without any of the exploration or experimental play we would have seen in United States babies" (p. 284). Similar observations during test administrations were made by Goldberg (1977) and Kaplan and Dove (1987).

These qualitative differences in object exploration and play could be due to a number of factors, including the effects of fear aroused in the testing situation, by either strangers, materials, or the physical setting; the effects of chronic malnutrition and poverty on the ability or motivation to perform; and differences in the degree to which these infants have had opportunities for object exploration and play, and to which the manipulation of objects has been encouraged or discouraged by their caregivers or playmates (Anderson & Mitchell, 1984; Finley & Layne, 1971; Goldberg, 1977; Kaplan & Dove, 1987; Super, 1981). It is likely that how much desirability or importance members of a given culture ascribe to the manipulation of objects by young children relates to larger cultural beliefs and socialization goals that influence the degree to which such behaviors are encouraged or discouraged (Ogbu, 1981; Whiting & Child, 1953). Without further research in this area, the relative importance of each of these factors is unknown. Studies comparing infant behavior in non-Western cultures that vary in the amount of object-related experience they offer their children would be particularly informative.

Finally, in a study in urban Senegal, Valantin (1973) found that, during testing sessions, many infants employed buccal prehension in the procurement of objects. Although she attributed the existence of this behavior pattern to the prolonged periods of breastfeeding these children receive, buccal prehension was not mentioned by any of the other researchers who studied infants in traditional, non-Western cultures where long periods of breastfeeding are also the norm. Given this unusual finding, further research on the extensiveness and correlates of this behavior pattern in humans would be informative.

GENDER DIFFERENCES IN CHILDREN'S SOLITARY OBJECT-DIRECTED BEHAVIOR

Gender differences in object-directed behavior have been studied much more extensively in humans than in other animals. However, as was the case for the animal literature, many investigators do not report analyses of gender differences even when a sufficient number of subjects is available for such comparisons to be made. Despite the limitations in the database, several general conclusions can be drawn.

First, there is little evidence that males and females differ in the amount of object manipulation they engage in during the first 2 years of life. Studies in Western cultures that have found no gender differences during this period include Belsky and Most (1981), Gibson and Walker (1984), Palmer (1989), Ross (1974), Ross et al. (1972), Ruff (1984), Steele and Pederson (1977), Streri and Pecheux (1986), van den Boom (1994), and Yarrow et al. (1983). Although this issue has received little attention in studies of non-Western cultures, two studies that did address it also found no gender differences during this age period (Bakeman et al., 1990; Sigman et al., 1988).

Studies of older children are less consistent. Of seven studies of preschoolers, five found no gender differences in the amount of object manipulation (Ali, 1973; Ball & Hagen, 1978; Henderson & Moore, 1979; McLoyd & Ratner, 1983; Schneider et al., 1983), whereas two found that boys engaged in more manipulation of a novel object than did girls (Daldry & Russell, 1982; Hutt, 1970a). For children of elementary school age, two studies reported no gender differences (Cohen, 1974; Henderson & Moore,

1979), whereas three did—again, with higher levels for boys (Coie, 1974; Kreitler et al., 1975; Schneider et al., 1983).

Studies of the nature of object manipulation, although fewer in number, highlight more consistent differences. Although most researchers point out that the nature of object manipulation in boys and girls is much more similar than it is different, some minor stylistic differences have been identified. An often-reported finding is that infant boys show greater vigor in their object play than do infant girls (Goldberg & Lewis, 1969; McCall, 1974; Wenckstern, Weizmann, & Leenaars, 1984). Research by Pomerleau and colleagues (Cossette, Malcuit & Pomerleau, 1991; Pomerleau, Malcuit, & Seguin, 1992) shows that this tendency can be seen in the behavior of infants as young as 5 months. In these studies, boys showed more tapping and more high-intensity arm movements toward objects than did girls. This gender difference observation is related to a now well-established gender difference in activity level that becomes greater with increasing child age (see the meta-analysis by Eaton & Enns, 1986).

Along with the gender difference in vigor, a second difference concerns the diversity of object manipulation. Studies of preschool children have found that boys score higher than do girls on measures of depth of exploration (Vandenberg, 1984), inventive exploration (Hutt, 1970a), variety of play (Goodfader, 1982), and configural richness of play (Cheyne & Rubin, 1983). Moreover, in a study of infants and toddlers, Zelazo and Kearsley (1980) found that girls engaged in more "appropriate uses" with toys than did boys.

Other studies of measures of play style during the first 2 years show no gender differences, including studies of attention span (Czerniecki et al., 1993; Ruff et al., 1990) and play tempo (Johnson & Brody, 1977).

Little research has been conducted on gender differences in constructive play. Although Goodfader (1982) identified differences between preschool boys and girls in the nature of their play constructions, Stiles-Davis (1988) did not. Regarding pretense, studies consistently find no gender differences before the age of 24 months. These studies include Belsky and Most (1981), Jackowitz and Watson (1980), McCune (1995), Russell and Russnaik (1981), and Watson and Fischer (1977). Studies of older children sometimes, but not always (e.g., Overton & Jackson, 1973; Robinson & Jackson, 1987), report differences in pretend play. For example, Largo and Howard (1979) found that 24- to 30-month-old girls were more likely than boys to show sequential pretend play, and Fenson and Ramsey (1980) found that 24-month-old girls engaged in more multischeme pretend than did same-aged boys. A study by Lowe (1975) suggests that these differences may be specific to the types of toys used. For example, girls showed a variety of more complex pretend behaviors with dolls than did boys, whereas boys engaged in more complex pretend behaviors with a toy truck than did girls. Numerous studies have shown that boys prefer playing with different types of toys than do girls; differences in preferences for gender-typed toys apparently begin around 12 months. Moreover, boys are less likely than girls to play with toys generally used by members of the opposite gender (for a review, see Rubin et al., 1983).

Studies of tool-use and problem solving show few significant differences in the behavior of boys and girls. Girls, however, have been found to show greater levels of success on haptic exploration tasks (Abravanel, 1968; Wolff, 1972). One study (Kleinman, 1979) found that girls engaged in more haptic exploration during these tasks than did boys. Studies of cross-modal transfer in infancy (Gottfried, Rose, & Bridges,1978; Rose, Gottfried, & Bridges, 1981) and studies using lure or similar tasks usually show no gender differences in performance (Smith & Dutton, 1979; Yarrow et al., 1983).

In summary, given the wide range of object-directed behaviors that have been studied in young children, the few consistent gender differences that have been identified (i.e., greater vigor and diversity in the object manipulation of boys, greater success in haptic exploration tasks for girls), show that the behavior of girls and boys with objects is much more similar than it is different. Measures of the *nature* of object manipulation have revealed more consistent gender differences than have simple measures of the *amount*. This is not surprising, given that the amount of object manipulation is influenced by a number of factors, including the type of behaviors being studied (e.g., exploration, play, tool-use), the nature of the objects available, and the setting in which the observations occurred. Moreover, given that by 12 months of age, boys and girls begin to differ in the kinds of objects they prefer to manipulate, certain gender differences in the nature of object-directed behavior may be an artifact of the objects boys and girls choose (Carpenter & Huston-Stein, 1980).

SOLITARY OBJECT MANIPULATION: SUMMARY AND CONCLUSIONS

The evolution of behavior patterns supporting object exploration, play, and tool-use appears to have been a major way in which *Homo sapiens*—among the most generalist of species—developed the flexibility to adapt to a range of environments. These behavior patterns likely evolved as a result of their survival advantages for both foraging and defense in early terrestrial habitats. The results of the cross-species analysis presented earlier are consistent with this interpretation. Phylogenetic proximity to humans was associated with a range of manipulatory abilities related to extractive foraging and defensive tool-use behaviors. Moreover, analyses of both human and nonhuman primates revealed striking similarities in the specific object-directed behaviors exhibited, and in the developmental order in which they emerged. As would be expected, the developmental patterns most similar to those of humans were those of the great apes, and the major differences between apes and humans emerged in the second year of life—a time of increasing human symbolic abilities.

As discussed earlier, object-directed behavior takes on at least three forms: exploration, play, and tool-use. Exploratory behavior occurs in the widest range of species (both vertebrates and invertebrates), whereas object play and tool-use are found mostly in birds and mammals, especially those with the longest periods of immaturity before adulthood. Although there are little experimental data to support this hypothesis, it is possible that object play provides the young of some altricial species with specific opportunities to practice and refine the motor skills required in later tool-use contexts.

Given the cross-cultural similarities of the basic forms of object manipulation in humans, as well as the patterns of similarities and differences in these behaviors between primate species, it is likely that the behaviors supporting object exploration, play, and tool-use in humans have a significant evolutionary basis. These appear to be highly canalized behavior patterns (Waddington, 1962) that emerge with a minimal amount of specific environmental support. However, because the *amount* and *quality* of object manipulation vary considerably across cultures, it is possible that providing opportunities for object manipulation, or encouraging it in other ways, may lead to more elaborated forms of object-directed behavior. Although the current cross-cultural data are consistent with such an interpretation (i.e., object manipulation is the most elaborate and complex in Western cultures where parents provide their children with numerous toys and frequently engage in object-mediated play—see chap. 9), alternative explanations, such as those involving problems in the assessment of object manipulation cross-culturally, need to be addressed.

Analyses of the development of object-directed behavior after the first year of life show a number of changes related to the child's developing cognitive abilities. Although much has been made about the differences among object exploration, play, and tool-use, similar developmental progressions occur in all three of these contexts. A review of the previous sections shows that, regardless of the area, with increasing age the child's object-directed activity involves a greater degree of coordination between motor systems, a focus on and coordination of multiple objects, a consideration of multiple aspects and dimensions of the situation, a move from simple repetition to more complex, hierarchically organized behavior patterns, and sustained attention and longer sequences of behavior. All of these changes are consistent with various theories of cognitive development that stress the child's developing ability to decenter and focus on multiple aspects of problems, as well as the increasing planfulness of child behavior (e.g., Case, 1985; Piaget, 1952; Siegler & Jenkins, 1989; Sternberg, 1984). Together, these developing cognitive abilities enable children to use tools and other objects in much more complex, flexible, and effective ways than any of the other primates.

Consideration of some of the other age differences in object-directed behavior highlight differences among exploration, play, and tool-use. For example, with increasing age and experience, behavior in specific tool-use contexts becomes more efficient, more predictable, and less variable. One would not expect to find similar changes for most forms of play, because behavior during play functions to increase, not decrease, diversity in responding. However, for types of play that provide practice and mastery of motor skills, decreases in diversity might be expected.

A problem with many current conceptualizations of play is that behaviors having what appear to be opposite functions (i.e., practice vs. the generation of novelty) are considered to be examples of the same general category. Practice play is in fact very different from the more generative forms of play and, as such, should probably be treated separately. Sutton-Smith and Kelly-Byrne (1984) made a similar point when they described the "bipolarity" in most major play theories.

Presented in Table 3.10 is an expanded conceptualization of exploration, play, and tool-use that includes this distinction, as well as several others that may be useful for future research and theory. The scheme is an elaboration of frameworks offered by Hutt (1967) and Nunnally and Lemond (1973). Following the work of Baldwin and Baldwin (1978b), the modes of object-directed behavior are listed in the order that they usually occur in response to novelty. Also provided is the likely goal of the behavior, the accompanying emotions, and the nature and stereotypy of the behavior patterns.

As shown in the table, avoidance is the most common behavioral response in situations where stimulus novelty is very high and the child is

TABLE 3.10

Expanded Conceptualization of Exploration and Play

Behavior	*Novelty*	*Arousal*	*Emotion*	*Goal*	*Behaviors*	*Stereotypy*
Avoidance	Very high	Over	Fear	Prevent harm	Escape	High
Exploration:						
Cautious	High	Over	Fear-interest	Investigate dangerousness	Approach-avoid	High
Structural	Moderately high	Optimal	Interest	Gain information about structure	Examining	Moderately high
Functional exploration/ practice play (mastery play)	Moderate	Optimal	Interest	Investigate functional properties	Mean-end experimentation	Moderate
Tool-use	Low	Optimal	Goal-directed drive	Achieve specific functional goal	Persistent goal-directed behavior	Moderate
Generative play	Low	Under	Interest joy	Increase arousal	Create novelty	Low

overaroused by fear. Following avoidance are two kinds of exploratory behavior: cautious exploration and exploration of structural properties. Cautious exploration, referred to as "early exploration" by Baldwin and Baldwin (1978b), occurs in response to high degrees of novelty and usually involves approach–avoidance sequences. The child is curious about the object, but also fearful.

Exploration of structural properties describes the kinds of behaviors that fall into Ruff's (1989) examining category. Through systematic holding, looking, rotating, pressing, and squeezing the object, the child attempts to uncover its structural characteristics. This is the initial kind of exploration that children engage in as their fear subsides. The goal of this behavior is to gain information about the object's structural properties, such as its size, shape, weight, hardness, texture, rigidity, and consistency.

Next comes a category somewhere between "pure" exploration and play, labeled *functional exploration/practice play*. This behavior pattern involves exploring the functional capabilities of the object by examining the effects of various actions on object responses, as well as practicing and refining the skills involved in producing these responses. Other researchers have referred to this behavior pattern as "mastery play" (e.g., Yarrow et al., 1983). Behavior in this category defies easy placement into either exploration or play. In the context of repeating and experimenting with various actions (play), children gain information about the relationship between their own behavior and characteristics of the object and its functions (exploration). Thus, stimulus seeking and information gathering are two sides of the same coin—when children repeatedly try to get the object to "do something," their understanding of the self in relation to the object increases.

Tool-use comes next in the scheme, because it is similar to functional exploration/practice play. The major difference is that, in tool-use situations, the child is trying to achieve a specific functional goal through the use of a mediational tool. Thus, whereas the focus during tool-use is on achieving specific *ends*, the focus during functional exploration/practice play is more on exploring the relationship *between* means and ends. Tool-use is application and functional exploration/practice play is practice; tool-use is more goal directed, and the latter focuses more on process. Finally, tool-use is a much narrower category of behavior, in that it involves specific situations in which objects are used as means to an end. Functional exploration/practice play is much broader because it includes a wide range of behaviors involved in manipulating a single or multiple objects.

The final mode, generative play, refers to situations in which the child is underaroused and tries to increase his or her arousal through actively creating novelty. This behavior is less predictable from object characteristics than is functional exploration/practice play, and is more a function of the child's ability and motivation to impose a framework or structure on the situation (e.g., the child places a toy figure into a toy cup and makes noises by shaking the cup). Although a specific end may be achieved during generative play (e.g., the noise produced in the example), the end is more a function of the child's creation of a unique situation than an end inherent in the specific object (e.g., the noise produced by a dial on a busy box). Many pretend behaviors are examples of generative play. Generative play serves to create behavioral diversity. Note that, according to this scheme, generative play is not behavior without a goal—instead, its purpose is to increase arousal in the face of familiarity.

Although the modes are listed in Table 3.10 in inverse order of novelty and level of arousal, their order is not fixed, and they can occur in any sequence during the course of object manipulation. For example, a child may initially find an object quite approachable and show no fear, but then, upon manipulation, produce an unexpected effect that leads to avoidance and then cautious exploration. Similarly, when engaging in tool-use to achieve a specific goal, the child may discover a new means to the end, which may lead to subsequent means–end experimentation (functional exploration/practice play).

The current conceptualization is offered in the hope that, by providing more homogeneous categories of behavior, that future research on the determinants and consequences of object-directed behavior might lead to more consistent results. To date, although there are considerable data supporting the information-gathering function of exploratory behavior, the unique contributions of play have yet to be convincingly demonstrated. By being more specific about the kinds of behaviors that are being studied, as well as by selecting outcome measures tied to the specific functions of different kinds of object manipulation, further progress in addressing this important question is likely. Such research would also be useful in further developing what is, admittedly, a preliminary attempt to classify the various forms of object-directed behavior.

II

Physical Activity Play

Anyone who has spent time on a playground, at the local swimming pool, or on a children's athletic field has witnessed physical activity play. In such settings, children often engage in vigorous, high-energy play that, at least on the surface, displays many similarities with serious predator chases and aggression. Although such activities have the potential for injury (usually when the partners get a bit too carried away), they are often accompanied by signs of enjoyment and positive affect, such as smiling, laughing, and bouncy, exaggerated motor movements. Parents, too, can get into the act—chasing their child around the house or rolling on the floor in a bout of "Big Time Wrestling" with a partner a third their size.

Physical activity play is common in other animals as well. In fact, play-fighting and play-chasing are the major forms of social play observed among most mammals that play (Fagen, 1981). Although biologists have had a hard time defining play-fighting or play-chasing, observers usually have little trouble identifying them and agreeing about when they occur (e.g., Miller, 1973). Because there are many similarities between play-fighting in animals and in humans, it is an excellent topic to begin our consideration of *social* play.

This section is made up of four chapters: two on play-fighting and two on locomotor play. The review is organized as in the previous section: The literature on nonhuman animals is considered before the literature on humans, and research on structure is reviewed before research on functions and effects.

4

Play-Fighting in Animals

Aggression is found throughout the animal kingdom and occurs between members of the same and different species. Depending on the species and the setting, intraspecies aggression can serve several functions, including obtaining food and other resources, gaining access to mates, defending territories, and establishing and maintaining position in a dominance hierarchy. Interspecies aggression can function to subdue prey and fend off their counterattacks, defend against predators, and defend others—especially the young. Selection pressure for the development and refinement of fighting skills has likely been significant, given the importance of the previously mentioned activities for survival and reproductive success.

Because fighting skills undoubtedly improve with experience and practice, many biologists (e.g., Farentinos, 1971; Symons, 1978) have argued that the young of many species engage in play-fighting as a way of developing and refining their fighting skills. The main basis for this argument is that play-fighting is made up of elements of adult aggression that differ in structure, vigor, or intensity—presumably in ways that facilitate learning (e.g., Aldis, 1975; Fagen, 1981). Others, in contrast, have argued that the similarities with serious aggression are superficial (e.g., Pellis & Pellis, 1998; Poole, 1966), and that play-fighting might serve other functions, such as social bonding or establishing dominance hierarchies. This issue has been the subject of much discussion and debate, but before turning to functional issues, we review the literature on the structure of play-fighting. This information—along with information on play partners and gender differences in play-fighting (see later discussion)—provides data for evaluating some of the numerous functional hypotheses that have been developed concerning play-fighting.

Although some theorists prefer to refer to play-fighting behaviors as immature forms of aggression (e.g., Havkin & Fentress, 1985; Lazar & Beckhorn, 1974), most argue that play-fighting and serious aggression are independent behavior patterns, each under the control of separate neurophysiological mechanisms. Pellis and Pellis (1991a) supported the view that play-fighting has its own motivation, using five types of evidence: Isolation studies show that deprivation of play-fighting leads to increases in its probability of occurrence; animals will learn tasks and engage in specific acts to get the opportunity to play; many species appear to have evolved separate play signals that are rarely found outside of the play context; some forms of play are interchangeable (e.g., cats deprived of social partners may engage in structurally similar object play); and in many species, the specific body parts attacked and tactics employed during play and serious fighting attacks differ. Play-fighting in animals is not viewed as "pretend" fighting, in that there is little evidence that animals are engaging in behavior symbolic of something else; instead, they are engaged in a particular kind of play activity that has many of the same components of serious aggression but differs from aggression in its apparent goals and structure. Given the ubiquity of play-fighting among mammals across the phylogenetic spectrum, to infer that members of many species are "pretending" to fight is inconsistent with our knowledge of species differences in symbolic abilities as demonstrated by analysis of animal behavior in other contexts (e.g., Call & Tomasello, 1996; Parker & Milbrath, 1994).

Despite the widespread occurrence of play-fighting behaviors across species, it is not without its risks. Fagen's (1981) listing of the risks of play contain five risks that are particularly relevant to play-fighting: eliciting damaging or fatal aggression, falling down or otherwise getting injured, alerting predators, becoming separated from the caregiver, and using up important time and energy. Given these risks, the evolution of a separately motivated play-fighting system suggests that its benefits likely outweigh its risks.

Biologists have long acknowledged the existence of play-fighting (e.g., Darwin, 1898; Groos, 1898). Naturalists often refer to play-fighting in their descriptions of the behavior of young animals in natural settings (e.g., Altmann, 1962; Rowley, 1929; Spencer, 1943; for a review, see Fagen, 1981). These descriptions make up the "secondary literature" on animal play. With only a few exceptions (e.g., Brownlee, 1954), research studies of the social play of animals did not appear until the late 1960s and early 1970s. Most of the papers on this subject published prior to this period were reviews that questioned whether play was a useful topic for scientific investigation (e.g., Beach, 1945; Lorenz, 1956; Schlosberg, 1947). Three major books and one major journal article reviewed the literature conducted during the 1960s and 1970s (Aldis, 1975; Fagen, 1981; Smith, 1982; Symons, 1978). After this initial flurry of activity, interest in the topic waned somewhat during the 1980s, but there has been continued interest ever since.

A review of Fagen's (1981) natural history of play behavior shows that play-fighting is found in almost all orders of mammals and in a few avian orders. Within orders, however, there is considerable variability, with members of closely related species sometimes showing very different amounts of play-fighting. A commonly mentioned example is that of laboratory rats (Norway rats, *Rattus norvegicus*), which show high levels of play-fighting compared to house mice (*Mus musculus*), which apparently show none (Fagen, 1981; Pellis, Pellis, Manning, & Dewsbury, 1991; Poole & Fish, 1975; Wolff, 1981). Unfortunately, the database on mammalian social play is so varied in depth and in quality that it is difficult to draw any firm conclusions about species variation in play-fighting patterns. This is partially due to the fact that the most detailed studies of play tend to involve (not surprisingly) the most playful species, whereas other, less playful species have not been systematically studied.

Although a few bird species exhibit play-fighting (e.g., Pellis, 1981; Skeate, 1985), it is rare and has been the focus of limited empirical study. Therefore, only mammalian play-fighting is considered in this text.

Given the large number of species in which play-fighting has been observed (see Fagen, 1981), the current discussion is restricted to a review of the primary play literature on those species for which sufficient empirical information about play patterns is currently available. Therefore, as shown in Table 4.1, research on 60 species from seven orders will be considered. These include animals from the following orders: marsupials, carnivores (felids, hyenas, canids, bears, and mustelids), pinnipeds, ungulates (two orders), rodents, and primates. These are species for which sufficient information on the nature of animal play could be found in the "primary" literature on animal play (i.e., empirical articles written specifically on play behavior). Only play in infants and juveniles is considered; research on adult play-fighting (e.g., Fischer & Nadler, 1978; Zucker, Dennon, Puleo, & Maple, 1986) is not reviewed here.

STRUCTURE

Much of the primary literature on animal play-fighting has been devoted to describing its structure. The underlying assumption of many of these studies is that, by providing detailed information on the structure of play-fighting across species, hypotheses about the evolution and functions of play-fight-

TABLE 4.1

Species Included in the Current Review of Animal Play-Fighting, and Studies Used in Constructing Tables 4.3, 4.4, and 6.1

Order marsupiala

Family macropodidae

Macropus rufugriseus, red-necked wallabies (Watson & Croft, 1993, 1996)

Macopus rufus, red kangaroos (Croft & Snaith, 1991)

Order carnivora

Family felidae

Acinonyx jubatas, cheetahs (Caro, 1995)

Felis catus, domestic cats (Barrett & Bateson, 1978; Caro, 1981; Martin, 1984a, Mendoza & Ramierez, 1987; West, 1974)

Panthera leo, lions (Schaller, 1972; Schenkel, 1966)

Panthera tigris, Siberian tigers (Wasser, 1978)

Family hyaenidae

Crocuta crocuta, spotted hyenas (Drea et al., 1996; Pedersen et al., 1990)

Family canidae

Canis aureus, golden jackals (Feddersen-Petersen, 1991)

Canis familiaris, domestic dogs (Bekoff, 1974, 1977; Feddersen-Petersen, 1991)

Canis latrans, coyotes (Bekoff, 1974, 1977)

Canis lupus, timber wolves (Bekoff, 1974, 1977; Feddersen-Petersen, 1991)

Cerdocyon thous, crab-eating foxes (Biben, 1983)

Chrysocyon brachyurus, maned wolves (Biben, 1983)

Speothos venaticus, bush dogs (Biben, 1983)

Family ursidae

Ursus americanus, black bears (Henry & Herrero, 1974)

Family mustelidae

Amblonyx cinerea, Oriental small-clawed otters (Pellis, 1984)

Mustela furo, ferrets (Biben, 1982b; Lazar & Beckhorn, 1974)

Mustela putorius, polecats (Poole, 1966, 1978)

Order pinnipedia

Family otariidae

Arctocephalus australis, South American fur seals (Harcourt, 1991)

Eumetopias jubatus, Stellar sea lions (Farentinos, 1971; Gentry, 1974)

Family phocidae

Mirounga angustirostris, northern elephant seals (Rasa, 1971)

Phoca vitulina, harbor seals (Renouf & Lawson, 1986; Wilson & Kleiman, 1974)

Table 4.1 continued

Table 4.1 continued

Order perissodactyla (ungulates)

Family equidae

Equus caballus, domestic horses (Araba & Crowell-Davis, 1994; Crowell-Davis et al., 1987; Fagen & George, 1977)

Order artiodactyla (ungulates)

Family tayassuidae

Dictyles tajacu, collared peccary (Byers, 1984)

Family cervidae

Odocoileus hemionus, black-tailed deer (Miller, 1975)

Family bovidae

Bos indicus, Boran cattle (Reinhart & Reinhart, 1982)

Capra hircus, domestic goats (Chepko, 1971)

Capra ibex, ibex (Byers, 1977, 1980)

Gazella cuvieri, Curvier's gazelle (Gomendio, 1988)

Hippotragus niger, sable antelope (Thompson, 1996b)

Oryx dammah, scimitar-horned oryx (Pfeiffer, 1985)

Ovis aries, domestic sheep (Sachs & Harris, 1978)

Ovis canadensis, bighorn sheep (Berger, 1980; Hass & Jenni, 1993)

Order rodentia

Family sciuridae

Mamota flaviventris, yellow-bellied marmots (Jamieson & Armitage, 1987; Nowicki & Armitage, 1979)

Spermophilus columbianus, Columbian ground squirrels (Steiner, 1971; Waterman, 1988)

Tamiasciurus hudsonicus, red squirrels (Ferron, 1975)

Family cricetidae

Mescocritetus auratus, Syrian golden hamsters (Pellis & Pellis, 1988b)

Microtus montanus, Montane voles (Pellis et al., 1989; Pierce et al., 1991)

Microtus ochrogaster, prairie voles (Pellis et al., 1989; Pierce et al., 1991)

Phodopus camphelli, Djungarin hamsters (Pellis & Pellis, 1989)

Family muridae

Mus musculus, house mice (Poole & Fish, 1975)

Rattus norvegicus, Norway rats (laboratory rats; Meaney & Stewart, 1981; Pellis et al., 1989; Pellis & Pellis, 1983, 1987, 1990; Poole & Fish, 1975)

Order primates

Family lemuridae

Lemur catta, ring-tailed lemurs (Pellis & Pellis, 1997b)

Table 4.1 continued

Table 4.1 continued

Family callitrichidae

Callithrix jacchus, common marmosets (Chalmers & Locke-Haydon, 1981; Stevenson & Poole, 1982; Voland, 1977)

Saguinus oedipus, cotton-topped tamarians (Goedeking & Immelmann, 1986)

Family cebidae

Alovatta palliata, mantled howling monkeys (Zucker & Clarke, 1992)

Ateles geoffroyi, black-handed spider monkeys (Pellis & Pellis, 1997b)

Saimiri sciureus, squirrel monkeys (Baldwin & Baldwin, 1978a; Biben, 1986; Biben & Symmes, 1986)

Family cercopithecidae

Cercopithecus aethiops, vervet monkeys (Fedigen, 1972; Govindarajulu et al., 1993)

Macaca fuscata, Japanese macaques (Hayaki, 1983; Imakawa, 1990)

Macaca iris, crab-eating macaques (Fady, 1976)

Macaca mulatta, rhesus macaques (Gard & Meier, 1977; Janus, 1989; Smith, 1978; Symons, 1978; Tartabini & Dienske, 1979)

Macaca speciosa, stumptail macaques (Bertrand, 1976)

Papio anubis, olive baboons (Bolwig, 1964; Chalmers, 1980; Owens, 1975b)

Papio cynocephalus, yellow baboons (Cheney, 1978)

Papio hamadryus, hamadryas baboons (Le Resche, 1976)

Presbytis pileatus, capped langurs (Islam & Husain, 1982)

Family ponidae

Gorilla gorilla, gorillas (Brown, 1988; Hoff, Nadler, & Maple, 1981)

Pan troglodytes, chimpanzees (Hayaki, 1985; Mendoza-Granados & Sommer, 1995; Merrick, 1977; Parquette, 1994; Spijkerman, Dienske, van Hooff, & Jens, 1996; van Lawick-Goodall, 1968)

Pongo pygmaeus, orangutans (Poole, 1987)

---

ing could be generated and/or tested. As already mentioned, the main purpose of structural analysis in this area has been to determine the degree to which play-fighting might be an important training ground for the development of serious fighting skills. In this section, the structure of play-fighting is reviewed in terms of the signals that indicate that "this is play"; the general characteristics of play-fighting; the behaviors involved in the initiation, conduct, and termination of play-fighting; and the specific ways in which play-fighting differs from serious fighting. Next we consider developmental changes in the structure of play-fighting, followed by information on the play partners that are usually chosen. Finally, differences between the play-fighting of males and females is compared. The choice of play-partners and gender differences in play-fighting have been examined to test functional hypotheses as well.

Before turning to a consideration of the specific characteristics and components of play-fighting, it is useful to consider some qualitative descriptions of these interactions from a range of species. These examples illustrate both the similarities in play-fighting across species, as well as how play-fighting interactions differ as a function of anatomical differences.

Red-Necked Wallabies (Watson & Croft, 1993, pp. 224–225):

Play fights usually began with a minimum of preliminaries. Initiation typically involved one partner rising into a high-stance posture or grabbing at the other.... Wallabies mostly used their forearms to attack or fend off their partners with the hindlegs used less often. Play-fighting was best described as mild to vigorous uni- or bilateral bouts of pushing, 'boxing' or wrestling usually interspersed with periods of nonplay activity. Termination occurred when one wallaby moved or oriented away from its partner.

Domestic Cats (Fagen, 1981, pp. 163–164):

A typical play encounter begins when one kitten crouches, holds its head near or against the ground, brings its hindlegs underneath its body, extends its tail straight behind or curls it around its body, and treads with its hind feet or sways its hindquarters from side to side. The kitten then springs forward, pushing off with its hindlegs. A second kitten, at whom the spring may have been more or less accurately aimed, falls or rolls belly-up with all four limbs held in a semi-vertical position, and with its tail held straight back and possibly moving back and forth. The hindlegs reach and the front legs reach at or paw at the first kitten, who stands up near or over its playmate, its head oriented toward the playmate's head and neck region. While in this stand-over, belly-up position, the two kittens hold their mouths open, exposing their teeth. They may paw at and attempt to mouth each other.... Two kittens at play may exchange roles, or both may lie on their sides, pawing and kicking at each other.

American Black Bears (Henry & Herrero, 1974, pp. 382–383):

The typical play-fighting sequence usually exhibits four phases: initiation, pushing-pulling, biting-clawing, and finally, termination. One bear approaches another bear and initiates play by rearing, pawing, biting, head butting, or, infrequently, lunging at that animal. Frequently, both animals rear and an extended pawing match follows. The purpose of pawing appears to be to knock the partner off balance so that a biting or clawing action can be delivered. Biting intention movements and face-pawing actions are repeatedly observed during these extended rearing-pawing matches. Once the social partner is pushed or pulled into a vulnerable position, a play-bite or neck-bite-hold is frequently delivered. The bitten animal then rolls over and may hind-leg-claw in order to break the neck-bite hold. Once freed, both bears frequently return to the rearing-pushing-pulling phase until one of the bears once again becomes vulnerable to a bite or clawing action. This alternation between the pushing-pulling and biting-clawing phases continues on the average for about 20 sec...until one of the bears closes its mouth, frequently gives a cut-off display, pivots, and flees. Flight behavior by one of the bears usually terminates the play-fighting sequence.

Northern Elephant Seals (Rasa, 1971, p. 86):

The "wrestling" type of interaction is usually initiated by one animal approaching another, either head on or from the side, uttering low volume "yipping" vocalizations.... At the same time, the head may be nodded up and down. Once the animals come into contact, one grasps the other slowly and deliberately on the side of the neck. This usually results in the animal which is being nipped raising itself on its fore flippers and pressing its throat against that of the other individual. The two then remain in this position, alternating in the delivery of the slow nips to the neck, or stay for some seconds with the throats pressed together and the heads pointing upwards.... Apart from bites to the neck region, bites may also be directed towards the lower jaw, nose and fore flippers. Weaners of 2 to 3 months of age showed, in addition, 'rearing and slamming' behaviour, raising the front two thirds of the body high off the ground and slamming down onto the other individual with the chest, usually with the mouth closed. Continual rearing and slamming by one individual usually results in the other's moving away and terminating the interaction.

Semi-Wild Boran Cattle (Reinhart & Reinhart, 1982, pp. 1–2):

The typical sequence of this behaviour was the following one:

1. *Solicitation.* With bouncing body and jerky head movements one animal approaches another with head lowered, chin stretched forward and tail raised.... These signals are generally expressed with more emphasis in young calves then in subadults or adults.

2. *Horning*. The solicited partner lowers its head whereupon both animals get their foreheads in contact with one another and gently rub and twist each other's crown and horns....
3. *Pushing*. The opponents push each other back and forth.... When their crowns or horns slide off they briefly circle and bunt each other's neck and shoulder, but quickly get into pushing contact again with their crowns. The partners break off only to solicit each other again. The roles of pushing respectively being pushed are changed and none of the opponents gains ground. The intensity of pushing is generally higher in young as compared to old animals.
4. *Termination*. After a few seconds or several minutes the participants become distracted and abruptly stop pushing each other. Termination has no evident consequences. Incidences which developed into routs, with one animal being chased away by the other, were rate; they have not been considered as mock fights.

Commonly mock fights were interspersed with running games and playful mounting, and all three activities were combined in free sequences.

Laboratory Rats (Panksepp et al., 1984, p. 466):

When such rat pups are placed together in a non-threatening environment, they rapidly begin to exhibit vigorous social interaction which outwardly appears like play-fighting: animals chase and pounce on each other, sometimes unilaterally, sometimes mutually with rapid role reversals. They repeatedly poke and nip each other, often at the nape of the neck but also on the ventral surface when one animal is pinned, and the consequent thrashing about may have more than a passing resemblance to human tickling. Animals also tumble over each other, frequently sustaining "dominant" or "submissive" postures momentarily. If animals are especially evenly matched, they frequently chase one each other in tight circles, apparently vying for a competitive edge.

Rhesus Monkeys (Symons, 1978, pp. 26–37):

A play-fight begins when one monkey approaches another or when they approach each other. The approach may be at a walk, run, gambol, stagger, or leap.... Either the approaching or the approached monkey may leap, and leaping may also occur during the play fight... In relatively nonvigorous play fights one or both partners may maintain a sitting position (especially likely when the partners differ substantially in body size), but in almost all vigorous play fights, both monkeys rear onto their hind legs. This posture varies from

> a squat to a bipedal stand with legs almost fully extended. Rearing frees the hands for reaching and grasping and increases mobility....
>
> Reaching toward the partner with the hands accompanies rearing at the beginning of a play fight and is a prominent feature of play-fighting.... Rhesus sometimes hit during play, striking with the palm side of the open hand, but this is surprisingly infrequent.... At the beginning of a play fight monkeys usually grasp each other with their hands, and mutual grasping continues throughout most of the interaction.... As the play fight develops, monkeys constantly release and regrasp in response to the fluid situation....
>
> Monkeys move to throw their partners to the ground. A monkey thrown to the ground rolls onto its back to face its partner.... The on-top monkey has the advantage over the on-bottom partner in leverage, mobility, and consequently in opportunity to play-bite.... Play-biting, usually accompanied by eyelid closure just before contact, occurs when one monkey places its open mouth on the partner during play-fighting.... Bites are generally aimed at the neck or the ventral surface (the ventral surface usually is the one exposed because of the tendency to face the partner)...but bites can be made anywhere on the partner's body.... Frequently in vigorous play-fighting monkeys feint with their heads while attempting to pull or push each other off balance, each apparently attempting to obtain a low-risk neck-biting opportunity. Once a neck bite is made, the biter attempts to keep its weight into the bite to prevent the partner from breaking free....
>
> Play-fighting may end when one monkey flees...although playing monkeys frequently simply stop and sit. (When one monkey stops play-fighting, its partner almost always stops as well, and they may separate or simply relax and hold each other).

Play Signals

As mentioned earlier, many species appear to engage in specific behaviors that are usually found only in the play context and appear to signal that "this is play" (Bekoff, 1995; Fagen, 1981). These behaviors apparently are highly stereotyped, making their interpretation easy and unambiguous (e.g., Bekoff, 1977). Therefore, by preceding or following fighting behaviors with play signals, the initiating animal signals that the behavior is playful not agonistic, and therefore the recipient is likely to respond accordingly. Without such play signals, it is likely that miscommunications would occur and play-fighting would be difficult to initiate or sustain. Although many candidates for play signals have been identified (e.g., the play face, play "bows," play vocalizations), it is often difficult to demonstrate that a particular behavior serves a play signaling function (Pellis & Pellis, 1996).

An assumption of many early approaches to the study of play-fighting was that one could differentiate between play and serious fighting by the presence or absence of play signals and threat gestures (e.g., Aldis, 1975; Bekoff, 1975; Fagen, 1981; Smith, 1982; Symons, 1978). It was assumed that serious fighting could be identified by the presence of threat gestures and the absence of play signals, whereas play-fighting could be identified by its accompanying play signals and lack of threats.

Although this approach may be useful for differentiating between some of the most obvious cases of play and serious fighting in some species, it has several limitations that were summarized by Pellis and Pellis (1996): play signals have not been identified for many species that play; in species that *do* play, many or even most play interactions are initiated without play signals; it has been difficult to identify play signaling behaviors that are *not* found outside of the play context; and many species *do* indeed engage in threat gestures during play. To further complicate matters, Pellis and Pellis (1996) concluded that the evidence for the existence of play signals is not that compelling. To illustrate their point, they considered published descriptions of one of the most common play signals in primates—the open-mouthed play face (van Hooff, 1978). As shown in Table 1 of their article, descriptions of this play signal in primates show wide variations in their specifics, including descriptions of the eyes (e.g., wide open, outer corners pulled up, drawn together, normal), the mouth (e.g., side open, partially open, open in oval or rounded shape, closed), the brows (raised, depressed, normal), and so on.

So, what is the current status of the concept of play signals in ethology? Some researchers now believe that animals frequently do not require a play signal to determine if a partner's behavior is playful or aggressive. In most situations, the context or the identity of the play partner is sufficient. Because young animals usually play with familiar partners of the same age and gender (see later discussion), a behavior initiated by such an animal is likely to be play. On the other hand, if an unfamiliar animal initiates a struggle, or a familiar animal initiates a struggle to gain access to food or water, its behavior is likely to be agonistic. When situational or partner characteristics are insufficient, stylistic differences in the behavior (e.g., its timing or intensity) may be employed (Pellis & Pellis, 1996).

Bekoff (1995) argued that play signals may be used mostly in situations of ambiguity, in which the attacking animal needs to communicate that a behavior is playful and not aggressive. In a study of coyotes, gray wolves, and domestic dogs, Bekoff (1995) demonstrated that play bows (a play signal common in canids) were more likely to occur before or after "biting with head shaking" (a generally agonistic behavior pattern) in coyotes than in the other two, less aggressive species. Therefore, animals did not simply use play signals to initiate play sessions, but instead employed them in close contact to behaviors likely to be interpreted as aggressive. By communicating the message "I want to play despite what I am going to do or just did—I still want to play" (p. 419), the attacking animal helps maintain the play mode and helps to prevent escalation to aggression.

A final issue concerns the degree to which play signals are indeed that—signals unique to *play*. Pellis and Pellis (1996) argued that because many such signals are found in both play and nonplay contexts, a behavior may not have to be unique to play in order to signal that a given interaction is agonistic or playful. Instead, the use of general amicable behaviors in the play-fighting context may be sufficient for communicating playful intention.

Despite the current controversies about the nature and contextual specificity of play signals, it is useful to summarize the types of play signals that have been identified in play-fighting interactions. This makes it possible to understand the range of ways that animals apparently communicate play intention, as well as the cross-species variability in the occurrence of these behaviors. A review of the literature indicated that play signals in mammals fall into six general categories: head or facial movements, vocalizations, body postures, tail positions, ambulatory patterns, and odors. The most common examples of each are listed in Table 4.2. Play interactions are often initiated with tactile patterns as well (e.g., play nips, pouncing, pushing, touching, or slapping—see Fagen, 1981), but these are not considered here because most researchers place them in the category of play-initiating behaviors, not as play signals—they are not unique to the play context, nor do they represent general amicable signals.

Table 4.3 presents the occurrence/nonoccurrence of these play signals by order or family for the 60 species listed in Table 4.1. The studies used in constructing this table are listed in Table 4.1. Information from Fagen's (1981) natural history of play was used as well. Data are not presented separately by family for the pinnipeds and ungulates, because very little descriptive data are available for many of the families making up these orders. Moreover, because play-fighting patterns differ considerably for the three pongid species, these are presented separately. For ease of presentation, however, the word *family* is used in the remaining discussion of this table.

A plus sign in a column in Table 4.3 indicates that at least one of the species in that family was described as showing the specific type of signal. Thus, it is important to note that a plus sign does not indicate that all members of the family show that behavior. Given large differences in the

TABLE 4.2
Common Play Signals

Head or facial movements	
	Head shaking (Drea et al., 1996; Fagen, 1981; Gentry, 1974; Pellis & Pellis, 1987; Watson & Croft, 1996)
	Ear position (Araba & Crowell-Davis, 1994; Drea et al., 1996; Henry & Herrero, 1974)
	Eye movements or play glance (Feddersen-Petersen, 1991; Hayaki, 1985; Spijkerman et al., 1996; Symons, 1978)
	Open-mouth play face (Bolwig, 1964; Chevalier-Skolnikoff, 1974; Drea et al., 1996; Gentry, 1974; Thompson, 1996a; van Hooff, 1978)
Play vocalizations	
	(Bekoff, 1974; Biben & Symmes, 1986; Goedeking & Immelmann, 1986; Masataka & Kohda, 1988; Rasa, 1971; Stevenson & Poole, 1982; van Lawick-Goodall, 1968)
Body postures	
	Play bow (Bekoff, 1977; Feddersen-Petersen, 1991)
	Roll onto back (Biben, 1983, 1986; Fagen, 1981; Schaller, 1972; Schenkel, 1966; Symons, 1978; Waterman, 1988)
	Looking between legs (Fagen, 1981)
	Crouch and stare (Symons, 1978)
	Hold out hand (Spijkerman et al., 1996)
Tail positions	
	(Drea et al., 1996; Kiley-Worthington, 1976)
Ambulatory patterns	
	Gamboling or staggering play gait (Bekoff, 1974; Biben, 1983; Henry & Herrero, 1974; Schaller, 1972; Symons, 1978; van Lawick-Goodall, 1968)
	Exaggerated approach–withdrawal (Aldis, 1975; Drea et al., 1996; Feddersen-Petersen, 1991; Henry & Herrero, 1974)
	Stalking (Caro, 1985; Schenkel, 1966; Stevenson & Poole, 1982; Wasser, 1978)
	Acrobatics and vertical leaping (Barrett & Bateson, 1978; Bekoff, 1974; Biben, 1986; Fagen, 1981; Feddersen-Petersen, 1991; van Lawick-Goodall, 1968; West, 1974)
	Shaking (Watson & Croft, 1996)
Odors	
	(Fagen, 1981; Wilson & Kleiman, 1974)

amount of detail available on play-fighting in different species of the same family, it is premature to assume that just because a particular behavior pattern is not mentioned in the few articles written on a given species, it does not exist in that species. Therefore, the table provides no more than a preliminary description of our current state of knowledge in this area. Still,

TABLE 4.3

Play Signals by Family or Order

Signal	*Macropods*	*Felids*	*Hyaenids*	*Canids*	*Ursids*	*Mustelids*	*Pinnipeds*	*Ungulates*	*Sciurids*
Head	+	+	+	+			+		
Ears		+	+		+			+	
Eyes		+		+					
Mouth		+	+	+	+	+	+	+	+
Vocalizations				+			+		
Postures		+		+					+
Tail		+	+	+				+	
Locomotion	+	+	+	+	+				
Odors									
Signal	*Cricetids*	*Murids*	*Lemurs*	*Callitrichids*	*Cebids*	*Cercopithecids*	*Orangutans*	*Gorillas*	*Chimps*
Head					+				
Ears									
Eyes						+			+
Mouth			+	+	+	+	+	+	+
Vocalizations				+	+				+
Postures					+	+			+
Tail									
Locomotion				+	+	+			+
Odors	+								

examination of the table reveals some general trends that may be useful in generating hypotheses for future research.

One of the most obvious features of the table is that the play face and locomotion cues are found in the greatest number of families. Second, of all of the families studied, two families of monkeys (the cebids and cercopiticines), the chimpanzees, and the carnivores showed the greatest variety of signals, whereas the rodents showed the least.

Several factors may be responsible for family differences in the number of play signals. The variety of play signals used by carnivores may reflect a greater need for members of these species to quickly provide unambiguous information about their play intentions. Given that carnivores have evolved a significant number of fighting skills used in the capture and consumption of prey, and given that some carnivores have the capacity for intense aggression from an early age (e.g., coyotes, spotted hyenas), it may be particularly important for members of these species to have fast and effective ways of communicating their intentions when play-fighting with such potentially dangerous partners. The results of several studies support this hypothesis.

First, in a comparative study of play signals in three canid species (timber wolves, coyotes, and domestic dogs), Bekoff (1974) found that coyotes, the species showing the most aggression from the earliest age, were the least successful in initiating play, the most likely to precede their play with play invitation signals, and the most likely to use unambiguous play signals. Thus, the aggressiveness of the species predicted its reliance on play signals in facilitating play-fighting.

Evidence that play signals are well developed in aggressive carnivores comes from a study of spotted hyenas by Drea, Hawk, and Glickman (1996). In this species, where the young show high levels of aggression in competing for nursing position just minutes after birth, play intention can be inferred from a wide variety of signals, including head movements, ear positions, facial configurations, tail positions, body movements, and ambulatory patterns. For example, play invitations often involved "jerky body movement directed at another animal, including tilting or rocking of head, with open mouth and lowered ears, and repeated approach/withdrawal sequences" (p. 1326).

Henry and Herrero (1974) identified an even more elaborated play-signaling system in their study of a large, powerful carnivore—the American black bear. They identified five separate ear postures, two of which (crescent and partially flattened) occurred during social play. They also identified seven separate facial expressions. One of them was only seen during play (relaxed open mouth), three others (relaxed, alert, and pucker-lip) were seen in both play and nonplay contexts, and the remaining three (jaw-snapping, jaw-gape, and biting face) were reserved for serious aggression. The existence of at least two elaborate visual systems for readily signaling play intention suggests that, for some species where play occurs between potentially dangerous partners, evolution may have resulted in redundancy in play-signaling systems.

Rodents, in contrast, rely on a more limited number of play signals. In many rodent species, play is preceded by few preliminaries—one animal simply pounces on another (e.g., Meaney & Stewart, 1981). It is not entirely clear why this would be the case because, like carnivores, rodents can inflict serious injuries on one another with their teeth. Several possibilities to consider are that rodents may rely on other, nonvisual clues, such as odors in signaling play intention (Fagen, 1981; Wilson & Kleiman, 1974); despite their ability to bite one another, the potential for injury from partners is not as great as for carnivores, due to differences in their overall aggressiveness; and structural differences in the nature of play-fighting as compared to serious fighting may be more pronounced for members of these species (see Pierce, Pellis, Dewsbury, & Pellis, 1991; see also later discussion in this chapter), making it easier to distinguish one from the other without the use of specific play signals.

An interesting question is why some primates show so many ways of signaling play. Table 4.3 shows that members of two monkey families (the cebids and the cercopiticines), as well as the chimpanzees, employ a variety of cues. The wide variety of signals identified may be a consequence of the fact that these are the families of primates that have received the most research attention; species from other families may use a variety of signals as well. Unlike the carnivores, the variety of play signals used by primates probably is not a consequence of the dangerousness of the play partners. Despite the omnivorous diet of many primates, the primary source of animal protein in most species is usually insects. Thus, primates have not evolved the predatory specializations of canids, hyenas, felids, or bears.

An alternative explanation for the variety of play signals seen in primates is the complexity and flexibility of their play behavior. Numerous authors have commented on the variability in primate play behavior (for a recent review, see Fagen, 1993), and play-fighting appears to be no exception (e.g., van Lawick-Goodall, 1968). If variability in play-fighting behavior is the norm, then one would expect a greater variety of play signals, as well as a greater need for play signals to clarify the numerous ambiguous situations that arise as a result of this variability.

Clearly, additional research on play signals is necessary. Such research should involve investigating a variety of possible play signals across a wide range of species, documenting not only the presence or absence of particular play signals in particular species, but also the situations in which they are used and not used, in both play and nonplay contexts. Only then will it be possible to identify some of the factors (e.g., dangerousness of likely play partners, structural similarity between play and serious fighting, reliance on visual vs. olfactory vs. auditory cues, complexity of play) that might predict the number, nature, variety, and utilization of play signals within a given species.

General Characteristics

According to some formulations (Poole, 1966; Schaller, 1963), play is described as behavior without a definite purpose. However, researchers have long agreed that play-fighting is goal-directed, with the apparent goal of seeking advantage over one's partner. The most commonly described goal is to "bite without being bitten" (e.g., Aldis, 1975; Fagen, 1981; Gentry, 1974; Symons, 1978); other goals include to sniff or touch a specific body part, to push one's opponent to the ground, or to wrestle so as to get on top (e.g., Fagen, 1981; Symons, 1978). Given the amount of self-handicapping and role-reversal in play, Pellis and Pellis (1998) provided the following modification: "Bite, allow yourself to be bitten, then defend yourself, and bite again." Referring to goals does not require inferring cognitive plans or intentions. Pellis (1988) put it this way: "One animal attempts to gain advantage over another...the word attempt is meant only as a descriptive term, capturing the observed behavior where repeated movements towards a body target by one animal are counteracted by evasive movements by the opponent" (p. 86).

When examining the behaviors that animals use in achieving their play-fighting goals, the behaviors—as in all forms of play—are similar to those in other functional contexts. Some of the early studies of play-fighting identified remarkable similarities with serious aggression. For example, Gentry (1974), in a study of Stellar sea lions, found that out of 98 "play components" identified, 18 were adult fighting behaviors (e.g., biting the neck, pushing chest to chest), and 18 were behaviors that adults used in boundary defense (e.g., ceremonial boundary display posture; walking parallel to the border after a boundary display). Other species showing parallels with adult boundary defense behaviors include South American bush dogs, who guard cavity openings during social play (Biben, 1983), and smooth-coated otters, who "defend particular spots from which other otters try to pull or push them" (Fagen, 1981, p. 153).

Bertrand (1976) noted many similarities between play-fighting and serious fighting in stumptail macaques. Out of 34 play behaviors he identified (e.g., crouch, push, kick, lunge), 20 were directly comparable with behaviors observed during adult aggression and the other 14 were modified versions thereof.

Researchers have shown, however, that behaviors during play-fighting sometimes come from contexts other than aggression. Pellis (1993), for example, noted that there are sexual components in the play-fighting of rodents, birds, and primates (e.g., mount, precopulatory behaviors), and Byers (1984) included "mounts" as one of the most common behaviors seen during play-fighting in ungulates. Aldis (1975) noted that the play behaviors of some carnivores apparently are predatory behaviors (e.g., stalking, vertical pouncing). Others have noted that predatory behaviors are particularly prevalent in felids such as domestic cats (Caro, 1981), lions (Schenkel, 1966), and tigers (Wasser, 1978).

Although the behaviors during play-fighting are similar to those in functional adult contexts, their form differs across the two contexts. For example, bites during play-fighting are inhibited (e.g., Aldis, 1975; Symons, 1978), the intensity of the behaviors is lower (e.g., Aldis, 1975; Pellis & Pellis, 1988b, 1989, 1998; Poole, 1966), and individuals often engage in self-handicapping when paired with a partner of lesser strength or abilities (e.g., Araba & Crowell-Davis, 1994; Bertrand, 1976; Croft & Snaith, 1991; Hayaki, 1985; Poole, 1966). This results in numerous cases of role-reversal and dominance reversals between attacker and defender (e.g., Aldis, 1975; Feddersen-Peterson, 1991; Jamieson & Armitage, 1987; Miller, 1975; Poole & Fish, 1976; Reinhart & Reinhart, 1982). Although it was once thought that behavior during play-fighting was exaggerated and randomly ordered (e.g., Aldis, 1975; Miller, 1973), there is little empirical support for this notion (Havkin & Fentress, 1985; Henry & Herrero, 1974; Hill & Bekoff, 1977; Symons, 1978).

Play Initiations

Play-fights are usually initiated with an attack that may or may not be preceded by some preliminary behaviors. The nature of the attack varies as a function of the "weapons system" of the animal. For example, laboratory rats and polecats initiate by pouncing on their partner and attempting a bite (e.g., Meaney & Stewart, 1981; Poole, 1966), whereas bears initiate by pawing (Henry & Herrero, 1974), and chimpanzees may initiate by tickling (van Lawick-Goodnow, 1968).

Whether an animal precedes the attack with any preliminary behaviors appears to vary considerably by species. Laboratory rats apparently engage in few or none (Meaney & Stewart, 1981), whereas cats often engage in prolonged locomotor patterns (Caro, 1981; Fagen, 1981). Aldis (1975) pointed out that "maneuvering" behaviors are seen in predators that attack large prey. He gave as examples dogs and terrestrial primates who precede their play attacks with circling and in-and-out locomotion. Similar behaviors have been found in hyenas (Drea et al., 1996). Stalking is a form of preliminary behavior that is seen in the play of larger felids (e.g., cheetahs—Caro, 1995; lions—Schenkel, 1966; tigers—Wasser, 1978).

Less agonistic forms of preliminary behaviors include: (a) rolling on one's back to a belly up position (e.g., canids—Biben, 1983; domestic cats—Fagen, 1981; lions—Schenkel, 1966; primates—Biben, 1986; ground squirrels—Waterman, 1988); (b) playwalking such as gamboling, swaggering, and exaggerated approach (bears—Henry & Herrero, 1974; canids— Bekoff, 1974; Biben, 1983; rhesus monkeys—Symons, 1978; chimpanzees—van Lawick-Goodall, 1968; lions—Schaller, 1972); and (c) leaping, bouncing, and acrobatics (e.g., canids—Bekoff, 1974; Feddersen-Petersen, 1991; Fagen, 1981; felids—Barrett & Bateson, 1978; West, 1974; primates—Biben, 1986; van Lawick-Goodall, 1968).

Finally, the diversity of play initiations varies by species as well—from simple initiations such as approach and bite in Stellar sea lions (Farentinos, 1971), and pounce and bite in rats (Meaney & Stewart, 1981), to a wide range of initiations in chimpanzees. In a field study of chimpanzees in Tanzania, Hayaki (1985) identified the following initiation behaviors: stamp; stare; shake, bite, or fold branch; play face; slap; hand up; bipedal; roll; move with object in the mouth or the hand; play-pant; lie with legs up; leaf-clip; and turn back. In her field work in the Gombe Stream Reserve, van Lawick-Goodall (1968) described the play initiations of infant chimpanzees as follows:

> Thus a 2-year-old infant usually walked slowly to a younger one and gently reached out to pat or tickle it, whereas when initiating play with a peer or older animal it often ran up with a boisterous play walk or gambol and flung itself onto the other, hitting, mock-biting, flailing or kicking. Infants often bounced after mature or adolescent males grabbing hold of their feet or ankles as they walked; often the male concerned kicked backwards, sometimes quite hard, but apparently, playfully. Sometimes infants approached a chosen playmate with a play walk whilst holding some "toy" and, if the other reached

to grab this, the initiator ran off looking back over its shoulder. (van Lawick-Goodall, 1968, p. 259)

In summary, species vary widely in the nature of their play initiations. It is an issue for further research whether these variations represent differences in the need to signal play intent due to factors such as potential dangerousness of the partner; complexity and/or flexibility of the animal's behavior patterns; or the need to practice functional locomotor behavior patterns, such as maneuvering and stalking, and whether these behaviors likely vary by species.

Play-Fighting Tactics

Presented in Table 4.4 is a summary of the fighting behaviors listed, by order, for the 60 species considered here. As was the case in Table 4.3, a plus sign indicates that at least one published report of a species in that order described that behavior as occurring during play-fighting in the species. Nine of the behavior categories in the table come from Byers' (1984) review of play-fighting in ungulates. Byers' "push/neckfight" category was broken into its two components to more effectively describe differences in the play-fighting strategies of the non-ungulate species. The remaining categories were added to include play-fighting behaviors mentioned by other authors in published studies on other mammals. The studies used in constructing this table are listed in Table 4.1. Fagen's (1981) natural history of play was used as well.

Examination of the table shows several commonalities in play-fighting tactics across a diverse set of species. Species from all of the orders studied demonstrated pushing, biting, and lunging or pouncing, and all but the macropods exhibited chasing. Moreover, all of the land-living animals showed pawing, swatting, or kicking. Pinnipeds cannot engage in these behaviors because they use their front flippers to stand up and support their weight (Aldis, 1975). The other behaviors varied as a function of the species' anatomy and available weapons systems. The two orders with the most limited use of their limbs (i.e., pinnipeds and ungulates) rely almost exclusively on their heads for play-fighting. Ungulates are the only order considered here to engage in head-butting or sparring contests with horns and antlers; both ungulates and pinnipeds engage in neck fights and mouth-to-mouth contact. Both primates and carnivores engage in grabbing, holding, and pulling—likely a result of primates' well-developed fine-motor capabilities, and carnivores' well-developed ability to use the mouth in prey

TABLE 4.4

Play-Fighting Behaviors, by Order

Behaviors	*Marsupials*	*Carnivores*	*Pinnipeds*	*Ungulates*	*Rodents*	*Primates*
Lunge, pounce*	+	+	+	+	+	+
Butt, clash, spar*				+		
Bite, slash*	+	+	+	+	+	+
Neck fight*			+	+		
Push*	+	+	+	+	+	+
Naso-naso contact*		+	+	+		
Paw, swat, kick*	+	+		+	+	+
Grab, hold	+	+				+
Pull		+				+
Wrestle, pin		+			+	+
Mount*				+	+	+
Chase*		+	+	+	+	+

*From the review by Byers (1984).

capture and consumption. Primates, carnivores, and rodents engage in wrestling—again reflecting differences in anatomy, probably the ease at which they can engage in rolling and pinning behaviors. Finally, mounting was reported only for ungulates, rodents, and primates.

Comparison of individual species within the orders revealed few systematic differences in the occurrence of these behavior patterns. One notable exception concerned the carnivores. Mouth-to-mouth fighting was only reported for hyenas, canids, and bears. A possible explanation for this findings is that mouth-to-mouth fighting may be most likely in those carnivores with long muzzles.

It is important to note that Table 4.4 presents only the most superficial kind of cross-order comparisons of play-fighting strategies. Although it is useful in demonstrating that diverse animals use a similar set of behaviors to achieve their play-fighting goals, it says nothing about similarities and differences in their relative use of the various strategies, or about differences in the nature, timing, and sequencing of these behaviors. Given the limited descriptive database on play-fighting in most species (Fagen, 1981; Pellis & Pellis, 1998), these aspects of behavior cannot currently be compared systematically across species. Several authors have presented sequential analyses of behavior patterns during play-fighting interactions (e.g., Biben, 1983; Croft & Snaith, 1991; Le Resche, 1976; Nowicki & Armitage, 1979; Poole & Fish, 1975, 1976; Stevenson & Poole, 1982; Wasser, 1978; Watson & Croft, 1993); however, given the relatively small number of studies and differences in the coding systems employed, few general conclusions can be drawn at this point.

Another limitation of the current database on play-fighting is that considerably more attention has been devoted to offensive than to defensive tactics. Pellis and Pellis (1991a) argued that species differ in both of these areas, and that a thorough description of play-fighting across species requires consideration of both. Pellis and colleagues (e.g., Pellis & Pellis, 1987, 1990, 1998; Pellis, Pellis, & Dewsbury, 1989) provided a wealth of such descriptive information in several species of murid rodents. They also published such information on birds (Pellis, 1981), otters (Pellis, 1984), and primates (Pellis & Pellis, 1997b).

The following excerpt from Pellis et al. (1989, pp. 305–306) illustrates how careful description of defensive strategies during play-fighting can illuminate significant differences in the play-fighting behaviors of closely related species. Pellis and colleagues demonstrated that, in play-fighting of several species of murid rodents, the apparent goal of play-fighting attacks is to contact the nape of the neck of the partner:

> Indeed, the four species studied had all the subcomponents of play-fighting: attack, defense, and counterattack. What produced the species differences was the frequency of combining these subcomponents.... The importance of the combination of these subcomponents is evident by the fact that any one element in itself could not replicate the pattern present in rats. For example, montane voles had rates of facing defense (i.e., turning to face the attacker, thus blocking access to an area such as the nape) similar to those of rats, and prairie voles had rates of counterattack similar to those of rats...but in each case, the pattern of play-fighting appeared different from rats. In rats, most of the defense involved turning to face the attacker, and over 40% of these were followed by counterattack.... It is counterattacks that elicit defense in the original attacker (Pellis & Pellis, 1987, 1988a) and thus produce the role reversals so typically described in the play-fighting by mammals (e.g., Fagen, 1981; Poole, 1985). In deer mice, most of the defense involved evasion (i.e., swerving away from the attacker), and counterattacks were infrequent.... Therefore, play-fighting by deer mice rarely looked like that of rats. The high rates of facing defense and low rate of counterattacks by montane voles made their play-fights appear more "agonistic," while the high rate of evasion and high rate of counterattack in the prairie voles made their play-fighting appear more like "mutual nosing" than like fighting. In all these species, therefore, the principle differences in the form of play-fighting arose from differences in the responses of the defenders, not of the attackers.

Although usually not as detailed, other sources for information on defensive strategies include Havkin and Fentress (1985—timber wolves); Poole (1966, 1978—polecats); Symons (1978—rhesus monkeys); Steiner (1971—Columbian ground squirrels); and Waterman (1988—Columbian ground squirrels).

Termination

Although a few authors identified specific behaviors in which animals engage before terminating a play-fight (e.g., Henry & Herrero, 1974; Poole, 1966), most stated that play-fights end when a partner breaks away or leaves (e.g., Henry & Herrero, 1974; Poole & Fish, 1975; Steiner, 1971). Play-fights appear to end when one or the other partner loses interest or the fight becomes too rough. In the latter case, the interaction may end with an aggressive threat (e.g., Poole, 1966). A important point made by most authors, however, is that most play-fights end amicably, and that the partners usually remain in close proximity to one another, unlike in the case of serious aggression (e.g., Hole, 1991; Watson & Croft, 1993).

Differences Between Play- and Serious-Fighting

Given all of the similarities between play-fighting and aggression, an important issue to consider in some detail is whether or not play-fighting can be considered as simply immature aggression (e.g., Havkin & Fentress, 1985; Lazar & Beckhorn, 1974). If this is the case, then arguments that play-fighting evolved as a separate system to serve important evolutionary functions become less convincing. Instead, the questions become: Why does aggression appear in an immature form, and why does it appear at the ages that it does?

As already mentioned, most play researchers believe that play-fighting is distinct from aggression and is under the control of different neurophysiological mechanisms (e.g., Panksepp et al., 1984). In fact, several observers have argued that a closer inspection of aggressive and play-fighting behavior patterns reveals that any similarities between behavior in these two contexts is superficial (e.g., Pellis & Pellis,1998; Poole, 1966).

Because the previous sections focused on the similarities between play-fighting and aggression, this section focuses on the differences. Presented in Table 4.5 is a summary of the documented differences between play-fighting and serious fighting. As can be seen in the table, these include differences in developmental course, presence or absence of threats, contexts in which the two behaviors occur, the behaviors themselves (i.e., intensity, duration, tactics, and sometimes targets), as well as the consequences from the interactions involved (i.e., harm, access to resources). Moreover, during play-fighting, animals do not respond in terms of the dominance relationships that govern aggressive interchanges (e.g., dominance reversals, self-handicapping, and role reversals). Additionally, aggressive interchanges terminate differently than do play-fighting interactions in terms of the reactions and proximity of the participants. Finally, studies that have investigated the relationship between the dominance hierarchies involved in play-fighting and serious aggression show no significant correlations.

Therefore, although play-fighting and aggression resemble each other in terms of their major behavioral components, the form and function of these behaviors in the two contexts differ substantially.

Developmental Course

Given the variation in the structure of play-fighting across species, and the few species for which longitudinal studies of play-fighting have been

TABLE 4.5

Differences Between Play- and Serious-Fighting

Characteristic	*Play-Fighting*	*Serious-Fighting*
Developmental course	*Developmental course of two behavior systems independent* (Bekoff, 1974; Byers, 1984; Caro, 1981; Drea et al., 1996; Reinhart & Reinhart, 1982)	
Threats	No threats (Farentinos, 1971; Symons, 1978)	
	Threats rare (Henry & Herrero, 1974; Poole, 1966; Reinhart & Reinhart, 1982)	
	Threats common (Feddersen-Petersen, 1991; Pellis & Pellis, 1996, 1998)	Threats and dominance behaviors common
	No dominance behaviors (Pellis & Pellis, 1998)	
Functional contexts	No obvious immediate function	Gain access to estrous females, food, water, and or grooming partners, and responses to intruders (defend territory; Hole, 1991; Reinhart & Reinhart, 1982; Watson & Croft, 1996)
Intensity	Gentle, inhibited (Croft & Snaith, 1991; Drea et al., 1996; Pellis & Pellis, 1988b, 1998; Poole, 1966, 1978)	Intense (Pellis & Pellis, 1989, 1998; Poole, 1966)
Duration	Longer (Croft & Snaith, 1991; Drea et al., 1996; Owens, 1975a; Poole, 1966)	Stand over and bite longer (Bekoff, 1977)
Tactics employed	No simultaneous defense during attacks—attacks only (Pellis & Pellis, 1998)	
	Defensive strategies more similar to serious aggression than are offensive strategies (Pellis et al., 1989)	Simultaneous defense during attacks
	Fewer attempts to immobilize partner (Poole, 1978)	
	No sideways attacks (Poole, 1966)	

Table 4.5 continued

Table 4.5 continued

Characteristic	*Play-Fighting*	*Serious-Fighting*
Targets	Agonistic: magpie, ground squirrels, polecats	Agonistic
	Amicable/precopulatory: golden hamsters, deer mice, voles, Norwegian rats (Pellis & Pellis, 1998)	
Consequences	No wounds (Pellis & Pellis, 1987)	Wounds
Dominance:		
Reversals	Yes (Biben, 1986; Watson, 1993)	No
Self-handicapping	Yes (Araba & Crowell-Davis, 1994)	No
Who usually initiates play	Independent of dominance status (Bertrand, 1976; Drea et al., 1996; Miller, 1973; Watson, 1993; Zucker & Clarke, 1992)	Older individuals (Watson, 1993) Dominant individuals (Drea et al., 1996)
Reactions to threats	No submission (Bekoff, 1975)	Submission
Termination	Dependent on who initiates (Watson, 1993)	Dependent on age and dominance status (Watson, 1993)
	Terminates with grooming bouts (Fagen, 1981)	Object of aggression withdraws (Reinhart & Reinhart, 1982)
	Ends amicably, in close proximity (Hole, 1991)	
Individual differences	*Individual differences in the overall rates of play-fighting and serious fighting, as well as wins and losses in the two contexts uncorrelated* (Araba & Crowell-Davis, 1994; Watson, 1993)	

conducted, it is difficult to make many firm conclusions about the developmental course of play-fighting. However, two general conclusions can be drawn. First, the frequency of play-fighting usually follows an inverted U function, with play-fighting starting sometime during infancy, peaking in the juvenile period, and then decreasing thereafter (see references in Table 4.6). The developmental stage at which play-fighting starts and stops differs significantly across species, and species vary, for example, in the degree to which play-fighting is observed in adults (e.g., Fischer & Nadler, 1978; Zucker et al., 1986). A second general trend is that, in some species, the vigor and roughness of play-fighting often increases with age (Baldwin & Baldwin, 1978a; Meaney & Stewart, 1981; Poole & Fish, 1976; van Lawick-Goodall, 1968). In fact, in some species, the likelihood of play-fighting escalating into serious fighting increases with age (Mendoza-Granados & Sommer, 1995; Pellis & Pellis, 1988b; Poole & Fish, 1976). Although advocates of the motor training hypothesis have argued that the refinement and coordination of the motor components of play-fighting should increase with age (see later discussion in this chapter), only a few researchers have noted such changes (e.g., Farentinos, 1971), and just as many reported few changes in the structure of play-fighting from its onset (e.g., Meaney & Stewart, 1981; Poole, 1966).

Play Partners

Research on who play-fights with whom has demonstrated that, when a choice is available, animals tend to play with partners similar to themselves. Specifically, as shown in Table 4.7, animals spend the most time play-fight-

TABLE 4.6

References Documenting Developmental Changes in Play-Fighting

Felids—Barrett & Bateson, 1978; Caro, 1981, 1995; Mendoza & Ramirez, 1987; Wasser, 1978
Hyenas—Drea et al., 1996
Canids—Bekoff, 1974; Feddersen-Petersen, 1991
Pinnipeds—Gentry, 1974; Renouf & Lawson, 1986
Ungulates—Crowell-Davis et al., 1987; Gomendio, 1988; Hass & Jenni, 1993; Pfeiffer, 1985; Reinhart & Reinhart, 1982
Rodents—Hole, 1988; Meaney & Stewart, 1981; Panksepp, 1981; Pellis & Pellis, 1990, 1993, 1997a; Waterman, 1988
Primates—Bramblett, 1978; Cheney, 1978; Govindarajulu et al., 1993; Imakawa, 1990; Koyama, 1985; Mendoza-Granados & Sommer, 1995; O'Neill-Wagner et al., 1994; Poole, 1987; Symons, 1978; van Lawick-Goodall, 1968; Zucker & Clarke, 1992

TABLE 4.7

Studies of Play-Fighting Partners: Age and Gender Differences

Animals play more with partners of same or similar age	
	Marsupials (Watson, 1993)
	Mustelids (Poole, 1966)
	Ungulates (Berger, 1980; Byers, 1980; Hass & Jenni, 1993; Pfeiffer, 1985; Reinhart & Reinhart, 1982; Thompson, 1996b)
	Primates (Cheney, 1978; Fady, 1976; Govindarajulu et al., 1993; Imakawa, 1990; Janus, 1989; Owens, 1975b)
Animals play more with partners of same gender	
	Ungulates (Byers, 1980; Reinhart & Reinhart, 1982)
	Rodents (Jamieson & Armitage, 1987; Meaney & Stewart, 1981)
	Primates (Biben, 1986; Cheney, 1978; Hayaki, 1983; Imakawa, 1990; Owens, 1975b)
Females avoid playing with males or terminate play initiated by males	
	Ungulates (Byers, 1977; Reinhart & Reinhart, 1982)
	Rodents (Jamieson & Armitage, 1987; Takahashi & Lore, 1983)

ing with partners that are close in age and are of the same gender. Most theorists interpret these tendencies to reflect animals' preference for play-fighting partners that are evenly matched in size and fighting abilities. This interpretation is supported by the finding that male–male play is often more intense than male–female play (e.g., Pfeiffer, 1985), and that the females of many species avoid or terminate play with males, presumably because of its greater intensity (see Table 4.7).

A second area of research concerns the biological relatedness of the partner. If play indeed has benefits for the players, an evolutionary analysis would predict that animals should play more with kin than with nonkin. Although this topic has not been extensively studied, most studies that have been conducted on the subject show that this is indeed the case (e.g., Siberian ibex—Byers, 1980; bighorn sheep—Berger, 1979; Japanese macaques—Glick, Eaton, Johnson, & Worlein, 1986; Koyama, 1985).

Studies of the dominance status of the partner are much less consistent. Although some studies show that animals are more likely to play-fight with partners of the same rank (wallabies—Watson, 1993; Boran cattle—Reinhart & Reinhart, 1982; Japanese macaques—Imakawa, 1990; rhesus macaques—Tartabini & Dienske, 1979), both Biben (1986) and Fady (1976) found that the primates they studied most frequently played with partners of similar rank or those who were submissive to themselves. Other studies find that the offspring of high-ranking females are the most frequently chosen play partners (e.g., Cheney, 1978; Gard & Meier, 1977). The results are no less confusing for the effects of dominance status on play *initiations*. Although several studies show that subordinate animals are more likely to initiate play with dominant animals than vice versa, this varies with the age and gender of the animal, as well as the stability of the dominance hierarchy (Garnd & Meier, 1977; Hayaki, 1983; Parquette, 1994; Pellis & Pellis, 1991b, 1992).

Because there has been much less research devoted to the relatedness and dominance status of the partner than to partner age and gender, it is difficult to draw conclusions about the importance of these factors. Moreover, because dominance hierarchies are more important for governing aggression than for play (recall that dominance reversals often occur during play), dominance status might have less relevance for play-fighting. It is also important to appreciate that the choice of play partners is under the influence of many factors: the number, age, and gender of potential partners available; the social structure of the animal group; the parents' dominance position in the group; the influence of the environmental setting; and so on. However, given the consistent tendency for animals to select partners of similar size and ability, one would expect that when choices are limited, animals will do their best to choose partners most similar to themselves. Byers (1980), for example, found that young male ibex usually found evenly matched partners by play-fighting with either same-aged males or older and larger females.

Gender Differences

Gender differences in play-fighting have been documented for a wide range of species (see reviews in Fagen, 1993; Meaney, Stewart, & Beatty, 1985; Thompson, 1996a). A widely held assumption, however, is that gender differences are not always found, but tend to occur for those species where there are differences between males and females in the importance of fighting skills for adult roles (Fagen, 1981). Byers (1980) and Symons (1978) argued that the existence of gender differences in play-fighting are tied to the reproductive strategy of a species. Specifically, where a major function of aggression is competition between males for mates at breeding time, there exist selection pressures for "male structures, behaviors, and dispositions that are useful in male–male competition" (Symons, 1978, p. 180). Thus, the young of species exhibiting such sexual dimorphism are expected to engage in more play-fighting in that play-fighting may function to increase males' fighting abilities and subsequently their reproductive success. The development of fighting skills in such species:

> may increase a males' reproductive success by increasing the probability that he will: (1) avoid being killed or wounded; (2) eliminate competitors by killing or wounding them; (3) succeed in joining a social group; (4) succeed in remaining in a social group; (5) defeat rivals and achieve high rank, and avoid being defeated and losing rank, which result, on the average, in more frequent copulation, more frequent copulation with older, more fertile females, and more frequent copulation in the fertile phase of the female's estrous cycle. (Symons, 1978, pp. 180–181)

Fagen (1981) argued that one would expect few differences in play-fighting in the young of those species where aggression serves the same functions for males and females. Such is the case for play among many carnivores, where both adult males and females use aggression for capturing and subduing prey, defending territories, and protecting offspring. Ferrets appear to be an intermediate case, in that adults of both genders hunt and defend territories, but males are larger and show greater aggressiveness than do females. The young of this species showed more gender differences in play than did canids, but less than did some ungulates and primates (Biben, 1982b).

According to Smith (1982), because the amount of intraspecific competition varies with reproductive strategy, gender differences in play-fighting should be greatest for highly polygynous species and less so for monogamous and solitary species. Presented in Table 4.8 is a summary of the relationship between monogamy and gender differences in play-fighting for the species considered in this chapter. Included in the table are only nondomesticated species for which data on gender differences in play-fighting are available. A plus in the table indicates (with one exception—see hyenas) that males are greater than females in the initiation, duration, and/or roughness/intensity of play-fighting. As can be seen in the table, the hypothesis was supported for all but one species (i.e., South American fur seals). Regarding the exception, Harcourt (1991) argued that he may not have found gender differences in play because observations were conducted only in the presence of the mother, and that maternal presence often suppresses play-fighting in this species. It should be noted, however, that few monogamous species have been studied, and that further studies employing larger samples of monogamous species (for a partial listing, see Kleiman, 1977) would generate more confidence in this hypothesis.

Probably the most compelling support for the relationship between adult roles and play-fighting in the young comes from a study of spotted hyenas

TABLE 4.8

Monogamy and Gender Differences in Play-Fighting of Young Mammals

		Gender Differences?	
Species	*Monogamous?*	*Present*	*Absent*
Marsupials			
Red-necked wallabies	No	Watson, 1993	
Red kangaroos	No	Croft & Snaith, 1991	
Carnivores			
Spotted hyena	No	Pedersen et al., 1990	
Coyotes	Yes		Bekoff, 1974; Hill & Bekoff, 1977
Timber wolves	Yes		Bekoff, 1974
Maned wolves	Yes		Biben, 1983
Crab-eating foxes	Yes		Biben, 1983
Bush dogs	Yes		Biben, 1983
Ferrets	No	Biben, 1982b	
Pinnipeds			
Stellar sea lions	No	Farentinos, 1971; Gentry, 1974	
South American fur seals	No		Harcourt, 1991
Northern elephant seals	No	Rasa, 1971	
Ungulates			
Black-tailed deer	No	Miller, 1975	
Curvier's gazelles	No	Gomendio, 1988	
Bighorn sheep	No	Berger, 1980; Hass & Jenni, 1993	
Siberian ibex	No	Byers, 1977, 1980	
Scimitar-horned oryx	No	Pfeiffer, 1985	
Rodents			
Yellow-bellied marmots	No	Jamieson & Armitage, 1987	
		Nowicki & Armitage, 1979	

Table 4.8 continued

Table 4.8 continued

Species	Monogamous?	Gender Differences? Present	Gender Differences? Absent
Columbian ground squirrels	No	Waterman, 1988	
Norway rats	No	Hole, 1988; Meaney & Stewart, 1981; Olioff & Stewart, 1978; Pellis & Pellis, 1990; Poole & Fish, 1976; Takahashi & Lore, 1983; Thor & Hollaway, 1983	
Primates			
Common marmosets	Yes		Stevenson & Poole, 1982
Squirrel monkeys	No	Baldwin & Baldwin, 1974;	
		Biben, 1986	
Japanese macaques	No	Hayaki, 1983; Imakawa, 1990; Koyama, 1985	
Rhesus macaques	No	Harlow, 1965; Hinde & Spencer-Booth, 1967; Janus, 1989; O'Neill-Wagner et al., 1994; Symons, 1978	
Yellow baboons	No	Cheney, 1978	
Olive baboons	No	Chalmers, 1980; Owens, 1975a, 1975b	
Hamadryous baboons	No	Kummer, 1968	
Vervet monkeys	No	Bramblett, 1978	
Orangutans	No	Nadler & Braggio, 1974	
Chimpanzees	No	Mendoza-Granados & Sommer, 1995; Nadler & Braggio, 1974; Spijkerman et al., 1996	

(Pedersen et al., 1990). In this sexually dimorphic species, adult females are larger and more aggressive than are adult males, so one would predict higher levels of play-fighting for young females than for young males. This was indeed the case.

FUNCTIONS/EFFECTS

There has been no shortage of hypotheses about the possible functions and/or effects of play-fighting, but, as was the case for solitary object play, there is little experimental evidence to support these claims. Originally, biologists who studied play-fighting argued that it either provided the motor training necessary to become an efficient and effective fighter (e.g., Farentinos, 1971; Symons, 1978) or served a socialization function in establishing social bonds and fostering group cohesion (e.g., Bekoff, 1974; Cheney, 1978; Poole, 1966). These possible functions are still debated today, but in the company of a much larger group of hypotheses. As shown in Table 4.9, hypotheses about the functions and effects of play-fighting fall into five general categories: aggression training, aggression control or prevention, socialization, cognitive stimulation and learning, and other. Hypotheses in the final category come from three recent reviews of the play-fighting literature, and have yet to generate much published research. It is likely that any or all of these might lead to separate functional categories in a future review.

The vast majority of hypotheses in Table 4.9 are not the result of simple armchair speculation. Instead, most were offered after a systematic observation of the structure of play-fighting in one or several mammalian species. In the sections that follow, we consider the empirical evidence for the major functional hypotheses. Three types of evidence are considered in separate sections, including evidence from analyses of play-fighting structure, correlational studies, and controlled experiments. As can be seen, there is little correlational or experimental evidence for the functions or effects of animal play-fighting. Instead, our knowledge comes almost exclusively from inferences drawn from observations of play-fighting structure.

Evidence From Analyses of Play-Fighting Structure

Aggression Training. In reviewing the hypothesized functions and effects of play on the development of motor skills, Bekoff and Byers (1981) suggested the catchall phrase "motor-training hypothesis" to refer to any

TABLE 4.9

Hypothesized Functions of Play-Fighting

Aggression Training	
	Experiences general physical training, which improves later fighting abilities, such as strength, endurance, flexibility, weapons systems (e.g., thicker horns; Aldis, 1975; Byers, 1980; Croft & Snaith, 1991; Fagen, 1981; Fontaine, 1994; Symons, 1978; Wasser, 1978; Watson & Croft, 1993)
	Learns specific fighting skills later used in gaining access to mates; territorial defense; establishing and maintaining position in dominance hierarchy; predatory aggression; defense against predators, and defending others (Aldis, 1975; Biben, 1986; Brownlee, 1984; Byers, 1977, 1980; Caro, 1981, 1995; Croft & Snaith, 1991; Crowell-Davis et al., 1987; Fagen, 1981; Farentinos, 1971; Govindarajulu et al., 1993; Hass & Jenni, 1993; Jamieson & Armitage, 1987; Owens, 1975a, 1975b; Parquette, 1994; Pfeiffer, 1985; Potegal & Einon, 1989; Reinhart & Reinhart, 1982; Schenkel, 1966; Steiner, 1971; Symons, 1978; Wasser, 1978; Watson, 1993; Watson & Croft, 1993, 1996).
	Engages in self-assessment of fighting skills (Thompson, 1996b; Wasser, 1978)
	Learns when to act in self-defense (Fagen, 1995)
	Reduces fear of launching attacks; self-confidence (Fagen, 1995; Pellis & Pellis, 1998)
Aggression prevention or control	
	Learns escape skills, such as speed and agility in flight (Brownlee, 1984; Wilson & Kleinman, 1974)
	Sublimates aggressive tendencies (Reinhart & Reinhart, 1982)
	Develops tolerance of other animals approaching (Einon & Potegal, 1991)
	Learns to assess ability or motivation of opponent (Fagen, 1981; Owens, 1975a; Reinhart & Reinhart, 1982; Symons, 1978; Wasser, 1978)
	Can prevent escalation of play-fighting into serious aggression by accepting pain as part of social interaction or learning restraint, self-handicapping, control, and modulation (Bekoff, 1974; Drea et al., 1996; Fagen, 1981; Pellis & Pellis, 1998; Potegal & Einon, 1989)
	Prevent aggression by establishing bonds with dominant animals (Pellis, 1993; Pellis & Pellis, 1991b)
Socialization	
	Learn identity of play partners (Symons, 1978; Wilson & Kleinman, 1974)
	Establish and reaffirm affiliative social relationships (Aldis, 1975; Bekoff, 1974; Cheney, 1978; Drea et al., 1996; Harcourt, 1991; Hayaki, 1983; Koyama, 1985; Nowicki & Armitage, 1979; Owens, 1975b; Poole, 1966; Reinhardt & Reinhardt, 1982; Symons, 1978)
	Establish, test, and maintain dominance relationships (Aldis, 1975; Meaney et al., 1985; Parquette, 1994; Tartabini & Dienske, 1979)
	Learn both submissive and dominant roles (Symons, 1978)
	Learn other social skills (e.g., communication, negotiation, conflict resolution; Aldis, 1975; Brown, 1988; Fagen, 1981; Feddersen-Petersen, 1991; Haas & Jenni, 1993; Symons, 1978)
Cognitive stimulation and learning	
	Cognitive stimulation (Fagen, 1993)
	Learn to approach novelty, not fear it (Fagen, 1993)
	Produce novelty (Fagen, 1981; Fedigen, 1972; Nowicki & Armitage, 1979; Symons, 1978)
Other	
	Sex and mounting skills (Byers, 1984; Pellis, 1993; Wasser, 1978)
	Signal well-being to caregivers and companions (Fagen, 1992, 1995)
	Energy loss: lower risk of predation, susceptibility to pathogens and cold stress (Barber, 1991)

hypothesis predicting that play activities might improve motor performance in any way (e.g., exercise, physical training, or practice). They argued that to talk about the effects of motor training in a general way is appropriate, because the various aspects of motor functioning are so interrelated. For example, if play-fighting serves the function of strengthening muscles, it does not strengthen all muscles equally, but only those involved in executing particular motor sequences. Once these specific muscles are strengthened, then it is likely that the animal can execute a specific motor sequence in a more efficient and economical way. Moreover, as more muscle mass is developed, it may change the animals' metabolism, which might influence weight gain, and therefore indirectly, endurance.Thus, in this example, play does not simply perform a strengthening function, but may have an indirect influence on motor efficiency and endurance as well.

As seen in Table 4.9, the motor-training hypothesis has been popular, and many authors have interpreted their findings on the structural characteristics of play-fighting as supporting some form of this hypothesis. This hypothesis has also been popular among researchers who study locomotor play (see chap. 6).

One of the most common arguments about how the structure of play-fighting supports the motor training hypothesis concerns observations of gender differences in both play-fighting and the selection of play partners. Studies that document the association between gender differences in adult aggression and in play-fighting are often cited as evidence of the motor training hypothesis (e.g., Biben, 1986; Byers, 1977, 1984; Caro, 1988; Crowell-Davis et al., 1987; Jamieson & Armitage, 1987; Pfeiffer, 1985; Reinhart & Reinhart, 1982; Symons, 1978; Watson & Croft, 1993). The basic assumption is that gender differences in play-fighting have evolved because differences in adult roles make practice during the infancy and juvenile periods more important for one gender than for the other. The finding that young animals spend most of their time playing with well-matched partners is viewed as supportive as well, because this would likely maximize motor training (Byers, 1980; Govindarajulu, Hunte, Vermeer, & Horrocks, 1993; Jamieson & Armitage, 1987; Pfeiffer, 1985; Reinhart & Reinhart, 1982; Watson, 1993).

The other type of structural evidence offered in support of the motor-training hypothesis concerns the specific ways in which the aggressive components of play-fighting differ from behaviors during serious aggression. The assumption of this argument is that if play-fighting facilitates the development of fighting skills, then the aggressive components of play-fighting should have structural characteristics that facilitate learning.

What is learned during motor training? Symons (1978) identified several consequences of motor practice, including increases in speed and economy of movement; decreases in reaction times and extraneous movements; and increases in strength, vigor, intensity, timing, precision, and automaticity. Practice also increases the ability to choose quickly among an array of possible actions.

In reviewing the literature on motor training, Fagen (1981) identified several characteristics of practice sessions that lead to the greatest motor learning: distributing practice across multiple sessions and time; engaging in social as well as solitary practice; breaking movements down into their components, and practicing them separately; slowing down behaviors when learning, and then increasing speed over time; paying little attention to details early on, and fine-tuning later; isolating the "bugs" that impede performance, and focusing on them in training; creating and "fooling around" with easier but related problems early on, and reattacking more difficult problems later; approaching tasks in several ways so that one does not get "stuck" into particular ways of behaving; and generalizing new skills as they are learned to new contexts, to see how well they "fit." As amply demonstrated in the literature, generalization of motor patterns is by no means automatic.

A comparison of these recommendations for practice with the general characteristics of play-fighting reviewed above reveals numerous similarities. Play-fighting (corresponding to the same order of items in the previous paragraph): is made up of bouts that are usually of short duration and occur at regular intervals (e.g., Hole, 1988; McDonald, 1977); is social; is made up of the components of serious agonistic behavior (e.g., Bertrand, 1976; Gentry, 1974); is reduced in intensity and may be slowed down in time (e.g., Aldis, 1975; Fagen, 1981; Poole, 1966); could be described as "fooling around" with the components of serious fighting; may show greater variability in behaviors and sequences than does serious fighting (e.g., Fagen, 1981; but also see Hill & Bekoff, 1977); and is generalized across locations, time, and partners. Although Symons (1978) and Fagen (1981) interpreted these similarities as supporting the motor-training hypothesis, Pellis and Pellis (1998) argued that a careful comparison of the structure of play-fighting and serious fighting leads to the conclusion that: "The practice hypothesis cannot account for the form of play-fighting in most species" (p. 116). They made this argument for two reasons. First, as shown in Table 4.10, the targets of "attacks" sometime differ for play-fighting and serious fighting, specifically for many species of rodents. Pellis and Pellis argued that, for these species, play-fighting may provide more practice for later

TABLE 4.10

Attack Targets During Play- and Serious-Fighting

Species	*Play-Fighting*	*Serious Fighting*
Carnivores		
Felids		
Tigers	Bite—all parts of body	Bite—Head
	Paw—head, shoulders, hind leg (Wasser, 1978)	Paw—Head (Wasser, 1978)
Canids		
Timber wolves	Neck and rear torso (Havkin & Fentress, 1985)	Nape, muzzle, throat (Fox, 1969)
	Thigh & Shoulder (Feddersen-Petersen, 1991)	
Ursids		
Black bear	Allogroom—lateral neck	Biting—Side of face or lateral neck
	Clawing—side of face, forehead, shoulder region	Clawing—mostly shoulder (Henry & Herrero, 1974)
Mustelids		
Polecats	Back of neck (Poole, 1966, 1978)	Back of neck (Poole, 1966, 1978)
Pinnipeds		
Stellar sea-lions	Head, neck, face, front and back flippers, side and back of body (Farentinos, 1971)	Neck (Gentry, 1974)
Northern elephant seals	Neck, lower jaw, nose, front flippers (Rasa, 1971)	Neck, chest, throat (Rasa, 1971)
Ungulates		
Bovids		
Domestic cattle	Forehead, horns (Reinhart & Reinhart, 1982)	Head or flank
Rodents		
Sciurids		
Columbian ground squirrels	Cheeks, throat, muzzle, forelegs (Steiner, 1971)	Cheeks, throat, muzzle, forelegs (Steiner, 1971)
Cricetids		
Prairie voles	Nape, anogenital (Pellis et al., 1989)	Rump and flanks (Pellis et al., 1991)

Table 4.10 continued

Table 4.10 continued

Species	*Play-Fighting*	*Serious-Fighting*
Montane voles	Nape (Pellis et al., 1989)	Nape and flanks (Pellis, 1993)
Djungarian hamsters	Head, face, rump (Pellis & Pellis, 1989)	Rump and lower torso (Pellis & Pellis, 1989)
Syrian golden hamsters	Cheeks (Pellis & Pellis, 1988a)	Lower torso, rump (Pellis & Pellis, 1988a)
Murids		
Norway rats	Nape, anogenital (Pellis et al., 1989; Pellis & Pellis, 1987)	Lateral and dorsal surface lower torso, rump, lower back, flanks (Pellis et al., 1989; Pellis & Pellis, 1987)
Primates		
Prosimians		
Ring-tailed lemurs	Body—dorsum, middle back (Pellis & Pellis, 1997b)	Body—dorsum, middle back (Pellis & Pellis, 1997b)
Callitrichids		
Common marmosets	Head, neck (Stevenson & Poole, 1982)	Limbs, face, ears (Stevenson & Poole, 1982)
Cercopithecids		
Rhesus monkeys	Neck or ventral surface (Symons, 1978)	Any accessible body part (Symons, 1978)
Olive baboons	All parts (Owens, 1975a)	Male–Male: All parts
		Others: neck, back, tail (Owens, 1975a)
Hamadryous baboons	Face or neck (Le Resche, 1976)	Shoulder, upper back, hands, head hair, neck (Le Resche, 1976)
Patas monkeys	Hands, arms, head, & neck (Pellis & Pellis, 1997b)	Hands and neck (Pellis & Pellis, 1997b)

sexual than for later agonistic behaviors (see also Pellis, 1993). Second, citing a review of the motor training literature written by Stamps (1995), Pellis and Pellis (1998) maintained that, during motor-skill learning, the transfer of skills across closely related behaviors is low. Therefore, if play-fighting is practice for serious fighting, then animals should spend most of their time practicing the most difficult components of serious fighting sequences. However, Pellis and Pellis stated that this is not the case. Instead, it appears that the easiest—not the most difficult—components of fighting dominate play. Consider the following example:

> During escalated fights in male deer, head-to-head slamming occurs, which at its most dangerous involves the animals jumping towards their opponents from a distance of up to 2.5 m away (Alvarez, 1993). Successful execution involves precise timing and placement of the blows. Errors can lead to injury (Geist, 1971). Yet these tactics are not the ones used during play-fighting, where gentler head butting and head-to-head pushing (Alvarez, 1993) are the tactics used.... (Pellis & Pellis, 1998, p. 121)

Moreover, Pellis and Pellis argued, the most difficult fighting tactics during serious fighting involve combinations of offensive and defensive tactics (i.e., the animal is trying to place an attack while simultaneously prevent being attacked), but these combinations are rarely seen during play. Clearly, analyses of the structure of play-fighting are open to many interpretations.

Byers and Walker (1995) refined the motor-training hypothesis and argued that it is possible to consider the different consequences of physical training separately. Specifically, they maintained that there are three distinguishable motor-training consequences: increased endurance, increased strength, and increased motor skill (i.e., increased energetic economy of movement). They stipulated that, because play is common in the juvenile period and then either disappears or occurs infrequently thereafter, any short-term motor-training benefits of play probably have little to do with long-term functions. That is, if a particular motor-training benefit of play disappears once the animal stops playing, then it probably does not serve a long-term motor training function. According to Byers and Walker's analysis, this appears to be the case for two of the three consequences of motor training: increased endurance and increased strength. These benefits apparently can be obtained at any age (infant, juvenile, or adult), and their effects are temporary—once the animal discontinues the activity, the motor-training benefits are lost.

In contrast, the effects of physical training on increased motor skill for juveniles are long-lasting, in that they result in specific physiological changes (i.e., muscle fiber differentiation and cerebellar synaptogenesis) that lead to long-term, irreversible changes in physical structures. Because the processes of muscle fiber differentiation and cerebellar synaptogenesis are largely complete (and therefore permanent) at the end of the juvenile period, any play experiences that impact these processes are likely to have long-term effects on development. In support of their hypothesis, Byers and Walker (1995) reviewed research on house mice, Norway rats, and house cats that show similar periods of development for play, cerebellar synaptogenesis, and muscle fiber differentiation.

Despite all of the speculation about how play-fighting might be important for the development of fighting skills, most ethologists believe that fighting skills are too important for survival to be *totally dependent* on play-fighting experiences for their acquisition. This has been demonstrated by research showing that animals who have been raised in social isolation (and obviously deprived of social play-fighting experiences) still develop aggressive behaviors (see the section on experimental studies later in this chapter). What play-fighting facilitates, most researchers maintain, is the refinement of fighting skills so that an individual's play-fighting background may give an animal certain selective advantages over others (Fagen, 1981; Symons, 1978). A problem with this assumption is that, in some highly aggressive species (e.g., spotted hyenas—Drea et al., 1996; collared peccaries—Byers, 1984), aggression appears in functional form shortly after birth, sometimes even before the young begin to play (Meaney et al., 1985; Pellis & Pellis, 1998).

Aggression Prevention and Control. Although most researchers have argued that play-fighting serves to increase serious fighting proficiency, it is sometimes argued that play-fighting operates in the service of aggression prevention and control. Six such hypotheses are listed in Table 4.9. The first, that aggression may function to help in the learning of escape skills, has been offered because play-fighting in some species has been characterized as having many locomotor or chasing elements (e.g., Pellis & Pellis, 1983; Wilson & Kleiman, 1974). Thus, in the course of play-fighting, animals often escape being pinned and may run at high speeds. The problem with arguing that such behavior is consistent with an escape function, however, is that once animals free themselves from being restrained they usually counterattack and come back for more. As already mentioned, play-fights usually end with the participants in close proximity, so little practice in successful escape seems to be offered.

Another hypothesis derived from observations of the structure of play-fighting is that it gives animals practice in restraint and self-control. Numerous observers have noted that biting, clawing, and other aggressive behaviors are inhibited during play-fighting (e.g., Henry & Herrero, 1974;

Pellis & Pellis, 1988b; Symons, 1978), indicating that it is a context in which restraint is exercised. A problem with interpreting such restraint in terms of the practice hypothesis is that restraint is usually present at the youngest ages at which play-fighting occurs, and there is little evidence that the ability to regulate or inhibit aggressive behavior develops over time. In fact, the exact opposite is true in that, at least in some species, the likelihood of play-fighting escalating into serious aggression increases, not decreases, with age (Pellis & Pellis, 1988b; Poole & Fish, 1976).

It could be that play-fighting may not develop the ability to control or regulate behavior per se, but instead increases the animal's tolerance for physical contact or pain during social interaction (Einon & Potegal, 1991; Pellis & Pellis, 1998). Animals with greater play-fighting experience might therefore be less likely to respond to physical contact with serious aggression.

A final way in which play-fighting may prevent aggression was suggested by Pellis and Pellis (1991b). Based upon their observation that subordinate rats often initiated play-fighting interactions with dominant rats postpuberty, they argued that this might serve the function of establishing and maintaining amicable relations with more dominant animals, so that the less dominant might be less likely to suffer aggression from the more dominant animals at a later point in time. This hypothesis suggests that play-fighting might serve more social than motor-training functions; thus, to it is to this category that we now turn.

Socialization. Several researchers, skeptical of the role of play-fighting in motor training, have argued instead that it serves in the establishment and maintenance of social relationships. Five such hypotheses are listed in Table 4.9. These hypotheses appear to be based less on analyses of the structure of play-fighting than on the fact that play-fighting so frequently occurs in the young of group-living species. The argument apparently is predicated on the evolutionary assumption that it is important for animals to form bonds with other animals in their social group (especially relatives or those with whom they are likely to cooperate later), and play-fighting is one form of amicable interaction through which this can occur. Thus, play-fighting may function like grooming in the formation of social bonds (e.g., Koyama, 1985).

Advocates of motor training, however, have argued that gender differences in play-fighting often work against social cohesion hypotheses. If play-fighting serves to establish social bonds and encourage intragroup harmony, then—in species where females remain in the same social group into adulthood and males disperse upon maturity—one would predict that play-fighting would be more common for females than for males. However, just the opposite is true. In such species, males usually show the most play-fighting (e.g., Owens, 1975a, 1975b; Symons, 1978; Waterman, 1988; Watson, 1993).

Research that has been cited in support of the socialization functions of play-fighting includes studies of play-fighting and dominance status. The basis assumption of these studies is that play-fighting has a role in the establishment and maintenance of dominance relationships. However, given the inconsistent relationship between dominance status and play-fighting (see the earlier review), it is not that surprising that just as many researchers have argued that they have evidence supporting this hypothesis (e.g., Koyama, 1985; Parquette, 1994; Tartabini & Dienske, 1979) as do researchers stipulating that they have evidence refuting it (e.g., Byers, 1980; Pellis & Pellis, 1989). Given the high levels of role and dominance reversals during play-fighting (see the earlier review), it is possible that play-fighting enables animals to learn about both submissive and dominant roles (Araba & Crowell-Davis, 1994; Biben, 1986; Symons, 1978).

Meaney et al. (1985) maintained that a consideration of dominance issues shows why the aforementioned gender differences in play-fighting are consistent with a socializing function. Gender differences in play-fighting, they argued, are related to gender differences in both the degree of difficulty in "integrating into a social structure and in achieving status within that structure" (p. 43). For example, in many primates among whom gender differences in play-fighting are common, it is males, not females, who must earn their dominance status through physical aggression (the dominance status of females is more often inherited directly from the mothers). Moreover, because males in many primate species must leave their natal group in order to integrate into another (females usually remain), males must rely on dominance behaviors in this context as well. This situation varies significantly from that of many canids, in which both males and females inherit their mothers' dominance status, and both males and females are equally likely to be dispersed from their natal groups and subsequently have difficulty integrating into a new group. Therefore, the existence of gender differences in play-fighting are dependent on the degree to which dominance relations for a given gender are important in both dispersion and integration. Comparing this argument with some of those offered earlier makes it clear that the existence of gender differences in play-fighting have been offered to support or to discredit a socializing function.

Cognitive Stimulation. When watching two animals engaged in a play-fight, it is obvious that such interactions provide a great deal of social and likely cognitive stimulation. Whether such stimulation is greater or less than what is received during other forms of social interaction, however, has not been empirically determined.

Play-fighting may indeed function to produce novelty. Compared to the research on object play, however, there is limited discussion in the play-fighting literature on its novelty-generating properties. Although play-fighting has been described as showing greater variability in behaviors and sequences than serious fighting (Fagen, 1981), there is little empirical evidence to support this notion (Hill & Bekoff, 1977). A few researchers, however, have commented on the novelty-generating possibilities in play-fighting. Nowicki and Armitage (1979) asserted that the novelty generated during the social play of yellow-bellied marmots aids in the adaptation of this species to a wide range of environmental settings. Fedigen (1972) made a similar argument regarding the social play of vervet monkeys. Both Aldis (1975) and van Lawick-Goodall (1968) made the point that play-fighting in chimpanzees shows high levels of diversity. Finally, Fagen (1993) maintained that play-fighting may play an important role in influencing animals' responses to novelty. Social play (including play-fighting), he argued, might lead animals to approach rather than to fear novelty.

Other. Several papers have proposed other possible functions of play-fighting that are consistent with the structural data as well. Pellis (1993) argued that, because behavior during play-fighting in several rodent species (both in play-fighting targets and tactics) is more similar to precopulatory than to agonistic behavior, play-fighting, at least in these species, may function as practice for behaviors in sexual reproduction. For example, in a study of adult sexual behavior, juvenile play-fighting, and juvenile serious fighting, Pellis and Pellis (1989) found that Djungarian hamsters licked and nuzzled the mouth of females at the beginning of sexual encounters, whereas bites during serious fighting were on the top of the head and the rump. Play-fighting targets, like the precopulatory ones, were directed toward the mouth.

Fagen (1992, 1995), in reviewing the factors that influence play, asserted that animals' social play (including play-fighting) might function to signal an animal's well-being to its caregivers and companions (see also O'Neill-Wagner, Bolig, & Price, 1994). Specifically, because play is most likely to occur when animals are well fed, healthy, and safe from danger, the amount and quality of play may signal to caregivers when a young animal is in need of assistance or care, and may also signal to potential companions and mates the desirability of that individual as a reproductive partner.

Finally, Barber (1991), reviewing much of the same literature as Fagen (1992, 1995) and Burghardt (1988), argued that play may serve an important energy-loss function. That is, animals play so that they burn up energy that might otherwise be stored as fat. Barber stipulated that by engaging in energy-burning play, animals remain lean and fit, making them less susceptible to predators. If excess calories were not burnt off in play, then the resulting obesity might increase the risk of predation by impeding escape ability through increasing balance problems, fatigue, muscle strain, inability to enter narrow spaces, and amount of nonpropulsive tissue. Moreover, because play activity raises basal body temperature, it could decrease the young animal's susceptibility to cold stress and pathogens. Barber (1991) cited evidence supporting his theory that the amount of play engaged in varies with the amount of food available, and showing that even young animals living in an environment with a surplus of food rarely develop obesity—they simply play more. By focusing on the risks of predation, cold stress, and pathogens, Barber (1991) considered those risks most relevant to young animals, and, by so doing, offered one of the few theories of play that focuses more on its short-term than its long-term benefits (although, obviously, short-term benefits on survival rates have long-term benefits for future reproductive success).

Summary and Conclusions. As this brief review indicates, the structure of animal play does not point more strongly to one function of play-fighting than another. Over the years, researchers have seen support for a wide range of possible play functions in their observations of play-fighting behavior. It is therefore difficult to use structural analyses to draw any firm conclusions about the functions or effects of play-fighting. Other kinds of correlational and especially experimental data would provide stronger evidence. It is to these types of studies that we now turn.

Correlational Studies

Surprisingly, besides the studies of play structure discussed in the previous section, few correlational studies of the effects or functions of play-fighting have been published. The term *correlational study* is used in this section to refer to investigations in which researchers examine within a single species the correlation between individual differences in the play-fighting of individuals or groups of individuals, and some separate measure of functional behavior that play is thought to impact. Presented in Table 4.11 are the only correlational studies of play-fighting in nonhumans that could be located, and a brief summary of their findings. From an examination of these studies, two general points can be made: the sample sizes for most studies were usually too small for any firm conclusions to be drawn (with sample sizes under 20, only very large effect sizes can be detected, and, because of potential violations of normality, the possibility of spurious findings is high), and usually no more statistically significant associations were found than would be expected by chance alone (with a p value of .05, one would expect on the average that one out of every 20 correlations would be significant due to chance alone). The one exception was the recent study of play in cheetahs by Caro (1995). Here the sample size was sufficiently large ($n = 51$), and the number of significant relations sufficient. As summarized in the table, Caro (1995) found that both stalking and crouching during play were correlated with these same behaviors during hunting, and that the amount of play-fighting (which he labeled *social contact play*) was positively correlated with rates of patting, biting, or grasping live prey that mothers released for the infants to play with after a maternal capture.

Given the current state of the literature, two possible conclusions can be drawn: (a) Play-fighting does not appear to have any effects on the development of fighting, predation, and other functional skills; or (b) insufficient correlational research has been conducted to date to fairly address this question. At the present time it is sensible to accept the latter conclusion, given the promising findings of the one study with a sufficient sample size (Caro, 1995) and that few studies to date have really provided a fair correlational test of these hypotheses. If play-fighting leads to a slight advantage in the development of fighting skills, one would have to design studies that made it possible to detect a small effect size, necessitating considerably larger samples than most of those employed to date. Finally, even if correlational relationships are found, one still must collect experimental data to examine the direction of causality.

Before turning to a consideration of the experimental work in this area, one additional correlation is worth considering. Although this study did not examine the statistical significance of relations between quantitative measures as did the studies in Table 4.11, it provides some interesting qualitative data on a group of animals in a natural setting that apparently never engaged in social play. Baldwin and Baldwin (1973), in their fieldwork in Central and South America, observed 2 of 44 troops of squirrel monkeys that apparently never engaged in social play due to a food shortage in their

TABLE 4.11

Correlational Studies of Play-Fighting Functions/Effects

Reference	*Species*	*Sample*	*Method*	*Findings*
Vincent & Bekoff (1978)	Coyotes	*n* = 4 infants	Play observations Observations of prey capture	Two out of 13 correlations significant: prey killing success positively correlated with frequency of play solicitations and pouncing during play (pouncing low frequency)
Caro (1981)	Domestic kittens	*n* = 7 litters	Play observations Observations of prey capture	Between 4 and 7½ weeks—8 out of 72 correlations between measures of play and predatory behaviors significant; between 8 and 12 weeks—4 out of 72 correlations significant; no coherent pattern of results; only one of the relationships was the same at both ages
Davies & Kemble (1983)	Northern grasshopper mice	*n* = 29 mice	Play observations Observations of prey capture	No correlation between play behaviors and cricket killing and consumption
Chalmers &Locke-Haydon (1984)	Common marmosets	*n* = 10 infants	Play observations Several Skills Tests	Correlated scores on 14 play measures with scores on seven skill measures collected no more than 3 weeks later for four different timepoints. Twenty out of 392 correlations significant
Caro (1995)	Cheetahs	*n* = 51 litters	Field observations of play and predatory behaviors	Four out of 12 correlations significant: stalking and crouching during play positively correlated with these same behaviors during hunting; amount of object and contact social play positively correlated with rates of patting, biting, and grasping live prey released by mother

environment. Based on their qualitative impressions, Baldwin and Baldwin (1973) described differences between members of these two (rather small) troops and the other troops they had studied. In comparison to the other troops, the adults in the troops in which the juveniles did not play showed: an absence of friendly behaviors (e.g., grooming; consort relations between males and females; travel and resting in groups); no contact aggression, although they observed chases and threat displays; a very small and simple sexual repertoire involving the minimal behaviors required for reproduction; and extreme social tolerance for close interpersonal contact. Despite these differences, Baldwin and Baldwin noted that "85% of adult females were accompanied by infants, which indicates a normal rate of reproductive success for that species" (p. 369). Baldwin and Baldwin (1973) interpreted these data to support the notion that social play is not *necessary* for the development of "an adaptive modicum of competence" (p. 379) in terms of sexual reproduction, troop cohesion, and the control of aggression. However, the differences in friendly, agonistic, and sexual behaviors summarized here suggest that "the opportunity to play provides learning experiences in which young animals can develop more complex, varied social interaction patterns and stronger habits for engaging in frequent overt exchanges" (pp. 379–380).

Although the Baldwin and Baldwin (1973) findings demonstrate that play is not *necessary* for the development of many adaptive behavior patterns, one needs to be careful about concluding that the differences in social behavior of the two monkey groups they studied were the result of the absence of play. Given the small sample on which these conclusions were based (the two troops they studied were both residents of the same forest in Panama), it is possible that other unique characteristics of these troops may have accounted for these findings. As acknowledged by Baldwin and Baldwin (1973), both troops were small in size and each had only two adult males—possibly accounting for the relatively low levels of aggression. Other factors, such as genetic factors or characteristics of the environment, may have been responsible for these differences as well. One of the most parsimonious explanations for their results is that, given the relative low level of food available, the monkeys were forced to devote most of their time to foraging and food consumption. Thus, they did not waste the limited available energy on social play, grooming, contact fighting, and extended sexual interchanges. Further support for Baldwin and Baldwin's hypotheses would have been obtained if they could have provisioned one of these troops after their initial periods of observation to see if the adults would begin to engage in more complex and more typical social interchanges, or if indeed they still had deficits in these areas that may have resulted from their early experiences (including the lack of play).

Experimental Studies

Three major methods have been used in experimental studies of animal play: social isolation, drugs, and surgery. In isolation studies, animals raised in social isolation are compared to controls in order to examine the effects of social experience (and therefore play-fighting) on development. These studies have shown that social experience is not necessary for the development of aggression. Studies of a range of species—including mice, gerbils, wild rats, chow dogs, sheep, and rhesus monkeys—show that animals raised in social isolation exhibit even higher levels of aggression once paired with conspecifics than do controls (see review in Cairns, 1979). In reviewing this research, Cairns (1979) argued that isolates "are jumpy" (p. 188), and that their social overreactivity to the approaches of and contacts with others may be responsible for their higher levels of aggression. Therefore, play-fighting apparently is not necessary for the development of aggression, but may in fact decrease its probability.

Isolation studies have also indicated that play-fighting is not necessary for the development of predatory behavior. Several studies have shown that rats raised in social isolation will engage in muricide (mouse killing). Once again, social isolation may have its effects—not on whether or not isolated animals engage in predatory behavior, but instead on the nature of the behavior itself. Although Potegal and Einon (1989) found no effects of social isolation on muricide, in their literature review they cited several studies suggesting that there may sometimes be such effects, with isolated rats engaging in greater muricide than controls.

The problem with isolation studies, however, is that no one has yet developed an experimental method for isolating animals from only play or play-fighting experiences. Therefore, it is impossible to determine what aspects of social isolation account for the effects on social behavior, for example, changes in all types of social interaction, exercise, sensory stimulation, arousal, and possibly nutrition (Bekoff, 1976).

Einon and colleagues (Einon, Morgan, & Kibbler, 1978; Einon & Potegal, 1991; Potegal & Einon, 1989) have come the closest to developing procedures for examining through isolation the effects of play-fighting experience. In these studies, laboratory rats that were raised in social isolation were compared to socially reared rats (either mother-reared, peer-reared, or both), as well as to rats that were given 1 hour of social experience per day over the same time period (partial isolation). This procedure provided information related to play-fighting, in that when rats between the ages of 3 and 7 weeks were given only 1 hour of social experience a day, they play-fought "more or less continuously for 45- to 60-minutes a session" (Potegal & Einon, 1989, p. 162). By using this method, Einon and associates found that many of the effects of social isolation on rat behavior (e.g., differences in habituation of locomotor activity and object contact, aggressive responses to shock; and responses to the attacks of an intruder) were alleviated by 1 hour of social contact (mostly play-fighting) per day. The most relevant findings for the hypotheses regarding play-fighting discussed earlier were that although total isolates showed greater aggression in response to shock and were more likely to get into serious fights that resulted in wounds, animals in the partial isolation group did not differ from controls on either of these measures (Potegal & Einon, 1989). The various groups did not differ, however, on the probability of launching an offensive attack (Potegal & Einon, 1989), nor of defending against a painful attack (Einon & Potegal, 1991). Einon et al. (1978) asserted that total social isolates had "difficulty in inhibiting their responding, or in switching between different patterns of behavior" (p. 224). This tendency was not found in rats in the partial isolation condition. Therefore, the studies of Einon and associates are consistent with the theorized aggression control functions of play-fighting. Unfortunately, however, it is still impossible to determine if it was the play-fighting experience itself that accounted for these effects.

Given the limitations of isolation studies, other researchers have attempted to conduct experimental studies involving drugs or surgery. The purpose of many of these studies has been to examine the neurophysiological processes involved in the play-fighting of rats (for a review, see Panksepp et al., 1984). Surgery studies show that play-fighting in rats appears to be precortical in that it survives decorticalization (Pellis, Pellis, & Whishaw, 1992). Studies involving the injection of hormones show that it is also under some hormonal control (e.g., Pellis & McKenna, 1992, 1995; Smith, Field, Forgie, & Pellis, 1996). Other studies have been conducted to examine the role of the senses in play-fighting (see the review in Pellis et al., 1996).

Another experimental strategy used in studies of play-fighting is to inject drugs into the play partner. In this way, one can experimentally manipulate the social responsiveness and playfulness of the partner, and then explore its effects on play-fighting behavior. For example, Pellis and McKenna (1995) demonstrated that attack targets differed during play and serious fighting, even when the defensive capabilities of the partner were modified by drugs. Similar studies have shown that the amount of play-fighting exhibited by a subject is a function of the experimentally manipulated playfulness of the partner (e.g., Einon et al., 1978; Pellis & McKenna, 1992).

MODELS OF THE EVOLUTION OF PLAY-FIGHTING

Given the level of disagreement between theorists regarding the functions of play-fighting, it is not surprising that several very different models of the evolution of this behavior pattern have been offered over the years. Besides Burghardt's (1984, 1988) theory that focuses on the role of ectothermy, endothermy, and parental care (see chap. 2), Byers (1984) offered a theory that focuses on motor training, Barber (1991) put forth a theory that focuses on energy loss, and Pellis (1993) provided a theory that focuses on sexual behavior. In most of these theories, it is argued that play evolved initially to serve a single function, but subsequent selection pressures led to specializations within species and to the development of other functions and effects. It is interesting to note that after reviewing much of the same literature (although different authors have focused on different species), Burghardt (1984) stipulated that play-fighting evolved from various precocial behaviors, Pellis (1993) maintained that it evolved from immature sexual behavior, Byers (1984) asserted that it evolved from locomotor rotational play, and Barber (1991) argued that it evolved as a way to expend excess nonprotein energy. Fortunately, most of these authors made specific, testable predictions derived from their theories about the nature of play and the factors that should influence it. Given the relative recency of these formulations, few empirical tests of these predictions have yet to be published. It is likely that such work is forthcoming.

ANIMAL PLAY-FIGHTING: SUMMARY AND CONCLUSIONS

In conclusion, play-fighting appears to have evolved as an independent system of behavior in many mammals and some birds. Given the considerable time devoted to play-fighting, as well as the potential risks involved, it is likely that play-fighting serves some important evolutionary functions. However, to date, students of animal behavior, have yet to arrive at any consensus about what might be the functions and effects of play-fighting, and empirical confirmation of any of the functional hypotheses is lacking. One possibility is that play-fighting may serve different functions for different species, or that play-fighting may have initially evolved to serve one function, but now serves several. Such hypotheses, however, are speculative and await future research.

Despite the lack of a consensus about the functions and effects of play-fighting, considerable progress has been made in describing its structure. Play-fighting, at least on the surface, is made up of the components of serious aggression, but differ in their timing and structure. Species from a wide range of mammalian orders show considerable similarities in their basic play-fighting tactics (see Table 4.4) and in the ways that these play-fighting behaviors differ from behavior during serious aggression. Species vary widely in the degree to which play-fighting is preceded by or accompanied by elaborate preliminary behaviors or play signals. The degree to which this is the case may be a function of such factors as the dangerousness of the partner, the complexity of the play patterns, or the degree of difference between play-fighting and serious-fighting behaviors. Across species, play-fighting begins in infancy, peaks during the juvenile period, and either stops or decreases to low frequency during adulthood, and in some species the vigor or roughness of play-fighting increases with age. Animals tend to choose play partners that are similar to themselves, usually those that are of the same gender, close in age, and with whom they are biologically related. Gender differences in play appear to be closely linked to gender differences in adult roles.

Despite the considerable amount of research devoted to play-fighting during the last 30 years, there is still much to be learned. Detailed studies of play-fighting have been limited to a rather small number of species, and with but a few exceptions (e.g., Pellis et al., 1989), comparative research on closely related species are difficult to find. Moreover, studies on the timing and sequencing of behavior during play-fighting are rare, as are detailed studies of offensive and defensive strategies. We still know little about the specific risks of play-fighting, the environmental and experiential factors that impact its occurrence, and patterns of individual differences in its nature and occurrence within specific species. Detailed analyses of play-fighting signals have yet to be conducted in many species, as have analyses of its initiation and termination. The relationship between dominance structures and play-fighting is still unclear, and little is known about the factors that lead to self-handicapping and role-reversal versus exploitation of one's advantages. The degree to which deception and manipulation of other animals occurs during play-fighting is an area with important implications for evolutionary analyses (Fagen, 1981). Finally, given the debate over the functions of play-fighting, more specific tests of the predictions of the various evolutionary models of play-fighting need to be conducted.

5

Play-Fighting in Children

Studies of children show that aggression can be as much a part of life for humans as it is for many other mammals. Although the evolution of language and complex social customs and skills have provided alternative means for humans to resolve conflicts over food, territories, mates, and other resources, humans often resort to aggression when alternative methods fail. This is especially true for young children, who lack many of the social skills necessary for conflict resolution (e.g., Laursen, Hartup, & Koplas, 1996; Shantz, 1987). Moreover, when one considers social interaction among children and adults in many hunting and gathering cultures (the cultures most similar to those in which many human behavior patterns evolved), aggressive threats and actions are very much a part of life—in hunting, in defense, and in achieving and maintaining status within the social group (Archer, 1995; Eibl-Eibesfeldt, 1989). Therefore, some of the same selection pressures that operated for the development and refinement of fighting skills for other animals likely operated in the evolution of human behavior as well.

As is true for other mammals, play-fighting behaviors in children are similar to behaviors during adult aggression, but differ in structure, vigor, and intensity (Aldis, 1975; Boulton, 1991a). Researchers who have studied play-fighting in humans generally agree that play-fighting is distinct from aggression and should be treated as a separate behavior system (e.g., Boulton, 1991a; Smith & Lewis, 1985).

Although the existence of play-fighting in humans has long been recognized (e.g., Groos, 1898; Hall, 1904), it was not until the mid 1960s that it received empirical attention. Blurton-Jones (1967), in an ethological study of children's behavior in a preschool classroom, was the first to apply Harlow's (1962) concept of rough-and-tumble play (play-fighting and play-chasing) to the behavior of humans. Since this time, most studies of play-fighting in children have been conducted by a small number of researchers, including Aldis (1975), Smith and colleagues (e.g., Smith &

Lewis, 1985), and Pellegrini (e.g., Pellegrini, 1988). Compared to other kinds of play, play-fighting has been the focus of little research in humans, so the descriptive base is limited (Pellegrini & Smith, 1998). Moreover, because many of the early studies employed Harlow's (1962) concept of rough-and-tumble play (which combines both play-fighting and play-chasing), the specific nature and functions of play-fighting are sometimes difficult to determine. Recently, however, researchers have begun to separate the fighting and chasing forms of rough-and-tumble play (e.g., Pellegrini, 1995).

STRUCTURE

Analysis of the structure of play-fighting in humans reveals many similarities with play-fighting in other mammals, especially the primates. However, the descriptive base on which this conclusion is drawn is very limited. Even though it has been more than 20 years since its publication, the work of Aldis (1975) remains one of the few sources of detailed descriptive analyses of play-fighting in humans. Based on analyses of films made in schoolyards, Aldis (1975) identified two forms of play-fighting: "wrestling for superior position" and "fragmentary wrestling." In "wrestling for superior position," found almost exclusively in boys, "One boy will make strenuous efforts to throw another to the ground, to get on top of him, to hold him down, to flatten him, and sometimes to pin him to the ground. Pinning constitutes a symbolic play victory" (p. 178).

In "fragmentary wrestling" (a less vigorous form of wrestling found in both boys and girls), "Neither child will attempt to achieve a clear-cut superiority but will merely grapple or push and pull in various directions" (p. 174). In this form of play-fighting:

> A great variety of responses may sometimes be seen in sequence. The following is an abbreviated description of a filmed sequence between two boys, each about 7-years old, who used their jackets in a play attack. The whole sequence was accompanied by much laughter. At the onset, boy A held both jackets. B then ran over to A, grabbed one jacket, and hit him with it. A proceeded to throw his jacket at B, but he missed; A then grabbed at B's jacket, but B held on to it. A kicked B, and then he made an exaggerated punching gesture towards B. Next he grabbed onto B's jacket and ran away pulling on it. B hung on, went over and picked up A's coat, and threw it at A. A threw the same coat at B and kicked at it on the ground; B ran away. A chased him, kicking at him. B stood over his own coat, which was by now

> lying on the ground, and A approached him and pushed him away from the coat, grabbed the coat, and ran off with it. B chased after him. A threw the coat away (pp. 194–195)

Play Signals

To date, the only play signals identified during the play-fighting of humans are signs of positive affect—specifically, smiling and laughing (e.g., Aldis, 1975; Blurton-Jones, 1967; Boulton, 1991b; Smith & Lewis, 1985). Thus, play in humans is signaled by "general amicable behaviors" (see Pellis & Pellis, 1996) rather than by play-specific signals such as the play-bow in canids (Bekoff, 1974). An examination of Tables 4.2 and 4.3 shows that the limited number of play signals in humans is at variance with the wide range of signals used in other primates, especially chimpanzees (our closest relative), who show the greatest variety of play signals among the primates. Although it is possible that human children actually do show a smaller variety of play signals than do many other mammals, it seems more likely that, due to the limited amount of descriptive work in this area, other human play signals have not been identified. A cursory examination of Table 4.2 suggests that this may be the case. One can imagine that if a child engaged in many of the behaviors in Table 4.2 at the onset of or during a break in play-fighting (e.g., gamboling or staggering gait, exaggerated approach–withdrawal), his or her partner would likely interpret the behavior that followed as play. Future research needs to consider such behaviors as possible signals. Moreover, human children undoubtedly have verbal ways of communicating play intention that have yet to be documented. Aldis (1975), for example, noted that children often apologize when they accidentally hurt one another during play-fighting, possibly as a way of communicating that "this is still play" and preventing escalation into serious aggression. Such conciliatory behaviors are not restricted to humans. Wasser (1978) found a similar tendency in tigers—in situations where the receiving cub showed an aggressive threat in response to being bit during play-fighting, the initiating cub would often lick the affected area and then resume play-fighting once it was clear that a serious fight would not ensue.

Most research on human play-fighting to date has used affective signals alone (positive vs. neutral or negative affect) and/or outcome factors in identifying instances of play-fighting (e.g., Boulton, 1991a; Pelligrini, 1993; Smith & Lewis, 1985). Play-fighting is generally assumed to occur when the child shows signs of positive affect and no negative affect. This contrasts with recent research of animals showing that many cases of play-fighting occur in the absence of play signals, and that animals apparently use contextual factors (e.g., identity of partner, presence or absence of limited resources) in determining if an aggressive initiation is playful or serious. As mentioned earlier, play signals appear to be used in situations of ambiguity (Bekoff, 1995; Pellis & Pellis, 1996). By defining play-fighting interactions as those in which children show aggressive behaviors that are accompanied by signs of positive affect, the frequency of play-fighting may be underestimated. On the other hand, play signals may be more important in humans than in other animals, because contextual information is often less available to humans. In contemporary human settings, for example, serious aggression in children rarely occurs in situations involving access to food or water, but instead occurs in conflicts over objects, personal space, or challenges to a child's self-esteem (Smith & Boulton, 1990). Thus, human children may be more likely than the young of other mammalian species to be the recipients of aggressive behaviors that are unprovoked, "out of the blue," and not tied to observable contextual factors. If this is the case, then play signals may be more necessary for children's play-fighting than for the play-fighting of other mammals.

General Characteristics

As was the case for play-fighting in other animals, play-fighting in children is goal-directed behavior as well. Although the goal of play-fighting in most mammals involves trying "to bite without being bitten," other common goals include trying to touch or sniff a specific body part, to push one's opponent to the ground, or to wrestle so as to get on top of another. Although biting is rare during children's play-fighting (see later discussion), the other three goals apply to children's play-fighting (Aldis, 1975). As was also the case for other animals, many of the behaviors that children exhibit during play-fighting come from the serious fighting context. Boulton (1991a) found that all of the behaviors he observed during serious fighting occurred during play-fighting, but that play-fighting was made up of some additional behaviors as well. These included behaviors that appeared to be stylized imitation of karate moves children likely observed on television (karate chop, back kick at, scissor kick at, knee at), rubbing with an open hand, running past the other child, and pushing and pulling. Play-fighting in children is most similar to aggression. Unlike many other primates, human children do not show sexual behaviors such as mounting during play-fighting, and, like other primates, they do not show the predatory or hunting behaviors found in the play-fighting of felids such as lions, tigers, and domestic cats.

At least on the surface, the differences between behaviors in the play-fighting and serious fighting contexts are similar for both humans and other animals: The intensity of the behaviors is reduced, aggressive attacks are restrained, and individuals engage in self-handicapping and role reversal (Aldis, 1975). Given the very limited descriptive base on play-fighting in humans, however, there is limited quantitative support for these conclusions.

Play Initiations

Play-fighting in other animals is usually initiated with an attack that may or may not be preceded by some preliminary behaviors. As discussed in the section on animal play-fighting, the extent of these preliminary behaviors varies considerably across species. No descriptive data on the nature of play-fighting initiations in humans could be found, although it is possible that play initiations in humans are as varied and complex as are those in chimpanzees. Humphreys and Smith (1987) made the point that play-fighting initiations differ from initiations of aggressive interactions, in that the former are offered as an invitation but the latter usually involve some kind of challenge.

Play-Fighting Tactics

Presented in Table 5.1 is a summary of the behaviors identified during play-fighting in five studies of children. To facilitate comparison with the animal data, the behaviors are listed in the same order as in Table 4.4; only three new behaviors had to be added. Comparison of Table 5.1 with the primate column in Table 4.4 shows many similarities between play-fighting in humans and in other primates. Of the nine behaviors listed for primates in Table 4.4, seven were found in studies of human children. The remaining two primate behaviors (i.e., biting and mounting) were not seen in children's play-fighting. Three additional behaviors had to be added to the table to describe children's play-fighting: leveraged throw/carry, trip/tackle, and "shoot" with a pretend gun (e.g., a finger, a stick). The first two are probably unique to humans because of the bipedal nature of human locomotion (a child who is standing needs to be thrown or tripped in order to wrestle and pin); the latter behavior is a function of the incorporation of symbolic pretense into children's play-fighting interactions.

TABLE 5.1
Children's Play-Fighting Behaviors

Behaviors	*Aldis (1975)*	*Boulton (1991a)*	*Humphreys & Smith (1987)*	*Neill (1976)*	*Smith & Lewis (1985)*
Lunge, pounce	+	+	+		+
Butt, clash, spar					
Bite, slash					
Neck fight					
Push	+	+	+	+	+
Naso-naso contact					
Paw, swat, kick, knee	+	+	+	+	+
Grab, hold	+	+	+	+	+
Pull	+	+	+	+	+
Wrestle, pin	+	+	+	+	+
Mount					
Chase	+	+	+	+	+
Leveraged throw, carry	+		+		
Trip, tackle	+		+		
"Shoot"					+

As was also the case for the animal literature, data on defense strategies during human play-fighting are limited. One exception to this comes from Aldis (1975). In his book, he described the various ways that one child underneath another may attempt to avoid being pinned or to extricate him or herself after being pinned (see Aldis, 1975).

Although comparing the results of studies of humans with those of other animals reveals numerous similarities in the basic behaviors used in play-fighting, there may be subtle differences in the timing, sequencing, intensity, and form of these behaviors that are not captured by these gross categorizations. Clearly, further descriptive research is in order.

Termination

Human children apparently terminate play-fighting in the same way that other animals do: One partner breaks away and stops the fight. It should be noted, however, that as was the case for other animals, play-fights usually end with children remaining in close proximity to one another (Aldis, 1975; Blurton-Jones, 1967; Smith & Lewis, 1985), in contrast to serious aggression.

Differences Between Play-Fighting and Serious Fighting

Presented in Table 5.2 is a summary of the documented differences between play-fighting and serious fighting in human children. This table is organized in the same way as Table 4.5, to facilitate comparison with the animal literature. (Note: one of the studies in the table—reported in Fry, 1987, 1990—was conducted in a non-Western culture and is considered in more detail later in this chapter). Comparison of the two tables shows that most of the documented differences between play-fighting and serious fighting of animals have also been documented for humans. For three of the differences that have been documented in animals—developmental course, tactics employed, and dominance status of initiators—insufficient data have been collected on humans to make a determination.

Some minor differences between animals and humans are evident. First, regarding contextual differences (as already mentioned), aggression in human children is more likely to involve possession of objects and use of space compared to animals' competition for food, water, and mates. Second, Aldis (1975) mentioned one difference between play-fighting and serious fighting in humans not found in other animals: Children often rest in the middle of play-fighting bouts, something Aldis did not observe in other animals. Finally, although research on animals has shown no correlation between rates of play-fighting and serious fighting, several studies of humans find positive correlations. For example, Frey and Hoppe-Graff (1994) found that the frequency of play-fighting correlated positively with both instrumental and serious fighting in 2- to 4-year-olds. Two studies of older children conducted by Pellegrini (kindergartners, second, and fourth graders—Pellegrini, 1988; sixth and seventh graders—Pellegrini, 1994) showed that although there was no significant correlation between serious and play-fighting overall, this correlation was significant and positive for rejected children (children receiving negative sociometric ratings from their peers). This correlation may be a consequence of the tendency of rejected children to respond to play-fighting initiations with serious aggression (see related discussion in the section on functions and effects).

TABLE 5.2

Differences Between Play-Fighting and Serious Fighting in Children

Characteristic	*Serious Fighting*	*Play-Fighting*
Developmental course	*No comparative data available*	
Threats	No threats	Frown, fixate, shout (Blurton-Jones, 1967; Boulton, 1991a, Smith & Lewis, 1985)
Functional contexts	No obvious immediate functions	Possess objects, use space (Smith & Boulton, 1990)
Behaviors:		
Intensity	Low intensity (Aldis, 1975; Fry, 1987)	High intensity (Aldis, 1975; Fry, 1987)
Duration	Longer (Fry, 1987)	Shorter (Fry, 1987)
Tactics employed	*No detailed descriptive data available*	
Targets	Trunk (Aldis, 1975)	Face (Aldis, 1975)
Consequences	Injury rare (Humphreys & Smith, 1987)	
Dominance:		
Role reversals	Common (Aldis, 1975; Fry, 1987)	Rare (Aldis, 1975; Fry, 1987)
Self-handicapping	Yes (Aldis, 1975; Fry, 1987)	No (Aldis, 1975; Fry, 1987)
Who initiates?	*No comparable data available*	
Termination	Close proximity (Aldis, 1975; Blurton-Jones, 1967; Fry, 1990; Smith & Lewis, 1985)	Child leaves (Aldis, 1975; Blurton-Jones, 1967; Fry, 1990; Smith & Lewis, 1985)

Table 5.2 continued

Table 5.2 continued

Characteristic	*Serious-Fighting*	*Play-Fighting*
Individual differences: Correlations between play-fighting and serious-fighting	2- to 4-year-olds (Frey & Hoppe-Graff, 1994)	
	instrumental aggression, $r = .38$	
	serious aggression, $r = .45$	
	Kindergartners, 2nd, and 4th graders (Pelligrini, 1988)	
	rejected children, $r = .47$	
	popular children, nonsignificant	
	6th and 7th graders (Pelligrini, 1994)	
	rejected children, $r = .54$	
	popular and average children, nonsignificant	

Developmental Course

Because most studies of play-fighting in children have employed cross-sectional designs (to date, only a single, 1-year longitudinal study has been published—Pellegrini, 1995), it is difficult to draw any firm conclusions about the developmental course of play-fighting in children. This is further complicated by the fact that most cross-sectional studies of play-fighting confound classroom with child age. For example, when classes of second and fourth graders are observed on a playground, differences between the classes in the rate of play-fighting are interpreted to reflect developmental differences. Because rates of play-fighting vary as a function of numerous contextual factors (e.g., size of the play area, number of toys available, hardness of the playground surface, teacher permissiveness—see Smith & Connolly, 1980), comparing children from different classes may be problematic, because such contextual factors often vary and are not controlled (e.g., fourth graders in a particular school may show less play-fighting than do second graders, because the fourth grade teachers in that school are less permissive of play-fighting during free play time).

Despite the limitations of the database, what little data on the developmental course of play-fighting have been collected are consistent with the inverted U function found for animals: starting sometime in infancy, peaking in the juvenile period, and then decreasing thereafter. Humphreys and Smith (1984) and Pellegrini and Smith (1998) pieced together data from various sources to describe changes in play-fighting with age. According to these reviews, play-fighting is first observed in 2-year-olds, is common in early and middle childhood, and then decreases during adolescence.

The other age trend that has been observed in the play-fighting of animals is that play-fighting becomes more vigorous and rougher with age. Data on this issue are not currently available for humans, although one study of 12- and 13-year-olds by Neill (1976) suggests that this might be the case. Whether play-fighting in humans is more likely to escalate into aggression with increasing child age has not been established. Finally, several studies have shown that, with increasing age, children watching videotapes of peer interactions are better able to correctly differentiate between instances of play-fighting and serious aggression (Boulton, 1993; Costabile et al., 1991).

Play Partners

As mentioned earlier, members of many mammalian species, when given a choice, tend to initiate play-fights with same-gender partners that are close in age to the initiator. Animals also tend to choose play partners with whom they are biologically related. Because most studies of children's play-fighting have involved observations of classes of same-aged children playing on playgrounds, researchers rarely have studied the effects of age or biological relatedness on partner choice. Regarding gender of the partner, numerous studies have shown that, starting in the preschool period, children usually play with same-gender partners when given a choice (Maccoby, 1988). This tendency has been documented in many studies and for many kinds of play, and is considered in detail in chapter 8. Suffice it to say here that children tend to choose play-fighting partners of the same gender as themselves. Moreover, because male–male play is more intense than male–female or female–female play, females often avoid or terminate play with males, presumably because of its intensity (Aldis, 1975; Di Pietro, 1981). However, when girls do play with boys, they show greater amounts of play-fighting than when they play with other girls (Aldis, 1975).

Although studies of the effects of dominance status of the partner on play-fighting were inconsistent in the animal literature, the literature for humans does not share that problem. For the most part, children tend to initiate play-fighting with children of the same or slightly lower dominance status (Boulton, 1991b; Humphreys & Smith, 1987; Pellegrini, 1993, 1994, 1995). In several of these studies, however, the relationship between dominance and partner choice was significant for some ages and not for others.

Finally, using sociometric methods, researchers have demonstrated that human children usually engage in play-fights with children whom they like (Boulton, 1991b; Humphreys & Smith, 1987; Smith & Lewis, 1985). Obviously, this is an issue that can only be studied in humans. However, is likely that this tendency is not unique to play-fighting: Children undoubtedly choose playmates they like for all types of play.

Gender Differences

The literature on gender differences in play-fighting of animals demonstrated that gender differences varied as a function of male–female differences in the importance of fighting for adult roles, the extent of sexual dimorphism, and reproductive strategy. According to this analysis, humans should show gender differences in play-fighting because humans are sexually dimorphic. Also, there are clear gender differences in the importance of aggression in adult roles among hunter–gatherer cultures (the cultural setting in which most of human evolution has occurred). On the other hand, as a mostly monogamous species, few gender differences in play-fighting might be expected. Examination of the available data on children's play-fighting show clear gender differences (e.g., Aldis, 1975; Blurton-Jones, 1967; Di Pietro, 1981; Frey & Hoppe-Graff, 1994; Hertz-Lazarowitz, Feitelson, Zahavi, & Hartup, 1981; Humphreys & Smith, 1987; Maccoby, 1988; Moller, Hymel, & Rubin, 1992; Pellegrini, 1989b; Smith & Connolly, 1972), with boys showing more play-fighting than girls. Moreover, Aldis (1975) reported that girls very rarely engage in wrestling play-fighting that involves attempts to pin the partner, and Di Pietro (1981) found that there were no girls in the high end of her distribution of rough-and-tumble play. Although no study presents data specifically on gender differences in the intensity of play-fighting interactions, the findings of Aldis (1975) and Di Pietro (1981) suggest that the genders may differ in intensity of play-fighting as well.

Studies of Non-Western Cultures

Numerous ethnographic studies of children growing up in non-Western cultures have shown that play-fighting in children is not unique to Western, industrialized cultures. Rough-and-tumble or wrestling play has been observed in a wide range of hunting-and-gathering and agricultural village cultures located in Africa (Konner, 1972; Whiting & Edwards, 1973), India (Whiting & Edwards, 1973), Japan (Takeuchi, 1994; Whiting & Edwards, 1973), Mexico (Whiting & Edwards, 1973), the Philippines (Whiting & Edwards, 1973), South America (Holmberg, 1969), and the South Pacific (Hogbin, 1946; Martini, 1994).

Unfortunately, most studies of non-Western cultures provide little information on such play activities other than that they have been observed or are common childhood behaviors. There are two exceptions, however. Whiting and Edwards (1973) conducted systematic observations in five non-Western cultures and found that, in four of the five cultures, boys engaged in more rough-and-tumble play than did girls. Fry (1987, 1990) published the only careful descriptive study of play-fighting in a nonindustrialized culture, and in it he found many results consistent with the studies of Western cultures described. Specifically, for Zapotec children in rural Mexico, Fry (1987, 1990) identified several differences in the nature of serious fighting and play-fighting that have already been discussed (see Table 5.2). Analyses of play partners showed that during play, partners were closer in age and more likely to be of the same gender (at least for boys—girls usually had same-gender partners for both play and serious aggression). Data from play initiations were more complicated, with play more likely to be initiated by a partner slightly older in age, except when males initiated play with females—in this case, the males initiating play were slightly younger than the recipients. There were no consistent age differences in the initiation of serious aggression.

FUNCTIONS/EFFECTS

As was the case for play-fighting in other animals, researchers have discussed many possible functions/effects of play-fighting in humans. Only a few hypotheses unique to humans have been offered, and the major hypotheses can be classified under the same categories used in Table 4.9. The hypotheses for humans are listed in Table 5.3. The main hypotheses unique to children concern the role of play-fighting in the development of certain social skills, such as perspective taking (e.g., Pellegrini, 1988, 1989b).

Unlike animal research, where the primary data on the functions/effects of play-fighting have come from analyses of play-fighting structure, the

TABLE 5.3

Hypothesized Functions of Child Play-Fighting

Aggression training
Experience general physical training that improves later fighting abilities: strength, endurance, flexibility (Aldis, 1975)
Learn specific fighting skills to be used later (Aldis, 1975; Boulton, 1991a; Humphreys & Smith, 1984, 1987)
Aggression prevention or control
Prevent escalation of play-fighting into serious aggression by accepting pain as part of social interaction or learning restraint, self-handicapping, control, and modulation (Dodge et al., 1990; Hartup, 1974)
Socialization
Establish and reaffirm affiliative social relationships (Blurton-Jones, 1967; Humphreys & Smith, 1987; Smith & Boulton, 1990)
Establish, test, and maintain dominance relationships (Pellis, 1995; Smith & Boulton, 1990)
Learn other social skills (e.g., communication, negotiation, conflict resolution—Pellegrini, 1987)
Cognitive stimulation and learning
Produce novelty/flexibility (Pelligrini, 1987)

major data for children involve examining the correlation between individual differences in play-fighting and various aspects of child development. Besides several studies on the effects of ecological factors on play-fighting (Smith & Connolly, 1980), no experimental data are available.

Given the lack of experimental data and the limited structural analyses conducted with humans, the following discussion of the possible functions and effects of play-fighting is organized by hypothesis rather than by method. Within each section, however, both structural and correlational evidence are considered.

Aggression Training

Support for the hypothesis that play-fighting provides training in specific aggression skills in humans is mixed at best. Although children usually play-fight with partners of the same gender, they tend to choose partners who are of equal dominance status or of a status slightly lower than themselves. Such a strategy may be effective in maximizing the vigor and challenge of play-fighting, while still leading to a high probability of play-fighting success (e.g., pinning, exhibiting dominance over one's partner). If children were only concerned with skill training, it seems that a strategy of choosing partners of equal or slightly higher dominance status would be more appropriate.

Second, although the aggressive behaviors that occur in play-fighting in humans show many of the same modifications found in other animals, it is not clear that these modifications facilitate the development of fighting skills (Pellis & Pellis, 1998). As already discussed in the section on animal play-fighting, repeated practice of restrained or inhibited behaviors may have little impact on the development of proficiency in high-intensity behaviors.

Finally, if play-fighting serves an aggressive training function, then one would expect a positive correlation between play-fighting and proficiency at aggression. Unfortunately, such a study has not been conducted with children, so it is necessary to examine the next closest thing—the correlation between the frequency of play-fighting and the overall level of aggression (it is likely that children who are proficient at aggression may be the most likely to use it). As described earlier, this correlation is positive and significant only for 2- to 4-year-old children (Frey & Hoppe-Graff, 1994), and for older children who are socially rejected (Pellegrini, 1988, 1989b, 1994). Because one of the most consistent predictors of social rejection in children is the child's level of aggression (Coie, Dodge, & Kupersmidt, 1990), it is possible that the positive correlations referred to earlier support the hypothesis that play-fighting serves an aggressive training function only for highly aggressive boys.

Several studies support the argument that play-fighting functions differently for rejected and for popular children. Pellegrini (1989a, 1989c) in a factor analytic study of elementary school children's behavior on the playground, found that although aggression and play-fighting came out as separate factors for popular children, these behaviors loaded on the same factor for rejected children. He interpreted these findings to suggest that play-fighting may serve an affiliative function for popular children and a dominance function for rejected children. Also, in an interview study of children's knowledge about play-fighting and serious fighting, Boulton and Smith (1989) found that rejected children had a much less differentiated conception of the two concepts than did other children.

Studies by Dodge and colleagues (e.g., Dodge & Frame, 1982) indicate that rejected children have a hostile attributional bias in their evaluation of social behavior that often leads to aggression. Specifically, when aggressive, rejected children are recipients of an ambiguous social behavior that could be interpreted as hostile, they assume it is hostile and respond with aggression. In the context of play-fighting, Dodge, Coie, Pettit, and Price (1990) argued that rejected children often misread the amicable intent of play-fighting and respond with aggression. This is not surprising, since such children come from homes with high levels of hostility and aggression (e.g., Weiss, Dodge, Bates, & Pettit, 1992), and a hostile attribution for ambiguous events is likely accurate in that setting. Consistent with this interpretation is the finding in three separate observational studies that only rejected boys showed a consistent tendency to respond to play-fighting interactions with aggressive behavior (Dodge et al., 1990; Pellegrini, 1988, 1994).

Aggression Prevention and Control

Given the amicable nature of play-fighting, some theorists have argued that instead of training fighting skills, play-fighting helps decrease the likelihood of aggression by preventing the escalation of aggressive behavior. Both Hartup (1974) and Dodge et al. (1990) asserted that play-fighting experience helps children learn to engage in aggressive behavior in a playful context, and therefore prevent rapid or reflexive escalation to serious aggression in response to physical contact. As discussed in the earlier section on animal play-fighting, because play-fighting requires a good deal of self-control and restraint of aggression, it might be good practice for exercising restraint in more serious contexts. Dodge et al. (1990) maintained that children who do not have sufficient positive play-fighting experiences show a greater likelihood of escalating play-fighting into serious aggression. It is possible that the relative amount of play-fighting and serious fighting experience a given child has may, in fact, be one of the best predictors of his or her interpretation of ambigous physical events, as well as the likelihood of aggressive escalation. Further data consistent with these arguments come from studies of the possible socialization functions of play-fighting, to which we now turn.

Socialization

In contrast to the animal research, considerable recent evidence exists in humans consistent with the various socialization hypotheses listed in Table 5.3. In support of the first hypothesis—that play-fighting aids in the establishment and reaffirming of affiliative social relationships—is the already-mentioned tendency of children to engage in play-fighting interactions with children whom they like (Boulton, 1991b; Humphreys & Smith, 1987; Smith & Lewis, 1985). Play-fighting thus occurs in a positive social context, and may in fact facilitate the development of social relationships.

Other findings consistent with the socialization functions are studies showing positive correlations between play-fighting and children's popularity—at least among boys in the early elementary school years (kindergarten through fourth grade—Dodge et al., 1990; Pellegrini, 1989b). Thus, young children who engage in play-fighting are liked by their peers—a finding consistent with a socialization function. One possible reason for the positive correlation between play-fighting and young children's popularity is that play-fighting may facilitate the development of specific social skills. Pellegrini (1988, 1989b) found that play-fighting was positively correlated with social problem solving in a group of kindergarten, second-, and fourth-grade students.

Pellegrini (1994, 1995) argued that, for older children, play-fighting may play a dominance rather than affiliative function, in that play-fighting in sixth to seventh graders was negatively correlated with peer popularity and positively correlated with dominance status as rated by peers. Moreover, the previously mentioned tendency of older children to choose play-fighting partners slightly lower in status than themselves may help children to maintain dominance patterns once they are established (Pellegrini, 1994, 1995). Other studies of boys in the later elementary school years (third through sixth grade) find that aggressive play is associated with peer rejection as well (Asarnow, 1983; Ladd, 1983). In a study of "horseplay" in college students, Gergen (1990) found that for both men and women the amount of play-fighting with male partners was positively correlated with the level of aggression.

Even among younger children, a study by Dodge (1983) indicates that the relationship between rough-and-tumble play and peer acceptance is a complex one. In this study of second graders, controversial boys (i.e., those who received high numbers of positive *and* negative nominations) showed the greatest amount of aggressive play, neglected and rejected boys showed an intermediate amount, and average and popular boys showed the least. Coie and Kupersmidt (1983), in a study of fourth graders, found that neglected boys (those receiving few positive *or* negative nominations) showed the least amount of rough-and-tumble play.

Novelty/Flexibility

Finally, it is possible that the variability inherent among play-fighting in humans provides opportunities to experiment with behaviors and sequences that contribute to behavioral flexibility. Recall that similar arguments were made for other primates, especially chimpanzees. A recent study by Pellegrini (1993) provides data on this issue. In a study of fifth-grade boys, Pellegrini found that flexibility in play-fighting patterns predicted the ability of children to generate multiple solutions to social problems, whereas temperamental flexibility did not. Clearly this is an important area for future research.

SYMBOLIC PLAY-FIGHTING IN HUMANS: THE CASE OF WAR PLAY

A form of play-fighting that apparently is unique to humans is "symbolic play-fighting," in which children take on aggressive roles in their pretend play (Sutton-Smith, 1988). Symbolic play-fighting differs from other forms in that an elements of pretense are at center of the activity. The most common examples of this behavior include symbolic play in which children act out aggressive roles (e.g., police officers, soldiers, or space warriors) either in their own behavior or through the use of toys such as action figures. The only reference to this type of play occurring in the play-fighting studies referenced earlier is the observation by Smith and Lewis (1985) that pretend "shooting" was identified as a play-fighting behavior.

Play with war toys is common in many non-Western cultures, with popular toys (usually made of sticks, reeds, or grasses) being bows and arrows, spears, or daggers (Hogbin, 1946; Jackson, 1964; Lancy, 1977; Rosenstiel, 1977; Shoemaker, 1964; Tonkinson, 1978; Turnbull, 1962; Whiting, 1963), and guns that shoot corn, peas, or thorns (Jackson, 1964; Shoemaker, 1964). Children also engage in other kinds of symbolic play-fighting in such cultures. For example, Hogbin (1946) described children from a New Guinea village pretending to be sharks; both Anderson (1982) and Storey (1977) described children engaging in pretend cockfights.

Whether or not it is desirable for children to engage in this type of pretend play has been the focus of much debate, especially with regard to the issue of using war toys. Professional opinions on this issue have ranged from the belief that children should be discouraged from engaging in such kinds of play (Nilsson, 1995) and a ban on the sale and production of war toys could allow for the "disarmament" of children (Andreas, 1969), to the belief that such behavior serves important developmental functions but should be channeled into more constructive activities (Carlsson-Paige & Levin, 1987, 1990), to the belief that interfering with a child's freedom to engage in war play amounts to "violence" against children (Wegener-Spohring, 1989).

In recent years, an equal amount of controversy has arisen over another form of symbolic play-fighting—the playing of "violent" videogames. In his 1982 addresses to the Western Psychiatric Institute and Clinic, the then-Surgeon-General of the United States, C. Everett Koop, stated that violent videogames were one the three root causes of family violence in the United States (Cooper & Mackie, 1986).

A brief history of war toys in the United States shows that the availability of such toys peaked in the mid-1960s and then declined thereafter—largely in response to antiwar sentiment generated by the Vietnam War (Andreas, 1969). In the late 1960s and the 1970s, numerous groups were formed in the United States and Europe that advocated banning war toys. These efforts eventually led to legislation and/or voluntary toy industry agreements in Sweden, Austria, Finland, and Malta that achieved that goal (Nilsson, 1995). However, with the early-1980s deregulation of children's television in the United States came a significant increase in the marketing of war toys, through both television advertising and children's television programming. This made TV and movie character toys easily and widely available for purchase in retail stores (Greenfield et al., 1990). As a result, sales for war toys increased 700% between the years of 1982 and 1989 (Beresin, 1989). As of this writing, this trend appears to continue. Moreover, with the increasing availability of home computers and increases in computer technology, the availability and realism of violent videogames has increased dramatically in the last 10 years.

Structure

Detailed studies of the structure of children's war play apparently have not been conducted. Studies have shown, however, that war play takes on two basic forms: acting out combat or fighting roles with toy or pretend weapons, and playing with action or combat figures (Costabile, Genta, Zucchini, Smith, & Harker, 1992; Wegener-Spohring, 1989). Children engage in such play both alone and with others, and it apparently begins between the ages of 2 and 3 years (Costabile et al., 1992). Numerous studies show that boys show greater interest and involvement in these activities than do girls (Little, 1989, cited in Smith, 1994; Sutton-Smith, Gerstmyer, & Meckley, 1988; Watson & Peng, 1992; Werebe & Baudonniere, 1991); this gender difference is found for videogames as well (Cooper & Mackie, 1986; Dominick, 1984; Fling et al., 1992; Kubey & Larson, 1990; Lin & Lepper, 1987; Sakamoto, 1994; Shimai, Masuda, & Kishimoto, 1990). Although Carlsson-Paige and Levin (1987, 1990) argued that war play is less complex and more repetitive than other kinds of pretend play, Sutton-Smith and colleagues (1988) presented data that suggest this may not be the case.

Functions/Effects

Although rarely couched in the language of biological function, there has been much discussion about the possible effects of war play on children's development. As was evident from the previous discussion, views of its hypothesized effects vary widely, from its having a very negative to a mildly positive impact on child development. Proponents of war play argue that such play helps young children expend energy; attain high levels of arousal and excitement; reduce anxiety; separate good from evil; experience power, control, and mastery; discharge aggressive energy; understand and/or defy death; practice strategic planning and problem-solving; and facilitate gender-role and friendship development (e.g., Beresin, 1989; Carlsson-Paige & Levin, 1987; Goldstein, 1995; Sutton-Smith et al., 1988). In contrast, opponents have argued that it desensitizes children to violence; encourages violence over other forms of problem solving, such as negotiation and cooperation; and interferes with the development of empathy and compassion (Nilsson, 1995).

The same debate has raged over children's playing videogames with aggressive content. In addition to the benefits listed previously, proponents argue that such games give children a sense of mastery and accomplishment; facilitate adolescent independence and autonomy; develop hand–eye coordination, attentional, and cognitive skills; and keep adolescents busy and away from other, potentially harmful activities (e.g., Egli & Meyers, 1984; Hull, 1985; Kesterbaum & Weinstein, 1985). Opponents add the additional arguments that such games result in wasted time and money, addiction, escapism, social isolation and alienation, poor grades, inflamed tendons, and an overemphasis on competition (e.g., Selnow, 1984; Soper & Miller, 1983).

Despite the intensity of these debates, only a limited amount of empirical data are available to address these issues. Regarding war play, Sutton-Smith (1988) critically reviewed the eight studies conducted on the topic as of 1988. These studies (three of which were unpublished theses or dissertations) were experiments in which the "aggressive" behavior of young children was observed either while they were playing with or just after they had played with war versus other toys. As pointed out by Sutton-Smith (1988), these studies are a heterogeneous group, with some examining more than just the effects of toy play. For example, in one study (Etaugh & Happach, 1979), children were given explicit instructions about how to behave (they were instructed to punch a bobo doll); whereas in another (Feshbach, 1956), children were read stories and then observed playing with toys from the stories. In such studies, it is not clear whether any subsequent toy play behavior was the result of the toys provided or the directions from the adult. Another problem Sutton-Smith (1988) noted concerns the measurement of child aggression in these studies. None of the researchers attempted to differentiate between serious fighting and play-fighting, and Sutton-Smith argued that most of the significant effects on child aggression were not effects on aggression on all but instead on playful behavior (see also Archer, 1995). Moreover, because it appears that none of the researchers used blind observers (the children appeared to be observed by persons aware of the experimental hypotheses), it is likely that differences in aggression might be an artifact of experimenter bias. Finally, Sutton-Smith (1988) argued that, when placed in a permissive environment, the play of young children becomes more aggressive over time regardless of the toys present, and that the results of many of the studies he reviewed could be an artifact of this tendency. Overall, three of the eight studies he reviewed showed no effects of war toys on aggression, whereas the others did show effects. However, given their methodological problems, Sutton-Smith (1988) maintained that the studies were "all unreliable pieces of work" (p. 64) and, "It is not possible on the basis of this body of work alone to conclude anything for certain about the relationship between war toys and aggressive behavior" (p. 64).

Only one additional experimental study published since this review could be located. Using a repeated measures design, Kaiser, Snyder, and Rogers (1995) observed a group of thirty-one 4- to 5-year-olds playing with neutral, prosocial, and antisocial toys. Although the researchers found differences in child behavior as a function of the toys available, the authors acknowledged that the observers "knew the hypotheses and the theory which contributed to the hypotheses" (p. 185), and there was no attempt made to differentiate between serious and play aggression. Smith described an unpublished senior thesis conducted by Little (cited in Smith, 1994) that examined both playful and serious aggression. Fifteen 3- to four-year-old children in a play group were observed playing with "TV war toys" (war toys advertised on television), "non-TV war toys" (e.g., knights, soldiers), and neutral toys (e.g., farm animals). Although the amount of "pretend aggression" (actions such as fighting, hitting, and shooting accompanied by positive or neutral facial expressions) varied across conditions, the amount of "real aggression" did not. Unfortunately, in the brief description of the article provided, it was not possible to determine whether researchers blind to the experimental hypotheses coded child behavior.

Finally, Watson and Peng (1992) conducted a correlational study of young children's play with war toys. Parents of thirty-six 3- to 5-year-olds completed a questionnaire about their child's and their own behavior at home, and the children were videotaped during free play at school. After the observations, two undergraduate coders (likely blind to the questionnaire responses of the parents) coded the frequency of pretend aggression, real aggression, rough-and-tumble play, and nonaggressive pretend behavior in the classroom. Results showed that parent reports of the amount of children's gun play in the home was positively correlated with the amount of real aggression children showed in the classroom (r = .37), negatively correlated with the amount of nonaggressive pretense (r = -.38), and uncorrelated with the amount of pretend aggression (r = –.04). Parent reports of children's gun play, however, were also positively correlated with parent reports of the use of physical punishment and independent ratings of the aggressiveness of the children's television preferences. Although Watson and Peng (1992) ran regressions in an attempt to sort out the independent contributions of each of the variables in predicting children's aggression, the sample was too small to produce reliable results. Moreover, Smith (1994) argued that their definition of real aggression, "whenever there was either an aggressive physical action toward a peer or toy (e.g., hitting, kicking, pushing, fighting, biting, grabbing a toy, throwing a toy) or an aggressive verbal action toward a peer (e.g., quarreling, shouting at a peer, threatening a peer, name calling)" (p. 378) was too broad, and might include many instances of play-fighting.

The limited experimental literature on "violent" videogames is plagued by many of the same problems. Of the six experimental studies that could be located using children or adolescents as participants, two found that, when compared to controls, children who played violent videogames showed more aggression during free play (Schutte, Malouff, Post-Gorden, & Rodasta, 1988; Silvern & Williamson, 1987); one found that children were less likely to donate nickels they had earned to poor children (Chambers & Ascione, 1987); and two found that they were more likely to play with an aggressive versus a nonaggressive toy (Cooper & Mackie, 1986; Schutte et al., 1988). In contrast, two studies found no differences between children who played violent versus nonviolent games when they engaged in laboratory tasks involving interfering with another child's progress in a game (Grayhill, Strawniak, Hunter, & O'Leary, 1987); or taking money away from another "child" the subject could not see, in a Milgramlike teaching task (Winkel, Novak, & Hopson, 1987).

The studies varied considerably in both the nature of the games employed and the definition of aggression. The violent games ranged from those involving boxing or hitting or shooting criminals (Chambers & Ascione, 1987; Grayhill et al., 1987; Schutte et al., 1988) to games involving battles between spaceships (e.g., Cooper & Mackie, 1986; Silvern & Williamson, 1987). Child aggression ranged from projective assessments (Grayhill et al., 1987), to child self-reports (Grayhill et al., 1987), to giving out punishment to an unseen child in a laboratory task (Cooper & Mackie, 1986; Grayhill et al., 1987; Winkel et al., 1987), to observed aggressive behaviors in a free-play situation (Kaiser et al., 1995; Schutte et al., 1988; Silvern & Williamson, 1987). None of the studies employing observational assessments attempted to differentiate between play-fighting and serious fighting, and it is not clear from most articles that the observers were blind to the study hypotheses.

A greater number of correlational studies have been conducted on videogame use. Although such studies do not usually differentiate between violent and nonviolent games, these studies are reviewed here because most games have some level of aggression. The typical study involves having children or adolescents (and occasionally a parent) complete questionnaires about the child's videogame use and provide information about other aspects of his or her life. In general, heavy levels of videogame use tend to be associated with lower grades and less time spent on homework (e.g., Creasey & Myers, 1986; Dominick, 1984), and higher levels of self- and teacher-reported aggression and antisocial behavior (self-reports— Dominick, 1984; Fling et al., 1992; Kesterbaum & Weinstein, 1985; teacher reports: Fling et al., 1992; Lin & Lepper, 1987). Such children spend more time watching television (Dominick, 1984) and playing other kinds of games (Lin & Lepper, 1987), spend less time reading (Lin & Lepper, 1987), and have sometimes been found to show such personality characteristics as competitiveness (Kesterbaum & Weinstein, 1985) and impulsivity (Lin & Lepper, 1987). The findings for self-esteem, sociability, and empathy are inconsistent (see Dominick, 1984; Fling et al., 1992; Kesterbaum & Weinstein, 1985; Sakamoto, 1994; Shimai et al., 1990). These findings, although consistent with a causal model regarding the effects of violent videogames on aggression, say little about cause and effect. Given that aggression existed long before the development of videogames, a more parsimonious explanation is that violent children are drawn to violent videogames, rather than the videogames being the cause of violence. Some third variables may be operating as well, in that parents who allow the unsupervised play of videogames may provide less direction, supervision, and monitoring of their children in general than do parents who directly or indirectly regulate the amount of play.

In a chapter on war play and violent videogames, Goldstein (1995) argued that the research in this area has been limited in theory and method, and only a limited range of questions has been addressed. He noted: "Among the dependent variables rarely or never studied in war play are its effects on imagination, militarist attitudes, attitudes toward violence and weapons, and the ability to cope with violence and death, such as the loss of a relative or a pet. What are the long-term effects, if any, of such play on militarism, aggression, role taking, problem-solving, self-concept?" (pp. 140–141).

Summary: Symbolic Play-Fighting

In summary, many young children across cultures, especially boys, devote at least some of their playtime to symbolic play-fighting activities—with or without toys (including videogames and computers). Whether it is desirable or undesirable for children to engage in these activities has been the subject of much debate, fueled as much by political ideology as by developmental theory or empirical findings. Pacifist parents especially are concerned when, after going to great lengths to buy only nonviolent toys, they find that their children engage in aggressive behavior with dolls, transform other toys into swords or guns, or show what seems to be an obsession with the war toys of other children. In a study of parents in Italy and England, Costabile and colleagues (1992) found that approximately half of the parents surveyed reported that they discouraged war play, about one quarter said they allowed it within limits, and about one quarter reported allowing it unconditionally. Italian parents were more likely than English parents to discourage war play, and this paralleled cultural differences in child play behavior as well. Despite parental behavior, about one third of the parents who reported discouraging war play had children who engaged in it anyway.

Despite the prevalence of war play, the large amount of advertising dollars devoted to its promotion, and the levels of parental concern, it impossible to draw any firm policy implications because of the lack of high-quality research in this area. If war play is indeed simply a symbolic form of play-fighting, it may result in some of the same positive developmental effects for young children discussed in the previous section (at least for most children). On the other hand, it is possible that such play may be detrimental for overly aggressive or rejected children, in that it may give these children an excuse or context to be too rough or coercive in their play with others, and therefore further increase the likelihood of peer rejection and aggression. Clearly, further developmental research on this issue is necessary, especially given what are to some parents disturbing trends in the severity and realism of violence portrayed in war toys, especially violent videogames. However, given the current state of affairs, the conclusions of Smith (1994) seem reasonable:

> Where does this leave us? Personally, and in our present state of knowledge, I find little to support the extreme anti-war-play position of some parents who ban any kind of war play in their home. They have a right to do it, but I suspect it is unnecessary and counterproductive. As a parent, I would be worried only if my child was obsessed with war toys or with

violence in his or her play; I would then seek the reasons for this attitude, probably outside the toys themselves. (pp. 82–83)

PLAY-FIGHTING: SUMMARY AND CONCLUSIONS

The results of the research reviewed in this and the previous chapter suggest that play-fighting in children has an evolutionary basis. Play-fighting is the most common form of play in mammals, and is found in children in a wide range of cultural settings. Most of the behaviors observed during play-fighting in children and other mammals are similar to those found in serious aggression, and a comparison of Tables 4.4 and 5.1 shows numerous similarities between these behaviors in humans and in other species, especially the primates. The young of play-fighting species, including humans, modify agonistic behavior for play-fighting in similar ways. As shown in Tables 4.5 and 5.2, when children and other mammals engage in play-fighting, their behavior is accompanied by general amicable signals, is reduced in intensity, is of longer duration, is directed toward different targets, shows role reversals and self-handicapping, rarely results in physical injury, and is terminated with the partners in close physical proximity. Moreover, individual differences in play-fighting and serious fighting are generally independent.

Further similarities come from studies of the developmental course of play-fighting, gender differences in its occurrence, and the play partners chosen. Play-fighting in humans apparently follows the same inverted U developmental course identified for most other mammals, and its roughness may increase with age. Also, humans and members of many other species usually choose play-fighting partners of the same gender, and females are more likely to terminate play-fights with partners of the opposite gender than are males. Finally, like many primates, males in both Western and non-Western cultures engage in more play-fighting than do females.

Despite these similarities, there are a few documented differences. Studies of humans include no mention of biting or mounting behaviors—behaviors common in many other mammals, especially primates—and several human play-fighting behaviors appear to relate to the bipedal nature of human locomotion (i.e., trip, tackle, leveraged throw). Aldis (1975) observed that, unlike other mammals, children sometimes take a rest in the middle of play-fighting before resuming, and humans apparently do not show the wide range of play-initiating behaviors found in other primates (especially the chimpanzees, our closest relative). Finally, play-fighting in children often has a verbal and sometimes a pretend component. Studies of children in a wide range of cultural settings note the existence of symbolic play-fighting using pretend weapons. Across cultures, these activities are more likely to be seen in boys than in girls.

The verbal component of play-fighting may account for some of the differences between humans and other mammals, especially the differences in initiations and resting. The differences in biting and mounting may reflect the impact of cultural suppression of these behavior patterns in humans.

Although it is impossible to *prove* that any behavior pattern in humans has an evolutionary basis, the research on play-fighting provides the strongest empirical support for this notion in the area of children's play. As one would expect for a human behavior pattern with a long evolutionary history, play-fighting is found across cultures, is common in hunting-and-gathering settings (the context in which human behavior patterns evolved), and shows structural similarities with the behavior of our animal relatives as a function of our phylogenetic proximity to them (with the greatest number of similarities found with the primates, especially the great apes). As a comparison of Tables 4.4, 4.5, 5.1, and 5.2 shows, the structural similarities are significant, especially in how behavior during play-fighting differs from that of serious aggression. As already mentioned, the research on developmental course, gender differences, and play partners is consistent with an evolutionary analysis as well. Although it is possible that the similarities and differences between the play-fighting of human and nonhuman animals is the result of socialization and cultural practices that have arisen independent of biological evolution, this seems unlikely given the considerable continuities across human and nonhuman species. In this case, it may be incumbent on those opposed to an evolutionary viewpoint to describe what combination of cultural and socialization practices could account for these widespread (and apparently coincidental) cross-species similarities. The evolutionary hypothesis is the most parsimonious explanation of this phenomenon, and very little data are inconsistent with it.

There is currently a dearth of empirical evidence to make a case for the functions or effects of play-fighting in humans. It is possible that play-fighting may serve social bonding effects for most young children and may facilitate the development of fighting skills in rejected children, but this hypothesis is based on limited empirical evidence. There is limited reliable data on symbolic play-fighting in humans as well, making it difficult to speculate about its effects. However, given the increasing opportunities in recent years for children to engage in interactive and graphic forms of symbolic play-fighting (especially while playing videogames), future research in this area is needed.

Given the commonalities between play-fighting in humans and other animals, further research on the structure and functions/effects of play-fighting from an evolutionary perspective is warranted. Because of the superficial level of descriptive detail that usually has been employed (especially in studies of humans), it is still possible that the similarities between play-fighting in humans and other animals are superficial. It is necessary to conduct studies of specific offensive and defensive strategies, as well as on the sequencing, timing, and other qualitative characteristics of play-fighting behaviors. Because the effects of play-fighting likely vary as a function of the child's gender and age as well as the social setting, further longitudinal studies of children's play-fighting in school and non-school settings need to be conducted. Such studies should include assessments of multiple constructs (e.g., temperament, serious aggression, play-fighting, prosocial behavior, social skills, emotional development) using multiple information sources (e.g., child-, peer-, parent-, and teacher-reports; structured and unstructured observations). The pioneering research by Aldis, Blurton-Jones, Boulton, Fry, Pellegrini, Smith, and their colleagues has just scratched the surface. There is much more to be learned from future inquiry in this area.

6

Locomotor Play in Animals

With some notable exceptions (e.g., pinnipeds and bats), running is critical to the survival of most mammals. Running is the primary way in which most prey species (such as many rodents and ungulates) escape predators, and many carnivores (such as felids and canids) capture prey. Depending on the species, running can be important in other contexts, including agonistic encounters, mating, and the defense of one's young. Selection pressures for running abilities in some species have been so strong that some of the fastest animals on earth are large carnivores such as the cheetah, along with some of the ungulate species on which they feed.

Researchers have differentiated between two major types of locomotor play in animals: solitary locomotor-rotational play, in which young animals engage in active locomotor behavior such as running, jumping, and kicking on their own; and chasing play, in which animals engage in short or extended chases with one other. Although chasing play is sometimes combined with play-fighting to make up the category of social play in some studies of animals (and the category of rough-and-tumble play with children), it is considered separately from play-fighting in the present text because of numerous differences between the nature of social interaction in these two contexts.

The observation of locomotor play shows that it has many of the locomotor components of prey avoidance and prey capture (Aldis, 1975) but, as was the case for play-fighting, these elements are varied in their structure. For the 60 species considered for this volume, most species that engage in play-fighting also engage in chasing play. The young of these species often engage in solitary locomotor-rotational play as well, although such behavior appears to be particularly common in some ungulate, rodent, and primate species. Very few species appear to engage in locomotor play but not play-fighting. Some bird species also engage in locomotor play, but given differences between locomotor behavior in terrestrial versus airborne species, only locomotor play in mammals is considered here.

Although common, locomotor play is not without it risks. Comparison of these risks to those of play-fighting reveal many similarities, including eliciting damaging or fatal aggression, falling down or otherwise getting injured, alerting predators, getting separated from the caregiver, and using up important time and energy. Because members of some species engage in locomotor play on sloped or rocky surfaces (e.g., Byers, 1977, 1980), other possible risks summarized by Fagen (1981) include getting caught in rocks or mud and danger from rolling rocks.

STRUCTURE

The literature on locomotor play is reviewed here using the same basic outline as in the discussion of play-fighting. Before turning to the literature, a few qualitative descriptions are presented:

Locomotor-Rotational Movements (multiple species—Fagen, 1981, pp. 287–291):

Play is rich in brisk and likely body movements. These locomotor-rotational movements (S. Wilson & Kleiman, 1974) fall into three general categories: locomotion, rotation of the whole body, and rotation of parts of the body.... The single most frequent and phylogenetically widespread locomotor act of play must surely be a leap upward.... Hops, springs, bounces, and bucks are variations on the basic vertical leap. Animals often run in play, using characteristic galumphing or lolloping gaits. They may run repeated courses—back and forth, circles, figure-eights, or semicircles—and may zigzag. Sudden starts and stops are frequent, as are sudden, sharp turns (jinks). Rotational movement is characteristic of play. Animals may somersault, roll, flip forward or backward, spin, whirl, pirouette, make handstands, chase their tails, rear, and kick up their heels. Rotations of body parts often occur in combination with whole-body movement. Often a vertical leap is decorated with body-twists, rear-kicks, or head-shakes. These acrobatics can be spectacular.

Black-Tailed Deer (Miller, 1975, pp. 150–151):

Two mature does, two yearling females, one female fawn, and one male fawn began a game of tag which lasted for 12 min. The yearlings gamboled about the adult females. The dominant doe chased one yearling and when the yearling ran close to the other deer they scattered as though to avoid being "tagged." When the chased yearling stopped, the doe stopped. They stood side by side for a brief period before the doe became the pursued and scattered the other deer by running among them. The yearling broke from the doe's trail and began chasing the other yearling which was "tagged" and in turn chased and "tagged" the subordinate mature doe. The subordinate mature doe appeared to tire of the game after being unsuccessful in catching the yearling or any other deer. The fawns were only chased for short distances and were never "tagged" or pursued except by each other. They appeared to play their own game while attempting to be a part of the main game.

Lions (Schaller, 1972, pp. 159–161):

Chasing. One animal bounds after another, swats at its legs, and tires to grasp its rump with the forepaws. Occasionally it runs beside the other, puts a forepaw on its shoulder, and pulls it down.... In such a situation the fleeing animal sometimes collapses in an exaggerated manner by falling on its side with a thump. Wrestling and pawing often follow a chase, but some end without further contact when the following animal turns and faces its pursuer. Chasing is especially prevalent when the group travels.

Rhesus Monkeys (Symons, 1978, p. 80):

The direction of a playchase is determined by the chasee, but the required behaviors are determined by the substrate over which the monkeys travel. Playchasing monkeys move through their environment at high speed, climbing tree trunks, running along the ground and branches, and leaping from the ground into trees or shrubs, from trees or shrubs to the ground, from branch to branch, and from branch to water. The chaser often leaps at the chasee on the ground, on a branch, or in the water, but the chasee usually anticipated these leaps and flees before it is hit.

Chimpanzees (van Lawick-Goodall, 1968, p. 258):

Individual chimpanzees, both in the wild and in captivity (Loizos, 1969), display an almost endless ability to vary the order, intensity and manner of execution of the play patterns described above. Chasing took place either in the trees or on the ground, or both. Juveniles and older infants often chased wildly through the trees, leaping from branch to branch or from tree to tree. When the pursuer caught up there was sometimes a bout of wrestling when each individual hung down from a branch by one arm whilst sparring or grasping hold of the other with his free arm and legs.... Often one youngest chased another along the ground, up a tree, out along a low branch and then down to the ground again. At other times two or more individuals chased each other round and round the base of a tree or a clump of vegetation.

Play Signals

Although considerable research on play signals has been conducted on species that engage in play-chasing, the researchers have rarely considered separately the nature of play signals used during fighting as compared to chasing play. Therefore, the bulk of this research was reviewed in the previous chapter. The only play signal unique to chasing concerned a particular way the play-face is displayed during a chase. In her description of play-chasing in Hamadryas baboons, Le Resche (1976) stated that "The pursued animal may on occasion look back over his shoulder at this pursuer, displaying the play face. The pursuer less frequently *shows* this expression" (p. 193).

General Characteristics

Although no specific goal has been offered for solitary locomotor-rotational play, the goal of chasing play appears to be for the chasing animal to pursue (and sometimes touch or "tag") another animal (Aldis, 1975). It has been noted that, as in play-fighting, role reversal characterizes play-chasing. That is, at some time during a chase, the pursuer and pursued change roles. The degree of role reversal, however, varies by species. For example, Aldis (1975) noted that role reversal is rare or nonexistent in brown bears and sloth bears, whereas it is common in domestic dogs, California sea lions, capuchin monkeys, patas monkeys, savanna baboons, and gibbons. Miller (1975) states that it is common in black-tailed deer, and Wolff (1981) found that it was common in wild house mice.

Play Initiations

The initiation of play-chasing appears to be much the same across species. Play-chases are usually initiated with one animal touching or nipping another and then running off (Aldis, 1975; Baldwin & Baldwin, 1978a; Henry & Herrero, 1974). The physical contact involved in initiating a chase is inhibited, likely communicating its playful intention. Because such initiations are so brief and are followed by the animal running off, preliminary behaviors apparently rarely occur.

Locomotor-Rotational and Play-Chasing Behaviors

Presented in Table 6.1 is a list of 10 kinds of locomotor-rotational behaviors observed during the solitary locomotor play of animals from various orders

TABLE 6.1
Locomotor-Rotational Behavior During Solitary Locomotor Play, by Order

Behavior	*Carnivores*	*Pinnipeds*	*Ungulates*	*Rodents*	*Primates*
Run*	+	+	+	+	+
Leap, jump*	+	+	+	+	+
Rear, rear and pirouette*			+	+	+
Heel kick, torso twist*			+	+	
Head shake, head jerk, neck twist*	+	+	+	+	+
Fast turns and stops*	+		+	+	+
Whirl, twirl*			+	+	+
Wallow, roll	+	+		+	+
Somersaults, acrobatics*	+		+	+	+
Swinging, hanging	+			+	+

*From the review by Byers (1984).

(references for this table are listed in Table 4.1). Eight categories come from Byers' (1984) review of play in ungulates; two were added to cover behaviors observed in members of other orders. Based on the review of the literature on the 60 species listed in Table 4.1, a plus sign indicates that at least one species in the particular order has been observed to show the behavior. As can be seen in Table 6.1, rodents, primates, and ungulates show the greatest variety of locomotor-rotational behavior, carnivores show an intermediate amount, and pinnipeds show the least. The data for pinnipeds, however, may be misleading, because these were studies conducted on land, where pinniped locomotion is limited (cf. Wilson, 1974). The wide variety of species that show these behaviors led Wilson and Kleiman (1974) to conclude that locomotor-rotational movements "may be ubiquitous to mammalian play, whether or not they are a major component of the play repertoire" (p. 360).

According to Aldis (1975), solitary locomotor-rotational play is prevalent in young animals and involves short sorties from their mothers that are repeated again and again. Similarly, Crowell-Davis et al. (1987) noted that Welsh ponies often trotted in circles around their mother in play, and that the sizes of these circles increased as the ponies got older. Locomotor excursions in the mother's absence do not appear to be random either.

Ferron (1975), in a study of red squirrels, noted that the play excursions of young animals tended to be along relatively fixed paths.

Although locomotor-rotational play is often solitary, in some species it can be social. Both Crowell-Davis et al. (1987) and Gomendio (1988) noted that group-living ungulates often engage in these behaviors in a side-by-side fashion, a pattern that Gomendio (1988) termed "parallel play."

The most common form of social locomotor play is play-chasing. Aldis (1975) described its structure succinctly:

> The basic motor patterns of chasing play are quite simple. Chasing almost always involves two—and only two—partners. One partner pursues; the other flees. Chasing always involves active running or other forms of locomotion. Although running is generally vigorous, it is at less than maximum speed.... In some primate species, there are additional responses such as "tagging" and "dangling," that have been added to the basic locomotor responses of chasing games.... Tagging usually occurs at the onset of a chasing game; one player may tag the other and then run off with the other pursuing. It may also occur when one player catches the other after a chase. (pp. 13–14)

Symons (1978) agreed that although play-chasing has a "certain simplicity" about it, concerning rhesus monkeys:

> The complexity of these chases becomes apparent, however, when one considers the problems of moving rapidly through intricate arboreal pathways. During playchases rhesus run swiftly along, and occasionally under, branches that vary continuously in diameter and flexibility while avoiding other branches that loom to block their path and make rapid, precise, and sudden turns and leaps without losing momentum. They must judge distance accurately, since leaping too far is as ineffective and potentially dangerous as not leaping far enough. They must anticipate which branches will support their weight and which will not, and this varies according to the weight of the animal, its velocity, and the distance leaped. While springing from a flexible branch monkeys must utilize the whip of the branch to augment the jump and when landing on a flexible support they must ride the rebound of the branch as they continue the chase. If they fall, they must twist to fall feet first, grasp a branch, climb on top, and resume running. (pp. 81–82)

Play-chases in primates sometimes involve circular paths around a tree (rhesus monkeys— Symons, 1978; chimpanzees—Hayaki, 1985; van Lawick-Goodall, 1968).

Subtle forms of communication appear to occur during play-chasing, at least in baboons. As already mentioned, Le Resche (1976) discussed baboons' flashing the play-face over their shoulder. Additionally, Owens (1975b) noted a behavior pattern that females more 5 months old often directed to infants during play-chasing—"encouraging" infants to engage in chasing play. Owen described behavior, termed "dodging" as follows:

> Sometimes animals appeared to invite another to chase them by making a bouncy leaping movement away from the partner (often accompanied by a hit, especially to the head). This behaviour is referred to as "dodging," and is a sub-division of approach–withdrawal play. If the partner responded by chasing the animal that dodged away, the animal characteristically waited for it to catch up, then dodged away again, and so on. (p. 396)

Three variations of play-chasing have been described in the literature. These include the huntinglike stalking, crouching play found in large felids such as cheetahs (Caro, 1995), lions (Schaller, 1972; Schenkel, 1966), and tigers (Wasser, 1978); a variation of hide-and-seek found in yellow-bellied marmots (Jamieson & Armitage, 1987); and a form of noncontact social play termed "approach–withdrawal play" in some primates. Fedigen (1972) differentiated approach–withdrawal from chasing play as follows: "Approach–withdrawal—one animal deliberately provokes or attracts another animal to it, runs a short distance with the second following, stops when ahead to provoke the second animal again, and as the second is about to catch up, runs off again" (p. 350). Chasing play, she argued, "may be differentiated from approach–withdrawal by its quality of sustained locomotion" (p. 350).

Termination

Play-chasing is apparently terminated in one of two ways: either animal loses interest and stops running (e.g., Aldis, 1975), or the pursued animal runs to cover (Symons, 1978).

Differences Between Play-Chasing and Serious Flight and Pursuit

Aldis (1975) described several ways in which behavior during play-chasing differs from behavior during serious flight and pursuit. Serious flight, he maintained, is at maximum speed; is directed away from danger toward cover or to a protector (such as the mother); may be accompanied by flight-related behaviors such as warning calls, hiding, or freezing; and ends with the animal staying in its hiding place for a long period of time. In contrast, play-chasing, is at slower speeds; is only temporarily directed away from the pursuer and then the animal may circle back toward the chaser; is rarely directed toward a protector; and is not accompanied by warning calls, hiding, or freezing. Additionally, play-chases are often accompanied by play signals and may involve role reversals between the chaser and chasee. Moreover, in play-chases, when the animal being pursued stops running the chaser does as well.

Symons (1978), in his study of rhesus monkeys, cited two characteristics of play-chases that contradict Aldis's (1975) conclusions: At least in rhesus monkeys, "freezing" behaviors during play-chases were common, and the pursued animal usually ran for cover.

Developmental Course

Although species vary in the specific developmental timetable of play-chasing, it appears to follow the same general developmental course found for play-fighting: an inverted U function, with play-chasing starting sometime during infancy, peaking in the juvenile period, and then decreasing thereafter. This function would explain why researchers find that some species show increases in the occurrence of play-chasing with age (e.g., Caro, 1988; Drea et al., 1996; Poole, 1978; Waterman, 1988), some exhibit decreases with age (e.g., Brown, 1988; Caro, 1995; Crowell-Davis et al., 1987; Gomendio, 1988), some display an inverted U relationship (e.g., Chalmers, 1980; Gard & Meier, 1977), and some demonstrate no changes (Meaney & Stewart, 1981).

Although less studied, other aspects of play-chasing change with age. Caro (1995) found that, in cheetahs, the length of play-chases increased with age, as did the animals' success in patting or grasping the play partner at the end of a chase. Owens (1975b), in his study of baboons, discovered that with increasing age animals were more likely to be the chasers than the pursuers. Finally, Symons (1978) observed that with increasing age, play-chases and play-fights among rhesus monkeys in the trees would occur at greater distances above the forest floor.

Play Partners

As was the case for play-fighting, many studies of play-partner choice do not differentiate between play-fighting and play-chasing. However, based on the few studies that are available, animals appear to be less selective about the age and gender of their partners when play-chasing compared to play-fighting. Aldis (1975) argued that size and age of the partner was irrelevant for play-chasing, in that any animal can be a suitable partner. This was confirmed in Berger's (1980) study of bighorn sheep, which found size preferences for play-fighting but not for play-chasing. Although there was no clear gender preference in partner choice, two studies found that, in male–female pairs, males were more likely than females to be the pursuers (marmots—Jamieson & Armitage, 1987; laboratory rats—Meaney & Stewart, 1981). No studies on the dominance status of partners could be found.

Gender Differences

Studies of gender differences are less consistent for locomotor play than for play-fighting. Although not entirely consistent, there appears to be—at least in primates—a tendency for females to engage in more solitary locomotor-rotational play than for males (rhesus monkeys—Gard & Meier, 1977; chimpanzees—Mendoza-Granados & Sommer, 1995). Also noted is that males engage in more chasing play than do females (baboons—Young, Coelho, & Bramblett, 1982; chimpanzees—Mendoza-Granados & Sommer, 1995). The exception to this pattern is a study of baboons by Owens (1975b), which showed that males engaged in more chasing play before 10 months of age, but females engaged in more such play after this age.

FUNCTIONS/EFFECTS

In contrast to the amount of debate regarding the possible functions and effects of play-fighting, there is much more consensus regarding locomotor play. Two major hypotheses have been offered: that these activities provide general exercise (Fagen, 1981), and that they train specific motor skills required in predator avoidance and/or pursuit (Byers, 1984; Ferron, 1975; Pellis & Pellis, 1983; Symons, 1978; Wilson & Kleiman, 1974). Besides these long-term effects, several researchers (i.e., Ferron, 1975; Stamps, 1995; Symons, 1978) suggested a short-term effect as well—to facilitate the learning of "environment–player relationships" (Symons, 1978, p. 85; see also Stamps, 1995). That is, by engaging in repeated locomotor play in a familiar environment, the young animal becomes familiar with the terrain and "lay of the land," thereby increasing the likelihood of predator avoidance and survival.

Evidence From Analyses of Play-Chasing Structure

Examination of the structure of play-chasing provides support for its hypothesized functions. For example, some of the behaviors in Table 6.1 are components of flight (e.g., run, leap, fast turns, fast stops, swinging), and others are behavior patterns that facilitate escape from a predator's grip (e.g., rear, heel kick, torso thrust, head shake, head jerk, neck twist, wallow, roll). The remaining behaviors (i.e., rear and pirouette, whirl, twirl, somersaults, acrobatics), although rarely displayed during flight, may help develop the strength, speed, and flexibility to perform some of the other behaviors.

Fewer of the behaviors in Table 6.1 seem similar to behavior during pursuit, although some are clearly relevant (e.g., run, leap, fast turns, fast stops, swinging). A comparison of behavior during stalking play in felids (e.g., Schaller, 1972; Schenkel, 1966; Wasser, 1978) shows many similarities with behavior during hunting.

Support for the escape function of play-chasing comes from a comparison of the orders in Table 6.1 on the solitary-play characteristics. The three prey orders (i.e., ungulates, rodents, and primates) show a greater variety of locomotor-rotational play behaviors than does the predator order (i.e., carnivores).

Symons (1978) argued that developmental changes in the nature of arboreal play-chasing support the escape function as well. Infant rhesus monkeys play-fight and play-chase in shrubs, where a fall is not dangerous, whereas juveniles play in thickly branched trees. Thus, play-chasing skills develop gradually, with chases among juveniles being more similar to escape patterns. These developmental changes, Symons asserted, are consistent with a skill development hypothesis regarding the acquisition of escape skills.

Finally, Byers (1977) provided some correlational data consistent with the motor-training hypothesis. During his observations of Siberian ibex kids, he found that solitary, locomotor-rotational behaviors (i.e., neck twist, heel kicks, jumps, leaps, and running) occurred on sloped surfaces, whereas social, play-fighting behaviors occurred on flat surfaces. Because play on sloped surfaces is more difficult and increases the chance of injury, one would expect all types of play to occur on the flat surfaces. Therefore, Byers (1977) argued that the long-term benefits of such behavior (presumably motor training) must outweigh the risks.

There has been less support for the general exercise function of play-chasing. As already discussed in the play-fighting chapter, in their reformulation of the motor-training hypothesis, Byers and Walker (1995) asserted that the endurance and strength effects of exercise were only short term and may not serve long-term functions. Moreover, several researchers have maintained that, with the exception of some ungulates, the duration of locomotor play bouts are too short to have an effect on the development of endurance (e.g., Hole, 1988; McDonald, 1977). Bekoff (1989), however, argued that this issue is far from being resolved.

7

Locomotor Play in Children

Young children love to run, and do so in many of their games, such as tag and hide-and-go-seek. Running also is found in other forms of outdoor play, even when children are not playing a specific "game"—be it running to the swingset, running to the corner, or skipping and hopping on the sidewalk. Much to the frustration of parents, children often bring these outside activities indoors and run up the stairs, chase one another around the house, or jump on a bed.

Descriptive studies of locomotor play in humans are hard to find. In a commentary on Pellegrini and Smith's (1998) recent review of this literature, Byers (1998) wrote: "I find it incredible that there are no complete accounts of the rates at which play motor acts are performed across human postnatal development, and I hope that the Pellegrini & Smith review will prompt a lot of this important descriptive work" (p. 599).

The few studies that have been done, however, suggest that human locomotor play is in many ways similar to that of other animals (e.g., Aldis, 1975; Blurton-Jones, 1967). As already mentioned, one of the limitations of the research on children's locomotor play is that chasing play is often coded as part of rough-and-tumble play, with no differentiation in the research between play-chasing and play-fighting.

As with other animals, it is likely that the development of running ability was selected for in human evolution because it facilitated both escape from predators and capture of prey. Descriptions of its structure (see later discussion in this chapter) indicate that locomotor play is made up of some of the same components as serious flight and pursuit. It is also likely that locomotor play has some of the same risks for humans as for other animals, including eliciting aggression, sustaining injury, alerting predators, becoming separated from the caregiver, and using up time and energy.

STRUCTURE

In one of the few published descriptions of children's chasing play, Aldis (1975) noted that chasing play in humans is usually of shorter duration than chasing play in other animals. He described it as follows:

> In brief chases, one child will typically make a brief play attack on another and then run away. The second child will then chase after him. Often, the second child will give up after running only a few yards, and the first child, when he sees that he is no longer being pursued, will then also stop. Sometimes the second child will quickly catch up to the first, either because he is a faster runner or because the first child handicapped himself and allowed the second to catch up. The second may then playfully attack the first and this, too, usually terminates the chase. Only rarely are the two children evenly matched runners, trying equally hard so that a chase continues for over 30 yards or so. Most chases end within 5 or 10 yards. (p. 207)

Although numerous researchers have studied the development of motor skills such as running, skipping, and jumping in structured test situations (e.g., Haywood, 1986), no descriptive study of solitary locomotor play in naturalistic play situations could be located.

Play Signals

Given the paucity of descriptive research on children's locomotor play, no study of the play signals used in the locomotor play context could be found.

General Characteristics

As was the case for other animals, the apparent goal of play-chasing appears to be pursuing (and sometimes touching or taging) the play partner (Aldis, 1975). Aldis (1975) also noted that play-chasing in humans sometimes involves role reversal and self-handicapping, where faster runners run at moderate speeds in order to allow themselves to be caught (see also Blurton-Jones, 1967).

Play Initiations

Boulton (1991b) identified two forms of play initiation in children: verbal and "hit-and-run." In the first, the child initiates a chase with a verbalization; in the second, the child (just as described for animals earlier) lightly hits another and runs, often encouraging a chase. Boulton made no mention of any preliminary behaviors that commonly precede the hit-and-run.

Locomotor-Rotational and Play-Chasing Behaviors

Presented in Table 7.1 is a list of the behaviors noted in the solitary locomotor play of humans. Comparison with Table 6.1 shows that two categories had to be added to include the behaviors observed in humans: slides and dances. The first of these, however, is not unique to humans: Several observers have noted sliding in otters (Fagen, 1981). Dancing, although a socially defined category of human behavior, has many characteristics similar to the locomotor-rotational play of other species, so much so that Fagen (1981) advocated using dance notation systems for describing the locomotor-rotational play of other animals.

Four categories in Table 6.1 are not found in Table 7.1: rearing, heel kicking, head shaking, and fast turns and stops. Given the few descriptive studies of solitary locomotor play in children, however, it would be premature to assume that some of these (e.g., head shaking and fast turns and

TABLE 7.1

Locomotor-Rotational Behavior During Solitary Locomotor Play: Humans

Behavior	*Blurton-Jones (1967)*	*Humphreys & Smith (1987)*	*Smith & Connolly (1972)*	*Smith & Connolly (1980)*
Run		+	+	+
Leap, jump	+	+	+	+
Rear, rear and pirouette				
Heel kick, torso twist				
Head shake, head jerk, neck twist				
Fast turns and stops				
Whirl, twirl, rotate		+		
Wallow, roll				+
Somersaults, acrobatics		+		
Swinging, hanging, climbing		+	+	+
Sliding			+	
Dancing			+	

stops) are not found in humans. Heel kicking and rearing, in contrast, appear to be more relevant for quadrupedal animals.

The behaviors observed in play-chasing (as described in the earlier passage from Aldis, 1975) are the same as identified in play-chasing of other animals. Again, given the paucity of descriptive research, information on the complexity of play-chases and paths is not available.

Termination

As described in the passage from Aldis (1975) quoted earlier, play chasing in humans is apparently terminated when one partner loses interest and stops chasing.

Differences Between Play-Chasing and Serious Flight and Pursuit

The differences between play-chasing and serious flight and pursuit that Aldis (1975) described for animals appear to be the same for humans. As Aldis (1975) argued for other animals, Blurton-Jones (1967) observed that in humans, compared to play-chases, serious flight was faster, characterized by a lack of play signals (i.e., smiling), and accompanied by a flight-related vocalization he characterized as a "tremulous scream" (p. 359). It is likely that the other characteristics identified by Aldis (1975)—direction of flight toward cover or a protector and protected hiding or avoidance upon termination—are found in humans as well.

Developmental Course

Only one developmental study of children's locomotor play could be located (Pellegrini, 1995—see later discussion). Clues about the developmental course of locomotor play come from studies of children's activity level and of the development of large-motor skills. Because activity level studies usually involve assessing the extent of children's large-motor activity in free-play contexts, it is likely—at least among children—that much of the behavior recorded in these studies is some form of locomotor play. Routh, Schroeder, and O'Tuama (1974) found that children's activity level (defined as the number of times children changed location in a laboratory playroom—i.e., entered different quadrants of the room as laid out in a grid on the floor) decreased from ages 3 to 8 years. Milich (1984) found a similar decline between ages 6 and 11 using both a similar grid-crossing assessment and readings off actometers (a mechanical device for assessing children's large motor activity). Children's activity level apparently evens out sometime during the school years. For example, Hovell, Bursick, Sharkey, and McClure (1978) found no age differences in observation of activity level between the third and sixth grades; and Pellegrini (1995) showed no change in chasing play between Grades 6 and 8. Pellegrini (1995) noted, however, that chasing is of low frequency by this age.

Although most studies of activity level show a developmental decline, it is likely that it follows an inverted U function, starting in infancy and peaking in the preschool years. This is partially supported by Routh, Walton, and Paden-Belkin's (1978) follow-up study that showed an increase in activity level between 10 months and 5 years. In an experiment designed to reconcile the inconsistencies between this and the earlier 1974 study, Routh and colleagues (1978—Experiment 2) deserved that the mother's presence affected the pattern of age differences in activity level.

Studies of the development of specific large-motor skills (such as running and jumping), although usually not analyzed in a play situation, show clear age-related changes that may be observed during play. For example, Haywood (1986) asserted that, with increasing age and experience, extraneous movements in children's running are reduced, and other motor changes occur to further increase the efficiency of movement. Specifically, lateral leg movements and outtoeing are minimized so that forces are restricted to the forward-backward plane, and the base of support is narrowed. Stride length increases, as does rotation of the trunk, and there is better coordination between movements of the legs and arms. The transition from beginning to advanced forms, however, is not smooth and continuous. Children often regress to less mature movements in the process of graduating to more efficient adult forms (Burnett & Johnson, 1971).

These developmental changes in the nature of motor development are similar to our earlier discussion of tool use—with increasing age and experience, children's behavior becomes more efficient, less variable, and more predictable. Because play often leads to *increased*, rather than decreased, variability in behavior, it would be interesting to compare developmental changes in the nature of motor development in play and nonplay settings.

Play Partners

Although no studies could be found that directly examined the partners children chose for chasing, it is likely that, as demonstrated in the large literature on children's play (e.g., Maccoby, 1988), children choose same-gender and most likely similar-age partners. However, as was the case with other animals, it is also possible that children may be less selective about chasing partners than play-fighting partners.

A hypothesis about activity level and play partners that was popular in the 1970s stipulated that the activity level of boys' play was more influenced by the presence of same-gender partners than was the activity level of the play of girls (e.g., Halverson & Waldrop, 1973; Maccoby & Jacklin, 1974). In a carefully designed study to test this hypothesis, Eaton and Keats (1982) found that, when playing in same-gender triads, both boys and girls showed higher levels of activity than when playing alone.

Gender Differences

Although research on gender differences in locomotor play is hard to find, considerable research has been conducted on gender differences in activity level. In a meta-analysis of 90 studies in this area, Eaton and Enns (1986) found significant differences in all age ranges studied, with an effect size about as large as gender differences in other areas, including spatial abilities, aggression, and mathematical reasoning. In all comparisons, boys showed higher activity levels than did girls. However, the size of the effect varied with several factors. Specifically, gender differences in activity level were largest at older ages, in unrestricted versus structured situations, and when peers were present. In a subsequent study, Eaton and Yu (1989) found that gender differences in activity level were not simply a product of maturational differences between boys and girls.

Studies published since the Eaton and Enns (1986) review support their conclusions. For example, both Klesges, Malott, Boschee, and Weber (1986) and McKenzie, Sallis, Nader, Broyles, and Nelson (1992) found that preschool boys were more active than preschool girls during naturalistic observations of children's behavior in the home and/or playground. In one of the few studies that included a specific measure of locomotor play, Pellegrini and Davis (1993) found that third-grade boys engaged in more "solitary exercise play" on the school playground than did third-grade girls.

Besides mean differences in the activity levels of boys and girls, several researchers have argued that there may be qualitative gender differences in activity patterns that may make adults more aversive to high-level activity in boys than in girls. Fagot and O'Brien (1994), for example, pointed out that even when few differences in mean activity level are found, the large-motor play of boys may be "more disruptive because it involves more noise, more aggression, or more conflict between participants. There is little doubt that, although dancing and wrestling may be rated equally on observational scales or by actometers, the typical adult observer is going to react quite differently to the two activities" (p. 396).

In their study of the effects of peer presence on activity level, Eaton and Keats (1982) provided evidence for this assertion. Although they found that boys and girls were equally likely to be affected by peer presence (see earlier discussion), the combination of main effects for gender and peer presence on their activity measures led to boys in triads being the most active of all subjects in their study: "It certainly seemed to us in the course of conducting the study that boys were markedly more active when in triads. Indeed, we feared for the fixtures in the playroom on a few occasions when male triads were present" (p. 539).

Studies of Non-Western Cultures

Ethnographic studies (many of the same studies considered in the section on play-fighting) provide evidence of locomotor play in a wide range of hunting-and-gathering and agricultural village cultures located in Africa (Lancy, 1977; Leacock, 1971; LeVine & LeVine, 1963; Turnbull, 1962), India (Minturn & Hitchcock, 1963), Sri Lanka (Prosser et al., 1986), Nepal (Anderson & Mitchell, 1984), Indonesia (Hogbin, 1946; Storey, 1977), Japan (Maretzki & Maretzki, 1963; Takeuchi, 1994), Mexico (Maccoby, Modiano, & Lander, 1964), the Philippines (Nydegger & Nydegger, 1963), South America (Miracle, 1977), and the South Pacific (Martini, 1994). Actions observed include chasing, dancing, climbing, and various arobatics. No studies including detailed descriptions of these behavior patterns could be found.

FUNCTIONS/EFFECTS

Although it is possible that locomotor play in humans evolved to serve some of the same functions hypothesized for its evolution in other animals (e.g., predator avoidance, learning environment–player relationships), the major functions that have been discussed for humans concern physical fitness. Again, most of these hypotheses come from discussions of activity level and exercise in general, not from discussions of locomotor play in particular. In a review of the literature on physical activity and fitness in children, Baranowski et al. (1992) listed eight components of physical fitness that might potentially be influenced by physical activity: agility, balance, body composition (i.e., amount of fat), cardiorespiratory endurance, flexibility, muscular endurance, muscular strength, and anaerobic power.

Despite the conclusion from experimental studies that increasing the physical activity of children appears to have physical-fitness benefits (e.g., lower blood pressure, higher HDL cholesterol, less obesity—for a review, see Krahenbuhl, Skinner, & Kohrt, 1985), there is controversy about whether these benefits lead to any long-term effects on health, and whether naturally occurring children's play provides sufficient activity for such training benefits to occur. In a typical study, Hovell et al. (1978) found that the level of physical activity showed by third- to sixth-grade children on the playground did not approach the level of intensity and duration necessary to achieve fitness benefits. Despite the nonaerobic nature of much of child activity, however, most children appear to be physically fit (Simons-Morton, O'Hara, Simons-Morton, & Parcel, 1987).

Probably the most studied physical-fitness correlate of children's activity level is children's obesity. Numerous studies have found an association between low levels of activity and children's obesity (e.g., Berkowitz, Argas, Korner, Kraemer, & Zeanah, 1985; Klesges et al., 1984, 1986; Waxman & Stunkard, 1980), although some have not (e.g., Klesges, Eck, Hanson, Haddock, & Klesges, 1990). Because of the correlational nature of these findings, it is impossible to determine if physical activity helps prevent obesity, or whether children who are obese have more difficulty being active. It is possible that both may be true. If low levels of activity do in fact contribute to childhood obesity, one possibility is that chronic inactivity caused by high levels of sedentary activities such as television viewing or videogame playing may account for the association. Several studies have documented associations among television viewing, inactivity, and obesity in children (e.g., Deitz & Gortmaker, 1985; Shimai, Yamada, Masuda, & Tada, 1993).

Although there is a shared consensus among many health professionals that children are less active and more obese at the present time than at any time in our recent history, the data for secular changes in obesity are more extensive than that for activity level (Baranowski et al., 1992). If these secular changes in obesity are due in part to reduced activity level, the increased incidence of obesity is likely to continue, given the trend toward the availability of more sedentary activities for the entertainment of children (e.g., VCRs, videogames).

Pellegrini and colleagues recently argued that physical-activity play may have some cognitive benefits as well (for a review, see Pellegrini & Smith, 1998). In this research, children's attention to academic tasks in the classroom was higher after recess than before. It is still not clear, Pelligrini et al. asserted, what processes may account for these effects: Was it the physical activity per se, a break from academic work, or some other factor? Clearly this is an important area for future research

LOCOMOTOR PLAY: SUMMARY AND CONCLUSIONS

Although considerably less research has been conducted on locomotor play than play-fighting, it appears that this behavior pattern in children may also have an evolutionary basis. Solitary locomotor-rotational and chasing play are seen in many species of animals, including humans in Western and non-Western cultures. These behavior patterns, which are made up of components of serious escape and pursuit behaviors, likely evolved to provide training in these behaviors so important to survival. Examination of the behaviors making up locomotor play in human and nonhuman species show a number of similarities (see Tables 6.1 and 7.1). Given the limited amount of research in this area, some of the differences between Tables 6.1 and 7.1 may reflect actual differences in the locomotor play of humans and other mammals, whereas others may not. The differences that will probably survive further investigation are those that are a consequence of the nature of bipedal versus quadrupedal locomotion (e.g., rearing and heel kicks), and those that are a result of the cultural ritualization of locomotor movements found in children's dancing.

As was the case for play-fighting, the ways in which chase and pursuit behaviors differ in the locomotor play and "serious" contexts are similar across human and nonhuman species; among them are role reversal, self-handicapping, and the lack of warning calls. Although not yet studied in humans, several other characteristics of animal play (e.g., circling back to chasing partner, lack of hiding or freezing) are likely to be characteristic of human play as well. Cross-species similarities are found in the initiation and termination of chasing play, as well as in its developmental course.

Most species (humans included) initiate play with physical contact and a run (although this contact often takes the form of biting in other mammals), and play is terminated when one partner loses interest and stops. The same inverted U developmental course has been found for both humans and other mammals.

The limited amount of research on humans makes it difficult to draw any firm conclusions about play partners and gender differences. Although boys are more active in their play than are girls, it is not clear what forms of locomotor activity account for these differences. Also, it is not clear whether the tendency for some other animals to be less selective about their play-chasing partners than their play-fighting partners is found in humans as well.

Most discussions about functions and effects of locomotor play for humans focus primarily on the exercise hypothesis (although cognitive benefits may accrue as well). The amount of locomotor play may, in part, account for the development of obesity in children, although there is still much research that needs to be conducted on this front.

Given the paucity of research in this area, future studies of locomotor play need to provide information on the following: play-chasing initiations and signals; cross-species differences in the sequencing, timing, and qualitative aspects of locomotor behavior during chasing and solitary play; differences between locomotor play and serious chase and pursuit; the developmental course of play and nonplay locomotor patterns; partner choices; and gender differences. Research on the functions and effects of play need to investigate more fully the relation between various forms of locomotor activity (play and nonplay), diet, physical fitness, obesity, metabolic processes, and children's involvement in other activities (e.g., television viewing, videogames). As was the case for play-fighting, it is likely that these relationships vary as a function of child gender, temperament, and age. Thus, multimethod, longitudinal research is in order.

III

Social Object, Social Pretend, and Parent–Child Play

Although the vast majority of social play interactions in nonhuman animals take the forms of play-fighting and locomotor play, in his natural history Fagen (1981) identifies several less common types. In rodents and ungulates these include playful mounting and "riding," playful "crawling on," and playful nosing and sniffing. Some canids and primates have been observed to engage in social play with objects. Their actions include playful struggles over objects and the ensuing chases (canids and primates—Aldis, 1975; Pedersen et al., 1990; Starin, 1990; van Lawick-Goodall, 1968), object tug-of-war (group-hunting canids—Aldis, 1975; Biben, 1982a, 1983; Fox, 1969; Pedersen et al.,1990); and playfully waving or shaking objects at another (primates—Starin, 1990; van Lawick-Goodall, 1968).

Although human children engage in play-fighting, play-chasing, and some of the rarer forms of social play seen in other animals (with the exception of mounting and sniffing), much of children's social play is devoted to activities that, at least on the surface, share few commonalities with the behavior of other species. The sociodramatic play and rule-governed games of early and middle childhood reflect children's developing cognitive and communicative abilities, and as such appear uniquely human. Moreover, although parent–offspring play is observed in many species (see introduction to chap. 9), the forms observed in humans differ considerably from those observed in other animals. These differences also appear to be due to the cognitive and communicative competence of human parents as play partners. Therefore, the next two chapters focus primarily on the play of humans, although occasional references to other animals are made when appropriate.

8

Social Object and Sociodramatic Play in Children

The major nonphysical forms of social play observed in children (e.g., sociodramatic play, games with rules, language play) are a consequence of our well-developed symbolic and communicative abilities (especially language abilities), and as such appear to be uniquely human. Given the recency of the evolution of language in the genus *Homo* (most scholars estimate that conventional language emerged about .25 million years ago—see Bickerton, 1990; Dunbar, 1993; Liberman, 1991), the forms of play dependent on language have a much shorter evolutionary history than other types, such as play-fighting and locomotor play. This does not mean, however, that these forms have not been influenced by natural selection. Language was probably incorporated into children's play as language evolved, and may even have influenced the evolution of play. Moreover, given the importance of language in enculturation, play forms that incorporated language may have had a significant role in cultural evolution as well. For example, anthropologists have illustrated how various culturally transmitted children's games serve important enculturation functions (e.g., Sutton-Smith, 1977).

If play is seen as structurally modified functional behaviors taken out of context, then much of sociodramatic play certainly meets this definition. As is seen in the discussion that follows, much of role-taking play involves children acting out adult roles in a pretend context. Given the altricial nature of humans, young children in numerous cultures spend limited time in functional contexts, and considerable time involved in exploration and play. Because this is a crucial time for enculturation, role playing may be one way in which young children develop the skills necessary for their integration into the culture, and therefore may indirectly influence their subsequent reproductive success.

When compared to play-fighting and locomotor play, nonphysical social play in humans is much less risky. Because there is no fighting involved and it may involve limited energy expenditure, nonphysical social play may be much less costly than other forms of play. It does not require that children venture far from their parents, nor is the risk of accidental injury very high. Although there may be social risks of such play (e.g., certain forms of play may distress playmates or caregivers), the major risk seems to be loss of time that could be devoted to "more productive" activities. There apparently is, therefore, no strong selection pressure against it, making only minor benefits necessary for its survival in the behavioral repertoire—through processes of either natural or cultural selection.

The study of social play in young children has a long history in child development research. With the establishment of several child development institutes in the 1920s came a flurry of studies of young children's play in institutional and preschool settings. These studies explored many social play and peer relations topics that did not become the focus of empirical study again until the 1970s and 1980s. Topics included studies of interactions between infants or toddlers (Bott, 1928; Bridges, 1933; Maudry & Nekula, 1939; Mengert, 1931), social participation (Parten, 1932, 1933), play activities (Manwell & Mengert, 1934), gender differences in play (Hattwick, 1937), conflict (Dawe, 1934), dominance relations (Hanfmann, 1935), peer popularity (Koch, 1933), friendship (Green, 1933), and social skills (Mallay, 1935). With the exception of a few studies of sociometric status (e.g., Bonney, 1955; Bonney & Powell, 1953; Dunnington, 1957) and ascendance behavior (a concept related to dominance—see Anderson, 1940; Mummery, 1947; Stott & Ball, 1957), few studies of social play were published in the 1940s and 1950s. Instead, most play studies during this period were neopsychoanalytic investigations of solitary doll play (for a review, see Levin & Wardwell, 1962).

With a decline in interest in psychoanalytic perspectives and the influence of observational learning theory in the 1960s (e.g., Bandura & Walters, 1963) came an interest in the effects of nonparent models and reinforcement agents on early child development. In the 1960s, Hartup and colleagues conducted pioneering research in this area (Charlesworth & Hartup, 1967; Hartup, Glazer, & Charlesworth, 1967; Hartup & Keller, 1960). The ethological study of early peer relations began in the 1960s as well, with studies by Blurton-Jones (1967), Grant (1969), McGrew (1972), and Smith and Connolly (1972). Interest in play as a topic began to reemerge with several classic studies published during this time period (Lieberman, 1965; Smilansky, 1968; Sutton-Smith, Rosenberg, & Morgan, 1963).

Research on children's social play increased significantly during the 1970s. Many of the topics addressed in the 1930s were rediscovered and became the focus of intensive empirical investigation. These included studies of interactions between infants and toddlers (e.g., Becker, 1977; Eckerman, Whatley, & Kutz, 1975; Mueller & Lucas, 1975), social participation (e.g., Barnes, 1971; Rubin, Maioni, & Hornung, 1976; Smith, 1978), gender differences in play (e.g., Fagot, 1977; Langlois, Gottfried, & Seay, 1973; Lever, 1978), and children's dominance relations (e.g., Abramovitch, 1976; Hold, 1976; Sluckin & Smith, 1977; Strayer & Strayer, 1976). Also during the 1970s, researchers conducted some of the first observational studies of young children's prosocial behavior (e.g., Eisenberg-Berg & Hand, 1979; Friedrich & Stein, 1973; Yarrow & Waxler, 1976) and sibling interactions (e.g., Abramovitch, Corter, & Lando, 1979; Lamb, 1978a, 1978b). The late 1960s and early 1970s were a time of considerable optimism among those dedicated to developing interventions for socioeconomically disadvantaged children, and during this period many play researchers advocated the use of sociodramatic play-training methods (e.g., Freyberg, 1973; Lovinger, 1974; Rosen, 1974; Saltz & Johnson, 1974).

Although the enthusiasm for play as a research topic diminished somewhat by the mid-1980s—possibly in response to several methodological critiques of the area (e.g., Caro, 1980; Christie & Johnsen, 1985; McLoyd, 1982; Simon & Smith, 1983)—research on peer and sibling relations continued into the 1980s and 1990s, especially in the areas of infant/toddler peer interactions (e.g., Camaioni, Baumgartner, & Perucchini, 1991; Hay, Nash, & Pedersen, 1983; Howes & Matheson, 1992; Ross, 1982; Vandell, Wilson, & Buchanan, 1980) and sibling interactions (e.g., Dunn & Kendrick, 1981; Dunn, Slomkowski, & Beardsall, 1994; Stoneman, Brody, & MacKinnon, 1984). The 1980s and 1990s also witnessed a renewed interest in three topics studied in the 1930s: children's conflict (e.g., Hartup, Laursen, Stewart, & Eastenson, 1988; Hay & Ross, 1982; Kolominskii & Zhiznevskii, 1992), peer popularity (e.g., Coie & Dodge, 1983; Ladd, 1983; Putallaz & Wasserman, 1989; Rubin & Daniels-Beirness, 1983), and children's friendships (e.g., Gottman, 1983; Howes, Droege, & Matheson, 1994; Werebe & Baudonniere, 1991). Finally, a committed group of researchers has continued to study children's social play, with a particular emphasis on the development of children's social fantasy play (e.g., Black, 1992; Connolly, Doyle, & Reznick, 1988; Farver, 1992; Goncu & Kessel, 1988; Howes & Farver, 1987; Lyytinen, 1991a; Pellegrini, 1985).

Given that the forms of social play during the infancy and toddler period differ considerably from those at later ages, this chapter is divided into two parts: the first on infancy and toddlerhood (0 to 2 years) and the second on early childhood (3 to 7 years). As with the previous chapters, research on the structure of play, as well as research on gender differences and play partners, is reviewed before the research on the functions and effects of social play. Following this is a review of the literature on play in non-Western, hunting-and-gathering, and village cultures. Because this volume is concerned with the play of *young* children, we deal only briefly with children's formal games. These activities usually emerge in middle childhood, and have been the focus of numerous volumes (e.g., Opie & Opie, 1969; Sutton-Smith, 1973). Although children engage in social play with their parents as well as with other children, social play with parents is covered in the next chapter.

One difficulty in writing this chapter was considering which topics to cover. Traditionally, play researchers have not focused on topics such as aggression, conflict resolution, prosocial behavior, social skills, and dominance hierarchies in their reviews, because these are considered aspects of social *interaction* between children, not aspects of social *play*. Instead, play researchers have looked primarily at children's interactions during more prototypical play forms, such as sociodramatic play. However, a problem with excluding the former topics is that behaviors like aggression, conflict, prosocial behavior, and so on, usually occur in a free-play context, making their exclusion somewhat arbitrary and artificial. Thus, for example, play researchers may study how children resolve conflicts by looking at who serves in what pretend role, but might not consider an ensuing object struggle as a relevant object of study. Obviously, different researchers are interested in studying different aspects of children's interactions with their peers, and researchers need to pick and choose which aspects of social interaction on which they will focus. However, when social convention (and research tradition) leads to the parsing of ongoing social interactions in free-play contexts into mutually exclusive categories of behavior (of which only some are relevant to play researchers), important interconnections between the areas are missed. Therefore, for this chapter, a decision was made to define social play very broadly, and if the researchers observed the unconstrained social interactions of young children in a free-play *setting*, the behavior observed was considered to be part of social play. Thus, the context of the social interactions was the primary factor considered for inclusion in this review. Studies of peer and sibling interactions during academic tasks, functional activities (e.g., chores), and structured experimental tasks designed by experimenters (e.g., referential communication tasks) were not included. Hence, according to the view of many play researchers, much of the research reviewed in this chapter is on children's social *interactions* in free-play settings, not on play per se.

STRUCTURE

Infancy and Toddlerhood

Beginning in the mid-1970s, several descriptive studies on the early social interactions of infants and toddlers were conducted to explore the development of early social skills and social responsiveness. In the typical study, pairs of infants or toddlers were brought together in a laboratory or home setting and videotaped while playing with a set of toys. Mothers were usually present, but instructed to not intervene in the interactions. Examples of such interactions include:

> Object exchange (OE): one partner offers/gives and the other receives/takes an object.... Attention request (ATR): one partner requests attention, verbally or non-verbally, from the other who, in turn, watches and eventually labels/comments.... Point and Name: one member of the dyad points to objects/events, waiting for a vocalization by the other in the form of naming the object/event. Pointing is often accompanied by a questioning vocalization. (Camaioni et al., 1991, p. 20)

> Curtain running: The children are engaged in mutual curtain running, where each child both (a) runs in turn through the curtain, and (b) acknowledges the other's runs by stopping and watching and/or by positive affect.... Peek-a-boo: The children are engaged in peek-a-boo, where one child hides and appears suddenly, and the other acknowledges this by smiling or laughing. The acknowledging child may or may not himself hide. (Brenner & Mueller, 1982, p. 384)

Types of Play. Although considerable research has gone into describing the nature of peer interactions in infants and toddlers, it is important to point out that the frequency of social interactions between peers in this age range is low (Bronson, 1981). Although toddlers have been observed to engage in extended social interactions, the "...toddlers' schemas for dealing with agemates is markedly fragile. Just the right peer and just the right circumstances may be sufficient to permit sustained interaction. However, under any less-than-optimal conditions it most often fails to support engagement in reciprocal social exchanges" (Bronson, 1981, p. 111). Social skills develop gradually during the second year and, as pointed out by Bronson (1981), the toddler's most active play experiences are with older individuals (siblings, parents) or solitary play with objects in the physical environment.

Researchers who study early peer interactions differentiate between two kinds of behavior during play: socially directed behaviors and social interactions. Socially directed behaviors (SDBs—Mueller & Lucas, 1975) are facial expressions, vocalizations, gestures, or physical contact behaviors that co-occur with a look toward the social partner. (The individual SDBs that have been identified are described later in this chapter.) Social interactions refer to turn-taking sequences of SDBs between partners. The shortest social interaction is when the SDB of one partner is followed by the SDB of another and the sequence terminates. Social interaction chains are therefore two turns in length or longer. Social interactions need not share a common focus to qualify as social interactions. Thus, if one infant looks and offers a toy and the other looks and vocalizes, this is as much an interaction as if one looks and offers a toy and the other takes it.

Various systems have been developed to classify the kinds of social interactions seen in infants and toddlers, as well as for describing the content of these interactions. A common approach is to distinguish between interactions in which one individual imitates another and those in which the two individuals have complementary roles (e.g., Camaioni et al., 1991; Eckerman, Davis, & Didow, 1989). Goldman and Ross (1978) referred to these extended social interactions as early peer "games." Unlike the formal rule-governed games of early childhood (e.g., hide-and-go-seek, tag), toddler games are extended sequences of social behaviors in which the partners have imitative, complementary, or reciprocal roles. According to Goldman and Ross (1978), in complementary games one child completes the action begun by another. Reciprocal games involve complementary interactions in which the children switch or reverse roles.

Researchers have conducted observational studies of peer interaction in order to describe the content of these interactions. The results of three of these studies are summarized in Table 8.1. As can be seen in the table, social interactions fall into six general categories: "agonistic" interactions, locomotor play, imitation, social object use, language interactions, and conventional games. One type of social play not studied by these researchers is social pretend behavior. Howes and colleagues (Howes, 1985; Howes, Unger, & Seidner, 1989) found that social pretend behavior emerges at later ages than does solitary pretend, and that children reach higher levels of pretend play complexity in the solitary context before they show it in social pretend play. In one cross-sectional study, Howes (1985) found that 16- and 17-month-olds showed primarily solitary pretend play (only 1 out of 9 subjects showed social pretend), whereas by 21 to 23 months of age , 8 out of 14 subjects showed social pretend. By 32 to 33 months of age, all of the

TABLE 8.1

Content of Social Interactions: Infants and Toddlers

	Eckerman et al. (1989)	*Brenner & Mueller (1982)*	*Camaioni et al. (1991)*
"Agonistic" Interactions			
Threat and retreat	+		+
Hit each other	+	+	
Rough and tumble play		+	
Locomotor play			
Lead–follow	+		
Run–chase		+	
Imitation			
Facial expression			+
Vocal		+	+
Motor	+	+	+
Social object use			
Object possession struggles		+	+
Object exchanges	+	+	+
Joint use of objects			+
Language interactions			
Verbal directive—response	+		+
Shared reference (point and name)		+	+
Conversation		+	+
Language games			+
Conventional Games			
Peek-A-Boo	+	+	+
Pat-A-Cake			+

subjects showed social pretend. A content analysis of the types of social pretend play exhibited in a study of 14- to 38-month-olds showed that pretend play fell into the same general categories as did the solitary pretend play discussed in chapter 2: preparing and eating food, grooming, caretaking, talking on the telephone; and "animating" inanimate objects.

Researchers have observed that a large proportion of peer interactions in the infancy and toddler periods involves the use of objects. For example,

Mueller and Brenner (1977) found that at 12 months, 92% of all social interactions were embedded in object-focused contacts, and that at all of the ages they studied (12, 14, 16, 17, 18, 21, and 23 months), no more than 17% of the social interactions were not mediated by the use of objects. Although Mueller and Lucas (1975) originally hypothesized that the capacity for social interaction develops as a consequence of object play (i.e., social interactions first appear in the context of parallel play with common toys), studies of younger infants suggest that this is not the case. Jacobson (1981), in a longitudinal study of 10- to 14-month-olds, found that at 10 months the greatest number of social interactions occurred in the absence of toys. He also found that long social interactions emerged simultaneously in the object- and nonobject-centered contexts between 10 and 12 month of age. Moreover, both Eckerman and Whatley (1977) and Vandell, Wilson, and Buchanan (1980) found that infants between the ages of 6 and 10 months engaged in more complex social interactions in the absence of toys than in their presence. In these studies, however, there was an increase in the social use of toys by the end of the first year. Jacobson (1981) concluded that although "peer social interaction does not originate in an object-centered context, object-centered play may enhance its development during the second year" (p. 618).

Social Play Behaviors. The socially directed behaviors (SDBs) identified in human infants and toddlers—at least on the surface—appear to differ little from those of the other primates, especially the apes. Three observational studies with fairly differentiated coding systems are summarized in Table 8.2. Early social behaviors included looking, facial expressions, gestures, vocalizations, and motor behaviors. Both affiliative and agonistic behaviors were observed.

A comparison of the behaviors in Table 8.2 with the information on primates in Tables 4.2, 4.4, and 6.1 (the tables summarizing behavior during play-fighting and locomotor play in other animals) and the information on primate social object play in the introduction to this chapter reveals that few of these behaviors are unique to early peer interactions in humans. Specifically, if the primate play-face is seen as corresponding to the human smile, then all of the socially directed behaviors in Table 8.2 (with the exception of pointing) are found in the literature on animal play reviewed earlier. However, several recent studies (e.g., Gomez, 1996; Leavens & Hopkins, 1998) describe instances of pointing in chimpanzees.

Despite the similarities in the individual behaviors identified, the peer interactions of human infants and toddlers do not necessarily differ little

TABLE 8.2

Socially Directed Behaviors: Infants and Toddlers

	Maudry & Nekula (1939)	*Holmberg (1980)*	*Vandell et al. (1980)*
Affiliation			
Attention (look at, watch)	+	+	+
Signal			
Facial expression			
Smile/laugh	+	+	+
Gestures			
Point			+
Show object	+	+	+
Vocalizations	+	+	+
Approach	+		+
Contact			
Touch/grasp	+	+	+
Prosocial behavior			
Give object	+	+	+
Terminate			
Withdraw	+		
Agonism			
Attack			
Bite	+		
Hit	+	+	+
Push	+		+
Competition			
Attempt to take object	+	+	
Terminate			
Submission			
Cry	+	+	+

from those of other primates. Major differences between human and other primates emerge in the second year. For example, the play behaviors of humans in this time period are sometimes part of larger, rule- and/or language-governed games, such as pat-a-cake or peek-a-boo (see Table 8.1). A comparison of Table 8.1 with the animal literature helps identify where some of these differences lie. Other primates have been observed to engage in many of the social interactions listed in the table, including agonistic interactions (see chap. 3), locomotor play (see chap. 4), and the three forms of social object use listed in the table (object possession struggles, object exchanges, and joint use of objects—see introduction to this chapter). Other primates, especially the apes, engage in imitation as well (see Russon, Bard, & Parker, 1996). The differences lie primarily in the areas of social pretend, language interactions, conventional games, and likely differences in the nature of joint object use. Moreover, numerous differences in the qualitative nature of the behaviors and interactions listed in Tables 8.1 and 8.2, as well as in their sequencing and timing, are likely to exist. It is unfortunate that no comparative studies involving humans and other primates could be found. Clearly, this is an area for future research.

Play Initiation, Maintenance, and Termination. Research on the initiation, maintenance, and termination of early peer interactions is limited. This differs considerably from the research on topics such as play-fighting, for example, in which considerable attention has been devoted to identifying the behaviors that serve to signal that what is to follow is play. One reason for this difference, of course, is that such signals are not as necessary in nonphysical social play. Given that the child is not engaging in behaviors that may be construed as physically threatening to the recipient, the use of ritualized play initiations and signals may not be required. Moreover, social interactions between infants and toddlers are often so short that the study of maintenance and termination processes is meaningless. However, a limited amount of information is available.

Peer interactions in infants and toddlers are apparently initiated with a wide range of behaviors. Any of the socially directed behaviors listed in Table 8.2 (with the obvious exception of withdraw) apparently can serve as an initiator, although some are more successful in achieving interaction than are others. In their study of 7- to 31-month-old infants, Finkelstein, Dent, Gallagher, and Ramey (1977) found that the behaviors most successful in eliciting peer interaction were object related (i.e., give, take, or show object). The probability of success following these behaviors was $p = .46$. Almost as high was the probability of success following talking or babbling, $p = .38$. Touching was the least successful initiator, with the probability of success at $p = .01$. Comparisons of infants (7 to 14 months) to toddlers (15 to 31 months) in the Finklestein et al. sample showed that only initiations using talking increased with age of the child. Becker (1977), in a study of 9-month-olds, found that the probability of an infant behavior successfully eliciting a response varied as a complexity of the behavior. Specifically, the probability of success was $p = .24$ for looks only, $p = .41$ for looks plus an additional behavior, $p = .59$ looks plus two additional behaviors, and $p = .77$ for looks plus three additional behaviors.

In a detailed observational study of 26- to 36-month-olds, Wellman and Lempers (1977) demonstrated that young children do not initiate interactions haphazardly, but time them in such a way as to facilitate success. Specifically, each of the following was true for 90% or more of all child communicative initiations: the child was less than 5 feet away from the receiver and the referent, there was an absence of a visual obstacle between the receiver and the referent, the receiver was not involved in interaction with another person, the communicator had played with or attended to the referent in the preceding 90 seconds, and the communicator was looking at the referent.

Howes (1985) examined strategies that children used to integrate pretense into social play among 16- to 33-month-olds. Verbal recruitment was more effective than nonverbal recruitment in successfully eliciting pretend play in the partner.

Research on factors predicting the termination of peer interaction could not be found. Likely terminators include one or both children are distracted, a child does not achieve the outcome that he or she desires, a conflict escalates to the point that one or the other child leaves, a caregiver intervenes, and so on.

Developmental Course. Researchers have identified many developmental differences between 6 and 30 months in the behaviors and interactions listed in Tables 8.1 and 8.2. In general, infants and toddlers become more active social partners with increasing child age, their social behaviors become more complex, and social interactions become longer and increasingly connected. Evidence for growing sociability with age include increases in the frequency of socially directed behaviors (Bridges, 1933; Finklestein et al., 1977; Holmberg, 1980; Maudry & Nekula, 1939; Mueller & Brenner, 1977), as well as in the frequency of verbalizations (Eckerman & Didow, 1996; Finklestein et al., 1977), approaches (Vandell et al., 1980), and object-related social acts (Eckerman et al., 1975; Maudry & Nekula, 1939; Vandell et al., 1980). Behaviors that have been shown to decrease between 6 and 18 months include crying, touching, and large body actions (such as flapping arms or clapping—Brooks-Gunn & Lewis, 1979; Vandell et al., 1980).

Increases in the complexity of social behavior are evidenced by increases in the frequency of coordinated socially directed behaviors (i.e., two or more socially directed behaviors that occur simultaneously—Brownell, 1990; Mueller & Brenner, 1977; Vandell et al., 1980), and changes in the nature of vocalizations—that is, decreases in vocalizations without verbal content (i.e., expressive sounds, unintelligible utterances) and increases in verbalizations that mark the child's activity, regulate the behavior of the peer, or are unrelated to the ongoing activity (e.g., "I like pizza"—Eckerman & Didow, 1996).

The increasing connectedness of social behavior during this period is evidenced by increases in the frequency of contingent interactions (Finklestein et al., 1977; Gunnar, Senior, & Hartup, 1984; Mueller & Brenner, 1977; Ross & Lollis, 1989), and in the frequency of games (Camaioni et al., 1991; Maudry & Nekula, 1939; Ross & Lollis, 1989), thematically related interactions (Eckerman et al., 1989), social pretend play (Howes, 1985), and the frequency and diversity of shared meanings (Brenner & Mueller, 1982). Regarding age differences in the nature of shared meanings, Brenner and Mueller (1982; see Table 8.1) found that: "At 12 months only 'object exchange' and 'object possession struggle' occurred. At 15 months, 'motor copy,' 'vocal copy', and 'vocal prosocial' were also seen. Only at 18 months were all 10 meanings found in PG1 [play group 1] observed" (p. 387).

Although studies have consistently found a developmental increase in the frequency of games, complementary interactions do not show age-related changes (Camaioni et al., 1991; Eckerman et al., 1989), and the data for imitative interactions are inconsistent. Camaioni et al. (1991) noted that imitative interactions decreased in frequency between 18 and 36 months, whereas Eckerman et al. (1989) discovered that imitative relations increased between 16 and 28 months. Although the possible reasons for this inconsistency are many (e.g., sample characteristics, objects available, settings, persons present), a likely possibility concerns the way in which *imitative* was defined. In the Camaioni et al. (1991) study, it referred to simple repetition of an observed vocal, expressive, or motor behavior, whereas in the Eckerman et al. (1989) study, it referred to "performing the same action as peer, one or more of the distinctive elements of peer's actions...a thematic variant of peer's actions...or an elaboration of peer's actions" (p. 444). The inclusion of elaboration in Eckerman et al's. (1989) definition may have accounted for their increased frequency with child age.

Caplan, Vespo, Pedersen, and Hay (1991) found similar age-related differences in a study of children's free-play conflicts during this period.

Although the frequency of conflicts was no different for 12- and 24-month-olds, the nature of the conflicts was. Compared to the conflicts of 12-month-olds, conflicts among 24-month-olds were longer, had more turns, were more likely to begin with distal behaviors, and were more likely to contain verbal objections. The 24-month-olds were less likely to use instrumental force and more likely to resolve conflicts with prosocial overtones.

Based on their work and reviews of the literature, both Hay (1985) and Howes (1987) described the developmental course of peer interaction during the infancy and toddler period. Howes (1987) presented a four-stage model that summarizes much of the literature reviewed earlier in this chapter. Three of the stages cover the infancy and toddler periods. The first stage, "recognition of the peer as a social partner," covers the first 12 months of life. Children during this stage direct individual social behaviors toward peers such as those listed in Table 8.2. Hay (1985) noted that children show some evidence of contingent interactions during this period; both Jacobson (1981) and Vandell et al. (1980) reported that social sequences occur during this period, but the majority of interactions are two turns in length. In Howes' (1987) second stage, "complementary and reciprocal play structures" (13 to 24 months), children engage in chains of social interaction that involve exchanges, turn taking, and/or complementary activities such as run and chase, hide and seek, or offer and receive. Hay (1985) stated that this is a period of cooperation, conflict (over objects and space), modeling, and verbal communication. In general, it is a period during which the behavior of the partner becomes "a distinct context for the other's activities" (Hay, 1985). Jacobson (1981) found that the average sequence at 14½ months was three turns or longer. In Howes' (1987) third stage, "communication of meaning" (25 to 36 months), children begin to add and communicate content to social interaction. The most common form of content is nonliteral pretend meaning, but other types are observed as well. Children in this age period begin to engage in extended games involving communicative behaviors that signal when and how to play, when to switch roles, and how to communicate knowledge.

Gender Differences. Gender differences in the nature of peer interaction during the first 2 years of life are rarely reported. Although many papers do not address the issue of gender differences, those that do usually report that none were found (e.g., Brownell, 1990; Eckerman et al., 1975, 1989; Eckerman & Didow, 1996; Hay et al., 1983; Hay & Ross, 1982; Holmberg, 1980; Howes et al., 1989; Ross & Lollis, 1989; Vandell et al.,

1980). Of all the papers reviewed for this text, only two presented significant ($p < .05$) gender differences in peer interaction during the first 2 years of life. Jacobson (1981) found that 10- to 14-month-old girls showed a higher frequency of socially directed behaviors than did boys, and Caplan et al. (1991) found that triads of 1- and 2-year-olds made up of two girls and one boy engaged in more conflict episodes than did triads made up of two boys and one girl. Because others have not replicated these results, it is likely that they were unique to the samples studied.

Play Partners. Although researchers have not documented differences in the peer interactions of infants and toddlers as a function of the gender of the play partner, age of the partner appears to make a difference. Rothstein-Fisch and Howes (1988) found that toddlers (12 to 23 months) in family day-care homes preferred 2-year-olds as play partners over same-age, younger, or preschool-age children. Compared to play with other partners, when toddlers played with 2-year-olds their play was more likely to include imitation and language. Toddlers spent the most time simply watching preschool children, and when next to them were most likely to be involved in solitary play. Brownell (1990) also found that toddlers who interacted with a slightly older partner showed increases in the complexity of their social behavior. When 18- and 24-month-olds interacted in same- or mixed-age dyads, the children, regardless of age, showed more positive affect, imitation, and distal signaling toward a 24-month-old partner than toward an 18-month-old partner. In another study (Howes & Farver, 1987), in which 2-year-olds were observed on separate occasions with 5-year-olds and age-mates, the 2-year-olds used more verbal meta-communication, teaching behaviors, and attempts to direct play with same-age peers than with older children. With the 5-year-olds they were more likely to engage in cooperative, social pretend play.

Familiarity of the play partner appears to make a difference as well. Two studies that manipulated the degree of partner familiarity (Becker, 1977; Mueller & Brenner, 1977) found that children with familiar partners engaged in more complex patterns of social interaction than did age-matched controls. Toddlers interacting with familiar peers engaged in a greater number of total interactions, coordinated socially directed behaviors, and contingent social interactions than did those interacting with strangers. Moreover, in a social relations analysis (Kenny & La Voie, 1984) of toddler interactions, Ross and Lollis (1989) found that patterns of social interaction unique to individual relationships emerged over successive play sessions.

Early Childhood

Children's play has been studied most extensively in the early childhood period. Most of this research, however, has been conducted in preschool settings, with a few studies conducted in laboratory or home settings. In the typical study, one or two observers visit a preschool classroom and, over the course of several weeks or months, collect observational data on the nature of children's interactions during free-play periods. Observations are conducted in the classroom, on the playground, or both. As such, the types of interactions observed are largely influenced by the materials available in the typical preschool classroom and the behavior of the teacher. Therefore, most of what we know about play in early childhood comes from studies of how children play with materials deemed educationally stimulating and age appropriate. Little information in currently available on how children with play with nontoy objects (e.g., sticks, mud, household objects) or toys thought to have little educational value (e.g., action figures, toy weapons). We also know little about the nature of play in the home (i.e., inside and outside play with friends and siblings) or in settings with low levels of adult supervision (e.g., play involving dangerous activities, excessive aggression or teasing, unacceptable language; see Fine, 1988, for a discussion of four types of "dirty play" in preadolescent boys—aggressive pranks, vandalism, sexual play, and racist invective).

Types of Play. The major types of play seen in preschool settings were identified in classic studies by Parten (1932) and Smilansky (1968). The categories developed for these studies (and often a combination of the two) are still popular in observational studies of children's play (e.g., Lyytinen, 1991a, 1991b; Saracho, 1993). Parten (1932), in an observational study of 2- to 4-year-olds in a preschool setting, developed a method for assessing the level of children's social participation during free play. She defined six levels of increasing social involvement:

> *Unoccupied behavior*—The child apparently is not playing....
> *Solitary independent play*—The child plays alone and independently with toys that are different from those used by the children within speaking distance....
> *Onlooker*—The child spends most of his time watching the other children play....
> *Parallel activity*—The child plays independently, but the activity he chooses naturally brings him among other children. He plays with toys that are like

> those which the children around him are using, but he plays with the toys as he sees fit, and does not try to influence or modify the activity of the children near him....
>
> *Associative play*—The child plays with other children.... All the members engage in similar if not identical activity; there is no division of labor, and no organization of the activity of several individuals around any material goal or product....
>
> *Cooperative or organized supplementary play*—The child plays in a group that is organized for the purpose of making some material product, or of striving to attain some competitive goal, or of dramatizing situations of adult and group life, or of playing formal games.... (Parten, 1932, pp. 249–251).

Smilansky (1968), as part of a play-training study for low-income preschoolers, proposed the following four-stage classification scheme for assessing the cognitive complexity of children's play behavior, based on the work of Piaget (1962):

> *Functional play.* At first the play of a child consists of simple muscular activities based on his need to activate his physical organism.... He repeats his actions and manipulations, imitates himself, tries new actions, imitates them, repeats them and so on....
>
> *Constructive play.* This form of play introduces the child to creative activity and thereby to the personal joy of creation.... From the sporadic handling of sand or bricks the child moves to building something from these materials that will remain even after he has finished playing....
>
> *Dramatic play.* The next stage in play development is symbolic play, which appears in the dramatic play of the child. This relationship allows him to acknowledge the objective world situation and, at the same time, to substitute an imaginary situation that satisfies his personal wishes and needs....
>
> *Games with rules.* This form of play, according to theoreticians, is the highest stage reached in play development. Here the child had to accept pre-arranged rules and adjust to them...This is the principle form of play that tends to accompany us into our adult lives.... (Smilansky, 1968, pp. 5–7).

Given that these two play schemes correspond to different aspects of children's play, Rubin (1977) developed a method for nesting one within the other for a more complete description of children's play. Therefore, with the exception of onlooker and unoccupied behavior, the remaining four forms of social participation (Parten, 1932) were divided into functional, constructive, and dramatic play (games with rules were not included, because of their low frequency in the preschool years). Categories such as solitary-functional and group-constructive were used by Rubin and others in descriptive and experimental studies of children's play (e.g., Mounts & Roopnarine, 1987; Pellegrini & Perlmutter, 1987; Rubin, Watson, & Jambor, 1978).

Although these coding schemes describe the majority of play interactions observed in preschool classrooms, Takhvar and Smith (1990) pointed out that they neglect certain less-commonly seen forms, such as physical activity play, language play, and rough-and-tumble play. Some researchers have modified these coding schemes to include such interactions (e.g., Coplan, Rubin, Fox, Calkins, & Stewart, 1994; Pelligrini & Perlmutter,1987; Saracho, 1984; Tizard, Philips, & Plewis, 1976).

Little is known about how often children outside of the preschool classroom engage in these various types of play. A study by Bloch (1987) provides some preliminary data. Parents of eighty-three 0- to 6-year-old middle-class children were randomly telephoned at times between 8 A.M. to 8 P.M. 5 randomly selected days a week over a 2-month summer period. When contacted, parents were instructed to go and observe what the child was doing at that moment and then report back to the interviewer. Results showed that children at all ages were playing in about one third of the waking periods sampled. The other common activities were eating (16%), making transitions (10%), watching television (8%), interacting socially (7%), and receiving care from others (6%). Play was broken down into the following categories: gross motor (11% of total waking observations), functional (9%), pretense (3%), miscellaneous (3%), constructive (2%), exploratory (2%), music/art (2%), and rough and tumble (1%). About half of all play was social in nature. Similar results were found in a maternal diary study by Giddings and Halverson (1981).

The other source of information on the types of play found in preschool children is detailed analyses of the nature of children's social pretend play. Using methods of qualitative analysis in a study of 4- and 5-year-olds, Roskos (1990) developed a taxonomy of pretend-play activity with three superordinate categories: play with objects (specific and generic), play like someone or something (TV/movie, functional, family/animal, occupational), and making up (people, places, things). She also identified 28 pretend play schemes. Although she offered no higher order categories for pretend-play content, the 28 schemes could be grouped into the following categories: domestic (e.g., house, babies), occupational (e.g., police, doctor), transportation (e.g., boat, train), entertainment (e.g., theatre, beach, camp out), and running errands (e.g., store, bank, beauty parlor). Pretend-play distinctions made by other researchers include acting out realistic versus fantasy roles (e.g., McLyod, Warren, & Thomas, 1984), and various kinds of material and ideational transformations (Matthews, 1977).

Social Play Behaviors. Compared to the play of infants and toddlers, the play of preschoolers shows tremendous complexity. Largely due to the development of pretense and language, the social complexity of social object play skyrockets during the third year of life. This is illustrated in the following example from Ariel (1992, pp. 120–121), of two 5½-year-old girls:

"Dina: Ruttie! Come here! Sit on this horse! [The 'horse' is a red tube, raised above the ground on metal stands. Dina climbs up the tube, seating herself astride. She moves her straightened arms forward and backward, apparently denoting 'riding']

Dina: Halt! [Ruttie climbs up and sits on the horse in front of her]

Ruttie: Move on! [She moves her straightened arms backward and forward]

Ruttie: First, let's pretend the baby-girl is in front and the mummy is behind, and then let's pretend you allow your baby to get lost, because, after all, you have a dog. Children, when they get lost, the dog gets them back. Come on! Let's pretend you arrived home. You told your little girl: "Sweetie! You may get lost!" So, are you getting off? [Dina gets off]

Dina: Yes, I am going home. I want you to work for me. [Dina some distance away from the tube]

Ruttie: Mummy! Mummy! Get me this bag! I need it! Here is a blanket! [Ruttie gets off the tube. She picks up a metal frame from an old table and places it on top of the tube]

Ruttie: The bag. [She picks up a large piece of cloth and places it on top of the tube]

Ruttie: A blanket [Pause] and food. [She runs toward the center of the yard, picks up an old tea kettle, and puts various small pieces of junk inside it until it is full. She carries it to the tube and places it on top of it, then picks up an old plastic shopping bag and fills it with little stones]

Ruttie: I haven't finished yet. I am gathering food. [She places the bag with the stones on top of the tube, climbs on the tube, starts "riding" moving her legs inwards and outwards rhythmically, calling "move!" Dina gets into an old metal cot about three yards away]

Dina: Ruttie! Let's pretend you are a mummy-baby. You are the queen, but you are a baby-girl. You are my queen baby. All this [Sweeping the yard with her arm] is your nursery.

Ruttie: But let's pretend I didn't come to the nursery today. Move! [Dina comes close to her]

Dina: You should say "Halt!" too! [Pause] Sweetie, here's some more food for you. [She hands Ruttie a little metal net]

Ruttie: Now I have arrived at the village. Fantastic! I've arrived at the village!

Dina: Let's pretend we live together in this village too.
Ruttie: No, let's pretend you are at home and this is a far, far away village.

As was true for the play of toddlers, the children in this example show turn taking, social responsiveness, and complementary roles, but the addition of language and social pretense yields interactions that differ substantially from those seen in the first 2 years. Not only does the range of possible activities increase dramatically with the addition of pretense, but the incorporation of language allows for greater planning, scripting, and negotiation of play roles, as well as for more complex forms of social influence. The play of preschoolers also reflects children's developing social-perspective-taking skills and their developing understanding of social conventions and norms.

Presented in Table 8.3 is a list of behaviors identified in three observational studies of free play in young children. To allow for easy comparison with the research on infants and toddlers, this table is formatted in the same way as Table 8.2. Note that three types of social behaviors are listed: affiliation, agonism, and nonagonistic social control. The latter category, which involves both controlling and control-granting behaviors, is not mutually exclusive with the other two, but refers to behaviors (primarily verbal) that often occur in the context of social play. A comparison of this table with Tables 8.1 and 8.2 shows the same general categories of behavior, but more individual behaviors within each category. Besides the verbalizations listed under nonagonistic social control, the behaviors identified in preschoolers and not in infants and toddlers include wave, kiss, hold hands, hug, help, verbal threat, kick, throw at, verbal insult, supplant, and several submissive gestures. Some of these differences may be artifacts of the coding systems employed by the different investigators. For example, Strayer and Trudel (1984) used a differentiated system for coding threatening and submissive displays, and it is possible that some of the same displays may be shown by infants and toddlers as well. Moreover, Brenner and Mueller (1982) observed hugs in their study of 12- to 23-month-olds.

Considering the differences between the play behaviors identified for preschoolers and toddlers, however, points to the increased verbal and social skills found in older children. Besides the obvious differences in their communicative abilities (e.g., nonagonistic social control), preschoolers were found to employ conventionalized signs of affection, including waving, kissing, and hand holding, as well as the more complex prosocial behavior of helping. These differences reflect both the child's growing awareness of cultural practices and norms as well as his or her growing ability to take the perspective of another person. Along with growing

TABLE 8.3

Peer Social Behaviors During Free Play: Early Childhood

	Strayer (1980); Strayer & Trudel (1984)	*Smith & Connolly (1972)*	*Jones (1984)*
Affiliation			
Attention (look at, watch)	+		
Signal			
Facial expression			
Smile/laugh	+	+	
Gestures			
Point	+	+	
Show object	+	+	
Wave	+		
Verbalizations	+	+	+
Approach (move toward, follow)	+	+	
Contact			
Touch/pat	+	+	
Kiss	+		
Hold hands	+	+	
Hug	+		
Prosocial behavior			
Help	+	+	
Give object	+	+	
Share	+		
Play			
Locomotor		+	
Play-fight		+	
Language (play noise, sing, scream)		+	
Social object play		+	
Terminate			
Re-orient (look away)	+	+	
Withdraw (move away)	+	+	
Agonism			
Threat			
Facial expressions			
Threat face	+	+	+

Table 8.3 continued

Table 8.3 continued

	Strayer (1980); Strayer & Trudel (1984)	*Smith & Connolly (1972)*	*Jones (1984)*
Gestures			
Body display	+		+
Motor			
Fragment-hit	+		
Lack	+		
Throw	+		
Chase	+		+
Verbal threat		+	+
Attack			
Bite	+		+
Hit	+	+	+
Grab	+		
Kick	+		+
Push/pull	+		+
Throw at	+		
Assault	+		+
Verbal insult			+
Competition			
Object struggle	+	+	
Steal	+	+	
Supplant	+		
Terminate			
Submission			
Gaze avert	+		
Crouch	+		
Cringe/flinch	+		+
Cry	+	+	+
Retreat			
Step back	+		+
Flee	+	+	+
Object/position loss	+	+	
Help seeking	+	+	+
Nonagonistic social control			
Controlling			

Table 8.3 continued

Table 8.3 continued

	Strayer (1980); Strayer & Trudel (1984)	Smith & Connolly (1972)	Jones (1984)
Start/propose activity			+
Directs Play			+
Instructs			+
Praises			+
States/interprets/enforces rules			+
Distributes objects			+
Grants/denies play			+
Refuses to comply			+
Offers alternative			+
Ignores			+
Control granting			
Follows directions			+
Copies			+
Asks for evaluation			+
Asks for help			+
Asks for object or permission			+

perspective-taking skills, however, come more sophisticated ways of harming another (i.e., verbal insult).

Much has been written about the complexity of social pretend during the preschool period. In a review of the literature on make-believe play, Bretherton (1989) identified three conceptual underpinnings of pretense: multiple role taking, transformations of reality or counterfactual thinking, and blurring conceptual levels or playing with the playframe. Like Magee (1989), Bretherton (1989) argued that pretend play can be viewed as dramatic performance, with the participants frequently shifting between various roles such as "coplaywrights, codirectors, coactors, and vicarious actors, without getting confused about which of these roles they or a playmate are momentarily adopting. Children must (as playwrights) decide on the play, and as directors decide on role distribution and props in order to (as coactors and vicarious actors) perform the make-believe drama" (Bretherton, 1989, p. 384). In a study of 3- to 6-year-olds, Magee (1989) found evidence of children enacting of the following roles during social fantasy play: acting, directing, audience, stage manager, and parallel actor.

Because social pretend involves coordinating the behavior of two or more players, it is not always clear to all involved where the play is heading. Children, therefore, often must move in and out of the pretend "frame" in order to successfully manage its course (see the earlier example). Giffin (1984) identified a continuum of communicative acts ranging from within-frame to out-of-frame communications. These are (moving from without to within): formal proposals, implicit pretend structuring, prompting, storytelling, underscoring, ulterior conversation, and pretend enactment.

Bretherton (1989) asserted that another component of pretense is the occurrence of reality transformations or counterfactual thinking. During pretense, she maintained, children are not bounded by the rules of reality, but instead can transcend the everyday and create unique patterns of relating. Finally, she argued, pretense involves the "deliberate interweaving (as opposed to confusion) of what is play and what is reality" (p. 390). Although the boundary between reality and pretense is diffuse for young children, even as this boundary becomes more distinct children can incorporate aspects of reality into their pretend play as a way of manipulating it, understanding it, coping with it, and bringing it under control.

That pretense differs from other forms of social object play is demonstrated in three studies by Connolly, Doyle, and associates (Connolly & Doyle, 1984; Connolly, Doyle, & Reznick,1988; de Lorimier, Doyle, & Tessier, 1995). In these studies, the nature of preschoolers' social behavior was compared when they were engaged in pretend versus nonpretend play (e.g., constructive play, drawing, games). Compared to the nonpretend context, during pretend play children showed more positive and less negative affect; engaged in longer sequences of behavior; engaged in more activity talk and attempts to influence another's behavior; were more successful in their influence attempts; and showed more compliance, shared play focus, and coordinated/reciprocal interactions. One reason for the longer length of sequences during pretend play is that much of children's social pretend play takes the form of scripted discourse (Nelson & Seidman, 1984). In a study that examined conflicts during pretend and nonpretend play (de Lorimier et al., 1995), the frequency of conflicts arising did not vary across the two contexts but conflicts during pretend play were of longer duration.

Five Social Interactional Processes: Gottman (1983)

One of the most comprehensive frameworks for describing the play interactions of young children to date was developed by Gottman (1983).

In this research, mothers of 3- to 9-year-olds made audiotapes of their children's play interactions in the home. In the first study, mothers made audiotapes of their children with their best friends; in the second, children were randomly paired and audiotaped during a series of three visits to one of their homes. Gottman and colleagues then spent approximately 5,000 hours transcribing and coding the tapes for sophisticated quantitative analyses. Based on the results of sequential analyses of the data, Gottman (1983) identified three social processes (play, self-exploration, and repair) and two affective states (amity and conflict) that described the nature of the interactions. Because these processes emerged from a set of analyses designed to differentiate between children who did and did not become friends, Gottman (1983) referred to these characteristics as friendship-formation processes. A diagram representing the temporal relationships between these processes during play interactions is presented in Fig. 8.1. These processes organize the remainder of this discussion of play structure.

Play. Gottman (1983) saw social interaction during play as a balancing act, when children work together in managing the "continual escalation and de-escalation of the involvement demanded by play" (p. 76). Central to his concept of play is the establishment of "common ground activity" (see Fig. 8.1). In the play interactions he coded, he asserted that there was "a hierarchy of successful conversation in common-ground activities, organized in terms of the responsiveness of the [social interactional] demand" (Gottman, 1983, p. 52). This hierarchy, borrowing in part from Parten (1932), was defined as follows:

1. *Parallel play, collective monologue.* Unconnected ME codes [ME codes were self-focused statements such as "I finished it so fast"]
   A: I'm making this blue.
   B: Staying in the lines, there, there.
2. *Parallel play, connected dialogue.* ME—ME chains of connected statements.
   A: I'm making mine blue.
   B: I'm making mine brown.
3. *Narration of the other child's play....*
   A: You're using blue to color that.
4. *Asymmetrical exchange*, in which at least one child affects the other's activity, with demands, compliance, and noncompliance.
   A: I'm putting pink in the blue.
   B: Pass the blue.

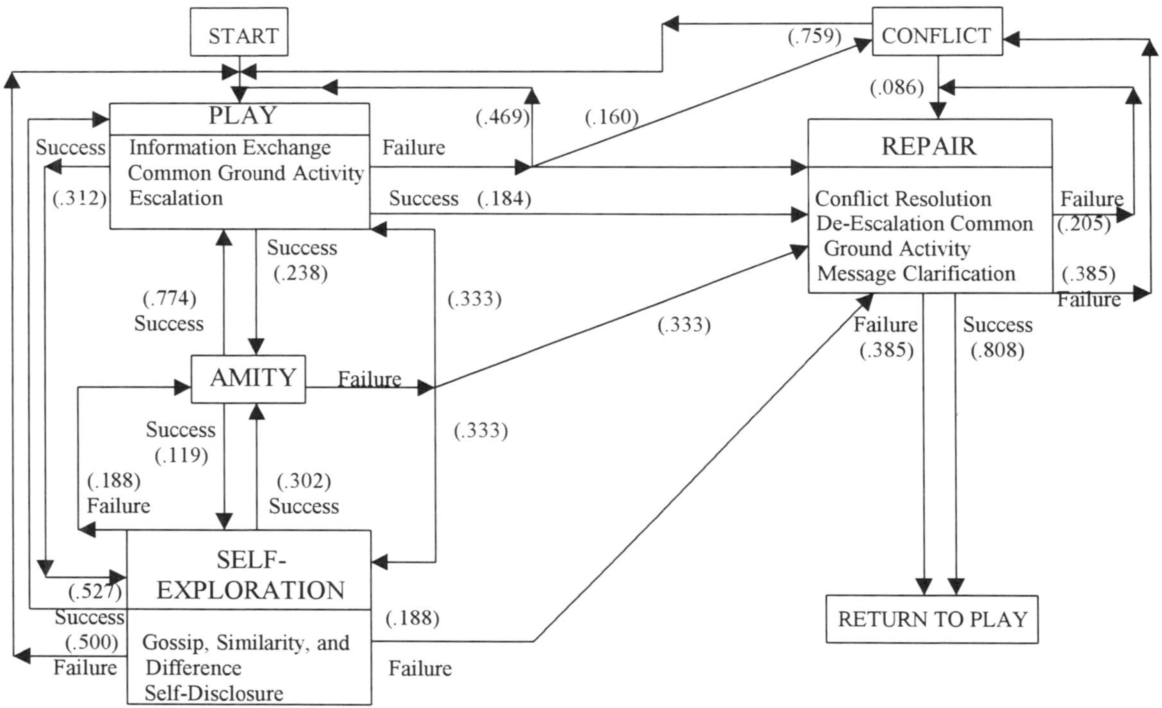

FIG. 8.1. Temporal flow model of friendship formation processes. From Gottman (1983), p. 69. Copyright © 1983 by Society for Research in Child Development. Adapted with permission.

A: I think I'll pass the blue.

5. *Symmetrical exchange* in which both children affect each other.
   A: And you make those for after we get in together, OK?
   B: 'Kay.
   A: Have to make those.
   B: Pretend like those little roll cookies, too. OK? Flat cookies, I mean.
   A: And make, um, make a, um, pancake too.
   B: Oh, rats, this is a little pancake.
6. *Joint activity* is successfully initiated....
   A: Yeah, let's play house.
   B: OK, play house. (Gottman, 1983, p. 53)

Escalation of social responsiveness is defined as moving up this scale; de-escalation involves moving downward. Information exchange, the other component of play in this model, refers to situations in which one child exchanges information with another. It was measured by the frequency of attention-getting and information statements, as well as the following sequences between the partners: an information question followed by an information statement from the partner, an attention-getting statement followed by an information statement, and an attention-getting statement followed by an agreement.

Self-Exploration. Gottman (1983) argued that a good deal of self-exploration occurs during play interactions (see Fig. 8.1). This occurs by talking about the self in relation to others during play interactions. In his model, these include similarity statements ("we both," "we against others," "me too," and "joining in" statements), contrast statements, reciprocated gossip, and self-disclosure (reciprocated statements about feelings, intimate personal statements about own feelings, and the sequence "question about feelings" followed by a "feelings statement"). Although many theorists have maintained that true self-disclosure does not occur until adolescence (e.g., Selman, 1980), Gottman (1983) asserted that early forms of self-disclosure occur, especially in the fantasy play context.

Other authors have also written about children's play as a context of social comparison. Both Mosatche and Bragonier (1981) and Chafel (1987) found that 3- to 5-year-olds spontaneously engaged in social comparison during play, making comparative statements about differences in ability, activity, possession, attitude, and status. Five functional categories were identified: "assessing self by comparison with a peer ('I can. I can do it, too'), besting ('I'm the general; that's higher than the captain, sorry'), cognitive uncertainty ('These are pretty, right?'), differentiating self from other ('You're making soapsud pie and I'm making a soap stew'), and establishing similarity between self and other ('We're both 4; we both had a birthday')" (Chafel, 1986, p. 52).

Amity and Conflict. Gottman (1983) identified two affective states that occur throughout play interactions: amity and conflict (see Fig. 8.1). Amity, which refers to interactions with a strong positive affective component, included the following statements: validation or approval of other, affirmation of relationship, sympathy, offers, affection, humor, reciprocated humor, and shared deviance. Conflict, in contrast, involved all "squabbles" and disagreement chains.

Although observational studies of free play in preschoolers indicate that prosocial behaviors such as sharing and comforting occur at relatively low frequencies (e.g., Eisenberg-Berg & Lennon, 1980), positive affect is one of the most commonly employed markers of play (e.g., Aldis, 1975). As for conflict, many studies have documented that it is common in the free play of preschoolers (e.g., Shantz, 1987), and several studies have been conducted to study its resolution.

Repair. Gottman (1983) referred to the process of conflict resolution during free play as "repair" (see Fig. 8.1). He identified three types of repair processes: message clarification (i.e., sequences of requests for repetition or clarification followed by repetitions or clarified messages), conflict resolution (weak demands, reciprocal joking regarding the conflict, and disagreements followed by clarified message), and de-escalation of the common-ground activity.

Other studies of conflict resolution in free-play settings provide information on the general nature of conflicts (Dawe, 1934; Hartup et al., 1988; Kolominskii & Zhiznevskii, 1992; Laursen & Hartup, 1989; Phinney, 1986; Sackin & Thelen, 1984). Although the ages, constructs, and operational definitions differ considerably across studies, it is still possible to draw some general conclusions. Conflicts during this period revolve primarily around objects, play roles, and the correctness of play actions (Dawe, 1934; Hartup et al., 1988; Kolominskii & Zhiznevskii, 1992). Children's reactions to such situations are generally to stand firm rather than to yield to the wishes of others (Hartup et al., 1988). Children in this age range are verbally or physically aggressive in about one third of all conflicts (Dawe, 1934; Hartup et al., 1988), with pushing, striking, and pulling the most common physical forms (kicking, pinching, stamping, and throwing things are rare; Dawe, 1934). In most cases, the conflict had a clear winner or loser. In the Hartup et al. (1988) study, 62% of the conflicts had a clear winner or loser, whereas in the remaining cases the outcome was coded as partial equality or some kind of compromise. Most conflicts are brief. Laursen and Hartup (1989) found that 63% of the conflicts they observed were 10 seconds in duration or less. The affective state of the participants after the conflicts is generally positive (in the study from Dawe, 1934, 75% of the participants' affect was positive), and children generally remain in close proximity with high probability of continued interaction (Hartup et al., 1988).

Little is known about the specific ways in which conflicts are resolved. In one of the few detailed descriptive studies, Kolominskii and Zhiznevskii (1992) provided the following list of the major ways in which successful resolution was achieved:

> introduction of additional elements into the content of the game (new roles, toys, play actions)—39% of the total number of favorably resolved conflicts by the children themselves; insisting on one's own claims by repeating the relevant declarations—15%; establishing an order in performing a role or using a toy—13%; emotional empathy with the peer who is "suffering" during the course of a conflict (in such situations children embrace each other, "feel sorry," excuse themselves—"I'm not unhappy")—11%; appeal to rules of the game—8%; compensation for an omission (the children give candy or toys in exchange for a concession)—5%; sanctions on the rival (for example, a threat to leave the game)—5%; proposal to play together—3%; resolution by an intermediary (i.e., solution of a dispute according to suggestions of other peers)—3%; certain algorithms for solving disputes (e.g., a counter)—2%; and finally, complaints as a means of achieving a concession from a peer—2%. (Kolominskii & Zhiznevskii, 1992, pp. 84–85)

Miller, Danaher, and Forbes (1986) differentiated between heavy-handed persuasion (physical force or the threat of such force) and moderate persuasion (simple proposition, check desire, check ability, give directions, entreaty, appeal to situational constraints, appeal to social norms, effusion of affect, appeal to past compliance, co-opting, justification by appropriateness, justification by desired outcome, offers exchange, de-escalation of request, clarification of social intent, clarification of reference). A final category, labeled "conflict mitigation tactics," involved strategies in which the child appeared to avoid addressing the conflict issue directly (i.e., clarification of other's feelings, changing the topic, peaceful acquiescence, proposal of compromise, indirect display of anger, avoidance).

Sackin and Thelen (1984) studied the behavior of the "losers" of a conflict, comparing situations in which children stayed together versus those in which they separated. Almost without exception, losers showed a submissive behavior just before a separation, and a conciliatory behavior if interaction was to continue (for a review and data on similar tendencies in other primates, see Aureli, 1997). Submissive behaviors, as identified by Strayer and Strayer (1976), included the appeasement gestures of cry-scream, flinch, cringe, withdraw, rapid flight, hand cover, and request cessation along with their object/position loss category. Conciliatory behaviors (identified by Sackin & Thelen, 1984) consisted of cooperative propositions, apologies, symbolic offers, object offer/share, and grooming.

Play Initiation, Maintenance, and Termination. Research shows that children initiate play either gradually or directly. When initiating gradually, children who are successful appear to engage in synchronous play behavior next to another child versus hovering and waiting (Bakeman & Brownlee, 1980; Johnson & Ershler, 1981; Ramsey & Lasquade, 1996). When initiating directly, children who are successful first get the other's attention and then make some sort of collaborative statement that includes or is responsive to the other child (Gottman, 1983; Krasnor & Rubin,1983; Mueller, 1972; Ramsey & Lasquade, 1996). These might include questions or comments about the other child's play or sociable statements that establish a friendly atmosphere with the other. Children who initiate with more forceful controlling statements appear to be less successful (Ramsey & Lasquade, 1996).

According to Gottman (1983), once play has been initiated, "The continual escalation and deescalation of the involvement demanded by the play is related to the amount of both conflict and amity that results. There is an intricate minimax problem the children have to manage. It is a sort of coordinated dance, and it must be, in part, the continual reinvention of this dance that makes play such an adventure for children" (p. 76). This alternation between play, self-exploration, and the affective processes of amity and conflict is illustrated in the transitional probabilities in Fig. 8.1.

Garvey (1974) argued that, as play evolves over time, children begin to abstract idiosyncratic rules for their interactions and maintain interest through theme and variation. Meckley (1994), in her 5-month ethnographic study of a nursery school classroom, found that members of a play group developed a unique shared knowledge of play events over time, and this knowledge facilitated the ongoing play interactions. Meckley listed five play event principles:

1. Each play event has distinct, consistent, and predictable patterns of actions, objects, and players....
2. Each play event has an ordinal place of occurrence, consistent within any given day in which that play event occurred....
3. The play events...are not only consistent in sequence of actions and within time but also show an evolution and interconnectedness with one another...
4. This group's social play reveals clear ownership of the play events by individual players and groups of players....
5. All of the young children in this group demonstrate *shared knowledge* of specific play event enactment, objects used in specific themes, and players' styles. (pp. 293–294)

Once play has begun, additional processes appear necessary for the initiation of pretend play. Doyle, Doehring, Tessier, de Lorimier, and Shapiro (1992) found that among children who pretended frequently there were three common activity modes through which they moved into pretend play: negotiation of pretend, "different social activities" (i.e., one child pretending or negotiating pretend and the other child involved in a literal activity), and solitary pretend enactment. The complexity of negotiating a pretend theme was illustrated by Verba (1993), who identified "cooperative formats" through which this process occurs: transmission–appropriation, consensual coelaboration, and conflictual coelaboration.

Researchers have commented on the difficulties of maintaining the pretend frame. Because children often go in and out of the pretend mode, children must develop both means for differentiating between the pretend and reality modes, and methods for the smooth transition from one to the other (Bretherton, 1989; Giffin, 1984).

No specific empirical studies on the termination of play could be found. As speculated in the earlier section on infants and toddlers, likely terminators include that one or both children are distracted, a child does not achieve the outcome that he or she desires, a conflict escalates to the point that one or the other children leave, a caregiver intervenes, and so on.

Developmental Course. Research using Parten's (1932) categories of social participation and Smilansky's (1968) cognitive play categories has yielded mixed support for these models. Rubin, Fein, and Vandenberg summarized the major findings of this research in their 1983 review. They concluded that, during the preschool period (in line with the predictions of Smilansky, 1968), functional play becomes less common with age, whereas interactive pretend and games with rules increase. These conclusions are consistent with the results of more recent studies (e.g., Cole & La Voie, 1985; Connolly, Doyle, & Ceschin, 1983; Field, De Stefano, & Koewler, 1982; Goldstein, Field, & Healy, 1989; Lloyd & Smith, 1985; Pellegrini, 1982, 1985), including Bloch's (1987) study using spot observations of child behavior in the home. Rubin et al. (1983) pointed out that research on constructive play, however, does not show any consistent changes with age, yielding questions about its utility as a measure of the cognitive maturity of children's play.

Regarding development into the elementary school years, Rubin et al. (1983) pieced together the results of several studies to provide preliminary evidence that pretend play peaks late in the preschool period. However, they cited only one study (Eifermann, 1971) that actually empirically documented a decrease beyond the preschool years. Two more recent studies support Rubin et al's. (1983) general conclusions. Doyle, Bowker, Serbin, Gold, and Sherman (1989) found that the duration of social pretend and the number of pretend object transformations during social pretend play decreased between Grades 1 and 6; de Lorimier et al. (1995) found that the proportion of time that children devoted to highly coordinated pretend play was lower for 6-year-olds than for 4-year-olds.

With a couple of exceptions (i.e., parallel play, solitary pretend play), the data on social participation are consistent with the Parten's (1932) belief that the level of social involvement should increase with child age. Associative (Barnes, 1971; Parten, 1932), cooperative (Barnes, 1971; Goldstein et al., 1989; Hold-Cavell, Attili, & Schleidt, 1986; Howes & Matheson, 1992; Parten, 1932; Pellegrini, 1982; Saracho, 1987), and group play (i.e., associative plus cooperative—Lyytinen, 1991a; Manwell & Mengert, 1934; Provost & La Freniere, 1991; Rubin & Krasnor, 1980; Rubin et al., 1978; Smith, 1978) have all been found to increase during the preschool years, whereas both nonplay (Goldstein et al., 1989; Hold-Cavell et al., 1986; Iijima, 1986; Lyytinen, 1991a; Manwell & Mengert, 1934; Parten, 1932; Provost & La Freniere, 1991; Rubin et al., 1978; Rubin & Krasnor, 1980; Smith, 1978) and solitary play (Barnes, 1971; Bloch, 1989; Goldstein et al., 1989; Iijima, 1986; Lyytinen, 1991a; Parten, 1932; Pellegrini, 1985; Provost & La Freniere, 1991; Rubin et al., 1978; Smith, 1978; Tizard et al., 1976) have been found to decrease. The data for parallel play are inconsistent. Some studies show an increase with age (Hold-Cavell et al., 1986; Lyytinen, 1991a), some show a decrease (Barnes, 1971; Goldstein et al., 1989; Howes & Matheson, 1992; Parten, 1932; Tizard et al.,1976), and others find no significant age differences (Provost & La Freniere, 1991; Rubin et al., 1978; Smith, 1978). Finally, despite an apparent overall decrease in solitary play with age, the results of several studies show that solitary *pretense* remains stable or increases with child age (Pellegrini, 1982; Rubin et al., 1983).

Many of the developmental changes in children's social object and social pretend play during the preschool years parallel, in the social realm, the same changes identified for solitary object play. They include the coordination of multiple aspects and dimensions of situations; a move from individual behaviors to complex, hierarchically organized behavior patterns; sustained attention and longer sequences of behavior; the ability to shift one's perspective from the self to another; and the ability to remove one's self cognitively from the current situation.

Children increasingly show evidence of coordinating multiple aspects of situations as their social play increases in complexity. Single-scheme play interactions decrease with age (Lyytinen, 1991a), whereas multiple-scheme events, differentiated scripts, and complex episodes increase (Duveen & Lloyd, 1988; Howes & Matheson, 1992; Lyytinen, 1991a; Pellegrini, 1985; Tizard et al., 1976). Similarly, with increasing age, children's social pretend is more likely to involve character portrayals than simple object transformations (e.g., Cole & La Voie, 1985; Field et al., 1982). Such behaviors are increasingly likely as the child's play becomes hierarchically organized around plans, goals, scripts, and game rules.

Unlike solitary play, however, this organization is co-constructed by the participants, requiring a range of strategies for making one's own plans known, uncovering the goals and plans of the play partners, and negotiating any differences. Therefore, with increasing child age, conflicts arising during children's play are less likely to be struggles over objects, and more likely to be conflicts over the composition of game participants, play roles, play rules, and the correctness of play actions (Dawe, 1934; Kolominskii & Zhiznevskii, 1992). Age-related changes in children's verbalizations during pretend play reflect the increasingly active nature of this co-construction process. With increasing age, children are more likely to announce in advance what they are going to do (Black, 1989; Field et al., 1982), make meta-communicative statements that are out of the pretend frame (Halliday-Scher, Urberg, & Kaplan-Estrin, 1995), and make references to the relationships between characters and between characters and their behaviors (Forbes & Yablick, 1984). Children's verbalizations during play also become increasingly responsive to the behavior of the play partner. Specifically, with increasing child age, verbalizations are less likely to be demands for action (Harris, 1985) or attention (Farver, 1992); more likely to be demands for information or agreement (Harris, 1985); and more likely to extend, build on, or repudiate the verbalizations of another (Farver, 1992; Garvey, 1974; Goncu, 1993; Goncu & Kessel, 1984; Harris, 1985).

Other studies further demonstrate the increasing social connectedness of children's verbalizations during play. Gottman (1983, 1986) identified several age differences between 3 and 7 years. Information exchange and common-ground activities were more common with increasing child age. Age was positively associated with attention getting and information statements, as well as these two sequences: questions followed by an information statement, and "we demands" followed by an agreement. Similarly, Harris (1985) found that information-giving statements increased between ages 2 and 4, although Farver (1992) found that attention-getting statements decreased between ages 2 and 5. Gottman (1983, 1986) also found that self-exploration increased with age, and noted positive associations with age for exploring differences (feeling statements followed by a disagreement), reciprocated gossip, and reciprocated feeling statements.

This increasing connectedness is also evidenced in children's entry strategies. Ramsey and Lasquade (1996) found that waiting and hovering became less common with age, whereas informational questions and statements became more common. Entry strategies that led to negative interactions or to parallel play were also less common with age.

When successful, the hierarchically organized, co-constructed nature of older children's play likely makes it possible for longer sequences of behavior to occur. Several researchers documented an increase in the length of play sequences with advancing child age (Farver, 1992; Goncu, 1993; Goncu & Kessel, 1984, 1988; Halliday-Scher et al., 1995; Kramer & Gottman, 1992; Reuter & Yunik, 1973). The ability to sustain social play for longer periods may help account for the increases in associative and cooperative play observed in preschool settings, as well as for decreases in time spent not playing.

Children's developing social perspective-taking skills and empathy probably also contribute to their ability to initiate and maintain longer sequences of play behavior. This apparently increases with age, as evidenced by increases in prosocial behavior. Research on prosocial behavior during free play in the preschool years shows that giving becomes less common with age, whereas comforting and sharing become more common (Bar-Tal, Raviv, & Goldberg, 1982; Eisenberg-Berg & Hand, 1979). Charlesworth and Hartup (1967) found that giving attention and approval became more common with increasing child age as well.

Despite the child's developing social skills, conflicts during play do not become less frequent with age (e.g., Shantz, 1987). This probably results from their increasing complexity—the topics, not the frequency, of conflicts appear to be what reflect age-related changes. Moreover, with increasing perspective-taking skills come new sources of disagreement and new ways to aggress against others, including verbal and relational aggression (e.g., Dawe, 1934; Kolominskii & Zhiznevskii, 1992).

Finally, the child's developing symbolic abilities and ability to remove one's self from the current situation also account for changes in play. With increasing age children are more likely to refer to objects that are not present, or to past and future events (Harris, 1985). Their play becomes less dependent on props (Black, 1989; Connolly et al., 1983), and the number of pretend themes increases, as do the number and complexity of self- and other pretend transformations, including complex forms of ideational, character, and fantasy-based pretend (Connolly et al., 1983; Doyle, Ceschin, Tessier, & Doehring, 1991; Johnson & Ershler, 1981).

Despite these parallels with the solitary-play context, it is unlikely that developmental changes in children's social object play are the result of cognitive factors alone. When playing in the social versus the object realm, emotional factors become particularly salient, and many of the age-related changes in children's play may reflect their developing ability to control and regulate their emotions. This may be particularly true of developmental changes in children's responses to conflict (e.g., Eisenberg, Fabes, Nyman, Bernzweig, & Pinuelas, 1994), as well as in their prosocial behavior (Eisenberg et al., 1996). Moreover, the degree to which children use other- versus self-focused play strategies depends, in part, on the ability to delay gratification and control impulses. Because much of the research and theory on developmental differences in children's play has had a cognitive focus, future developmental research needs to explore more of the affective aspects of young children's play experiences.

Gender Differences. In direct contrast to the research on infants and toddlers, numerous gender differences in peer interaction and play have been found in studies of preschoolers and school-age children. Many of these are summarized in Table 8.4. As seen in the table, active, physical, outdoor play (i.e., play-fighting, aggression, activity level, outdoor play, rate of social initiations/interactions) and functional play are more common in boys, whereas constructive play is more common in girls. Although boys and girls have not been found to consistently differ in the *total amount* of pretend play (see the table section on inconsistent findings), boys and girls differ in the roles they assume in this context (fantastic for boys and family for girls), and boys apparently engage in more object transformations.

TABLE 8.4

Gender Differences in Children's Play Behavior

	References
Well-established gender differences	
Boys greater than girls	
Play-fighting	See chapter 5
Aggression	For a review see Maccoby & Jacklin (1980)
Activity level during play	See chapter 7
Outdoor play	Lever, 1978; Sanders & Harper, 1976; Stoneman et al., 1984
Rates of social initiations/interactions during play	Austin, Salehi, & Leffler, 1987; Cameron, Eisenberg, & Tryon, 1985; Dodge, Schlundt, Schocken, & Delugach, 1983; Greenwood, Walker, Todd, & Hops, 1981; Hinde et al., 1993; Hinde, Titmos, Easton, & Tamplin, 1985; Mize & Cox, 1990
Functional play	Dunn & Herwig, 1992; Johnson & Ershler, 1981; Moller et al., 1992; Rubin et al., 1976; Urberg & Kaplan, 1989
Object transformations: pretend	Cole & LaVoie, 1985; Doyle et al., 1989; Field et al., 1982; Johnson & Ershler, 1981; Jones & Glenn, 1991; Matthews, 1977
Fantastic roles: pretend	Johnson & Ershler, 1981; McLoyd et al., 1984; Werebe & Baudonniere, 1991
Self-focused interpersonal behavior	
Verbal and nonverbal attention getters	Austin et al., 1987; Hazen & Black, 1989; Phinney, 1979
Asserting own desires	Cook, Fritz, McCornack, & Visperas, 1985; Haslett, 1983; Hazen & Black, 1989; Leaper, 1991; Putallaz & Wasserman, 1989
Protecting self-interest	Haslett, 1983
Other self-statements	Black, 1992; Black & Hazen, 1990
Information-giving statements	Cook et al., 1985; Haas, 1979; Phinney, 1979
Showing off	Cook et al., 1985; Turner, Gervai, & Hinde, 1993

Table 8.4 continued

Table 8.4 continued

	References
Girls greater than boys	
Constructive play	Dunn & Herwig, 1992; Johnson & Ershler, 1981; Moller et al., 1992; Pellegrini, 1982; Rubin et al., 1976, 1978
Family/domestic roles: Pretend	Black, 1989; Connolly et al., 1983; Doyle et al., 1989; McLoyd et al., 1984; Tizard et al., 1976a; Werebe & Baudonniere, 1991
Other-focused interpersonal behavior	
Acknowledge actions, vocalizations, or situation of others	Austin et al., 1987; Haas, 1979; Haslett, 1983; Leaper, 1991
Conversation facilitators that follow up on the play or conversation of the other	Austin et al., 1987; Black, 1992
Request clarification	Black, 1992; Phinney, 1979
Extended coherent play episodes with topically linked interchanges	Black, 1989; Black & Hazen, 1990
Mitigation strategies during conflict	Miller et al., 1986; Sheldon, 1992
Inconsistent findings	
Amount of dramatic play	
Boys greater than girls	Cole & LaVoie, 1985; Cook et al., 1985; Doyle et al., 1989; Flannery & Watson, 1993; Johnson & Ershler, 1981; Moller et al., 1992; Pellegrini, 1982; Rubin et al., 1976, 1978; Sanders & Harper, 1976
Girls greater than boys	Hinde et al., 1985; Jones & Glenn, 1991; Lloyd & Goodwin, 1993; Maguire & Dunn, 1997; Tizard et al., 1976a; Urberg & Kaplan, 1989; Werebe & Baudonniere, 1991

Table 8.4 continued

Table 8.4 continued

	References
Amount of parallel play	
Boys greater than girls	Harper & Huie, 1985; Lyytinen, 1991a; Tizard et al., 1976a
Girls greater than boys	Halliday & McNaughton, 1982; Johnson & Ershler, 1981
Amount of social play	
Boys greater than girls	Harper & Huie, 1985; Hinde et al., 1985; Ladd, 1983; Moller et al., 1992; Pellegrini, 1982
Girls greater than boys	Lyytinen, 1991a; Tizard et al., 1976a
Amount of talking during free play	
Boys greater than girls	Cook et al., 1985; Mueller, 1972
Girls greater than boys	Field, 1982; Ladd, 1983; Langlois et al., 1973; McLoyd, 1980; Urberg & Kaplan, 1989; Werebe & Baudonniere, 1991
Character roles: pretend play	
Boys greater than girls	Connolly et al., 1983
Girls greater than boys	Doyle et al., 1989; Jones & Glenn, 1991; McLoyd, 1980
Prosocial behavior: free play	
Boys greater than girls	Cameron et al., 1985; Marcus & Jenny, 1977
Girls greater than boys	Friedrich & Stein, 1973; Werebe & Baudonniere, 1988

Other studies have documented differences in the social interactional processes involved in children's play (see Table 8.4). Analyses of children's conversations during play show that girls and boys often approach establishing common ground in very different ways. Girls apparently establish common ground by considering the wants and needs of the play partner, whereas boys do so by focusing on the self. These gender differences in verbal style parallel later gender differences in the conversations of adults (e.g., Fasold, 1990; Smith, 1985)—research shows that, in some situations, women use a more indirect, other-focused verbal style than do men. Theorists point out, however, that this does not necessarily mean that girls are less assertive than boys, just that their means for achieving social influence are more oriented toward considering the viewpoint of others (Leaper, 1991; Sheldon, 1990). In a study examining the use of directives only during *pretend*, Lloyd and Goodwin (1993) found that girls showed a higher proportion of directives than did boys.

Although a review of the literature on children's games is beyond the scope of this volume, a commonly cited gender difference in children's play is that boys engage in more competitive games than do girls (e.g., Stoneman et al., 1984). In a classic study of gender differences in the nature of children's games, Lever (1976, 1978) found that, compared to those of girls, the games of boys were more competitive, were longer in duration, had a greater number of rules and roles, required more interdependence between players, had more explicitly defined goals, and were more challenging in terms of the ceiling of possible performance. These findings were replicated in a study by Borman and Kurdek (1987).

In addition to these behavioral differences in the nature of children's play, boys and girls also differ in the specific play activities they prefer. Presented in Table 8.5 are lists of play activities that have been shown to vary consistently as a function of child gender. Observational studies of preschool children from the 1920s to the 1990s have documented these differences (for a review, see Eisenberg, 1983). The studies used in constructing Table 8.5 are listed at the bottom of the table. Only differences found in at least two studies are listed. Most children begin to show these differences in play preferences during the second year of life (e.g., O'Brien & Huston, 1985; O'Brien, Huston, & Risley, 1983). As can be seen in Table 8.5, gender differences have been reported primarily in three kinds of activities: aesthetic, pretend, and large motor. An attempt was made in constructing the table to identify pairs of activities that were similar in terms of their general purpose and structure. Paired activities appear in the same rows in the table.

TABLE 8.5

Gender Differences in Children's Play Activities

	Girls	*Boys*
Aesthetic	Art	Blocks and construction toys
	Play dough	Sandbox
	Listen to music	
Pretend	Domestic—kitchen, laundry	Carpentry
	Dolls and accessories	War and fantasy toys
	Telephone	Transportation toys
Large motor	Dance	Rough and tumble
		Riding toys
		Balls

References: Barry and Barry (1976), Berenbaum and Hines (1992), Cameron et al. (1985), Carpenter (1983), Carpenter and Huston-Stein (1980), Clark, Wyon, and Richards (1969), Conner and Serbin (1977), Duveen, Lloyd, and Smith (1988), Eisenberg, Tyron, and Cameron (1984), Etaugh, Collins, and Gerson (1975), Fagot (1977, 1984, 1985), Fagot, Leinbach, and Hagan (1986), Fagot and Patterson (1969), Giddings and Halverson (1981), Halliday and McNaughton (1982), Lamb and Roopnarine (1979), Lloyd and Smith (1985), Parten (1933), Pellegrini and Perlmutter (1989), Powlishta, Serbin, and Moller (1993), Rosenberg and Sutton-Smith (1960), Rubin (1977), Smetana and Letourneau (1984), Stoneman et al. (1984), Tizard et al. (1976a), Turner et al. (1993).

Not only do children show these preferences, but a number of studies have documented that preschool children reinforce one another for engaging in gender-typed activity choice during free play (e.g., Fagot, 1977, 1984, 1985; Fagot & Patterson, 1969; Lamb, Easterbrooks, & Holden, 1980; Lamb & Roopnarine, 1979; Roopnarine, 1984). Although the results of individual studies vary, in general children receive more positive consequences from their peers when engaged in gender-typed activities than in cross-gender activities. Moreover, these positive consequences also predict the likelihood that a child will continue or terminate the activity (Fagot, 1985; Lamb et al., 1980; Lamb & Roopnarine, 1979). The patterns of differential reinforcement and reinforcement effectiveness are strongest for boys and for same-gender peers. Moreover, boys, but not girls, are likely to receive negative peer responses for engaging in cross-gender activities.

In a series of correlational and experimental studies, Carpenter and Huston (Carpenter, 1983; Carpenter & Huston-Stein, 1980; Carpenter, Huston, & Holt, 1986; Huston, Carpenter, Atwater, & Johnson, 1986) demonstrated that some of the behavioral gender differences found in children's play may actually be a consequence of the kinds of activities typically engaged in by members of the different genders. They asserted that activities in the preschool classroom vary in their level of adult provided structure, with male-stereotyped activities low in structure and female activities high. According to Carpenter (1983), "Activity structure is defined as the extent to which there are externally imposed rules, norms, or guidelines about appropriate ways of performing an activity or task. These guidelines can be provided by two sources: observation of adult models and direct feedback or instructions" (p. 120).

Using a parent questionnaire and observations of teacher directions and feedback, Carpenter and Huston-Stein (1980) completed structure ratings on 19 child activities seen in at least four of five preschool classrooms. The results of these ratings are summarized in Table 8.6. A comparison of Tables 8.5 and 8.6 should make the association between level of structure and gender differences in children's activity preferences clear. Once the structure of the activities had been identified, Carpenter and Huston-Stein went on to demonstrate (through both correlational and experimental methods) that child behavior during preschool activities was more a function of the level of the activity structure than of the child's gender. Specifically, when engaged in low-structure activities, children of both genders showed higher rates of leadership and other social behaviors to peers. However, in high-structure activities, they showed more compliance and teacher-directed bids

TABLE 8.6
Structure Classification of Child Activities
(Data from Carpenter & Huston-Stein, 1980)*

Low Structure	*Middle Structure*	*High Structure*
Blocks	Board games	Art
Rough and tumble	Dolls	Books
Sand table	Fantasy games	Carpentry
Tinker toys	Painting	Cooking
	Play Dough	Play House
	Playground equipment	Science
	Puzzles	
	Small motor toys	
	Trucks and trains	

*Because Carpenter and Huston-Stein (1980) provided 11 structure ratings for each activity—1 rating for model availability and 10 ratings for feedback (individual and group feedback ratings for each of five classrooms)—for this table the activities were classified into three general categories by combining across ratings.

for attention. Therefore, Carpenter and Huston-Stein asserted, previously identified gender differences in dependence, independence, and compliance may be a function of the nature of activities that boys and girls prefer.

The results of the Carpenter and Huston research raises the possibility that some of the gender differences in child behavior observed during free play may be an artifact of the types of activities in which boys and girls are involved during the free-play observations. Rubin et al. (1983) also argued that different child activities may elicit differences in child behavior: "The toys parents (and grandparents) buy for boys tend to 'pull for' gross motor, functional activities (balls, vehicles) and for fictional character role enactments (laser guns, light sabers). The toys purchased for girls tend to elicit familial role enactments (dolls, carriages) and school-related pretense and constructive play themes (art construction materials)" (p. 740).

In order to more directly consider the possibility that some of the gender differences in children's play behaviors described earlier may be a function of the activities in which boys and girls are usually observed, a more extensive (and admittedly tentative) analysis of young children's play activities is presented in Table 8.7. For each of the activities listed in Table 8.5, a preliminary attempt to classify it among multiple dimensions was undertaken. These dimensions correspond to commonly held gender-stereotypes regarding the behavior of young boys and girls. Some of the dimensions refer to areas in which clear gender differences have been found (e.g., activity level, aggression), and others refer to areas in which the data on gender differences are equivocal (e.g., prosocial behavior, competitiveness). In order to facilitate comparison of activities with similar structures, similar activities are listed on adjacent rows.

A consideration of the information in Table 8.7 suggests that gender differences in activity preferences may explain some of the aforementioned gender differences in children's play behaviors. First consider children's aesthetic activities. Note that of the five aesthetic activities listed in Table 8.5, three are favored by girls and two are favored by boys. Examination of Table 8.7 suggests some differences in the nature of these activities that might have implications for gender differences in child behavior. Specifically, according to this analysis, aesthetic activities preferred by boys are more likely than those preferred by girls to involve involvement in a three-dimensional versus a two-dimensional medium, and more likely to have functional goals. In contrast, aesthetic activities preferred by girls, are more likely to involve teacher modeling and assistance. They are also more likely to be conventional, nonactive, played inside, and appeal to children's feelings versus their actions. If these speculative gender differences in the

TABLE 8.7

Preliminary Classification of Child Activities Across Multiple Dimensions

Activities	*Two vs. Three Dimen-sions*	*Functional–Aesthetic*	*Teacher Assist vs. No Assist*	*Child vs. Adult Model*	*Non–conventional–conventional*	*Active–Not Active*	*Outside–Inside*	*Actions–Feelings*	*Antisocial–Prosocial*	*Self– vs. Other–Serving*	*Competitive–Noncompetitive*	*Noisy–Quiet*	*Messy–Neat*	*Large Motor–Fine Motor*	*Risky–Safe*
Aesthetic															
Art–G	2	A	N&T	A&C	C&N	N	I	A	—	S	N	Q	N	F	S
Blocks–B	3	F	N	C	N	N	I	A	—	S	N	Q	N	F	S
Play Dough–G	3	F	N	C	N	N	I	A	—	S	N	Q	M	F	S
Sandbox–B	3	F	N	C	N	A	O	A	—	S	N	Q	M	F	S
Music–G	—	A	N	—	C	N	I	F	—	S	N	Q	—	—	S
Pretend															
Domestic–G	—	—	N	A	C	A	I	A	P	O	N	Q	N	F	S
Carpentry–B	—	—	T	A	C	A	I	A	P	O	N	N	M	F & L	R
Dolls–G	—	—	N	A	C	A	I	A & F	P	O	N	Q	N	F	S
War/Fantasy–B	—	—	N	C	N	A	I	A & F	A	S	C	N	N	F & L	R
Telephone–G	—	—	N	A	C	N	I	F	P	S	N	Q	N	F	S
Transportation–B	—	—	N	A	C	A	I & O	A	P	S	C&N	N	N	F&L	S
Large Motor															
Dancing–G	—	—	N & T	A & C	C & N	A	I & O	A & F	P	S	N	N	—	L	S
Rough and Tumble–B	—	—	N	C	N	A	O	A & F	A	S	C	N	—	L	R
Riding Toy–B	—	—	N	C	C	A	O	A & F	P	S	C & N	N	—	L	R
Balls–B	—	—	N & T	A & C	C	A	O	A	P	S	C & N	N	—	L	S

nature of children's aesthetic activity preferences withstand empirical test, then one might expect higher levels of independence, activity level, outside play, and visual–spatial skills in boys than in girls. Moreover, the overall preference of art activities in girls versus boys might account in part for the finding that girls show more constructive play in preschool classrooms than do boys.

The nature of the preferred pretend activities of girls and boys (see Table 8.5) might also account in part for gender differences in behavior. As shown in Table 8.7, the pretend activities preferred by girls are more likely than those preferred by boys to be conventional and modeled by adults, and to emphasize feelings, prosocial behavior, cooperation, quietness, and neatness. In contrast, those preferred by boys emphasize self-serving actions, and are often competitive, destructive, noisy, and messy in nature. Once again, if research confirms these as general characteristics of boys' versus girls' activities, then they would have many implications for play behavior. For example, girls' tendency to consider the needs of the others during play may be a consequence of involvement in caregiving and nurturing interactions with dolls, whereas boys' greater use of criticism and self-focus may be a function of the competitive, self-serving nature of boys' play activities.

Some of the gender differences in preschool-classroom pretend play may be a consequence of the types of play materials *not* available. Given boys' interest in war toys (see chap. 5) and the fact that few preschools provide such materials for children to use in play (most preschool teachers and directors feel such objects should be left at home), it is possible that boys engage in more object transformations to provide themselves with the "weapons" they need for war play. Because one of the most common object transformations seen in boys is using a stick, a peg, a finger, a puzzle piece, or some other toy to represent a gun, the lack of war toys might account for many such transformations. This might also explain why boys show more pretend play that is independent of the props available (e.g., Black, 1989; Johnson & Ershler, 1981). Because girls prefer domestic themes and appropriate props are usually available, object transformations are less necessary. Furthermore, given that war play often involves superheroes, space monsters, and so on, boys' preference for war-play activities may also account for their greater involvement in fantastic themes during play—girls' domestic play activity preferences rarely require fantastic themes. Considering the high level of girls' involvement in art activities, the higher level of boy's dramatic play found in some studies may be due to war themes as well.

Finally, given that three of the four large-motor activities in Table 8.5 are preferred by boys, these activity preferences might account for some of the gender differences in activity level found in observational studies (see chap. 7). Moreover, because most of these are low-structure activities (most are not modeled by adults, or do not require adult instruction or assistance), aggression, risk-taking, and play involving high levels of physical activity may be more likely to occur.

Although these hypotheses are clearly speculative, future studies need to examine the degree to which gender differences in children's social object play remain when the nature of the specific activity is controlled. The only research located that has examined this hypothesis (Carpenter, 1983; Huston et al., 1986) suggests that the activity versus the child's gender is a more important predictor of child behavior. In two studies of *solitary* play with objects, the results are less consistent. When Karpoe and Olney (1983) compared the play constructions of boys and girls playing with stereotypically masculine or stereotypically feminine toys, they found that only toy type predicted child constructions; the effects of child gender were not significant. Liss (1981), in contrast, found that when 5-year-old boys and girls played with either male or female gender-typed toys their play differed in many ways, even when using the same toys. For example, girls showed more nurturance, talking, gentle, and active play than did boys when playing with dolls, whereas boys were more active when playing with trucks.

Play Partners. Despite the many gender differences in the play behaviors and activities of preschool children, several theorists have asserted that it is important to take a relational perspective on these gender differences (e.g., Deaux & Major, 1987; Maccoby, 1990, 1998). That is, one should not view gender differences in the behavior of boys and girls in individual terms, but instead in the context of specific social relationships. For example, consideration of the findings in the previous section might be misleading, because most involve comparing the play behavior of children in same-gender dyads—that is, dyads of boys compared to dyads of girls (Maccoby, 1998). As is seen later in this chapter, the nature of gender differences in children's play has been shown to vary as a function of the gender of the partner, with greater differences usually found in same-gender pairings. For example, recall from chapter 7 that Eaton and Keats (1982) found that the greatest gender differences in children's activity level occurred during play in all boy triads—so much so that they feared for the safety of their laboratory equipment.

As we have discussed, gender differences in children's play are usually found even in observational studies where the gender of the partner is not controlled by the experimenter. One factor that might explain this is that, beginning during the 3rd year of life (at least in structured child-care settings), children start to spend more time playing with same-gender children (for reviews, see Hartup, 1983; Maccoby, 1998; Maccoby & Jacklin, 1987). Researchers have found that girls start to show this preference at a younger age than do boys. La Freniere, Strayer, and Gauthier (1984) uncovered that girls began to spend more time with same-gender playmates at 27 months, whereas boys did not show this tendency until 36 months. These results are consistent with Howes (1988a), Powlishta, Serbin, and Moller (1993), and Moller and Serbin (1996), who found greater gender segregation for girls than boys at ages 2 and 3, and with numerous studies reporting gender segregation for both boys and girls between the ages of 3 and 5 (Clark, Wyon, & Richards, 1969; Fagot, Leinbach, & Hagan, 1986; Fishbein & Imai, 1993; Gottman, 1986; Jones & Glenn, 1991; McCandless & Hoyt, 1961; Parten, 1933; Phinney & Rotherman, 1982; Turner, Gervai, & Hinde, 1993; Urberg & Kaplan, 1989). Despite the fact that girls start earlier, La Freniere et al. (1984) found that, by age 5½, boys showed greater gender segregation in their play than did girls. Other researchers have found more gender segregation in preschool boys than girls as well (Iijima, 1986; McCandless & Hoyt, 1961). Although studies show that boys' play groups are larger than those of girls in middle childhood (Lever, 1976; Luria & Herzog, 1991), this difference apparently is not found in the preschool period (Beneson, 1993). The extent of gender segregation, however, varies across settings. Bianchi and Bakeman (1978) found more cross-gender interaction during free play in an open school that consciously minimized gender typing. Experimental studies in classrooms also showed increases in the level of cross-gender interactions (e.g., Lockheed, 1986; Serbin, Tonik, & Sternglanz, 1977).

Researchers have offered various explanations for why gender segregation in early play patterns occurs. One of the most popular (Maccoby, 1988, 1990, 1998) is that girls initially withdraw from mixed-gender play because boys dominate the interactions and are nonresponsive partners. This theory was recently supported in a study by Moller and Serbin (1996) who found that gender-segregating boys had a more disruptive play style than did girls or nonsegregating boys, whereas gender-segregating girls scored higher than other children on teacher ratings of social sensitivity (i.e., verbal skills, not shy, meets new situations well). Other theories emphasize the roles of modeling, differential reinforcement, activity preferences, and cognitive developmental factors (for reviews, see Hartup, 1983; Huston, 1983; Maccoby, 1998).

Direct comparisons between the nature of children's play in same-gender versus cross-gender dyads reveal a number of differences. Although the differences between same- and cross-gender interactions are often different for boys and girls, some differences are found for both genders. For example, children receive more positive consequences for their behavior from same-gender versus opposite-gender peers (Snyder, West, Stockemer, Gibbons, & Almquist-Parks, 1996), children in same-gender dyads are more likely to reinforce gender-typed behaviors (Charlesworth & Hartup, 1967; Fagot & Patterson, 1969; Lamb et al., 1980), and the reinforcement of same-gender peers appears to be more effective than reinforcement of peers offered by the opposite gender (Fagot, 1985; Roopnarine, 1984). Same-gender children are more likely to challenge each other during play (Phinney & Rotherman, 1982), assert their needs (Serbin, Sprafkin, Elman, & Doyle, 1982), and offer conciliatory gestures in terminating conflicts (Sackin & Thelen, 1984). In contrast, children talk more to one another during opposite-gender play (Liss, 1983), and are less likely to engage in gender-stereotyped activities (Lloyd & Smith, 1985; Smetana & Letourneau, 1984). However, in an experimental study in which children engaged in solitary toy play either alone or in the presence of a same- or an opposite-gender peer, children showed the greatest amount of gender-typed play in the presence of an opposite-gender peer (Serbin, Connor, Burchardt, & Citron, 1979).

Several studies have shown that, when interacting with an opposite-gender peer, both girls and boys act more like the opposite gender than when interacting in same-gender dyads. The net effect of this change is that, when with boys, girls become more direct and assertive, and, when with girls, boys become less so. Specifically, as compared to when they are playing in same-gender dyads, girls playing with boys speak in shorter utterances (Duveen & Lloyd, 1988), are more controlling (Leaper, 1991), make more offers and information statements (Phinney & Rotherman, 1982), are more boisterous (Smith & Inder, 1993), are less cooperative (Neppl & Murray, 1997), show more parallel play (Neppl & Murray, 1997), and are more physically aggressive (Howes, 1988b; Phinney & Rotherman, 1982; Smith & Inder, 1993). Conversely, they are less quiet (Smith & Inder, 1993), and less likely to smile, to engage in positive body contact (Langlois et al., 1973), or to use indirect requests (Serbin et al., 1982) or collaborative statements (Leaper, 1991). Overall, girls' direct requests are more successful in play with girls than with boys (Serbin et al., 1982). Boys, when engaged in cross-gender play, speak in longer utterances (Duveen & Lloyd, 1988); make fewer counterassertions (Serbin et al., 1982), imperatives

(Phinney & Rotherman, 1982), and direct requests (Serbin et al., 1982); and are less physically aggressive (Howes, 1988b; Langlois et al., 1973; Phinney & Rotherman, 1982). However, boys in cross-gender interactions are less responsive to the requests of their partners, less likely to withdraw in response to the partner's prohibitions (Jacklin & Maccoby, 1978), and less likely to respond positively to the partner's request for social interaction (Howes, 1988b). Moreover, mixed-gender dyads are those with the highest probability of one or the other partner exiting play (Smith & Inder, 1993).

Despite the tendency for gender differences in behavior to become less pronounced in the presence of an opposite-gender partner, not all gender differences in play behavior disappear. Examination of gender differences during cross-gender play yields mixed results. Probably the most consistent findings are that gender differences in aggression are typically not found in mixed-gender play (Langlois et al., 1973; Phinney & Rotherman, 1982; Serbin et al., 1982), that gender differences in vigorous play remain (Lloyd & Smith, 1985), and that girls are more likely than boys to withdraw from play in the mixed-gender context (Duveen et al., 1988; Jacklin & Maccoby, 1978). Studies of verbalizations are inconsistent (e.g., Haas, 1981; Langlois et al., 1973; Miller et al., 1986; Phinney & Rotherman, 1982; Serbin et al., 1982).

To further complicate matters, it appears that gender differences in play also vary as a function of the age makeup of the classroom. Field (1982) found that children only showed preference for same-gender peers when they were involved in interaction with same-age peers; when children were in a mixed-age group, these preferences disappeared. Although gender segregation has been found in other studies of mixed-age classrooms (Goldman, 1981; Gray & Feldman, 1997; Urberg & Kaplan, 1989), the reasons for the differences between these studies is not known. Gender segregation may also be a function of the availability of same-gender, same-age peers. In two studies of children's play partners in the home (Bloch, 1987; Ellis, Rogoff, & Cromer, 1981), same-gender child companions were no more common during the preschool years than were opposite-gender companions. This might be a function of the limited number of same-age peers available in this context. Ellis et al. (1981) found that children began to spend more time with same-gender companions only after the age of 6.

Other studies of the age of the play partner also yield mixed results. Although some studies show preferences for interaction with same-age peers (Gray & Feldman, 1997; Urberg & Kaplan, 1989), others do not (Goldman, 1981). Moreover, the preference for same-age playmates appears to increase with age (Ellis et al., 1981), at least up to adolescence

(Gray & Feldman, 1997). Although some researchers have found that children engage in more interactive play in same-age groups and more parallel play in mixed-age groups (Roopnarine, 1981; Roopnarine & Johnson, 1983; Urberg & Kaplan, 1989), both Goldman (1981) and Mounts and Roopnarine (1987) saw more parallel play in same-age groups. Two researchers found higher levels of play in mixed-age versus same-age groups: Mounts and Roopnarine (1987) noted more constructive play; and Field (1982) found more imitation, verbal interaction, and fantasy play. Both Lougee, Grueneich, and Hartup (1977) and Bailey, McWilliam, Ware, and Burchinal (1993) observed that play with younger children was associated with less positive (or more negative) social interaction, and play with older children with more positive (or less negative) interaction. Langlois, Gottfried, Barnes, and Hendrick (1978) found complex interactions among age of the child, age of the partner, and gender composition of the dyad.

In summary, the nature of gender differences in children's play is complex and varies with the gender and age of the partner. As already mentioned, one of the problems with naturalistic, observational research in this area is that, due to early patterns of gender segregation and the age homogeneous settings in which most studies are conducted, the gender and age of the children in these studies are confounded with the gender and age of the play partners. Thus, such studies tell us less about gender differences in the behavior of individual children than about gender differences in the behavior of same-gender dyads. When researchers have varied the gender and/or age of the play partner, the results are complex. However, some general conclusions can be drawn. First, and not surprisingly, children's behavior in their preferred context (same-gender, same-age interaction) is more positive, responsive, and conciliatory. Many (but not all) of the gender differences in children's behavior go away when children are playing with an opposite-gender partner; for example, with girls becoming more directive, and boys becoming less so. Research on this issue, however, is inconsistent, and studies examining the situational factors (e.g., type of play materials, structure of play setting, other people present) that moderate the effects of gender of partner on children's play behavior need to be conducted.

Gender differences in children's behavior vary with the age of the partner as well. Children appear to play with same-gender, same-age partners when they can, but when such partners are not available, the existing research suggests that age may be more important determinant than gender in partner choice. The differences between same- and mixed-age play vary considerably across studies, but these may be a function of the size of the age difference, the ages of the children, the gender composition and size of the group, and the setting in which the observations occur. For example, if the age difference is large, one might expect more parallel play, whereas if the age difference is small, more complex play (at least for the younger members of the group) would be expected. Finally, the effects of mixed-aged interactions may be different for younger and older children, with younger children showing more positive interactions and higher levels of play (as long as the age difference is not too great), but interactions for older children being more negative.

Research on the effects of partner familiarity (and especially friendship) is more consistent. In a meta-analytic review, Newcomb and Bagwell (1995) found that friends differed from non-friends in the nature of their social interactions. Most of the studies of young children involved observations of free play. Specifically, when interacting with friends, children showed greater social contact, talking, cooperation, positive affect, and conflict resolution than when interacting with acquaintances or strangers. Children have been found to engage in more pretend play in general, and more cooperative and/or elaborated pretend play in specific, with friends than with nonfriends (Doyle, Connolly, & Rivest, 1980; Howes et al., 1994; Vespo, 1991; Werebe & Baudonniere, 1991).

Starting in the late 1970s, a number of observational studies of young children's interactions with their siblings were published. In some of these studies, a few differences in the way in which children play with friends versus siblings were identified. For example, Abramovitch, Corter, Pepler, and Stanhope (1986) found that children engaged in more prosocial acts and play-related interactions with friends than with siblings; and Phinney (1986) observed that children were more likely to use compromise and reasoning in conflicts with friends, whereas they were more likely to use counterassertions with siblings. Stoneman et al. (1984) noted that boys' interactions with friends were more egalitarian than were their interactions with siblings. Stoneman et al. found that boys' activity preferences varied with partner as well: Boys were more likely to play competitive physical games with friends, and to play with toys with siblings. Girls, in contrast, were more likely to play ritualized games with siblings versus friends. One problem with comparing the behavior of children with their friends versus their behavior with siblings is that age differences between the children are usually confounded with partner status. That is, children's friends are usually closer in age to the child than are children's siblings. Although this could be untangled by observing interactions with a twin sibling or with a friend of the sibling's age, such studies could not be located.

The behavior of older and younger siblings differs during sibling play. Stoneman et al. (1984) summarized much of this research by describing the nature of the roles during sibling interactions. Younger siblings were found to spend more time in the learner or managee roles, whereas older siblings spend more time teaching and managing. Specifically, when toddlers interact with older siblings, they are more likely than their brother or sister to monitor and imitate the partner's behavior, to smile, to accept toys, to pick up toys abandoned by their partner, and to submit to aggression (Abramovitch et al., 1979; Lamb, 1978a, 1978b). Older siblings are more likely to vocalize, to offer and take toys, to be aggressive, and to respond to aggression with counteraggression (Abramovitch et al., 1979; Lamb, 1978a, 1978b). Similar patterns are found among preschool children and their older siblings. Younger siblings are more likely to approach the older sibling, to imitate his or her behavior, to submit to aggression, and to respond positively to the partner's prosocial behavior (Abramovitch et al., 1986; Abramovitch, Corter, & Pepler, 1980; Corter, Pepler, & Abramovitch, 1982; Pepler, Abramovitch, & Corter, 1981). Conversely, older siblings are more likely to initiate play, to engage in on-task and prosocial behavior, to be verbally and physically aggressive, and to respond to aggression with counteraggression (Abramovitch et al., 1980, 1986; Corter et al., 1982; Dunn & Munn, 1986a; Pepler et al., 1981; Vespo, Pedersen, & Hay, 1995). Abramovitch and colleagues (1986) observed that the tendency for younger siblings to imitate the behavior of older siblings disappeared by the time the younger sibling was 5 years of age, and imitation occurred infrequently. Although most studies do not report effects for the size of the age difference between siblings, two studies found that older siblings were more prosocial if there was a large versus a small age difference (Corter et al., 1982; Dunn & Munn, 1986a).

Research on sibling play confirms some of the same developmental trends identified in peer play described earlier. For example, both prosocial behavior and aggression increase with age (Abramovitch et al., 1986; Dunn & Kendrick, 1981; Dunn & Munn, 1985; Pepler et al., 1981), as does sibling cooperation (Dunn & Munn, 1986a; Volling, Youngblade, & Belsky, 1997). Sibling conflicts also become increasingly verbal in nature (Munn & Dunn, 1989; Vespo et al., 1995), and are more likely to be resolved through the use of such strategies as reasoning and explanation or references to social rules (Dunn & Munn, 1985; Phinney, 1986).

Fewer gender differences have been identified in the context of sibling play than peer play. The only gender difference reported by more than one researcher was that boy siblings use more aggression than do girls (Abramovitch et al., 1979; Hinde, Tamplin, & Barrett, 1993; Lamb, 1978b). Lamb (1978b) reported that girls initiated more social behaviors toward their younger siblings than did boys, and Abramovitch and colleagues (Abramovitch et al., 1979; Corter et al., 1982) noted more prosocial behavior in girls. No other consistent gender differences in sibling interaction have been reported.

Findings concerning the effects of the gender composition of sibling dyads on play behavior are complex and often inconsistent. For example, although Dunn and Kendrick (1981) found more positive behavior in same-gender than in opposite-gender dyads, Stoneman et al. (1984) found just the opposite. Other findings show that same-gender siblings display more prosocial behavior (Abramavitch et al., 1986), spend greater time in conflict (Vespo et al., 1995), are more involved in competitive physical activities (Stoneman et al., 1984), and spend more time in the learner role (Stoneman et al., 1984). Stoneman, Brody, and MacKinnon (1986) observed that opposite-gender dyads, in contrast, spent more time in art activities, and Abramovitch and colleagues (1986) found that, with increasing age, younger siblings in opposite-gender dyads were less likely to imitate the behavior of their older sibling. Abramovitch et al. (1986) speculated that this later finding may reflect changes in gender-role development.

Stoneman and colleagues (1984) also presented numerous findings regarding the differences between male and female siblings in same- and mixed-gender dyads. These findings are even more complex than the findings reported previously for peers, because not only does one need to consider the gender composition of the dyad, one also has to consider in mixed dyads whether the younger child is a boy or a girl. In general, the results of their analyses showed that same-gender dyads showed more gender-stereotypical behavior than did mixed-gender dyads, and that the older rather than the younger sibling in mixed-gender dyads had the greatest effects on the nature of play behavior.

Dominance Hierarchies. Besides friendship and affiliative relationships, dominance patterns in groups of young children also predict their behavior during free play. Several studies have shown that agonistic interactions during children's free play are governed by dominance hierarchies (Frankel & Arbel, 1980a, 1980b; Jones, 1984; La Freniere & Charlesworth, 1983; Pettit, Bakshi, Dodge, & Coie, 1990; Sluckin & Smith, 1977; Strayer & Strayer, 1976, 1978; Strayer & Trudel, 1984). Most studies employ two quantitative indexes that assess the degree to which children's agonistic interactions conform to a linear dominance hierarchy: indexes of linearity and rigidity. Linearity refers to the proportion of dyadic *relationships* that conform to a linear dominance hierarchy, whereas rigidity refers to the proportion of agonistic *episodes* where the outcomes are consistent with a linear model. These indexes are computed by making up a matrix of all agonistic interactions and then computing the following values:

$$\text{Linearity} = \frac{\text{Total Dyads} - \text{Relational Reversals}}{\text{Total Dyads}}$$

$$\text{Rigidity} = \frac{\text{Total Episodes} - \text{Episode Reversals}}{\text{Total Episodes}}$$

Therefore, a linearity value of .90 means that only 10% of the dyadic relationships observed violated linearity. For example, in a linear dominance hierarchy A, B, C, D, if D were dominant to B it would be a relational reversal. If rigidity were .90, that means that 90% of the actual agonistic interactions followed the identified dominance hierarchy. Thus, even though A was usually dominant to B, at least some of the time B might dominant A (an episodic reversal).

Despite differences in the samples studied, the behaviors coded, and the ways in which dominance is defined (see later discussion), studies of dominance hierarchies during the free-play interactions of young children have been remarkably consistent. Of the studies found for the present review (those listed earlier), all reported linearity values of .81 or higher, and all but the Frankel and Arbel (1980b) study of 2- to 3-year-olds reporting values of .91 or higher. With the exception of the work of Strayer and Strayer (1976, 1978), rigidity values are typically lower, with values ranging between .67 and .81. Strayer and Strayer (1976, 1978), probably for reasons considered later in this chapter, reported much higher values, ranging from .92 to .95. The linearity and rigidity values for children are in the same range as those found for other primates. For a study they conducted of captive macaque monkeys, Strayer and Strayer (1976) reported values of 1.00 for linearity and .90 for rigidity.

Dominance hierarchies have been identified for groups of children ranging from 25 to 28 months (Frankel & Arbel , 1980a) and first to third grade (Pettit et al., 1990). Dominance has been defined in a variety of ways, with the two most common being wins at object struggles, and aggressive behaviors or threats followed by submissive behaviors. Strayer and Strayer (1976, 1978) employed the most stringent definition of dominance, restricting their definition of dominance interactions to those episodes in which an aggressive behavior or threat by one child was followed by a submissive behavior by the other. Because such a definition excludes certain episodes from the analysis, this might explain their higher rigidity values. They maintained that the behavior of both partners must be considered in order to accurately assess dominance, because dominance is a characteristic of relationships, not individuals. Strayer and Strayer also found that dominance hierarchies using the results of object struggles yielded lower values for linearity and rigidity, probably because nonstatus variables, such as prior possession rules, govern the outcome of such encounters (see Frankel & Arbel, 1980a).

Besides levels of linearity and rigidity similar to other primates, detailed observational analyses of children's nonverbal behavior during agonistic interactions show similarities with the behavior of other primates as well. Camras (1977) identified aggressive facial expressions displayed by children during conflict that predict both the displaying child's resistance to subsequent attacks as well as the partner's hesitance in launching a counterattack. As was found by previous child ethologists (e.g., Blurton-Jones, 1967; Grant, 1969), these expressions (i.e., lowered brows, stare, face thrust forward, lips pressed together, lips thrust forward) have similarities to threat gestures in other primates. Similarly, Zivin (1977a, 1977b) identified a "plus" face that children often showed after winning conflicts, and a "minus" face that they bore after losing. These faces, Zivin argued, have similarities to threat and submissive gestures in many primates. Ginsburg, Pollman, and Wauson (1977) found that children who went on to lose conflicts reliably engaged in "diminution" of their body posture just prior to the cessation of aggressive behavior by their partner. This tendency, similar to an appeasement gesture shown in many mammals (Darwin, 1898; Eibl-Eisenfeldt, 1970), involves "making oneself smaller" as a sign of submission. Examples included bowing the head, slumping the shoulders, lying motionless on the ground, kneeling, and even shoe tying. Finally, Keating and Bai (1986) found that 4- to 7-year-old children labeled pictures of adult models with lowered brows and/or unsmiling faces as more dominant than those with neutral brows or smiles.

Although the early studies of dominance hierarchies focused on physical aggression or object conflict, other studies have shown similarly high linearity and rigidity values for hierarchies based on observational systems that combine verbal and nonverbal aggression, and for those that combine aggression with nonagonistic social control (e.g., Jones, 1984; La Freniere & Charlesworth, 1983; Pettit et al., 1990). When investigators have constructured dominance hierarchies separately for aggression and verbal influence, they are highly correlated. For example, La Freniere and Charlesworth (1983), observed that ranks from verbal and physical dominance hierarchies were correlated positively, with *r*s ranging from .60 to .74. Jones (1984) noted that the dominance hierarchy based on physical and verbal aggression was correlated positively with the dominance hierarchy based on "leadership organization" (i.e., directs play, instructs, states/enforces rules, distributes offers/objects, grants/denies play). This correlation was $r = .52$ for males and $r = .55$ for females. This is consistent with the view of many ethologists that dominance hierarchies maintain the social order in higher primates not simply through aggression and threats of aggression, but also via other, more subtle means.

Theorists often cite the low levels of actual aggression observed in groups of young children as support for the role of dominance hierarchies in controlling aggression. In two studies that examined the development of dominance hierarchies over time, it was found that aggression declined as the hierarchies developed (Frankel & Arbel, 1980a; Pettit et al., 1990). Moreover, Strayer and Strayer (1978) noted that the most dominant child was not necessarily the most aggressive. A child can be dominant without engaging in much overt aggression and, in their studies, the most aggressive children have been somewhere in the middle of the hierarchies.

Given the low levels of observed aggression in preschool groups, some theorists have argued that factors other than agonistic interactions are important for the social organization of these social groups (e.g., Vaughn & Waters, 1981). For example, Strayer (1980) found that the agonistic and affiliative structures in the classrooms studied were sometimes very different. It has been asserted, however, that high frequencies of overt aggression are not necessary for the maintenance of the social order in the higher primates. One alternative theory was offered by Chance (1967), who stipulated that social behavior in groups of primates is organized in part by an underlying "attention structure." In this analysis, dominance relations reflect the degree to which animals dominate the attention of others in the social group:

> A redefinition of dominance is now proposed on the assumption that the attention-binding effect of an animal in a group is essentially the quality which puts it in a behaviourally focal position, and which also tends to place it near the group's spatial centre. The dominant animal therefore may be said to dominate the attention of others at most, if not all, times, and usually without specific actions to do so. (p. 509)

Chance (1967) asserted that one can examine the attention structure of a primate group by analyzing the patterns of visual attention between its members.

Numerous studies of attention structure in children's groups have indeed found that dominance rank based on agonistic and/or assertive interactions is associated with attention structure, with the most dominant individuals receiving the greatest amount of attention (Abramovitch, 1976; Anderson & Willis, 1976; Frankel & Arbel, 1980b; La Freniere & Charlesworth, 1983; Melson & Dyan, 1987; Vaughn & Waters, 1981). Moreover, agonistic and/or assertive dominance status is positively associated with a number of other positive social behaviors, including the amount of conversation, affiliation, and social play (Jones, 1984; La Freniere & Charlesworth, 1983; Pettit et al., 1990; Strayer & Trudel, 1984); the number of play partners (Jones, 1984); leadership and success in social control (Jones, 1984; Pettit et al., 1990); the receipt of positive attention and/or prosocial behavior (Frankel & Arbel, 1980a, 1980b; Jones, 1984); being imitated by others (Frankel & Arbel, 1980a, 1980b); and peer ratings of social preference, leadership qualities, humor, aggression, and positive psychological adjustment (Pettit et al., 1990). Regarding dominance status and social influence, Keating and Heltman (1994) observed that dominance status was positively associated with the ability to encode credible, deceptive messages in preschool children.

Given the positive correlation between attention and dominance status, it should not be surprising that the positive social behaviors listed previously have been found to be associated with high attention rank as well (Abramovitch & Grusec, 1978; Hold, 1976; La Freniere & Charlesworth, 1983; Vaughn & Martino, 1988; Vaughn & Waters, 1981; Waters, Garber, Gornal & Vaughn, 1983). In some studies, attention received has actually been found to be a better predictor of these behaviors than has agonistic dominance status (e.g., Vaughn & Waters, 1981). Vaughn and Langlois (1983) noted that the correlation between attention rank and peer popularity was not an artifact of physical attractiveness: Physical attractiveness ratings were positively correlated with peer acceptance, but were uncorrelated with attention rank.

Given the correlations among dominance, attention, and social behavior, it is not clear why dominant individuals receive the most attention from their peers. Three possibilities to consider are that submissive children monitor, imitate, and initiate positive interactions with dominant children to prevent becoming victims of aggression, because dominant children engage in the highest levels of social influence and control, or because dominant children are the most socially active and/or involved in the most interesting activities. Despite the reasons for these associations, it is well established (at least in the preschool years) that dominance, attention, and level of social activity (both initiated and received) tend to co-occur. Hold-Cavell (1985) added another behavior to this list: Showing off—a behavior similar to display behaviors in other animals—was also positively associated with aggression and attention received.

Studies of Non-Western Cultures. The amount and nature of children's social object and social pretend play varies significantly across cultures. Social play varies so much, in fact, that anthropologists have used the study of children's play to analyze culture. By examining the structure and rules of children's games, researchers have gained insights into the social dynamics of the cultures they study (for reviews, see Schwartzman, 1976, 1986). Although games are but one form of play seen in young children, anthropologists have devoted considerably less attention to nongame forms (e.g., constructive play, sociodramatic play). Moreover, what little is known about these forms comes from ethnographic field notes and provides little information about the frequency, quantity, quality, representativeness, and sequential structure of children's nongame play patterns. Instead, researchers usually describe in a brief paragraph the kinds of play that are observed and may provide one or two examples.

Despite the limited database, research has shown that children from diverse hunting-and-gathering and agricultural village cultures around the world begin to engage in social object and social pretend play with other children as toddlers or shortly thereafter. The most commonly described play form in this age range is sociodramatic play (children's games generally emerge in middle or later childhood). Anthropologists sometimes call this *imitative play*, in that the play involves acting out the behavior of adults in the environment. The most common themes are those centered around domestic life (e.g., cooking, eating, cleaning, child care, sleeping, dressing up, going to market), adult work activities (e.g., hunting, gathering, planting, harvesting, fishing), and adult rituals (e.g., marriage, dancing). Children in many cultures have few toys for engaging in these activities, but show a great deal of resourcefulness in using natural materials and/or discarded items in constructing toys. Children construct houses, dolls, tools, weapons, canoes, and other vehicles (e.g., Bloch, 1984; Jackson, 1964; Prosser et al., 1986; Shoemaker, 1964). These basic forms of imitative play have been observed in hunting-and-gathering or agricultural village cultures in Africa (Bloch, 1984; Eibl-Eibesfeldt, 1989; Leacock, 1971; LeVine & LeVine, 1963; Liddell, Kvalsvig, Strydom, Qotyana, & Shabalala, 1993; Turnbull, 1962), India (Minturn & Hitchcock, 1963; Roopnarine, Hossain, Gill, & Brophy, 1994), Japan (Maretzki & Maretzki, 1963), Mexico (Romney & Romney, 1963), Nepal (Anderson & Mitchell, 1984), the Philippines (Nydegger & Nydegger, 1963), South America (Eibel-Eibesfeldt, 1989; Jackson, 1964; Shoemaker, 1964), the South Pacific (Hogbin, 1946; Martini, 1994), and Sri Lanka (Prosser et al., 1986).

Information on other forms of social object play is much less common. Two studies were located that collected quantitative, observational data on children's play behavior in traditional village settings: one in Senegal (Bloch, 1989) and one in South Africa (Liddell et al., 1993). The main methodological difference between the two studies is that Liddell et al. (1993) collected data both inside and outside of the home, whereas Bloch (1989) primarily collected data out of doors. Liddell et al. (1993) studied 5-year-olds; Bloch (1989) studied 2- to 6-year-olds. Both studies found that fantasy occurred in about 5% of the observation periods for boys and about 10% of the observation periods for girls. The most frequently occurring types of play varied considerably across studies. In the Liddell et al. (1993) study, engrossed manipulation of objects was most common (about 15% of observations), followed by fantasy (about 10%), children's games (about 8%), and acrobatic play (about 5%). In Bloch (1989), the most common activities were gross motor (about 12% of observations), followed by pretense (about 7%), and functional (about 4%). If one considers that the Bloch (1989) observations were conducted mostly outside, this might have accounted for the higher levels of gross motor and lower levels of functional play. When Bloch (1989) compared her observational data collected in Senegal to the data she collected on children in Wisconsin using parent phone reports (Bloch, 1987—see earlier discussion on play structure), she found only one significant difference: "While American children engaged in more gross motor play (total time) than did Senegalese children, both groups were virtually identical in time spent in pretense, functional, constructive, exploratory, rough-and-tumble, and music/art play" (pp. 139–140).

Data on qualitative differences in children's play across cultures is even harder to find. Several observers have noted that children's play in some traditional cultures is noncompetitive (e.g., Anderson & Mitchell, 1989; Draper, 1976). This was also found in a comparative study of children in North American and Chinese preschools (Navon & Ramsey, 1989). Others have commented that children's sociodramatic play in some traditional cultures is more routinized and less elaborated than in Western cultures (Gaskins & Goncu, 1992; Martini, 1994). Also, children's play in traditional cultures involves imitation of adult behavior in traditional roles and rarely involves fantastic transformations (i.e., character roles that the child will seldom, if ever, enact or encounter in later life; Feitelson, 1977; Gaskins & Goncu, 1992).

The limited data on developmental changes in children's play in traditional cultures is consistent with the research conducted in Western settings. In studies of traditional cultures, functional play decreases with age (Bloch, 1989), sociodramatic play begins in early childhood (see studies cited earlier), and games begin somewhat later (for reviews, see Schwartzman, 1976, 1986). In a study of two kindergartens in urban Taiwan (children between the ages of 3 and 6), Pan (1994) found that constructive and parallel play were more common with increasing child age, whereas functional and solitary play were less common.

Gender differences in children's play are mostly parallel with Western research. In the studies collecting quantitative data, girls were found to engage in more fantasy play than did boys (Bloch, 1989; Liddell et al., 1993; Prosser et al., 1986), as well as in more cooperative play and rhymes (Liddell et al., 1993). Boys, in contrast, engage in more negative actions (Prosser et al., 1986; Whiting & Edwards, 1973), more rough-and-tumble play (Whiting & Edwards, 1973), more gross-motor play (Bloch, 1989; Graves, 1978), and more solitary play and engrossed manipulation of objects (Liddell et al., 1993). In their qualitative descriptions of children's play activities, ethnographers (with few exceptions) reported that girls engage in domestic activities (e.g., cooking, eating, cleaning, going to market, playing with dolls) and dancing, whereas boys engage in hunting, fighting, or farming play (see references cited earlier). A comparison of these activity preferences with those in Table 8.5 for Western cultures reveals many similarities (especially for girls). Boys' play in Western cultures has not been found to emphasize hunting or farming themes, possibly because the studies of Western children are usually carried out in urban settings.

Research on play partners shows that, unlike children in most Western school settings, children in traditional cultures usually play in mixed-age groups (Anderson & Mitchell, 1984; Draper, 1976; Farver & Howes, 1988; Gaskins & Goncu, 1992; Liddell et al., 1993; Martini, 1994; Rogoff, 1981; Whiting, 1963). In most settings, this appears to be due to at least two factors: the high incidence of sibling caregiving in many cultures (Weisner & Gallimore, 1977), and the limited number of same-age play partners available for play (Draper, 1976; Gaskins & Goncu, 1992). Compared to children in Western cultures, children in many traditional cultures spend most of their days in mixed-age play groups, under little adult supervision, in which toddlers and young children are cared for by older children in the group (e.g., Anderson & Mitchell, 1984; Eibl-Eibesfeldt, 1989; Farver & Howes, 1988; Martini, 1994; Whiting, 1963). Children have specialized roles in these groups, with toddlers on the periphery, mostly watching and cared for by older children, but occasionally given "bit parts" in play activities; "initiate members" who are old enough to play a more active role in group activities, but who usually follow the lead of others and may care for the toddlers; and leaders who introduce activities, direct group play, invent new play, and also care for the younger children, making sure that they follow the group rules (Anderson & Mitchell, 1984; Eibl-Eibesfeldt, 1989; Gaskin & Goncu, 1992; Martini, 1994). As children grow older they move into new roles—usually from peripheral to follower to leader roles (Eibl-Eibesfeldt, 1989; Martini, 1994). In one study that compared the nature of play in same- and mixed-age play groups in Indonesia, Farver and Howes (1988) observed that children showed more positive affect and social play in same-age groups and more complementary and reciprocal play in mixed-age groups.

Although age segregation is rare in these cultures, gender segregation is more common (Farver & Howes, 1988; Liddell et al., 1993; Whiting, 1963)—usually because of apparent differences in preferred activities. Gender segregation, however, can only occur when a sufficient number of play partners is available. For children with small numbers of available partners, gender segregation is not reported (e.g., Draper, 1976; Gaskins & Goncu, 1992).

In a study of dominance relationships in a playgroup of G/wi San children in the Kalahari desert (Hold, 1980), attention rank was positively correlated with children's initiating and organizing behaviors. Older children in the group were more likely than younger children to be imitated and to receive various social behaviors (i.e., ask, tell, and show).

Because the level of children's social play varies so much across cultures, theorists have tried to identify those factors that might contribute to these differences. Based on a review of the literature, Feitelson (1977) listed four: extended periods of free time for play to occur, adequate play materials, sufficient space in the physical environment, and a favorable attitude of adults toward children's play. Bloch (1989) provided data relevant to the first factor. She found that the amount of time children in her study spent in housework was negatively correlated with time spent in play ($r = -.40$). Similarly, Harkness and Super (1983) noted that as children's time spent in work increased developmentally there was a corresponding decrease in time spent in play. Schwartzman (1986), however, questioned these "current assumptions about play." Children, she argued, find ways to integrate work and play, playing while they work; do not need much space to play; and, when insufficient play materials are available, make their own.

FUNCTIONS/EFFECTS

Play theorists have been remarkably resourceful in generating possible functions and effects for children's social object and social pretend play. Children's social play has been hypothesized to contribute to the cognitive, motor, and social development of children, including the development of perception, attention, memory, problem-solving skills, language, communication, creativity, logical operations, emotion recognition, self-regulation, social skills, gender roles, social relationships, conflict resolution, coping with stress, and so on (see chapters in Bergen, 1988, for a particularly comprehensive view). Given that children devote a good deal of time to social play, undoubtedly much learning occurs in this context. However, research on the functions and effects of social play has been limited to a much smaller number of areas. These are summarized in Table 8.8. Research on these functions/effects is reviewed here.

Cognitive Stimulation and Learning

Correlational Studies. As shown in Table 8.8, the most studied functions/effects regarding children's cognitive development fall into four areas: promoting general cognitive development, promoting linguistic

TABLE 8.8

Hypothesized Functions/Effects of Social Object Play in Children

Cognitive stimulation and learning
Promoting general cognitive development
Promoting linguistic skills
Providing information about the physical environment
Providing novelty and diversity in responding
Social development
Establishing and reaffirming social relationships
Facilitating the development of social perspective-taking skills
Facilitating the development of social problem-solving skills

skills, providing information about the physical environment, and producing novelty and diversity in responding. Correlational evidence for the first two functions/effects includes studies reporting positive correlations between the quality of children's social play interactions and standardized measures of cognitive functioning. In general, studies have found that play at higher levels of Smilansky's (1968) play classification is associated with higher scores on such measures. For example, Johnson et al. (1982) found that involvement in constructive play correlated positively with performance on the Peabody Picture Vocabulary and the Raven Progressive Matrices Tests, as did the number of symbolic transformations during play. Similarly, Johnson (1976) noted that social (but not nonsocial) dramatic play positively correlated with preschoolers' performance on the Peabody and the picture completion subtest of the Wechsler Primary and Preschool Scale of Intelligence. Moreover, Pellegrini and colleagues, in two separate studies (Galda, Pellegrini, & Cox, 1989; Pellegrini, 1980), observed that involvement in dramatic play was positively associated with performance on tests of reading and writing readiness in preschool and kindergarten children. These same play behaviors have been seen to correlate positively with assessments of children's understanding of the physical environment. Jennings (1975) found that the amount of constructive play correlated positively with children's performance on several physical knowledge tests (picture completion, block design, geometric design, and object classification), and Rubin and Maioni (1975) observed that dramatic play positively correlated with classification skills. Functional play, in contrast, was negatively correlated with classification performance (Rubin & Maioni, 1975).

Although the data are somewhat inconsistent, children's involvement in gender-typed play activities has also been shown to be associated with children's performance on verbal and performance measures. In general, involvement in gender-typed feminine activities is correlated with verbal skills, whereas involvement in gender-typed masculine activities is correlated with nonverbal performance. Although one study has found these relationships for children regardless of their gender (Serbin & Conner, 1979), some studies found significant results only for girls (Coates, Lord, & Jakabovics, 1975; Fagot & Littman, 1976), whereas others found results only for boys (Conner & Serbin, 1977).

Play, especially dramatic play, has been shown to be positively associated with performance on assessments of creativity and divergent thinking. For example, Johnson (1976) found that social, but not nonsocial, fantasy play was positively associated with divergent thinking in both a multiple-uses and a story-completion task. Moran, Sawyers, Fu, and Milgram (1984) observed positive correlations between imaginative play and divergent thinking as well. Fantasy-play measures in the preschool years predict creativity during middle childhood. Shmukler (1982–1983) found a positive relationship between observed imaginative play in the preschool period and creativity in a multiple-uses task in the third grade. Similarly, Griffing, Stewart, McKendry, and Anderson (1983) noted that social interaction and verbal communication during dramatic play in the preschool years positively correlated with fluency in a story-telling task 5 years later. These results are consistent with studies employing teacher ratings of children's imaginativeness or playfulness and children's divergent thinking (Barnett & Kleiber, 1984; Durrett & Huffman, 1968; Lieberman, 1965; Singer & Rummo, 1973).

Despite the number of studies that have shown relationships between children's play (especially constructive and dramatic play) and their convergent or divergent thinking, not all studies have obtained significant relationships (see Cole & La Voie, 1985; Doyle et al., 1991; Dunn & Herwig, 1992; Truhon, 1983). However, comparison across studies is often difficult given differences in the populations, settings, and measures employed. Even when associations are found, however, it is not clear what they mean. Does play contribute to cognitive development, does cognitive development contribute to play, or are these correlations due to some unmeasured third variable, such as verbal competence, temperament, or parenting? Without experimental data, the direction of cause and effect is impossible to determine.

Experimental Studies. To more directly test the hypothesis that children's social play contributes to their cognitive development, numerous experimental training studies have been conducted. The vast majority of these studies were conducted in the 1970s and early 1980s, when enthusiasm for play as a research topic was at its peak. Most studies were conducted with low-income children as an attempt to increase the level of play in these populations. In the typical study, children in a preschool setting serving low-income families were randomly assigned to either a play-training group or one or more control conditions. Play-training, as first developed by Smilansky (1968), involves having an adult facilitate pretend play in small groups of children (most studies have had four to six children per group). The nature and level of adult involvement has varied from study to study, ranging from facilitating dramatic play by providing appropriate play materials and suggestions to actively structuring pretend play by reading children a story and them helping them act out the story themselves. The length of the training intervention also varies from study to study, ranging from a single meeting to sessions over the course of 3 years. However, most studies of play-training involve several sessions a week extending for at least a 2-month period. The effects of play in these studies usually have been explored by obtaining assessments of children's cognitive and/or social functioning both before and after the training, in order to determine if performance on the outcome measures is affected by the experimental intervention (as compared to nonplay controls).

In the mid-1980s, numerous reviews of the play-training literature were published that raised questions about the interpretation of the results of these studies (e.g., Christie & Johnsen, 1985; McLoyd, 1982; Rubin et al., 1983; Smith & Simon, 1984). The main criticism was that it was not clear whether the cognitive and social gains demonstrated by subjects in these studies were due to the effects of play per se, or due to the cognitive, verbal, and/or social stimulation children received from the adult play tutor. Also, most studies did not determine whether or not children in the sociodramatic play conditions actually engaged in more sociodramatic play in the classroom than did the controls. Although researchers attempted to deal with these issues by employing nonplay tutoring attention controls, few studies directly examined differences between the conditions in the level of adult simulation provided or in children's subsequent play in the classroom. Other concerns were that the specific play-training techniques were rarely based on a specific theoretical perspective, that the samples and effect sizes of these studies were small, that usually only short-term outcomes were examined, and that the reliance on low-income populations might have exaggerated their effectiveness. Finally, it is not clear from most studies that the persons conducting the assessments were blind to the experimental conditions or the hypotheses, raising the possibilities of experimenter bias and experimenter effects. Although research could easily be done to address most of these criticisms (e.g., see Christie, 1983; Pellegrini, 1984b; Smith, Dalgleish, & Herzmark, 1981), the unanticipated effect of these critiques apparently has been to discourage future inquiry in this area.

In reviewing the results of the experimental studies in this and the next sections, it is important to keep these limitations in mind. However, given the consistency of several findings, unless they are the result of experimenter bias (a determination that cannot be made because most publications do not provide the relevant information), some of the correlational findings reviewed earlier may reflect the operation of causal factors. In a meta-analysis of 46 studies conducted between 1974 and 1987 (of which only 7 were published after 1984), Fisher (1992) found an average effect size of $r = .35$ across all areas of cognitive, linguistic, and social development. Effect size varied little as a function of dependent variable, ranging from $r = .26$ for conservation skill to $r = .39$ for perspective taking. However, when the specific subtypes of play were examined there was much more variance in effect size, with studies of sociodramatic play showing the largest value, $r = .60$. Unfortunately for our purposes, though, the meta-analysis combined the results of experimental and correlational studies, making direction of cause and effect impossible to determine.

Experimental play-training studies have provided information about all four cognitive outcomes listed in Table 8.8. Saltz and colleagues (Saltz, Dixon, & Johnson, 1977; Saltz & Johnson, 1974), in two separate studies, demonstrated positive effects of play-training on low-income children's scores on standardized IQ tests, including the Weschsler Preschool and Primary Scale of Intelligence and the Picture Test of Intelligence. Two methodological strengths of these studies is that play-training lasted over a significant period (three times per week for 4 or 6 to 7 months), and two play and two nonplay control conditions were examined in both studies. Although the authors did not present data on the level of adult stimulation and interaction as a function of condition, the control conditions in both studies were designed to engage the children in high levels of interaction with an adult in either a nondramatic play or an instructional context. At the end of training in both studies, children in the fantasy and/or sociodramatic play-training conditions scored higher on standardized intelligence measures than did those in the control conditions. In the first study (Saltz & Johnson, 1974), however, there was an independent effect of "dimensionality training" on standardized test performance as well (i.e., teaching children "to recognize various forms and changes of physical stimuli," p. 625). In contrast, in a study comparing sociodramatic play training to "skills tutoring," Smith et al. (1981) found no statistically significant differences between the two groups in Wechsler performance. This study was methodologically superior to the Saltz et al. studies, in that Smith and colleagues observed tutor behavior during the sessions and demonstrated that there were no differences between the conditions in the time the that children spent with the tutor, in the level of activity involvement, or in the amount of assistance that the tutor provided. The study was also similar in that it employed low-income subjects, although training was more intense (5 times a week vs. 3 times a week) over a shorter period of time (8 weeks vs. 4 to 7 months).

Most play-training studies have focused on linguistic outcomes. Twelve studies were located. Outcome variables included performance on standardized assessments of verbal performance such as the Peabody Picture Vocabulary Test or the Illinois Test of Psycholinguistic Abilities (Christie, 1983; Dansky, 1980b; Lovinger, 1974; Shmukler & Naveh, 1984–1985; Smith et al., 1981; Smith & Syddall, 1978) or performance on story memory or storytelling tasks (Dansky, 1980b; Marbach & Yawkey, 1980; Pellegrini, 1984a, 1984b; Pellegrini & Galda, 1982; Saltz et al., 1977; Saltz & Johnson, 1974; Silvern, Williamson, & Waters, 1982; Taharally, 1991). The studies examining general verbal performance have, with but one exception, studied children from low-income families (Smith & Syddall, 1978, studied children from working-class families). Also with one exception (Shmukler & Naveh, 1984–1985), the results of these studies are consistent with the verbal stimulation hypothesis—that is, it is not play per se but the opportunity to interact with a trained experimenter in verbally stimulating activities that predicts gains in verbal skills over controls. Evidence supporting this notion are: (a) Lovinger's (1974) finding that children given sociodramatic play-training showed increases in their verbal performance, whereas controls attending regular nursery school did not; (b) the findings of three separate studies (Christie, 1983; Smith et al., 1981; Smith & Syddall, 1978) showing no differences in the verbal performance of children who had received sociodramatic play versus skills tutoring; and (c) Dansky's (1980b) finding that children given "exploratory training" (i.e., spending time with an experimenter discussing similarities and differences between objects and playing object guessing games) showed gains in verbal performance that were as high as those of children receiving sociodramatic play training (in fact, even though the exploratory and sociodramatic play groups did not significantly differ from one another, only the exploratory training group showed significantly higher posttraining verbal performance than did free-play controls).

The only study supporting the hypothesis that play per se leads to greater increases in general verbal performance is one conducted by Shmukler and Naveh (1984–1985) in South Africa. Shmukler and Naveh found that children in their two play-training groups (having an adult lead the children in acting out fairy tales, and having an adult facilitate children acting out themes of their own) had higher Wechsler verbal scores 1 month after the completion of 12 training sessions than did children in their two control groups (adult leads children in nondramatic play activities and no-treatment controls).

Studies of the effects of play-training on performance in story-related tasks (story memory and storytelling) provide evidence for the benefits of sociodramatic play per se. These studies include subjects from a broader range of social classes than did the previously mentioned studies, and in some cases the length of the intervention was limited to three or fewer sessions (Marbach & Yawkey, 1980; Pellegrini, 1984b; Pellegrini & Galda, 1982). Despite these differences that should work *against* finding effects unique to sociodramatic play, in every study children given sociodramatic play-training scored higher on measures of story recall and/or production than did those in attention or nonplay controls (Dansky, 1980b; Marbach & Yawkey, 1980; Pellegrini, 1984a, 1984b; Pellegrini & Galda, 1982; Saltz et al., 1977; Saltz & Johnson, 1974; Silvern et al., 1983; Taharally, 1991). This was the case despite the fact that these studies included the widest range of nondramatic play controls, such as dimensional training, nondramatic play activities, reading/discussing stories, exploratory training, picture drawing, playing with puppets, studying pictures, information rehearsal, and accommodating questioning by adults. Compared to controls, children in these studies who were encouraged to engage in sociodramatic play or to act out stories showed higher performance on story recall or storytelling measures collected on separate stories after the completion of play-training. In a study that included measures of both general verbal abilities and story measures, Dansky (1980b) noted that children given exploratory training scored as high on the verbal expression scale of the Illinois Test of Psycholinguistic Abilities as did children given sociodramatic play training (see earlier discussion), whereas children in the sociodramatic training group did better than the exploratory training group in four different storytelling tasks (story recall, story sequencing, storytelling, and story retelling).

Studies that have examined the effects of sociodramatic play training on children's understanding of the physical environment are rare. Although Golomb and colleagues (Golomb & Bonen, 1981; Golomb & Cornelius, 1977; Golomb, Gowing, & Friedman, 1982) conducted studies on play-training and children's conservation, play-training in these studies was done via single-child interactions with an adult experimenter and did not involve peers. In the one study of conservation we located that involved sociodramatic play training with peers (Fink, 1976), no effects on conservation of number or continuous quantity were found. Moreover, Guthrie and Hudson (1979) failed to replicate the Golomb and Cornelius (1977) findings, and suggested that those findings may have been due to experimenter effects. Finally, in the Dansky (1980b) study cited earlier, exploratory training increased children's recall for information in a curiosity box task, whereas children receiving sociodramatic play-training did no better than children in the free-play condition.

A final area that has been of interest to researchers studying play-training is its effect on producing novelty and diversity of responding. Seven such studies were located, with four showing positive effects of sociodramatic play-training on tests of divergent thinking (Dansky, 1980b; Nash, 1975; Shmukler & Naven, 1984–1985; Udwin, 1983) and three showing no differences in divergent thinking between children given sociodramatic play and skills tutoring (Christie, 1983; Smith et al., 1981; Smith & Syddall, 1978). A closer look at those studies finding significant effects shows that Nash (1975) reported effects unique to play for only one of three divergent thinking variables, and Udwin (1983) involved a unique and particularly high-risk group of subjects—institutionalized abused children who had been removed from their families. Moreover, none of these studies provided data on the level of adult activities as a function of condition.

Socialization and Social Skills

Correlational Studies. Correlational studies of the social effects/functions of social play have been primarily in three areas: the establishment and maintenance of social relationships, and the development of both social perspective-taking and social problem-solving skills. Each of these is reviewed in turn.

Correlational evidence on the possible effects of play on the development of social relationships in children comes primarily from studies of the behavioral correlates of children's sociometric status. In the typical study, children are individually interviewed about their playmate or friendship preferences, and the sociometric information from these interviews is correlated with data collected during observations of children's free play in a classroom or playground setting. The procedures for collecting and analyzing sociometric data have evolved considerably over the last 10 years, and a review of these methods is beyond the scope of this chapter (for a review, see Newcomb, Bukowski, & Pattee, 1993). However, two important methodological conclusions in this area are that it is important to collect separate data on peer acceptance and rejection (these variables have separate correlates, and one is not necessarily the opposite of the other—e.g., controversial children are high on both); and one needs to differentiate between neglected and rejected children—although both are low on peer acceptance, they appear to be so for different reasons.

A review of the behavioral correlates of sociometric status reveals a number of consistent findings. Because the correlates of acceptance and rejection differ, these are considered separately. In general, children who show high levels of peer acceptance are socially active, and engage in positive, cooperative, responsive, cohesive interactions during social play, whereas rejected children are often domineering or disruptive. Socially neglected children often show extreme passivity.

An early view of the correlates of peer acceptance or popularity was that popular children were more socially active than were unpopular children. Although this view has been criticized and it has been argued that the quality of social interactions is more important than their quantity (Asher et al., 1981; Gottman, 1977), some data support the interaction rate hypothesis (Begin, 1986; Ironsmith & Poteat, 1990). Not only are frequently nominated children more socially active, but several studies have found that they engage in more frequent social conversations than do less popular children (Bonney, 1955; Ladd, 1983; Steinkamp, 1989).

Regarding interaction quality, the most commonly reported behavioral correlate of peer acceptance is the proportion or number of positive interactions that children have with their peers (Asarnow, 1983; Beck, Collins, Overholser, & Terry, 1985; Campbell & Yarrow, 1961; Coie & Dodge, 1988; Dodge, Coie, & Brakke, 1982; Gresham, 1982; Ladd & Price, 1987; Mize & Ladd, 1988; Roopnarine & Adams, 1987; Rubin & Daniels-Beirness, 1983; Rubin, Daniels-Beirness, & Hayvren, 1982). Other related commonly reported correlates are the frequency with which children dispense positive reinforcement (Gottman, Gonso, & Rasmussen, 1975; Hartup et al., 1967; Masters & Furman, 1981) and the amount of positive affect (Bonney, 1955; Bonney & Powell, 1953; Denham, McKinley, Couchoud, & Holt, 1990; Howes, 1988a; Ironsmith & Poteat, 1990; Steinkamp, 1989). Anger shows a negative relationship (Denham et al., 1990; Fabes & Eisenberg, 1992).

Several general conclusions can be drawn from research using the Parten (1932) and/or Smilansky (1968) categories: Peer acceptance is positively associated with degree to which children enter into cooperative, reciprocal, or group play (Asarnow, 1983; Bonney, 1955; Bonney & Powell, 1953; Coie & Kupersmidt, 1983; Dodge, 1983; Howes, 1988a; Ladd, 1983; Ladd & Price, 1987; Ladd, Price, & Hart, 1988; Rubin et al., 1982; Steinkamp, 1989), as well as time spent in sociodramatic play (Connolly & Doyle, 1984; Doyle & Connolly, 1989; Howes, 1988a; Rubin et al., 1982; Rubin & Maioni, 1975). Solitary play is sometimes negatively associated with peer acceptance (e.g., Bonney, 1955; Coie & Kupersmidt, 1983; Dodge, 1983; Rubin et al., 1982), whereas the relationship between parallel play and acceptance is inconsistent (Coie & Kupersmidt, 1983; Ladd, 1983; Rubin

& Daniels-Beirness, 1983; Steinkamp, 1989) and varies with the age and gender of the child.

More detailed descriptions of the nature of peer interaction during play reveal some additional correlates. For example, Doyle and Connolly (1989) found that peer acceptance was positively associated with the degree to which children engaged in negotiation during pretend, and Coie and Kupersmidt (1983) noted that norm setting was positively correlated with peer acceptance. Kemple, Speranza, and Hazen (1992) observed that the cohesiveness of children's discourse during triadic interactions predicted their level of peer acceptance. Popular children, in response to the conversational initiations of their peers, were more likely than less popular children to accept, acknowledge, or answer the initiation, and, if they rejected the initiation, they were more likely to do so in such a way that they extended the topic, "typically by providing a reason or rule for the rejection, or an alternative idea or compromise" (Kemple et al., 1992, p. 381). They were also less likely to engage in nondirected initiations themselves (initiations not clearly directed at a particular child) or to engage in noncontingent responses to the initiations of another.

A setting that has demonstrated its usefulness in identifying differences between children of varying sociometric status is the group entry situation. In a typical study, two children are allowed to play for a set period of time and then a third child is sent into the room to join them. Research of this type has found that popular children are more likely than unpopular children to make group-oriented (Dodge, Schlundt, Schocken, & Delugach, 1983) or relevant statements (Putallaz, 1983) in these situations. They are also more likely to engage in synchronous nonverbal behavior and to agree with statements of the "host" child (Dodge, Pettit, McCluskey, & Brown, 1987). Findings for information statements as entry strategies are inconsistent: Hazen and Black (1989) noted that they were positively associated with popularity, whereas Putallaz (1983) gleaned a negative association. In a detailed analysis of children's verbalizations during pretend play in an entry situation, Black (1992) found that peer acceptance was positively associated with extending statements, explanations, and request clarification. Both Black and Hazen (1990) and Black and Logan (1995) observed that, during entry situations, the play of popular children had more coherent episodes than did that of less popular children. Additionally, Black and Logan (1995) noted that popular children showed more alternating turn taking and were more likely to use open-ended questions and explanations.

Putallaz and Sheppard (1990) studied low and high social-status dyads of first-grade children in three play situations with "limited resources"—the children had only one toy available with which to play. Children in high-status dyads were more likely to collaborate or compromise with one another in this situation, whereas children in low-status dyads were more likely to compete. Similar results were found by Fonzi, Schneider, Tani, and Tomada (1997) in a study comparing friends to nonfriends in an object-sharing task.

A number of correlates of social rejection have been identified as well. Not surprisingly, the most consistent predictor is the number or proportion of negative interactions (Asarnow, 1983; French & Waas, 1987; Gottman, 1977; Ladd & Price, 1987; Rubin et al., 1982; Rubin & Daniels-Beirness, 1983). Other related positive correlates include the administration or punishment or negative reinforcement (Hartup et al., 1967; Masters & Furman, 1981), aversive or hostile verbalizations (Coie & Kupersmidt, 1983; Dodge, 1983), arguing or disagreements (Ladd, 1983; Ladd et al., 1988; Putallaz, 1987), exclusion of peers (Dodge, 1983), and noncohesive discourse (Kemple et al., 1992). Rejected children (as well as neglected children) are more likely to show play that is "inappropriate" for the situation (Coie & Kupersmidt, 1983; Dodge, 1983).

Solitary play is sometimes associated with peer rejection or neglect (e.g., Coie & Kupersmidt, 1983; Dodge, 1983; Rubin et al., 1982), as is onlooker and unoccupied behavior (French & Waas, 1987; Gottman, 1977; Ladd, 1983). Social isolates, however, are a heterogeneous group, and researchers have begun to make distinctions between various kinds of social withdrawal (e.g., Asendorphf, 1990, 1991; Coplan et al., 1994; Harrist, Zaia, Bates, Dodge, & Pettit, 1997; Rubin, 1982; Rubin, Coplan, Fox, & Calkins, 1995; Rubin, Hymel, & Mills, 1989). This research has identified at least three kinds of social withdrawal: an immature, active pattern associated with externalizing symptoms; an anxious, inactive pattern associated with internalizing symptoms; and a solitary, passive pattern associated with little psychopathology. The first pattern is most consistently associated with social rejection, whereas the other two are at greater risk for social neglect. Given the relationship between submissiveness and later victimization by peers (Schwartz, Dodge, & Coie, 1993), it is possible that children showing either of the latter two patterns may be at risk for victimization as well.

Studies of Smilansky's (1968) categories show that functional play is sometimes associated with social rejection (Rubin & Daniels-Beirness, 1983), and that group-dramatic play is negatively associated (Rubin et al., 1982). Studies of entry behavior show an additional number of correlates. Rejected children are more likely to use disruptive entry tactics (Dodge et al., 1983, 1987), whereas neglected children are more likely to wait or hover

(Dodge et al., 1983). Putallaz and Wasserman (1989) found more hovering in rejected children as well. Black and Hazen (Black, 1992; Black & Hazen, 1990; Black & Logan, 1995; Hazen & Black, 1989) and Dodge et al. (1987) noted several other correlates of social rejection during group entry. Rejected children were more likely to state their own feelings or attitudes; show unfocused and nondirective initiations, noncontingent responses, and irrelevant turns; and make demands or rejecting statements in negotiating pretend and nonpretend play interactions.

Although aggression tends to be associated with social rejection, the association between aggression and sociometric status is complex and varies by type of aggression and age and gender of the child. The relationship between aggression and peer rejection is stronger in the middle elementary school years than in preschool or early elementary period, probably due to differences in its severity (Coie, Belding, & Underwood, 1988; Coie, Dodge, Terry, & Wright, 1991). For young children, at least, it also is not clear that aggression per se is the reason for rejection. In a study of first graders, Shantz (1986) observed that although the relationship between aggression and peer rejection became nonsignificant when controlling for the level of conflict, conflict remained a strong predictor of rejection when the level of aggression was controlled. Given the relative frequency with which aggression occurs in young children, and given their limited knowledge of alternatives to aggression, conflict resolution through aggression may be viewed by children as more acceptable in early childhood than in the later years. Shantz (1986) asserted that children's judgments of the fairness of the partner's behavior may be a more important determinant of peer rejection than whether or not the partner uses aggression.

Because most research in this area has focused on physical aggression, it would be informative to conduct a similar analysis on relational aggression in preschoolers (a nonphysical form of aggression found more often in girls—see Crick, Casas, & Mosher, 1997). Even among older children, however, aggression is not always associated with consistent peer rejection. Controversial children (those who receive a high number of both positive and negative peer nominations), show the highest levels of aggression of all sociometric groups (see the meta-analytic review by Newcomb et al., 1993). These may be children who receive positive nominations from a network of aggressive peers (Cairns, Cairns, Neckerman, Gest, & Garieppy, 1988). Bierman, Smoot, and Aumiller (1993) compared rejected and nonrejected boys who had been identified by their peers as highly aggressive on various observational, self-report, peer rating, and teacher rating measures. Although the groups were equal in their level of physical aggression, rejected aggressive boys showed higher levels of argumentative/disruptive, insensitive, and atypical behavior.

Evidence for the role of play in the development of specific friendships (vs. popularity) comes from two sources: the literature on friendship and social interaction reviewed earlier, and the already described longitudinal study of friendship formation by Gottman (1983). As previously mentioned, studies on the interactions of friends versus nonfriends show that, when interacting with friends, children show greater social contact, talking, cooperation, positive affect, and conflict resolution than when interacting with acquaintances or strangers (Newcomb & Bagwell, 1995). In the Gottman (1983) study, unacquainted dyads of children who later "hit it off," compared to those who did not, showed higher levels of information success, gossip success, self-disclosure, and amity success in their play sessions. Children who did not hit it off showed higher levels of information failure, escalation success, conflict, and amity failure.

Moving to the relationship between play and the development of social skills, correlational studies are consistent with the hypothesis that play may contribute to the development of social perspective taking. In general, play interactions at higher levels of Smilansky's (1968) cognitive play scale tend to be associated with social perspective-taking skills. Specifically, Connolly and Doyle (1984) noted that social dramatic play was positively correlated with affective role taking in a task requiring identifying and justifying characters' emotions in a story, but was uncorrelated with performance in a gift-choice task. Rubin and Maioni (1975) found that the amount of dramatic play was positively correlated with performance on a spatial perspective-taking task. They also observed that functional play was negatively associated with perspective taking—both in a role-taking (emotion labeling) task and a spatial perspective-taking task. Maguire and Dunn (1997) discovered that the amount of sociodramatic play with a best friend at age 6 predicted children's understanding of conflicting emotions 7 months later. Finally, Slomkowski and Dunn (1996) noted that the mean length of play episodes, connected episodes, and pretend episodes were positively correlated with children's responses to an emotion-understanding and false-belief task.

Game complexity is correlated with social perspective taking, but apparently only for girls. In a study of second and fifth graders, Borman and Kurdek (1987) found that the complexity of girls' games (using definitions from Lever, 1978) was positively correlated with performance on an interpersonal understanding task. For boys, however, game complexity was negatively associated with interpersonal understanding. Borman and Kurdek interpreted their findings in regard to differences in the nature of boys' and girls' games. They asserted that in girls' games, which are usually small group or dyadic in nature, interpersonal understanding is rewarded; whereas in boys' games, which usually are structured, rule-governed, competitive, team-oriented activities, too much sensitivity to the needs of others might interfere with performance.

Research on the level of social participation (Parten, 1932) has shown that higher levels of social participation appear to be related to perspective taking as well. For example, Rubin (1976) found that whereas associative play was positively correlated with role-taking (emotion labeling) performance, onlooker-unoccupied and parallel play were negatively associated. Finally, although social dramatic play is positively associated with perspective taking abilities (see earlier discussion), Rubin (1982) noted that *solitary* dramatic play was negatively correlated with perspective taking in a penny-hiding task.

The final area that has been examined in correlational studies is the relation between children's play and social problem-solving skills. The typical study employs some version of a Spivak and Shure (1974) task in which children are given a series of hypothetical social problem-solving situations, and the number and rated friendliness of their social problem-solving strategies are assessed. Mize and Cox (1990) observed that the number of strategies children generated in such situations was positively correlated with the frequency of positive peer interactions during free play. In two other studies, the friendliness of the strategies generated in such tasks was positively associated with the frequency of positive peer interactions (Mize & Ladd, 1988) and reciprocal/complementary and cooperative social pretend play (Howes & Matheson, 1992), whereas simple social play (Howes & Matheson, 1992) and aggression (Mize & Ladd, 1988) showed negative relations. Eisenberg et al. (1994) used a more differentiated system for coding enacted strategies, strategies for friendliness, aggressiveness, and degree of assertiveness. The friendliness of enacted strategies was not associated with the amount of cooperative, onlooking, or solitary play. Instead, the assertiveness of enacted strategies for girls correlated with these variables, showing a positive correlation with cooperation and a negative correlation with onlooking behavior. Friendliness and aggressiveness of enacted strategies were associated with children's responses to situations that elicited negative emotions in the classroom, but only for boys. Enacted friendliness was correlated with verbal versus escape responses, and enacted aggression was positively correlated with defensive and escape responses, and negatively correlated with verbal responses.

Experimental Studies. Although the vast majority of correlational studies have focused on the relationship between play and social relationships, the experimental studies have focused mostly on social perspective-taking skills. Only two experimental studies of play and social relationships could be found. In a study by Furman, Rahe, and Hartup (1979), preschool children who showed low levels of social interaction in the classroom attended 10 dyadic play sessions over a 4- to 6-week period, with either same-age or younger children. The adult present in these sessions did not structure child behavior, but instead simply observed the children in their play. Compared to a no-treatment control, children who attended the play sessions increased their level of social interaction in the classroom—specifically, their level of positive reinforcement and neutral acts. Although the two treatment groups did not significantly differ from one another on any of the variables, the means for the younger-age partners were higher than those for the same-age partners.

Given the success of the Furman et al. (1979) study, it is surprising that follow-up studies could not be found. Many investigators have since conducted studies showing that social skills training interventions can impact children's social relationships (for reviews, see Hughes & Sullivan, 1988; La Greca, 1993; Ogilvy, 1994), but these studies provide little information about the role of play because a combination of strategies involving coaching, modeling, and so on are used in such interventions.

Parker (1986) reported the results of two experimental studies designed to experimentally test Gottman's (1983) predictions about friendship-formation processes in young children. Because preschool children are too young to serve as experimental confederates, Parker (1986) constructed a 2-foot-tall "talking extraterrestrial" with a hidden mike and speaker to serve as a playmate in his experimental studies of friendship formation. Four- and 5-year-olds were told that they "would soon meet 'another boy (girl) your age' who was 'special and from a place far away' and interested in meeting the child and playing in the room" (pp. 127–128). An experimenter watching from another room had the extraterrestrial converse with the child in either a skilled or unskilled manner and then, based on both child and mother reports, it was determined whether or not the child hit it off with the extraterrestrial during the experimental session. Social skills were defined in terms of six conversation processes identified by Gottman (1983): communication clarity and connectedness, information exchange, establishing common-ground activities, exploring similarities and differences, self-disclosure, and conflict resolution. Results showed that, in two separate studies, children (especially girls) were more likely to hit it off when the extraterrestrial conversed with them in a skilled versus an unskilled manner.

Eight studies were identified that examined the effects of play-training on social perspective taking (Burns & Brainerd, 1979; Fink, 1976; Ghiaci & Richardson, 1980; Rosen, 1974; Saltz et al., 1977; Saltz & Johnson, 1974; Smith et al., 1981; Smith & Syddall, 1978). All but three of these studies (Burns & Brainerd, 1979; Fink, 1976; Ghiaci & Richardson, 1980) were reviewed in the earlier section on cognitive effects. The studies included children from a range of social classes, and training ranged from six sessions over 6 consecutive days (Ghiaci & Richardson, 1980) to training three times a week for 6 to 7 months (Saltz et al., 1977). Social perspective taking was assessed with a number of measures, including perceptual role taking, gift choices, emotion labeling, social role conservation, understanding of kinship relations, hiding tasks, and a social repertory grid task. Because the major hypothesis of the studies was that sociodramatic play would facilitate perspective-taking abilities, in each study a sociodramatic play-training condition was compared to one or more controls. A range of controls was used, including dimensional training, nondramatic play activities, reading/discussing stories, free play, and skills tutoring. In every study, at least one of the social perspective-taking measures was higher in the dramatic play group than in the controls. Even in the previously discussed studies showing no differences between the dramatic-play and skills-tutoring groups on cognitive outcomes (i.e., Smith et al., 1981; Smith & Syddall, 1978), effects of dramatic play-training were demonstrated on at least one of the social perspective-taking measures. Across all of the studies, significant effects were found for all types of social perspective-taking tasks except for the hiding task.

SOCIAL OBJECT AND SOCIAL PRETEND PLAY: SUMMARY AND CONCLUSIONS

In contrast to play-fighting and locomotor play, social object play in children is, at least on the surface, very different than play in other animals. The major forms identified in humans—sociodramatic play and games with rules—appear to be uniquely human and are apparent consequences of our symbolic, communicative, and other cognitive abilities. Although elements of the play of other animals (object struggles, tug of war, shaking objects) are occasionally seen in the social object play of children, they are more often components of locomotor play (e.g., chasing games such as "keep away") or structured games (e.g., basketball, football, rugby, soccer) than of sociodramatic play.

Social behavior toward age-mates in infants and toddlers is not that different from that of other primates (especially the apes). With the exception of obvious differences in vocalizations, the individual affective and social play behaviors (e.g., look, smile, point, show, approach, touch, grasp, give, follow, joint object use, withdrawal), as well as the specific agonistic behaviors (threaten, bite, hit, push, attempt to take object), are much the same. (Comparative studies on qualitative differences in these behaviors and their sequential structure, however, have yet to be conducted.) Differences emerge in the second year, with children showing longer, more coordinated social interchanges, increasing verbal interaction, conventional rule-governed games, and social pretend. Conflicts between children in the second year show similar developmental changes as well. Whether these changes are found in children outside of Western, industrialized cultures is unknown and an area for future research.

As children move into the third and fourth years of life there is increasing divergence from the behavior of other primates, but some similarities are still evident. Children and other primates show many of the same affiliative and agonistic behaviors (see Table 8.3), but increasing communicative development in children brings a whole new dimension of social interaction—verbal social control. Also, one sees the increasing impact of culture in the development of conventionalized signs of affiliation (e.g., waving and hand holding), games, and social pretend, as well as the impact of the child's developing social perspective-taking skills (e.g., helping, teasing, relational aggression). Despite differences in the form and content of the play of preschoolers and other primates, aspects of children's agonistic behavior in group settings are governed by dominance hierarchies similar to those in other primate groups. Aspects of social play and affiliative behavior are impacted by dominance hierarchies as well, including attentional patterns, verbal social control, and conciliatory behaviors.

Research on the structure and content of play during early childhood has mostly been conducted in preschool settings. Therefore, we know very little about certain kinds of play in the preschool years, especially play in the absence of close supervision by adults and play forms discouraged in educational settings. As would be expected, the schemes used for describing children's social object play to date reflect the interests of play researchers, and a review of the research shows that the descriptive database on children's social object play is rather limited. Most research comes from a social cognitive or a social skills-training perspective. The social cognitive studies include those using variations of Smilansky's (1968) procedures (i.e., categories developed to discriminate among play types of varying cognitive complexity from a Piagetian perspective), as well as studies of children's pretend transformations (i.e., studies of how developing symbolic abilities contribute to the cognitive complexity of pretend play). The social skills-training approach is reflected in studies using Parten's (1932) procedures (i.e., categories reflecting the child's increasing ability to engage in cooperative social interactions with others), as well as the correlates of sociometric status (many of these studies were conducted to identify social skills deficits in socially isolated or rejected children so that effective training programs could be developed). Although this research provides a wealth of information about certain aspects of children's social object play, certain areas have been largely ignored, including the roles of emotional expression and regulation, temperament and personality factors, self-exploratory and social comparison processes, group dynamics and problem solving, play as a training context for later social roles, play in moral development and enculturation, and social play in creativity and divergent thinking. Although descriptive studies of these aspects of play have been conducted for other types of play (e.g., enculturation in parent–child play; divergent thinking in solitary object play), these areas (with but a few exceptions) have been the focus of little systematic descriptive research on children's social object play. Moreover, most of the research from the social cognitive and social skills-training perspectives, with but a few exceptions (e.g., Gottman, 1983, 1986), provide little specific descriptive information about how play provides experiences for the development of these skills. Thus, our understanding of the nature of children's social object and social pretend play is still very limited, and more research from a wide range of perspectives is in order.

Research on gender differences and partner effects on social object play reveal several general findings. First, when given the opportunity, most children appear to prefer to play with same-age and same-gender partners from an early age; and second, children in both Western and non-Western cultures spend a good deal of time in gender-related play activities based on adult roles. The reason behind this age and gender segregation is that children probably prefer to play with individuals like themselves. Research shows that in same-gender and same-age groups the interactions are more egalitarian, the partners are more matched in their interests and abilities, and children show the greatest levels of involvement and enjoyment. With older playmates, children often experience less freedom to play the way that they want, and younger playmates may be less stimulating.

A general characteristic that accounts for some of the gender differences and partner effects on children's play is the tendency to attend to and imitate the behavior of older or dominant individuals. In Western cultures this is true of the behavior of younger children toward their older siblings, of younger toward older children in mixed-age play groups, and of lower dominance-status children toward higher-status children in same-age play groups. This tendency is also found in the mixed-age play groups of some non-western cultures: Younger children imitate the behavior of older children in the group, and the group as a whole imitates the behavior of adults in the larger culture.

If the tendency to imitate the behavior of more dominant individuals was selected for in the evolution of humans, this might be a major mechanism (along with the evolution of language and communication) through which culture is transmitted from one generation to the next (see Boyd & Richerson, 1985). Because studies of traditional cultures suggest that children probably evolved in cultural settings where there was little age segregation, throughout most of our evolutionary history children likely spent considerable time playing with other children who were in a position to transmit important cultural information. However, when children are given greater freedom to choose their play partners (an opportunity unique to contemporary Western cultures), they are likely to base their choices not on the potential for learning but instead on the potential for fun, and therefore choose same-gender, same-age partners. Although this may not make the most sense for the transmission of culture, the environments in which many Western children find themselves today differ considerably from the environments in which social play patterns evolved. Therefore, parents and other adults (such as teachers) may need to take a greater role in the cultural transmission process than peers in contemporary Western cultures (see chap. 9).

As concluded by many other reviewers (e.g., Rubin et al., 1983; Smith, 1982), there is little experimental evidence for speculations about the functions and effects of children's social object and social pretend play. The research that has been conducted to date is primarily correlational in nature, showing that children who play frequently and at higher levels of complexity tend to show higher levels of cognitive and social functioning than do those who do not play in this way. As already discussed, such studies provide little information about cause and effect, in that competent children may demonstrate their abilities in both play and nonplay domains. Unfortunately, most experimental studies of play-training have methodological limitations. In studies employing more appropriate controls, it appears that many of the cognitive "effects" of play-training appear to be effects of adult verbal stimulation. The strongest data on the cognitive role of play per se come from studies of story recall, that is, giving children the opportunity to act out stories increases their memory for material in other stories more than simply hearing stories or discussing them with adults. Therefore, pretend play may be a useful way for children to familiarize themselves with the narrative structure of stories, thus facilitating story comprehension and retention.

The data on social functions and effects are no more compelling. Although researchers have demonstrated significant correlations between children's social behavior during play and their relationships with peers, it is not clear that play has a causal role. Do children develop social skills during play, or do socially skilled children simply play more and in more complex ways? Or is there some third variable (e.g., parenting practices, child temperament, or child psychopathology) correlated with social skills, play, and peer acceptance that accounts for these relationships? Although longitudinal and social skills-training studies indicate that social skills probably play a role in peer acceptance, it is not clear what role, if any, can be attributed to play. The strongest data for the social effects of play is for social perspective-taking skills—both correlational and experimental studies suggest that dramatic role play may make a contribution in this area.

Clearly, despite over 30 years of intensive work on the topic, there is little empirical support for the role of social object and social pretend play in development. Moreover, despite a limited amount of ethnographic work on this topic, virtually no quantitative data on the nature of children's social object play outside of Western, industrialized settings are available. Given significant cultural differences in the nature and amount of children's play, detailed and descriptive studies of children's play are needed to provide a base for the development and testing of hypotheses. Even in the area receiving the most empirical attention (the behavioral correlates of sociometric status), the level of description is usually global, examining variables such as the frequency of "positive" or "negative" interactions or the amount of "reciprocal" play. Although these global categories were appropriate for the purposes of the investigations in which they were used, the results of such studies tell us little about the nature of children's play interactions and their possible effects on development.

With the exception of the focus on creativity and divergent thinking, early play researchers may have been misguided in their search for the cognitive effects of social play. The only consistent experimental data on social play effects to date are for social perspective-taking skills, and a review of the descriptive data considered earlier suggests many ways that play with peers and siblings may effect the social development of young children.

Social object and social pretend play undoubtedly are important contexts in which social relationships are formed and strengthened. They also provide numerous opportunities for the practice of social skills. During play, children must successfully initiate interactions with others, communicate their wishes, take the perspective of their play partners, negotiate play themes and scripts, solve social problems, resolve conflicts, and assume both leadership and follower roles. Emotional development may be affected, in that children must learn to take turns; delay gratification; resist temptation; and control their frustration, anger, excitement, and activity level. Other possible areas include moral development and the development of self-understanding. When children play they frequently must deal with issues of equity and fairness in addressing conflicts and in making decisions about play rules and roles. Also, considerable self-exploration and social comparison goes on during play, as does the discussion of feelings. Finally, play may be an important context for the practice of later roles and behaviors, including gender roles, family roles, occupational roles, and so on. Although young children obviously learn little about how to put out fires when playing "firefighter," they may learn important lessons about social roles, rules, relationships, and the larger social structure in this context. In cultural settings where there is little segregation between adults and children, young children have considerable opportunity to observe adults engaged in their daily activities. In such cultures, a good deal of preparation for adult roles may occur during children's play. These are just a few of the possible social effects of children's play suggested by the descriptive work conducted to date, and should provide some directions for future inquiry in this area.

9

Parent–Child Play

One of the greatest joys of parenting is playing with one's children. When parents start a game of peek-a-boo or lift their infant into the air, they can put aside the stresses of everyday adult life and become a kid again. Although most children eventually play primarily with siblings and peers, their first playmates are usually their parents. Parent–infant play is a common phenomenon in Western cultures, and researchers have done much to describe its complexity.

Parent–offspring play is found in many other species as well, although much less attention has been devoted to this topic than to play between agemates. We therefore only briefly review the research on other animals (primarily the apes) before turning the research on humans. Two topics are considered: parent–offspring play and parental influences on peer or sibling play. The second topic is important, because long after parents lose the distinction of being the primary playmate they continue to have a significant impact on their children's play in more indirect ways.

ANIMAL RESEARCH

Compared to the research on the play of agemates, very little quantitative research has been conducted on parent–offspring play. The literature in this area is made up mostly of brief descriptions of parent–offspring play found in articles focusing primarily on play between infants or juveniles. Fagen (1981) described parent–offspring play for almost every major mammalian order, with the great apes apparently showing the highest levels of mother–offspring play—their play is characterized by the greatest inventiveness and diversity (e.g., Aldis, 1975; Bard, 1995; Biben & Suomi, 1993; Fagen, 1981). Consider the following examples.

Gorillas (Hoff et al., 1981, p. 67):

Reciprocal mother–infant play first appeared in the fourth month. At this point, it was initiated and usually terminated by the mother. It typically involved gently nudging with finger, arm, or leg by the mother, and wrestling with and mouthing of the mother by the infant. Play was often exhibited in a discontinuous, reciprocal fashion, with the infant and mother alternately exhibiting play behavior toward the other.

Chimpanzees (van Lawick-Goodall, 1968, pp. 244–246):

All chimpanzee mothers were seen to play with their infant.... Most mothers initiated play during the first few months of their infants' lives...by tickling them with their fingers, usually under the chin or in the groin. Flo frequently tickled Flint with small nibbling movements of her worn teeth.... Also, on many occasions, she lay on her back, held Flo above her with one of his ankles or wrists clasped in her foot, and tickled him.... Play behavior of mothers with their one-year-old infants included rolling the infants over, mock biting and tickling them, gently sparring with them or pushing them to and fro as they dangled from a branch.... The infants responded by grabbing and biting the mothers' hands, flinging themselves onto the mothers with flailing arms or tickling them.

A factor that might contribute to the amount of mother–infant play observed in the great apes and the low levels found in monkeys (Biben & Suomi, 1993) is the length of immaturity in young apes (Aldis, 1975; Bard, 1995). Because chimpanzees, gorillas, and orangutans are born in an immature state (especially chimpanzees, who cannot independently cling to their mother's fur until 2 or 3 months after birth), mothers must actively hold their infants and provide for their needs. Compared to many monkeys who can independently crawl or walk in the first days or weeks after birth, great ape infants are in almost continuous contact with their mothers for the first few months of life. Therefore, peer interaction for great apes does not occur until a much later age than for Old and New World monkeys.

Chimpanzee mothers spend a good deal of their time looking at and touching their infants (Bard, 1994; van de Rijt-Plooij & Plooij, 1987; Yerkes & Tomlin, 1935). Their infants are also in a better position to look back, because of differences in motor development. Higley and Suomi (1986) reported that, during the first month of life, chimps spent only 20% of their time in nipple contact, whereas this figure was 80% for macaques.

Chimpanzee mothers also begin initiating play in the newborn period. For example, Bard (1994) noted that among chimpanzee mothers, 3% of their time with newborns was spent playing and eliciting smiles from them. Mothers watch their infants closely and respond to their motor behavior and vocalizations (Yerkes & Tomlin, 1935). The level of maternal play and infant responsiveness increases over the first few months, with infants smiling and laughing in response to maternal tickles by 2 months (Plooij, 1979; Yerkes & Tomlin, 1935), and infants initiating tickle games with their mothers by the third month (Plooij, 1979; van Lawick-Goodall, 1968). Starting around the third month, chimpanzee mothers spend a good deal of time encouraging motor development with exercises such as stretching the infant's muscles, assisting standing, and so on (Yerkes & Tomlin, 1935). Mother–infant play becomes less common after the first year of life, as young chimpanzees spend a greater proportion of their time playing with siblings or peers (Bard, 1995; Biben & Suomi, 1993; van Lawick-Goodall, 1968). Aldis (1975) argued that this likely occurs because once infants develop the skills to engage in longer play interactions and the independence to venture farther from their mothers, they choose play partners that are more evenly matched in their play-fighting or play-chasing abilities—that is, same-age and same-gender peers.

One general characteristic of parent–offspring play that has been identified is the tendency of parents to inhibit their behavior or self-handicap when playing with their young (e.g., Baldwin & Baldwin, 1978a; Biben & Suomi, 1993; Fagen, 1981; Hoff et al., 1981). This is illustrated in the chimpanzee example cited earlier, and is particularly characteristic of parent–offspring play-fighting.

As mentioned earlier, few studies have examined in any detail the structure of parent–offspring play. Presented in Table 9.1 is a list of the maternal behaviors identified in three studies of mother–infant interactions in chimpanzees. As can be seen in the table, many of these behaviors are affiliative, although few are by definition playful. Empirical studies of parent–offspring play are rare. Presented in Table 9.2 are the various types of play that mothers and adult males directed toward infants in a field study of bonobos (pygmy chimps; Enomoto, 1990).

Although a few studies have examined gender differences in the nature of adult–offspring play interactions (e.g., Baldwin & Baldwin, 1978a; Enomoto, 1990; Fedigen, 1972; Hayaki, 1985; Suomi, 1977), too few have been conducted to draw many firm conclusions. One point that can be made is that the degree to which male and female adults play with their young varies considerably by species, and among primates these differences are

TABLE 9.1

Maternal Behavior During Mother–Infant Interactions in Chimpanzees

	Savage & Malick (1977)	*Yerkes & Tomilin (1935)*	*van Lawick-Goodall (1968)*
Affiliation			
Attention (look at, watch)	+	+	+
Signal			
Facial expressions			
Play-face			+
Gestures			
Head nod	+		
Hand beg	+		
Mouth cover	+		
Vocalizations		+	+
Approach	+	+	+
Contact			
Hold/support/reposition	+	+	+
Touch/grasp	+	+	+
Groom	+	+	+
Push/pull on	+	+	+
Kiss			+
Play			
Tickle	+	+	+
Play bite	+	+	+
Roll over			+
Spar			+
Terminate			
Disengage hand	+		
Withdraw	+	+	+
Caretaking			
Restrain/retrieve		+	+
Feeds		+	+
Restrict access nipple (weans)		+	+
Encourage motor development		+	+

TABLE 9.2

Types of Adult-Offspring Play Identified in Bonobos (Findings from Enomoto, 1990)

	Mothers	*Adult Males*
Shake/bob head	+	+
Suspend infant with its hand	+	
Suspend infant with its foot	+	
Shake branch infant is on	+	
Grasp infant by hand and swing	+	
Rub infant with sexual skin	+	
Hold infant between legs and shake	+	+
Play pant	+	+
Touch with mouth		+
Tickle with foot		+
Rock infant on legs		+
Place on abdomen		+
Lift in air while laying on back		+
Wrestle		+
Play bite		+
Infant climbs on	+	+
Infant jumps on	+	
Infant hangs on	+	
Infant hangs on and swings	+	
Infant touches adult's genitals	+	+
Infant kicks adult	+	+
Infant play face	+	+
Infant play pant	+	+

associated with the nature of the animals' living conditions. When few peers are available for play, adult–offspring play is more common (Aldis, 1975; Hinde & Spencer-Booth, 1967). This appears to be particularly true for play by adult males (e.g., Redican & Mitchell, 1974; Suomi, 1977; Zucker, Mitchell, & Maple, 1978).

Several studies suggest that mothers influence their offspring's play interactions with peers, primarily through intervention when the play-fighting gets too rough. In studies of howler (Baldwin & Baldwin, 1978a) and vervet monkeys (Govindarajulu et al., 1993), the mother's presence alone was enough to terminate play interactions. Govindarajulu et al. (1993) also observed that maternal interventions became increasingly less likely as the offspring got older. In a study of spotted hyenas, Drea et al. (1996) noted that mothers rarely interrupted play, but did break up instances of aggression. Mothers also apparently take a role in protecting their infants when they are playing. In an ingenious experimental study, Biben, Symmes, and Bernard (1989) found that mother squirrel monkeys became particularly vigilant toward a source of likely danger when their infants were engaged in play. Mothers may influence the content of their offspring's play as well. Based on an analysis of the diffusion of stone play in a group of Japanese macaques, Huffman (1984) concluded that this unique form of play was likely passed from mother to infant.

CHILD RESEARCH: STRUCTURE

In contrast to the research on other animals, researchers who study humans have devoted a significant amount of time to the study of parent–child play. Interestingly, the vast majority of these studies have involved parents and infants, with studies of parent–child play after the age of 2 hard to find. Therefore, the bulk of this section is devoted to reviewing the literature on parent–infant play. Because the nature of these play interactions varies significantly as a function of the child's developing cognitive, motor, emotional, and social abilities, it is necessary to review this literature developmentally, focusing on the major periods of parent–infant interaction. Despite the years since its publication, Sander (1962) provided one of the best descriptions of the nature of these changes to date. Based on longitudinal observations of infants over the first 18 months of life, he identified five distinct periods of mother–infant interaction: initial adaptation (0 to 2½ months), reciprocal exchange (2½ to 5 months), early directed activity (5 to 9 months), focalization on the parent (9 to 12 months), and self-assertion (12 to 18 months). Each period represents a qualitatively different phase of mother–infant interaction, largely because of the infant's developing social capabilities. In general, as the infant develops greater social skills, parents spend less time visually, auditorily, and/or physically stimulating their infant, and more time engaged in increasingly cooperative and collaborative activities. These stages are be used to organize the following review on the structure of parent–infant play. Although Sander (1962) studied only mothers, because these stages apply equally well to fathers they are discussed as stages of *parent*–infant play in the current discussion.

As was the case for the chapters on solitary object and social object play, a difficulty in reviewing the literature for this chapter was in deciding what kinds of studies to include. The typical study in this area involves either observing parents and infants when the adults are instructed to "play" with their infants in a home or a laboratory playroom setting, or observing "naturalistic" parent–infant interactions during unstructured observations in the home. Such studies are usually referred to as studies of parent–infant "interaction," although sometimes they are described as studies of "play." Given the difficulty in separating play and nonplay interactions in such contexts, a decision was made to cast as wide a net as possible and include studies in which parents engaged in social interactions with their infants when not involved in caretaking (e.g., feeding, soothing) or structured teaching tasks. Although parents often "play" with their infants in the context of such activities, much of the interactions are devoted to nonplay activities. Studies of parent–infant interaction during reading were excluded for this reason as well. Therefore, as was the case in the previous chapter, this review involves research on affiliative behaviors between parents and infants, not just "play."

Period of Initial Adaptation (0 to 2½ Months)

The primary adaptive task during Sander's first period of parent–infant interaction involves the sensitive meshing of parental activities with cues of the infant's state (Sander, 1962). During this period, the infant has a limited social repertoire. According to Ainsworth, Blehar, Waters, and Wall (1978), sensitive parenting during this phase involves learning to interpret and respond to the infant's signals of distress and interest, as well as providing the infant with an optimal level of early social stimulation in the forms of looking, smiling, talking, and touching. This is the period in which parents learn about the idiosyncratic needs and characteristics of their new infant, and in which first-time parents learn about parenting and infants in general.

Limited Infant Responsiveness

Because infants during this period show limited social responsiveness (e.g., sustained eye contact and social smiling do not appear until the second month—Sroufe & Waters, 1976; Wolff, 1963), parent–infant play in this phase does not show the complex, reciprocal patterns of interaction found in the second period of parent–infant interaction (see the next section). Although parents do play with their infants during this period, a greater amount of time is devoted to basic caregiving activities such as feeding, changing, dressing, soothing, and putting the infant down to bed (e.g., Moss, 1967). When Brazelton, Koslowski, and Main (1974) asked mothers to communicate with their 1-week-olds, they were:

> ...faced with the problem of communicating with infants who, if they were not crying or thrashing, were often simply hanging in the infant seat with closed or semiclosed eyes, or just as frequently, were 'frozen' motionless in some strange and uninterpretable posture—staring at nothing. While infants in the first weeks did seem to look directly into the face of their mothers, the communicative aspects of eye-to-eye contact sometimes were dampened when the mothers moved their heads and found the infants still staring off in the original direction. (p. 67)
>
> In the first week, the most definite "interactive" activities we noted on the part of infants were (1) withdrawal from a "noxious" stimulus such as contact in a sensitive area, (2) efforts to push away the noxious stimulation, (3) slight and quickly fading apparent effort to maintain eye-to-eye contact, (4) lulling to a caress or another stimulus from the mother, (5) brightening of the face and turning of the head to her voice or her face, and (6) responding to holding efforts from her by a gradual quieting and alerting to her. (pp. 67–68).

Despite the limited social responsiveness of their infants, mothers of newborns would "endow the smallest movements with highly personal meaning and react to them affectively" (Brazelton et al., 1974, p. 68). Mothers acted as if their newborns were intentionally communicating with them and would respond to and elaborate on infant behaviors "*as if* highly significant interaction has taken place when there has been no action at all" (Brazelton et al., 1974, p. 68). This integration of infant behavior into a familiar communicative frame becomes increasingly common during the next period.

Parental Behaviors

Although microanalytic descriptive studies of mothers' interactions with their newborns have been conducted in hospital settings (e.g., Bakeman & Brown, 1977; Dunn & Richards, 1977; Kaye, 1977), the observations in these studies were conducted during feeding rather than play. Listed in Table 9.3 are parental behaviors identified in studies that included observations of play as well as feeding. Comparing this table with Table 9.1 for primate mothers reveals a number of similarities and differences. Three differences (i.e., rocking, stimulating feeding, and burping) involve behavior during caregiving and not play. Many of the behaviors listed in the tables for chimpanzee mothers but not for humans (i.e., hand beg, push/pull on, tickle, roll over, spar, withdraw, restrain/retrieve, restrict access to nipple) are behaviors that chimpanzee mothers do not engage in until later in infancy, once the infant develops greater locomotor skills. Human parents engage in these same behaviors with older infants as well (see sections that follow). Only three of the behaviors (i.e., mouth cover, play bite, spar) have not been identified in studies of human parents. Aldis (1975), however, identified some variants on the play-bite in human mothers:

> Next, we shall consider the various kinds of mouth–body contact that may occur in play. Four general categories may be distinguished. Of these, kissing is the gentlest. Kissing may be defined simply as placing the mouth on the baby. Occasionally, kissing is accompanied by breathing, which consists of blowing air on the baby or sucking it in. Blowing air may tickle the baby and carries slight overtones of a playful attack. Nuzzling, which consists of placing the nose or mouth on the baby and making side-to-side or up-and-down movements, is generally more forceful than either kissing or breathing and, thus, carries stronger overtones of playful attack. Finally, nibbling consists of placing the mouth in an open position on the baby and making nibbling movements with the lips or teeth. Occasionally, the baby's hand or foot is placed inside the mouth and chewed on. Nibbling represents the strongest form of play attack. Nibbling may be considered to be a special form of the mouthing or mocking-biting responses that we have discussed in earlier chapters on animal play. (pp. 242–244)

Several behaviors that appear to be unique to human parents (besides the obvious differences in the nature of vocalizations) include making faces, imitating faces, pointing, imitating vocalizations, and playing with toys. It may be premature, however, to assume that these behaviors are not found in chimpanzee mothers. Chimpanzees show a wide range of facial expressions (van Lawick-Goodall, 1968). They also engage in social imitating and pointing (Gomez, 1996; Leavens & Hopkins, 1998). Whether these behaviors occur during mother–infant play in chimpanzees is a question for future observational research.

The results of a study of older infants suggest that chimpanzee and human parents vary considerably in their use of objects during parent–infant play (Bard & Vauclair, 1984). In this study of 8- to 11-months-olds, five

TABLE 9.3

Parental Behavior During Parent–Infant Interaction in the First 3 Weeks of Life

	Moss (1967)	*Parke & Sawin (1980)*	*Kaitz, Zvi, Levy, Berger, & Eidelman (1995)*
Affiliation			
Attention (look at, watch)	+	+	+
Signal			
Facial expressions			
Smile	+	+	+
Make face		+	+
Imitate face	+	+	
Gestures			
Empathic head move			+
Point			+
Vocalizations	+	+	+
Imitates vocalizations	+	+	
Approach	+		
Contact			
Hold/support/reposition/walk with	+	+	+
Rock	+	+	+
Affectionate touch	+	+	+
Kiss	+	+	+
Stimulating touch	+	+	+
Play			
Show toy		+	
Make noise with toy		+	
Touch with toy		+	
Terminate			
Caretaking			
Feeds	+	+	
Stimulates feeding	+	+	
Burps	+	+	
Wipe/clean/arrange clothes		+	+
Encourage motor development (stresses musculature)	+		

adult–infant dyads were videotaped playing with a standard set of toys: a human mother and her infant, a human female adult (caretaker) and a chimpanzee, a bonobo female adult (substitute mother) and a bonobo infant, a human adult (caretaker) and the same bonobo infant, and a chimpanzee mother and her infant. The adult apes almost never acted on objects with the apparent intention of engaging the infant's attention, whereas the human adult females frequently did. Although these results need to be replicated and extended to other primate species, they do point to a major difference in play style.

Period of Reciprocal Exchange (2½ to 5 Months)

Sander's second period of parent–infant interaction is characterized by reciprocal, affectively charged, face-to-face interactions between parents and infants. With the development of sustained eye contact, social smiling, and cooing, infant responsiveness to social stimulation increases dramatically. Not only are there changes in social behavior around the second month of life, but there are also changes in learning, attentional and motor control, cognitive development, EEG and hormonal activity, and sleep and activity cycles (Stern, 1985). According to Stern (1985): "The age of two months is almost as clear a boundary as birth itself.... Almost everything changes. And all observers of infants, including parents, agree on this" (p. 37).

The Social Infant

Several longitudinal studies have documented the increasing social responsiveness of infant behavior from the first to third month of life. These include observations of unstructured parent– infant interactions in the home (Belsky, Gilstrap, & Rovine, 1984; Blehar, Lieberman, & Ainsworth, 1977; Moss, 1967) as well as observations during brief, structured parent–infant play sessions (Fogel, 1982; Kaye & Fogel, 1980; Parke & Sawin, 1980). During this period, infants spend less time sleeping (Moss, 1967) and crying (Belsky, Gilstrap et al., 1984; Moss, 1967), and more time in alert states (Moss, 1967). Infants show increases in looking at the mother, smiling, vocalizing, and bouncing (Blehar et al., 1977; Fogel, 1982; Moss, 1967). They also spend more time mouthing and otherwise exploring objects (Belsky, Gilstrap et al., 1984; Moss, 1967). Between 6 and 15 weeks, Blehar et al. (1977) found that infants showed increasing intensity of response to face-to-face interactions with their mothers. As well, Belsky, Taylor, and

Rovine (1984) noted that between 1 and 3 months the number of three-step contingent social exchanges between mothers and infants became more common. Kaye and Fogel (1980) provided additional evidence for the infant's increasing social responsiveness in a molecular analysis of mother–infant interaction during face-to-face play. Specifically, they compared the amount of expressive behavior that infants showed when either looking at or not looking at their mother. At 6 weeks the amount of expressive behavior did not differ across these two contexts, whereas by 13 weeks the rates of vocalizations and smiling were two times higher when looking at the mother.

Increases in Parental Stimulation

As infants become more social, their parents become more socially stimulating. In these same studies, a number of infant age-related differences in parent behavior were identified. Over this age range parents spent less time holding their infants close to their bodies (Moss, 1967; Parke & Sawin, 1980), and providing routine physical care (Belsky, Gilstrap et al., 1984; Moss, 1967; Parke & Sawin, 1980) and more time looking at (Moss, 1967) and stimulating their infant. Forms of infant stimulation that became more common with age included affectionate contact; stressing the infant's musculature; showing toys or objects; and talking to, smiling at, and imitating the baby (Belsky, Gilstrap et al., 1984; Moss, 1967; Parke & Sawin, 1980).

Face-to-Face Interactions

Numerous descriptive studies of face-to-face interactions between parents and their infants during this period have been conducted. In the typical study, films or videotapes of parent–infant interactions are made in a laboratory or home setting under controlled conditions. The infant is usually placed in an infant seat and the parent sits in a chair. Parents are instructed to play with or talk to their infants as they would at home, and are told to feel free to touch their infants but not take them out of the seat to play. The interactions observed in this way are brief, usually about 5 minutes or shorter in length. Although one could argue that the constraints placed on parent and infant lead to play interactions that are not representative of parent behavior in more natural settings, researchers in this area argue that these interactions are representative of brief, socially engaging interactions that occur frequently in more natural settings. Stern (1977), who videotaped such interactions as they occurred during everyday mother–infant interactions in the home, described them this way: "They are fairly ordinary and common interpersonal interchanges occurring between a primary caregiver and an infant during the first half year of life. These are moments that are almost purely social in nature. They often occur at unlikely or unexpected times in the middle of or in the space between other activities" (Stern, 1977, p. 2).

As part of an observational study of 26 infants followed across the first year of life, Blehar et al. (1977) identified characteristics of naturally occurring face-to-face interactions in the home that occurred between 6 and 15 weeks. Fifty-five percent of the interactions they observed occurred when the mother was holding the baby, and 41% occurred when the mother bent over the infant when the infant was supine or propped up (3% involved holding by a third person). Mothers played a greater role in both the initiation and termination of face-to-face interactions. Ninety-one percent of these interactions were initiated by mothers, and mothers were more likely than infants to terminate these interactions as well: mothers terminating interactions (44%), infants terminating interactions (19%), unable to determine (37%). The interactions were usually brief, with 50% lasting fewer than 30 seconds.

One of the first papers to describe in detail the temporal structure of early face-to-face interactions was written by Brazelton et al. (1974). These authors identified seven segments of the typical interaction: initiation, orientation, state of attention, acceleration, peak of excitement, deceleration, and withdrawal or turning away. According to this scheme, initiation usually occurs when the infant looks at the mother, and orientation involves the infant orienting his or her body toward the mother and extending his or her extremities in her direction. The level of infant positive affective involvement increases through the segments labeled *state of attention* and *acceleration,* and reaches its highest level at *peak of excitement*. Throughout these segments the mother increases her level of visual, verbal, and tactile stimulation, encouraging the infant to higher levels of affective involvement. After reaching the peak of excitement, deceleration of infant and maternal behavior occurs until the infant looks down or turns away. After the infant "recovers," the cycle starts over again.

Brazelton et al. (1974) demonstrated that individual mothers differed in their sensitivity to their infant's cycles of attention and nonattention, with the most sensitive mothers adjusting their rhythm to the infant's by following cues of attention and withdrawal. Insensitive mothers appeared to either steadily bombard their infants with stimulation or to try to establish their own rhythm to regulate the infant–both patterns of insensitivity apparently decreased the level of infant positive involvement in the interactions.

Stern (1977) offered an arousal regulation model to account for the cyclical nature of early face-to-face interactions. Put briefly, he asserted that when mothers provide their infants with affectively arousing events, they increase the infant's level of involvement in the interaction up to the point that the infant needs to take a break. At that point, the infant deescalates and may turn away. Sensitive mothers, Stern argued, closely monitor attentional and affective cues from the infant and use them in guiding their level of stimulation. Moreover, by a process of repetition and variation, mothers encourage their infants to higher levels of involvement and thus prevent boredom. Stern maintained that mothers, by exaggerating their behavior and slowing it down in time, adapted their behavior to the infant's preferences and varied it in ways that hold the infant's attention.

Protoconversations

Since the early work of Stern (1971) and Brazelton et al. (1974), numerous investigators have described the specific patterns of maternal behavior shown during early face-to-face interactions. In general, they have confirmed Stern's (1977) view that maternal behavior in this context is often comprised of normal adult communicative behaviors exaggerated in both time and space. Mothers also, to a certain extent, act in accordance with the rules of adult conversations, engaging in "protoconversations" (Bateson, 1975) with their infants, treating them as if they are competent social partners and taking over those aspects of the infant's role that he or she cannot do for the self. As a result, a high degree of repetition and imitation characterize these interactions. Four aspects of maternal behavior have been examined in some detail: gaze patterns, vocalizations, facial expressions, and motor behavior. Each is considered in turn.

Gaze Patterns. Gaze patterns during early face-to-face interactions can best be described by Fogel's (1977) concept of *framing*. Mothers' gaze provides a "frame" for infant gaze, because although mothers gaze at their infants almost continuously, the infant alternates between gazing at the mother and gazing away. This pattern has been identified in several observational studies (e.g., Kaye & Fogel, 1980; Messer & Vietze, 1984; Stern, 1974). Stern (1974) noted that, based on the rules of adult conversation, when the parent continuously gazes at the infant, the child is treated as if he or she is talking and the parent is listening. During observations of unstructured mother–infant interactions in the home, Messer and Vietze (1984) observed that, on the average, such maternal frames occurred once every 4 minutes at 10 weeks of age, and lasted a mean of 3⅓ minutes. Mothers and infants appear to be mutually responsive to each other's gaze during this period. For example, Stern (1974) found that both mothers and infants were more likely to look away from their partner if their partner was not looking at them, and Messer and Vietze (1984) noted that "both mother and infant were more likely to start a gaze and less likely to stop a gaze when the other person was looking at them rather than looking away" (p. 167).

Vocalizations. Maternal vocalizations during early face-to-face interactions are exaggerated versions of the verbalizations used in conversations between adults, apparently modified in such a way as to elicit and sustain infant interest. In comparison to adult conversation, they have an expanded range of pitch and loudness; show sudden changes in pitch, rhythm, and stress; are of higher pitch; and have elongated vowel sounds (Aldis, 1975; Stern, 1974). Regarding content, maternal verbalizations during face-to-face play are characterized by high levels of repetition and nonsense words or sounds (Aldis, 1975; Penman, Cross, Milgram-Friedman, & Meares, 1983; Stern, Beebe, Jaffe, & Bennett, 1977). Repetition is not restricted to verbal content: Spectrographic analyses show high that melodic repetition is common as well (Papousek, Papousek, & Haekel, 1987). Moreover, in a study of 2-, 3-, and 5-month-olds, Papousek and Papousek (1989) noted that, with increasing child age, mothers matched a greater number of features of their infant's vocalizations, including pitch, contour, duration, rhythm, vowel sounds, and consonant sounds.

Although infants of this age are obviously incapable of conversation, mothers engage in "protoconversations" (Bateson, 1975) during face-to-face play by inserting their vocalizations between the infant's so that an early form of turn-taking occurs. Stern (1977) pointed out that even when the infant is not vocalizing, that the length of pauses between maternal verbalizations are of the duration one would predict if the infant had given an imagined vocal response. In addition, Stern (1977) noted, the mother may play the role of both participants by answering the questions she asks for the child. In a quantitative analysis of the length of pauses between mother and infant vocalizations, Beebe, Alson, Jaffe, Feldstein, and Crown (1988) found that, at 3⅓ to 4 months, mother–infant "conversations" during face-to-face interactions already showed the vocal congruence found in adult conversations. In a longitudinal study of three infants, Ginsburg and Kilbourne (1988) observed that all three mothers shifted, rather dramatically, from a primarily overlapping vocalization pattern prior to 12 weeks to a primarily alternating pattern shortly thereafter (the change occurred for two children between 15 and 17 weeks, and for the other between 10 and 12 weeks).

Mothers vocalize more to their infants than their infants do to them (Anderson, Vietze, & Dokecki, 1977) and, as in conversations between adults, sequential analyses of maternal and infant vocalizations have shown that both infants in this age range and mothers respond to their partner's vocalizations with vocalizations of their own (Anderson et al., 1977; Stevenson, Ver Hoeve, Roach, & Leavitt, 1986). This reciprocity, however, only occurs where the partners are looking at one another (Stevenson et al., 1986). Stevenson et al. (1986) also noted that the probability of vocalization is suppressed for both infants and mothers after a vocalization, probably to allow for a response. Anderson et al. (1977) found no reciprocity for vocal offsets.

Spectrographic analyses of maternal speech during face-to-face interactions have been conducted to identify the functions of parental pitch variations in this context. Stern, Spieker, and MacKain (1982) discovered five common pitch contours: sinusoidal, bell, rise, fall, and bell-right. Rising contours appeared to have an attention-getting function, in that they were common when the infant was looking away from the mother with neutral affect; and sinusoidal (up-down-up or down-up-down) contours appeared to serve in holding attention, in that they were common when the infant was gazing at the mother and smiling. Katz, Cohn, and Moore (1996) instructed mothers to direct their infant's attention, to show approval, and to act as if they were comforting him or her. Spectrographic analyses using multiple data analytic methods confirmed that attention-directing utterances were characterized by a rising intonation and a short duration. Approval utterances had waveforms high in fundamental frequency and of long duration (this similarity to Stern et al.'s sinusoidal utterances suggests the possibility that approval statements may be used for attention holding). Comforting utterances were of long duration but were not associated with any specific waveform. Both Fernald and Simon (1984) and Papousek et al., (1987) identified similar pitch patterns in research with German mothers and/or fathers interacting with their infants.

Researchers have conducted analyses of the verbal content of maternal vocalizations as well. In a study of mothers of 3-month-old infants during free play in a laboratory setting (the infants were in an infant seat but, unlike the other studies reviewed here, toys were made available), Penman et al. (1983) found high levels of nonsense vocalizations and questions. Of those statements that had a specific referent (39% did not), references to the infant's actions were the most common (35%), followed by references to the external environment (12%) and the infant's feelings (11%). References to the mother's behavior were rare (4%). When infants looked away from the mother, mothers made more statements regarding the infant's feelings or the mother's behavior.

Although the speech modifications made for interacting with infants have many of the same characteristics of "motherese" (the speech register that adults and children use in interact with language-learning children—see the section on toddlers later in this chapter), Kaye (1980) pointed out that the baby talk used with young infants differs from that used with toddlers. Speech to toddlers and younger infants probably differs because parents of toddlers likely simplify their language in response to early signs of language comprehension, whereas parents of young infants probably modify their language to elicit and sustain infant attention. Kaye (1980) noted that compared to maternal speech to 2-year-olds, maternal speech to infants was more rapid (more utterances/minute), shorter (fewer words/utterance), and more repetitive. Regarding the latter characteristic, 16% of utterances to infants were exact repetitions and 33% were partial repetitions. These same figures for mothers of 2-year-olds were 4% and 31%, respectively. Kaye (1980) also found that 21% of mother's speech to infants were "phatics"—a class of one-word greetings such as "Huh," "Yeah," "Well," and "Ooh." These were rarely used by mothers of 2-year-olds. Finally, lexical babytalk (e.g., "froggy") was rare. Papousek et al. (1987), in the previously cited study of German parents and infants, identified many of the same characteristics, including short utterances, high levels of repetition, and the frequent occurrence of "lexically ambiguous" utterances. Like Kaye's (1980) "phatics," approximately 60% of mothers' and fathers' utterances:

> were either interjections, emotional exclamations, and calls, or musical, rhythmic, imitative, and onomatopoeic sounds...such as calls for visual attention (e.g., calling by name, "*guckguck*," "*hallo*"); calls for an infant turn (e.g., isolated questions tags such as "hm?" "*gell?*" "*ja?*"); approvals (e.g., "*jaah!*" "*fein!*"); or readiness for playful interchanges. (Papousek et al., 1987, pp. 501–502)

Facial Expressions. Stern (1977) identified five maternal facial expressions that are common during face-to-face play: mock surprise, frown, smile, concern/sympathy, and neutral. Compared to facial expressions in interactions between adults, maternal facial expressions during mother–infant play are exaggerated in size and slowed down in time (slower tempo and longer duration). Stern suggested that these facial expressions play different roles in face-to-face interactions, with mock surprise more often serving to initiate interactions, frowns to terminate, smiles and concern/sympathy (the "Oh, you poor dear" expression) to maintain and modulate interaction, and neutral expressions used in avoidance situations.

Parents frequently imitate the infant's facial expressions during face-to-face interactions. Parental imitation is one of the easiest ways to establish a turn-taking frame. In two studies of mother–infant play, Malatesta and colleagues (Malatesta, Grigoryev, Lamb, Albin, & Culver, 1986; Malatesta & Haviland, 1982) conducted second-by-second coding of mother and infant facial expression to look for patterns of imitation. Mothers showed high levels of responsiveness to infant changes in infant facial expression. For example, in the Malatesta and Haviland (1982) study, 25% of all maternal facial expression changes followed a change in infant expression by 1 second or less (most responses were less than 1 second). Moreover, with just two exceptions (knit brow and pain), the most common form of maternal response to an infant expression change was to match the infant's expression (35% of all contingent responses were matches). This was found for interest, joy, surprise, sadness, anger, and brow-flash expressions. Rather than imitate, mothers were most likely to ignore pain and knit-brow expressions (knit brows reflect concentration and often immediately precede infant distress). Besides sequences that reflect the suppression of certain expressions, the only other significant sequences identified were for mothers to respond to interest with brow-flash expressions and to anger with a knit-brow expression.

In a follow-up study using both preterm and full-term infants (Malatesta et al., 1986), the results for full-terms largely replicated the Malatesta and Haviland (1982) study. As before, matching was the most common maternal response to interest, joy, surprise, sadness, and anger (data on brow flashes were not presented in this second study). Also, as before, ignore was only significant in response to knit-brow and pain expressions. Although mothers in the first study responded to anger with a knit-brow expression, mothers in the second study responded to anger with surprise. Also, mothers in the second study responded to pain expressions with a knit brow (a sequence that approached statistical significance in the first study).

Other studies provide evidence of maternal imitation of facial expression as well. Fogel (1982) found that mothers were three times as likely to smile at their 6-week old infant when he or she was smiling at her than when the infant was not. Similarly, Moran, Krupka, Tutton, and Symons (1987) noted that mothers imitated their infant's mouth opening, lip movements, smiling, and eye movements in the second after they occurred, whereas infants did not contingently imitate their mother's behavior. Instead, infant matching of mothers' facial expressions was more delayed, and tended to occur 2 or 3 seconds after maternal onsets. In a subsequent study of the relationship between maternal and infant smiling, Symons and Moran (1994) observed that mothers showed significant responsiveness to infant smiling (i.e., the probability of a mother smile following an infant smile was greater than the baseline probability of a mother smile), but their smiling did not show significant dependency on infant smiling (i.e., the probability of an infant smile preceding a mother smile was no greater than the baseline probability for an infant smile). The opposite was true for infants—they showed significant dependency, but not significant responsiveness.

Motor Behavior. Pawlby (1977) found that mothers do not restrict their imitation to facial expressions during early play patterns. Mothers frequently imitated the infant's hand and body movements and speech and nonspeech sounds. Although two toys were available for play during the mother–infant play sessions in this study, only two instances of mothers imitating infant behavior with objects were found for the age period of 17 to 24 weeks. As found by Moran et al. (1987), infant imitation of mothers during this age period was rare.

Although less attention has been placed on maternal motor behavior during face-to-face interactions, Stern (1977) described various head movements, including nodding up and down, wagging side to side, head cocking, and head averting. Like maternal vocalizations and facial expressions, these were exaggerated and slowed down in time. Some parents play touching and poking games in this context as well (Field, 1978; Yogman, 1981).

Temporal Aspects. Some researchers have gone beyond the description of individual maternal behaviors to identifying larger order units. For example, Fogel (1977) identified "runs" of maternal behavior that involved repetition of individual behaviors separated by short intervals and "time-outs" (which referred to longer intervals between runs). He also offered the concept of *framing* (see earlier discussion) for when an extended maternal behavior (e.g., a look or a hold) provides a stable context in which other behaviors can occur. According to Fogel (1977), frames can exist between or within individuals. Stern et al. (1977) also offered a system for larger order units, including "phrases" (short, near-continuous sequences of behavior), "runs" (strings of repeating phrases), and "episodes of maintained engagement" ("a longer series of phrases, usually containing several runs and bounded by shifts in maternal behaviors and often attention," p. 181). In their 1977 paper, Stern et al. illustrated how these concepts can be applied to the analysis of both vocal and kinesic behaviors.

Kaye and Fogel (1980)—in a longitudinal study of 37 infants at 6, 13, and 26 weeks of age—applied the concepts of runs to describe cycles of maternal activity during face-to-face interaction. Mothers were considered "on" when they where smiling, bobbing their heads, or making exaggerated facial expressions, and "off" when they were not. Maternal gazing and vocalizations were not used in defining cycles, because mothers looked and spoke throughout the sessions. At each of the three ages studied, the rate of maternal "on" cycles was 3.1 to 3.3 cycles per minute (about 6 seconds on and 13 seconds off). The rates of maternal "on" behavior were higher when the infant was looking at the mother than when he or she was not looking at the mother. At 13 weeks, mothers appeared to get their infant's attention by touching and bouncing the baby (unlike those in many of the previously mentioned studies, mothers in this study held their infants in their laps rather than put them in an infant seat), then greeted the infant with smiles and exaggerated head bobbing. The infants often responded with a smile, a vocalization, or a wide mouth. Mothers then apparently used talking and facial expressions to hold the infant's attention.

Because Brazelton et al. (1974) based their conclusions about the cyclic nature of mother–infant interaction on a small number of cases, in subsequent research they refined their coding and data analytic procedures so that they could conduct quantitative analyses on larger samples. In a 1980 paper, Tronick, Als, and Brazelton presented a system for coding mother–infant face-to-face interactions that yields a single quantitative index suitable for time-series analyses. This system makes it possible to assign the behavior of the mother and the infant to one of several "monadic phases" for each second of the interaction. The six monadic phases for infants and the seven for mothers are summarized in Table 9.4. The phases are listed in order from the lowest to highest level of infant and parent involvement (despite the ordering in the table, Tronick and colleagues argue that talk is not necessarily a higher measure of involvement than play). Note that increases from lower to higher phases correspond to increases in the levels of both positive, mother-directed involvement for infants and positive, infant-directed stimulation for mothers.

In a study designed to quantitatively examine the cyclical structure of early face-to-face interactions, Lester, Hoffman, and Brazelton (1985)

TABLE 9.4

Monadic Phases for Describing Parent–Infant Interaction

Phase	*Gaze*	*Head Orientation*	*Body Orientation*	*Facial Expression*	*Voice*	*Specific Handling*
Infant						
Avoid	Away	Away	Away	Negative to neutral	Cry and others	—
Avert	Away	Partially away	Partially away	Negative to neutral	Negative to neutal	—
Monitor	Toward	Toward	Toward	Negative to bright	Negative to neutral	—
Set	Toward	Toward	Toward	Neutral to positive	Neutral	—
Play	Toward	Toward	Toward	Positive	None, Single laugh	—
Talk	Toward	Toward	Toward	Positive	Repeated and coo	—
Adult						
Avoid	Away	Away	Away	Negative to none	Negative	None or negative
Avert	Away	Partially away	Away	Negative to none	Negative adult	Varied
Monitor	Toward	Toward	Toward	Negative to none	Little infantalization	Varied
Elicit	Toward	Toward/moving	Toward/moving	Varied	Rapid/none	Abrupt/touching
Set	Toward	Toward	Toward	Neutral to positive	Infantalized	Varied
Play	Toward	Toward/moving	Toward	Positive	Infantalized	Varied
Talk	Toward	Toward	Toward	Positive	Burst-pause	Varied

Note. From "Monadic Phases: A Structural Descriptive Analysis of Infant–Mother Face-to-Face Interaction," by Tronick, Als, and Brazelton (1980). *Merrill-Palmer Quarterly, 26,* pp. 9–10. Adapted with permission of Wayne State Unversity Press.

applied both the original and an expanded (13-point) version of the monadic phases system to the play interactions of 20 full-term and 20 preterm infants and their mothers at both 3 and 5 months of age. Spectral and cross-spectral time-series analyses were used to quantify periodicities in the behavior of both mothers and infants, as well as to estimate the synchrony or coherence of interactional rhythms between mothers and infants. Results showed that, for most infants and mothers, the cycles were between .022 and .10 Hz. This corresponds to periods of 10 to 45 seconds. The level of synchrony (or coherence) between the cycles of mothers and infants increased between 3 and 5 months, and there was evidence that both mothers and infants led the interactions, depending on the frequencies under consideration. Gottman and Ringland (1981) also noted evidence for periodicity and bidirectional influence in a reanalysis of the data in three dyads from an earlier study by Tronick and colleagues.

In a similar study of 3-, 6-, and 9-month-old infants, Cohn and Tronick (1988) found little evidence of periodicities. Only 5 of 18 infants at the age of 3 months, and 1 of 18 infants at the age of 6 months, showed a significant periodic component in their behavior, and even for these infants the periodic component accounted for little variance in behavior. Instead, the cycles identified were nonperiodic in nature. At 3 months, evidence for influence of mothers on infants and infants on mothers was found, not in terms of the coherence of cycles but instead in terms of how moment-to-moment changes in the behavior of one partner predicted changes in the other. Unfortunately, differences between the studies make comparison of the results difficult. The studies differed in the coding systems employed (13 points vs. 9 points), the length of the interactions modeled (3 minutes vs. 2 minutes), and major differences in the nature of the time-series analyses employed. Kaye and Fogel (1980), using very different analytic methods, also found little evidence of periodicity in infant behavior, but instead noticed aperiodic rises and falls. Although the issue of periodicity (i.e., cycles of regular, predictable lengths) is far from resolved, this does not mean that maternal and infant involvement does not increase and decrease over time, nor that mothers and infants do not respond to one another in predictable ways. The controversy concerns how well the increases and decreases can be modeled as a predictable function of time alone.

Experimental Variations. Several experimental studies of adult–infant interaction have been conducted that support Stern's (1977) model. In two separate studies, Pelaez-Nogueros and colleagues (Pelaez-Nogueros et al., 1997) had adult females interact with infants in a standardized manner, to examine the effects of different forms of contingent responsiveness on infant behavior. In both studies, experimenters responded contingently to infant looks with various forms of stimulation, and examined their effects on infant behavior over time. In the first study (Pelaez-Nogueros et al., 1996), contingent physical stroking paired with smiling and vocalizing at the infant led to more eye contact, more positive affect, and less negative affect over time than did contingent smiling and vocalizing alone. In the second (Pelaez-Nogueras et al., 1997), contingent stroking led to more eye contact, more positive affect, and less negative affect than did tickling and poking. Although these investigators interpreted their findings in an operant framework, an equally plausible interpretation is that the combination of smiling, vocalizing, and stroking was optimally arousing and elicited higher levels of infant involvement than did smiling and vocalizing alone, whereas tickling and poking may have been overstimulating and led to more infant distress.

Field (1977) also provided some support for Stern's (1977) model. In a study of full-term, preterm, and postterm infants, she noted that when mothers of full-terms were instructed to try to get their infant's attention, the mothers increased their level of stimulation over that of free play, and the amount of time that their infants looked at them actually *decreased*. When mothers were instructed to imitate the infant's behavior, the mothers reduced their stimulation level and the amount of infant looking increased. These findings are consistent with Stern's (1977) arousal regulation model if the high levels of maternal stimulation during attention getting were overstimulating and thus increased infant gaze aversion, and if the additional maternal responsiveness during imitation increased the interest value of maternal stimulation and thus led to more infant looking. However, Symons and Moran (1987) did not replicate Field's (1977) findings. They found that infant attention was highest in the attention-getting condition and lowest during imitation (although this difference was not statistically significant). However, the level of maternal stimulation during attention getting in their study did not differ from the level during free play, so it may have not been overstimulating to the child.

The most common type of experimental study in this area involves use of Tronick and colleagues' (Tronick, Als, Adamson, Wise, & Brazelton, 1978) "still face" procedure. In such studies (e.g., Cossette, Pomerleau, Malcuit, & Kaczorowski, 1996; Field, Vega-Lahr, Scafidi, & Goldstein, 1986; Kogan & Carter, 1996; Lamb, Morrison, & Malkin, 1987; Mayes & Carter, 1990; Murray & Trevarthen, 1985; Toda & Fogel, 1993; Weinberg & Tronick, 1996), parents play with their infant for a short period and then terminate normal social interaction and present a motionless "still face" (usually for about 2 minutes). Infants in these situations usually show decreases in smiling and looking at the parent, increases in motor activity, and in some studies fussiness or crying. If this disinterest or distress is the result of boredom from insufficient stimulation, then these results support Stern's (1977) model in that infant positive affective involvement is dependent on the presentation of interesting patterns of parental stimulation.

A problem with the still face situation, however, is that so much changes upon presentation of the still face that it is difficult to account for its effects on infant behavior. For example, parental behavior in a range of modalities (e.g., facial expression, looking behavior, vocalizations, touching, head movements) decreases, the overall level of stimulation decreases, the change in parental behavior is abrupt and unanticipated, parental behavior is no longer responsive to that of the infant, and the parent shows a pattern of behavior that the infant may not have seen before. Therefore, instead of supporting the arousal model, the effects of the still face could be due to expectancy violations, decreases in a single or multiple modalities, a loss of social control, the suddenness of the change, and so on. More likely, probably more than one of these processes is operating. Muir and colleagues (D'Entremont & Muir, 1997; Ellsworth, Muir, & Hains, 1993; Gusella, Muir, & Tronick, 1988; Hains & Muir, 1996a, 1996b; Stack & Muir, 1992) and others have conducted a series of studies aimed at exploring which aspects of the situation may be most influential in affecting infant behavior. These studies provide data supporting the arousal regulation, expectancy violation, and noncontingency explanations. For example, supporting arousal regulation was the finding that if adults (mothers or experimenters) provided physical stimulation during the still face sessions (even if the adult's hand was not visible), the infants showed more positive affect than if no physical contact was provided (Stack & Muir, 1992). The moderating effect of adult touch, however, was only significant for active, not passive, touch. Thus, infants who continued to receive stimulation in the form of active touch during the still face period probably experienced levels of arousal high enough that they did not become bored and distressed.

Findings supporting the expectancy violation hypothesis are those showing that more natural decreases in the level of maternal activity are associated with less distress than is the novel still face pattern. Thus, infants show less distress if their mother leaves the room, turns her head away, breaks eye contact, or offers a smiling versus a neutral still face (D'Entremont & Muir, 1997; Field, Vega-Lahr, Scafidi, & Goldstein, 1986; Hains & Muir, 1996b; Murray & Trevarthen, 1985).

Evidence for the role of contingency comes from studies showing that, when infants interact with an noncontingent adult, they show less attention and positive affect and sometimes more distress than when interacting with a contingently responsive one (Hains & Muir, 1996a; Murray & Trevarthen, 1985). Some of the earliest research on the role of contingencies comes from studies on infant operant learning conducted during the early 1960s. Although it was originally demonstrated that experimenters could condition the smiling or vocalizations of young infants (e.g., Brackbill, 1958; Rheingold, Gewirtz, & Ross, 1959), subsequent research showed that the presentation of either a contingent or noncontingent talking face appeared to serve as an eliciting stimulus for infant social behaviors as long as the adult established eye contact (Bloom, 1974; Bloom & Esposito, 1975). However, Bloom and colleagues later demonstrated that although both contingent and noncontingent adult vocalizing increased the frequency of infant vocalizations, contingent adult vocalizations were more likely to elicit infant vocalizations that showed pause patterns similar to those used in adult conversation, as well as vocalizations that were closer to spoken language in their structure (i.e., syllabic vs. vocalic sounds—Bloom, 1988; Bloom, Russell, & Wassenberg, 1987; see also Legerstee, 1991). Finally, Cohn and Elmore (1988) observed that, when mothers presented brief 5-second still faces contingent on their infant's vocalizations, the infant's behavior was found to be more responsive to the mother's behavior than when the still faces were presented randomly.

Summary and Conclusions

In summary, the increasing social responsiveness of infants during the third month of life is associated with both quantitative and qualitative changes in the nature of parent–infant play. As infants become more capable play partners, their parents spend more time playing with them, and the pairs engage in more complex, reciprocally satisfying interactions. During face-to-face play (the primary form of play studied for this age), parents engage their children in extended "proto-conversations" in which they provide sufficient support (or scaffolding; Bruner, 1978) so that the child can be a partner in a conversational interchange with limited active involvement. Parents treat infants as if they are the predominant speakers in these interactions, giving continuous gaze, waiting for or trying to elicit a response, and imitating (usually within a second) infant facial expressions, vocalizations, and motor behaviors. Parents also ask questions, wait for a response, and provide one of their own when the infant does not supply one.

Thus, parents incorporate their child's behavior into the only kind of conversational structure they know—a simplified version of that used with other adults.

Although parents simplify their language when playing with infants, their language differs from the child-directed speech of the second year of life, in that it is more rapid, shorter, and involves more repetition. Parents use a large number of phatics that have no lexical meaning, and a variety of prosodic variations that function to elicit and maintain the attention of a child with no receptive language abilities. In fact, several studies have demonstrated that many of the speech variations characterizing this period yield vocalizations that infants prefer (e.g., Fernald & Kuhl, 1987; Werker & McLeod, 1989). Most of these variations reflect the high levels of parental emotional involvement in play—although it is rarely mentioned in the literature, parents are actually having *fun* when they make faces, vary the pitch and loudness of their vocalizations, show sudden changes in pitch, rhythm or stress, and make nonsense words and sounds. Moreover, work by Malatesta and colleagues show that mothers discourage or ignore negative affect in this context as well.

Infants' social responsiveness increases throughout this period. With advancing age infants become more responsive to parental vocalizations, changes in parental looking behavior, and changes in the level of affective involvement. Face-to-face play is characterized by patterns of reciprocity and mutual responsiveness, and sensitive parents appear to play a role in the regulation and modulation of infant affect. Although other forms of play occur during this period (play with objects, physical play), little is known about their structure.

Period of Early Directed Activity (5 to 9 Months)

In Sander's (1962) third stage of parent–infant interaction, the infant begins to play a greater role in initiating social interactions with the parent. Although these initiations are often simple in nature (looks at parent, crawls to parent, etc.), they represent the infant's early attempts to engage the parent in play. An even greater difference between parent–infant interactions in this stage compared to those in the previous stages concerns the role of objects. Because this stage covers a period of rapid locomotor and fine-motor development (the infant changes from an immobile child who can barely grasp objects to one who can creep, crawl, and manipulate virtually everything in reach), the infant spends the vast majority of his or her time manually exploring an environment that was previously only available for visual inspection. Not surprisingly, most parent–infant interactions during this period are object centered.

Infant Initiations and Social Responsiveness

The infant's growing tendency to initiate interactions was addressed by Kaye and Fogel (1980), who found that by 6 months spontaneous infant greetings were as frequent as those elicited by their mothers (at 3 months, mother-elicited greetings were much more common). Although Kaye and Fogel (1980) noted that with increasing age infants spent less time gazing at their mother's faces (see later discussion), they determined that "those declining periods contain a richer and more balanced exchange of expressive acts between the two partners" (p. 463). This is also evidenced in the findings of other investigators, who observed increases in a number of infant-expressive behaviors between the period from 3 through 5 to 6 through 8 months, including increases in vocalizations (Bornstein & Tamis-LeMonda, 1990; Field, Vega-Lahr, Goldstein, & Scafidi, 1987), smiling (Field et al., 1987; Malatesta et al., 1986), and imitation of the mother's behavior (Pawlby, 1977). The frequency of negative affect expressions (i.e., knit brow and pain/discomfort expressions) decreased over this same time period (Malatesta et al., 1986; Malatesta & Haviland, 1982). Using sequential analytic techniques, Cohn and Tronick (1987) and Tronick and Cohn (1989) noted increasing social responsiveness to mothers between 3 and 6 months of age.

The Role of Objects

The results of numerous studies document the other major change in this period: the increasing role of objects in parent–infant play. As a result of their developing manipulatory abilities, infants spend a great deal of their time handling objects during this time period, and much of parent behavior during object play can be seen as parental attempts to facilitate the infant's exploration or play with objects.

Looking at Objects. Researchers who have compared infants in the 3 through 5 versus 6 through 9 month age ranges find that as age advances the amount of time that infants and mothers look at objects increases, and the time spent looking at the play partner decreases (Findji, 1993; Friedman et al., 1976; Malatesta et al., 1986; Messer & Vietze, 1984; Tronick & Cohn, 1989). This results in less total time spent in mutual visual regard (Friedman et al., 1976). However, because infants in the 6 through 9 month range and their mothers divide their time between looking at objects and looking at their partner, the frequency of partner-directed looks actually increases (Messer & Vietze, 1984). Messer and Vietze (1984) observed that, between 2½ and 6 months, the frequency of maternal frames increased, but their duration decreased. Messer and Vietze found an increase in the likelihood that mothers would look away from their infant after the infant looked away. These looks away likely involved mothers following their infant's gaze toward an object.

Talking About Objects. Besides looking at objects during parent–infant play, parents spend time talking about the objects as well. Maternal greetings and verbalizations during play that are comments on the infant's behavior or condition decrease between 3 through 5 and 6 through 8 months, whereas verbalizations that provide information about objects and the external environment become more common (Penman et al., 1983; Sherrod, Crawley, Petersen, & Bennett, 1978). Maternal directions, commands, and questions also become more common during play (Penman et al., 1983; Sherrod et al., 1978). These include both attention-directing utterances as well as those that involve trying to influence infant behavior toward objects. Statements with no clear referent and those that are affect related also decrease (Malatesta & Haviland, 1982; Penman et al., 1983). In general, maternal behaviors that involve social stimulation decrease over this period, whereas didactic interactions increase (Bornstein & Tamis-LeMonda, 1990).

There is a close connection between maternal language and current environmental objects or events during this period. Harris, Jones, and Grant (1983) observed that at both 6 and 9 months, about two thirds of maternal utterances began in response to something said or done by the child, and 70% of maternal utterances related to an object at which the child was currently looking. When mothers did change the topic (i.e., said something that was not in response to an infant vocalization or behavior), 95% of the time they accompanied their verbalizations with an action (compared to only 61% of the time for utterances in response to the child). Overall, the content of maternal utterances was most likely to refer to objects and their attributes, or to potential actions the child could perform with the objects.

Motor Behavior With Objects. Maternal motor behaviors change during this period as well, with more time devoted to object play (Tronick & Cohn,

1989), and imitation of infant behavior with objects (Pawbey, 1977), and less time to tactile stimulation (Crawley et al., 1978; Field et al., 1987). As opposed to 3 months, at which Cohn & Tronick (1988) found similar proportions of mothers and infants to be responsive to moment-to-moment changes in the other's behavior, at 6 months the proportion of mothers was higher. Cohn and Tronick attributed this tendency "to infants' increased interest in objects at this age" (p. 388). Also, probably as a function of the greater diversity of maternal behaviors that result when mothers engage in both object- and nonobject-mediated play, the diversity of behaviors imitated (Pawlby, 1977) and games played (Crawley et al., 1978) increases during this period as well.

Parent–Infant Games

A form of parent–infant play that becomes common during this age period—likely in response to the infants' increasing social responsiveness and developing motor abilities—is the parent–infant game. Crawley et al. (1978) identified four kinds of parent–infant games in a laboratory study of nontoy play between mothers and their 4-, 6-, and 8-month-olds: tactile, gross body movement, limb movement, and visual stimulation games. Tactile and gross body movement games simply involved physical stimulation and did not require an active motor role on the part of the infant. Limb movement and visual stimulation games were classified as those in which the infant did or did not have a conventional motor role. For both limb movement and visual stimulation games, the proportion of games in which the infant had an active motor role significantly increased between 4 and 8 months. These investigators identified the four games with a conventional motor role for infants: pat-a-cake, peek-a-boo, clapping, and waving bye-bye. Gustafson, Green, and West (1979) provided a more comprehensive list of games involving an active motor or vocal role, including those identified by Crawley et al. (1978) and the following: gonna get you, horsie, ball, vocal games, build tower–knock down, give and take, point and name, so big, telephone, and "overboard" (where the parent turns the infant upside down and says "little girl/boy overboard").

A microanalytic analysis of mothers playing peek-a-boo with their 5-month-old infants (Stifter & Moyer, 1991) highlights the importance of affect regulation in the game-playing context. Using molecular coding of facial affect, Stifter and Moyer (1991) observed that infants showing moderate- and high-intensity smiles during peek-a-boo exhibited more gaze aversion during peek-a-boo than did those showing low-intensity smiles. Moreover, smile duration was positively correlated with the frequency of gaze aversions overall. Sequential analyses of individual smiles showed that long smiles tended to be followed by long gaze aversions, and the smiles followed by long gaze aversions tended to be of greater intensity and longer duration than were smiles followed by no gaze aversion. Moderately active mothers elicited the most frequent, most intense, and longest infant smiles as well as the highest level of gaze aversions. In contrast, infants of low- and high-intensity mothers showed more nonpositive gaze aversions. These findings are consistent with Stern's (1977) arousal regulation model of play interactions, particularly his suggestion that gaze aversion serves an arousal regulating function. Although Stifter and Moyer (1991) expected the highest level of infant arousal for highly stimulating mothers, they asserted that their results were more consistent with the hypothesis that infants of mothers showing low levels of stimulation were not positively aroused by the interactions and thus showed less gaze aversion. Stifter and Moyer stipulated that mothers who show high levels of activity may have increased their level of stimulation in response to infant disinterest.

A Taxonomy of Play Patterns

Power and Parke (1982) conducted a laboratory study and a home observation study of play between 7- to 10-month-old infants and their mothers and fathers in an attempt to develop a taxonomy of parent–infant play patterns during this age period. Rather than focus on the coding of individual parent and infant behaviors, they identified 11 parent–infant "bouts" that corresponded to larger units of social interaction during mother– and father–infant play, and three kinds of managerial interactions that regulated infant exploration in the home. Microanalytic coding was conducted on typical examples of the bouts identified in the laboratory setting to identify their sequential structure (such analyses could not be conducted in the home, because videotape records were not obtained in this setting). As shown in Table 9.5, parent–infant bouts fell into two general categories: social/affective and exploratory). Social/affective bouts did not involve toys or objects, whereas exploratory bouts did. Consistent with the high degree of object-mediated play during this age period, analyses of the home observation data revealed that exploratory bouts accounted for about half of all parent–infant play interactions, with managerial interactions accounting for another 12%. Only about one third of parent–infant play interactions were made up of the nonobject-mediated social/affective bouts.

TABLE 9.5

Parent–Infant Interaction Bouts: 7 to 10 Months (Power & Parke, 1982)

	Description
Social/affective bouts	
Face	Parent positions face in front of infant's face → Eye contact → Infant looks Away → Parent Moves own Face away
Vocal	Parent asks question answerable by one word → Infant utters word *or* Parent labels object → Infant labels object
Minor physical	Parent touches, tickles, bounces, or shakes infant
Lifting	Parent lifts infant in air → Parent positions Infant in face-to-face orientation → Face bout parent shakes infant → Parent lowers to lap
Exploratory bouts	
Watching	Parent presents toy → Make toy salient (make noise, move) → Present toy
Toy touching	Parent presents toy → (Move toy → Touch with toy)[a] → Present toy
Grasping	Parent presents toy → Infant reaches → Infant grasps toy → Parent vocalizes
Retrieving	Parent presents toy out of reach → Infant crawls and reaches → Infant grasps toy
Imitating	Parent demonstrates motor behavior with or without object → Parent gives object to infant → Infant imitates motor Behavior
Searching	Parent hides object → Infant searches → Infant finds object *or* Verbal instructions to find object → Infant searches → Infant finds object
Large-motor assisting	Parent assists infant in crawling, walking, climbing, etc.

[a]The sequence in the parentheses was often repeated a number of times.

Period of Focalization on the Parent (9 to 12 Months)

Infant Socially Directed Behaviors

Sander's (1962) next stage, the focalization on the parent, coincides with a number of significant developments in the infant's ability to initiate social interactions with others. During this period, the infant develops a number of social skills that greatly expand its initiating repertoire. These include such diverse behaviors as socially directed vocalizations (Bates, Camaioni, & Volterra, 1975; Harding & Golinkoff, 1979), communicative pointing (Murphy, 1978), and showing and giving objects (Rheingold, Hay, & West, 1976). An important characteristic of this period "concerns the extension of initiative in the child to achieve something of a manipulation of the mother, especially a focalization on her as a person who meets his needs" (Sander, 1962, p. 142). Although Mosier and Rogoff (1994) demonstrated that these conventional means for achieving "instrumental use of the mother" develop gradually between 6 and 10 months, their frequency of occurrence doubled between 9 and 10 months, and then increased by half again between 11 and 12 months. Not only do infants develop a number of new ways to influence adult behavior during this period, they also become more likely to direct these behaviors toward parents or significant others. This is a period of the intensification of attachment behavior (Ainsworth et al., 1978) and, in contrast to the previous periods, it is a time characterized by a large number of infant-initiated interactions.

Both Bates et al. (1975) and Masur (1983) provided longitudinal data on the development of communicative behaviors in the parent–infant play context between 9 and 12 months of life. The developmental sequence in most infants is open hand reaching followed by showing, then giving, and then pointing. Although pointing as a conventional communicative gesture does not usually appear until the end of the first year (Masur, 1983 found the mean age of first pointing to be 12 months), infants show hand positions with an extended index finger during face-to-face interactions as early as 2 to 3 months (Fogel & Hannan, 1985; Hannan, 1987). However, an analysis of the form of these "prepoints" and the contexts in which they occurred (Blake, O'Rourke, & Borzellino, 1994) suggests that these early finger extensions have little in common with later communicative pointing.

The increase in infant-initiated play interactions characterizing this period was documented in a longitudinal, observational study of unstructured mother–infant interactions in the home (Green, Gustafson, & West,

1980). Green et al. found that, between 6 and 12 months, the frequency of child-initiated interactions increased. Two kinds of child interactions showed an increase: those in which the infant directed behavior toward the mother that elicited a maternal response, and "nondirected" initiations in which an infant behavior not directed toward the mother elicited a maternal response. Directed infant initiations that showed an increase included fussing and directing the mother's attention; nondirected initiations that increased included locomotor actions and manipulating objects. Infant initiations in which the infant simply looked at the mother decreased over this time period. Observations of infants who could and could not engage in independent locomotion (as well as observations of infants in and out of walkers) suggested that two factors accounted for these increases in infant initiations: the infant's developing locomotor and communicative skills.

Decreases in Physical Stimulation; Increases in Coordinated Joint Attention

Concurrent with the development of these communicative abilities comes a decrease in the amount of physical play and an increase in toy play (Lamb, 1977). Parents of children in this age range appear to show high levels of responsiveness to infant communicative behaviors, especially in the context of object play. In Masur's (1982) home observations, mothers responded verbally or nonverbally to over 95% of all showings of objects (infant extends object toward mother) and about 80% of all points. Mothers often responded to such behaviors (especially points) with object labeling. Similarly, in a study of mother– and father–infant interactions between 6 and 12 months, Yarrow and associates (Hunter, McCarthy, MacTurk, & Vietze, 1987; Yarrow et al., 1984) noted an increase in attention focus (present object, set problem, offer object, take object, demonstrate) and a decrease in sensory stimulation (toy touching, prompt gross motor act, tactile stimulation, physical stimulation) with increasing child age. They also found increases in the frequency of joint vocalizations, joint object play, parent vocalizations when infants played with an object, and infant vocalizations when parents presented an object. During this same time period, Bakeman and Adamson (1984; Adamson & Bakeman, 1985) observed increases in coordinated joint play and positive affective expressions, and decreases in unengaged behavior and looking at the mother.

As was the case for the previous period, maternal verbalizations during toy play are tied closely to objects in the physical environment. In a laboratory study of toy play, Messer (1978) found that at 11 months, 86% of all maternal utterances referred to the toys in the playroom, especially the toy or toys with which the infant or mother were currently playing. Using information theory analyses, Messer (1978) noted that although the frequency with which the various toys were mentioned reduced uncertainty about the referents of mothers' utterances by only 1% to 4%, knowing the identity of the toy with which the child was currently playing reduced uncertainly between 57% and 89%. Even when there was a lack of correspondence between the toy currently being played with and the content of mothers' utterances, the toy mentioned frequently became the next object to be manipulated. In a study of mother's play with familiar and novel objects, Masur (1997) found that, when mothers first mentioned the name of a novel object, it was almost always (92% or more of the time) accompanied by physical designation of the object as well. Moreover, in virtually every case, the name of the whole object was given before references to its parts or other characteristics.

Kruper and Uzgiris (1987) revealed age-related changes in the nature of parent verbalizations. In a cross-sectional study of mother– and father–infant interaction, they found that parents of 9-month-olds showed a greater number of attention-seeking utterances, game-related statements, and statements "for" the child than parents of 3-month-olds. In contrast, parents of 3-month-olds showed more interpretations, comments, explanations of actions, interpretations of affect, and requests for permission.

Increasing Infant Role in Parent–Infant Games

Several studies have shown that both the frequency (Gustafson et al., 1979; Power, 1985) and the complexity of mother–infant games continued to increase during this age period. With increasing age, infants are more likely to play and to initiate games. For example, Gustafson et al. (1979) found that although infants initiated only 7% of the mother–infant games at 6 and 8 months, they initiated 48% of these games at 12 months. As was also true during the last period, infants take on a more active motor and vocal role with increasing child age (Camaioni & Laicardi, 1985; Crawley & Sherrod, 1984; Gustafson et al., 1979; Hodapp, Goldfield, & Boyatzis, 1984; Ratner & Bruner, 1978). Moreover, as infants take on a more active role, their behavior becomes more routinized and organized around the game's structure (Ratner & Bruner, 1978). Thus, for example, the number of game-related infant gestures and the frequency of infant smiles increase with age, as does the co-occurrence of infant vocalizations with game-related gestures (Rome-Flanders, Cossette, Ricard, & Decarie, 1995; Rome-Flanders & Richard, 1992).

Parents' behavior, in turn, serves as a social "scaffold" (Bruner, 1978) for infant behavior during games, by incorporating the infant's behavior into the game structure and providing support and structure for infant behavior when the infant is unable to perform certain behaviors independently (Hodapp et al., 1984; Rome-Flanders, Cronk, & Gourde, 1995). However, as the infant develops new skills the parent allows the infant to take a more and more active role. In mother–infant ball games, for example, Hodapp et al. (1984) noted that mothers of infants who had yet to successfully return the ball often invented "non-return variants" on the game so that the games could continue. These variants, along with attention-getting and stage-setting behaviors, became less common with advancing infant age (see also Rome-Flanders, Cronk, et al., 1995) as infants became more active and competent participants. In contrast, mothers holding out their hands as a target for the ball and reinforcement became more frequent with age. Interestingly, the mean age at which infants first returned the ball (and when infants uncovered themselves in a game of peek-a-boo) was around 10 months—the same age as the development of many other communicative behaviors in some of the previously reviewed studies.

Summary and Conclusions

In summary, parent–infant play interactions during this and the previous age period are more frequent, of shorter duration, and more likely to involve objects and be initiated by infants. Moreover, interactions between parent and infant become increasingly collaborative in nature, with infants taking on more active motor and social roles. However, the increasing focus on objects does not mean that the interactions become less positive in nature. Cohn and Tronick (1987; Tronick & Cohn, 1989) observed that, in the absence of objects, the amount of time mothers and infants spent in social play increased between 6 and 9 months, as did the probability of the co-occurrence of mother and infant positive affect. Cohn and Tronick (1988) also noted that although mothers primarily responded to infants at the 6-month period (see previous section), a more balanced pattern of social influence was found at 9 months. Given their high levels of social responsiveness, it appears that in the transition from the period of early directed activity to the period of focalization on the parent, infants learn new ways of incorporating their increased interest in exploring and manipulating objects into complex, rule-governed, social interactions with their parents.

Period of Self-Assertion (12 to 24 Months)

During Sander's (1962) final period of parent–infant interaction, the primary issue is the development of autonomy. As the infant begins to walk and more actively explore the physical environment, parent–infant interactions come more and more under the infant's control. During this period, parents begin to shift from the role of stimulator into a more managerial role (Power & Parke, 1982). The parent spends less time engaged in extended interactions with the infant, and more time organizing and arranging the infant's home environment, and setting boundaries on the range of the home setting to which the child has access (Parke, 1978; Power & Parke, 1982). The onset of walking can impact the affective quality of mother–infant interactions as well, with both more positive exchanges and more "testing of wills" (Biringen, Emde, Campos, & Applebaum, 1995). Moreover, the infant's developing cognitive and communicative abilities transform the nature of parent–child play. The development of language greatly increases the infant's capacity to play an active role in initiating and directing parent–infant interactions, and the child's developing symbolic abilities (especially pretend play—see chap. 3) broadens the number of possible themes for parent–child play. Together, these developments allow infants to show a greater amount of independence than during the first year of life, and sensitive parents respond to these changes by granting their children more autonomy during the second year.

Autonomy Granting

Studies of changes in parent–infant play during the transition from the period of focalization on the parent to the period of self-assertion confirm increases in the parents' granting of autonomy. For example, Power (1985) found that parents of 13-month-olds are less likely than parents of 10-month-olds to direct infant attention or perform motor behaviors for the child. They are also less likely, with advancing child age, to engage in directive speech acts (e.g., direct hearer's attention, negotiate mutual attention, negotiate immediate activity), and are more likely to engage in child-centered speech acts (e.g., discuss joint focus, discuss clarification of communication, discuss hearer's thoughts/feelings; Pan, Imbens-Bailey, Winner, & Snow, 1996). Clarke-Stewart (1973) observed over this time period increases in maternal responsiveness as well.

Verbal Mediation

Related to greater autonomy granting is the increasing tendency for parents to use verbal means alone to direct and regulate infant behavior during play (Power, 1985). Researchers have reported increases in the frequency and duration of parent and infant vocalizations during the second year (e.g., Clarke-Stewart & Hevey, 1981; Stern, Spieker, Barnett, & MacKain, 1983; Vandell, 1979), as well as increases in the number of vocal turns during parent–infant play (Rutter & Durkin,1987). During this period of rapid language development, parents, when playing with their toddlers, are more likely to ask questions, praise the child, point, label objects, imitate the child's utterances, and engage in didactic utterances and prohibitions (Halliday & Leslie, 1986; Vandell, 1979).

Parental Speech Simplification

As already mentioned, considerable research has demonstrated that parents use a simplified language register with children in this age range (for reviews, see DePaulo & Bonvillian, 1978; Pine, 1994). Differences in the nature of adult-to-child versus child-to-child speech are summarized in Table 9.6. Blount and Padgug (1977), who focused more on prosodic, paralinguistic, and interactional features of parent–child speech (in both English and Spanish), identified some of the same and many additional features as well (see Table 9.6).

Child Initiations

Clarke-Stewart and Hevey (1981) provided data on changes in the initiation of mother–infant interactions during the second year. They found a decrease between 12 and 30 months in the ratio of mother-initiated to child-initiated interactions during unstructured mother–infant interactions in the home. According to these authors, "Over this age period, mothers and children were clearly becoming equal partners in initiation of interaction, and by 2½ years they had achieved equality in the frequency of verbal initiation and were approaching it for affectionate play and responsiveness" (pp. 133–134).

Developing Gaze Patterns

Studies of the relation between gaze and vocal behavior show that infants begin to approach adult patterns of social interaction by the end of the second year. In contrast to earlier time periods, when mothers gaze almost continuously at their infants and this gaze behavior serves as a frame for

TABLE 9.6

Characteristics of Parent–Child Speech

DePaulo & Bonvillian (1978)	*Blount & Padgug (1977)*
Short sentence length Clear enunciation Slow rate of speaking High pitch, exaggerated intonation Phonological simplification: Consonant clusters, initial word sounds, distinctive consonant-vowel combinations, lengthened vowels, reduplication Semantic simplification Unique lexicon: Terms of endearment, diminutives, words with concrete referents, less diverse vocabulary Syntactic simplification: Fewer subordinate clauses, verbs, verb forms, compound verbs, modifiers, function words, coordinate clauses, that- and wh- complementizers, transformations, embeddings, conjoinings, more content words	Volume: air Breathiness, breath held Volume: sound Whisper, lowered volume, raised volume Pitch/intonation: Falsetto, high pitch, low pitch, exaggerated intonation, singing Paralinguistic: Alteration, creaky voice, tenseness, nasality, rounding, phonetic substitution Duration: Lengthened vowel, lengthened consonant, shortened vowel, shortened consonant Tempo: Slow tempo, fast tempo Lexicon/forms: Special lexical items, nonsense forms, personal pronoun substitution Grammar: Grammatical deviations, tag questions Interaction: Instructional, hmm?, attentionals, turn substitution, repetition, imitation, interpretation

Note. From DePaulo and Bonvillian (1978), p. 193. Copyright © 1978 by Plenum. From Blount and Padgug (1977), p. 72. Copyright © 1977 by Cambridge University Press. Both adapted with permission.

infant behaviors (see earlier discussion), patterns of infant gaze become more balanced during the second year. For example, Farran and Kasari (1990) noted that, between 6 and 20 months, the duration of mother looking and infant's terminating mutual gaze decreased with age, whereas the frequency and duration of infant gazes at the mother increased. Time spent in mutual visual regard increased over this period as well, as did the "simultaneous" initiation and termination of mutual regard (i.e., infant and mother changed looking behavior within the same 1-second interval). Rutter and Durkin (1987) examined the timing of infant looks away from their mothers during vocal behavior, and observed increases between the ages of 12 and 24 months in looks away at the end of the infant's and the mother's vocal turns. The former is evidence of the development of floor-yielding behavior seen in adult conversations; the latter is another adultlike pattern of signaling readiness to take the floor.

Increasing Social Coordination

In general, play interactions between parents and infants become more complex and coordinated during the second year. For example, Vandell (1979) observed increases in the following between 16 and 22 months: the percentage of time in interaction, the mean interaction duration, the mean number of action changes per sequence, the mean number of actor changes in the longest interaction, the frequency of social acts the infant directed toward his or her parent, the number of coordinated social behaviors directed toward the parent, and the percentage of social behaviors that were the child's. Vandell found decreases over this time period in the number of both infant and parent isolated social behaviors (i.e., social behaviors for which there was no partner response). Interestingly, the frequency of simple parent social behaviors increased with child age, suggesting that parents may at least initially simplify some aspects of their social behavior once the infant develops the ability to become a more active partner in the interaction. Similarly, Biringen, Robinson, and Emde (1994) noted that minimally elaborated mother–toddler interactions decreased between 18 and 24 months, whereas mutually controlled interactions increased; and Farver and Wimbarti (1995b) found that mutual involvement between fathers and toddlers increased between 18 and 24 months as well.

Social Pretend Play

Finally, pretend play between parents and infants becomes increasingly common during the second year. Power (1985) noted an increase in pretend play between 10 and 13 months, and several other investigators have documented increases in the frequency, duration, and/or complexity of pretend play between parents and toddlers during the second year (Dixon & Shore, 1993; Haight & Miller, 1992; Kavanaugh, Whittington, & Cerbone, 1983; Slade, 1987a; Tamis-LeMonda & Bornstein, 1991). Although infants sometimes engage in solitary pretend, Haight and Miller (1992) found that, between 12 and 24 months, 75% to 80% of pretend play was social, and about 90% of the time the play partner was the mother (pretend play with other children was rare in this age range, and observations were conducted when fathers were not at home). Research has shown that mothers initiate (and to a large extent structure) pretend-play interactions early in the second year, whereas by 18 and especially 24 months infants take a much more active role in the process (Haight & Miller, 1992; Kavanaugh et al., 1983). When engaged in pretend-play interactions, mothers spend a good deal of time both verbally and nonverbally prompting and elaborating on child behavior (Dixon & Shore, 1993; Dunn & Dale, 1984; Haight & Miller, 1992; Kavanaugh et al., 1983; Miller & Garvey, 1984), and several studies have shown that toddlers' pretend play in interaction with the mother is of longer duration and/or of higher level complexity than in solitary pretend (Fiese, 1990; Haight & Miller, 1992; O'Connell & Bretherton, 1984; Slade, 1987b). In a recent study, Damast, Tamis-LeMonda, and Bornstein (1996) noted that mothers prompted pretend play more than exploratory or nonsymbolic play (as did Dunn & Wooding, 1977), and that, when they prompted, they usually encouraged pretend play at a developmental level equal to or higher than the behavior in which the infant was currently engaged.

Summary and Conclusions

Together, the data on autonomy granting, child initiations, verbal mediation, social coordination, and gaze patterns all show that parent–child play becomes increasingly egalitarian during the second year. Infants take a much greater role in initiating and sustaining play interactions with their parents, and aspects of their communicative behavior start to approach more adult forms. The incorporation of pretend broadens the range of play topics considerably, and provides a new context for the parental scaffolding of developmentally complex behavior patterns.

Early Childhood (2 to 5 Years)

During the preschool years, the child's developing language and symbolic skills have a major impact on the nature of parent–child play. Unfortunately, however, in contrast to the research on parents and infants and toddlers, few descriptive studies of the nature of parent–child play interactions during the preschool years have been conducted. Although numerous observational studies of parent–child interactions have been conducted during this period, many involve observing parents and children in structured teaching contexts rather than in play (e.g., Bee, Van Egeren, Streissguth, Nyman, & Leckie, 1969; Frankel & Rollins, 1983; Hess & Shipman, 1965; Laosa, 1978).

The majority of studies of free play during this period involve descriptions of the kind of language that parents use during their interactions with language-learning children. Besides documenting many of the differences between adult- and child-directed speech (see Table 9.6), these studies have documented a number of developmental changes in the nature of parent–child speech. In general, parental language during this period becomes more complex and collaborative, and less explicitly didactic. Developmental changes identified include increases in the overall amount of verbal stimulation and the length of verbal interchanges (McLaughlin, White, McDevitt, & Raskin, 1983; Phillips, 1973); MLU, linguistic diversity, and syntactic complexity (Lipscomb & Coon, 1983; Longhurst & Stepanich, 1975; McLaughlin et al., 1983; Phillips, 1973; Rondal, 1980; Seitz & Stewart, 1975); references to objects not present (Wanska & Bedrosian, 1986); and more "collaborative" language forms that set the stage for more extended conversations such as information questions, complex expatiations, and modeled questions (Lipscomb & Coon, 1983; McLaughlin et al., 1983; Reichle, Longhurst, & Stepanich, 1976). Parental language characteristics that become less common with increasing child age include directives and imperatives (McLaughlin et el., 1983; Rondal, 1980); and clarifications, repetitions, corrective feedback, and other forms of direct tuition (Lipscomb & Coon, 1983; McLaughlin et al., 1983; Moerk, 1974, 1975; Rondal, 1980; Seitz & Stewart, 1975).

The few studies that have examined more than the verbal aspects of parent–child play after the age of 24 months generally find that with, increasing child age, parent–child play becomes more interactive; that the play is more likely to include pretend themes; and that it is less physical in nature. MacDonald and Parke (1986) conducted one of the few studies of parent–child physical play beyond the infancy period. They made phone interviews of 390 families with children ranging in age from under 1 year to 10 years, regarding their physical play activities. Most forms of physical play (i.e., swing child, toss child, roll on soft surface, bounce on knee, pattycake, horsey, and piggyback) peaked during the period of 1 to 2 years and declined thereafter, although other forms peaked at 3 to 4 years (wrestle, tumble, and playground activity). When comparing the ways that mothers and fathers played with their 12-month-olds versus their 3- to 4-year-old siblings, Stevenson, Leavitt, Thompson, & Roach (1988) found that parents engaged in more functional and physical play with infants and more pretense play with the preschoolers. Finally, in a study of parents and 3- to 6-year-olds exploring a new set of toys, Henderson (1991) noted that parent-directed learning decreased with age, whereas interactive exploration increased.

In a recent paper on pretend play in the home, Haight, Masiello, Dickson, Huckeby, and Black (1994) observed that pretend play between mothers and children during the preschool years apparently can serve many functions and occur in many different contexts. Children and mothers often engaged in pretend play in nonplay contexts, including chores, caregiving, reading, and transitional periods. Pretend play was used in comforting the child, in teaching, and in negotiating problems.

Gender Differences in Parent–Child Play

An area that has received considerable attention is that of gender differences in parent–child play. Two areas have been studied are differential treatment of the genders by parents and mother– father differences in play style. Each is considered in turn.

Differential Treatment of the Genders

The degree to boys and girls are treated differently by their parents has long been an area of debate in developmental psychology. The major motivation for conducting this type of research is to understand the degree to which early gender differences in child behavior are the result of early socialization versus genetic factors. Because this has been an area of much disagreement and debate, the major literature reviews in this area are considered, along with a review of studies published since the most recent large-scale review.

The history of this debate during the last 3 decades starts with the publication of two books by Eleanor Maccoby (Maccoby, 1966; Maccoby & Jacklin, 1974). In the latter book, Maccoby and Jacklin (1974) argued that besides the reinforcement of involvement in gender-stereotyped play activities, there was very little empirical evidence supporting other forms of differential treatment of the genders by parents, although they acknowledged that this issue had not been the focus of intensive study. Again using limited evidence, both Birns (1976) and Block (1976) responded to Maccoby and Jacklin's (1974) conclusions by arguing that differential treatment did occur and was responsible for many (if not all) early gender differences in behavior. This debate led to an increased interest in this topic, and throughout the 1970s numerous studies of the differential treatment of boys and girls by their parents (especially boy and girl infants) were conducted.

In a review of the infancy studies conducted during the 1960s and 1970s (most of which were studies of parent–infant play), Power (1981) concluded that although the evidence for differential treatment of the genders by mothers was equivocal, there was clear support for differential treatment by fathers. Specifically, of 24 studies examining mother–infant interaction that were located, only 16 publications considered infant gender as a factor and only 7 identified statistically significant patterns of differential treatment. Moreover, only 2 of 15 studies that were published after 1975 found evidence of differential treatment by mothers, and both of the significant effects could be explained by a tendency for mothers to verbalize more to their girls than to their boys. This contrasted greatly with the situation for fathers, where all but 1 of the 15 studies of father–infant interaction located found statistically significant differences in treatment of the genders. The patterns of differential treatment were summarized as follows:

> While fathers actively encourage the development of manipulatory and visual-spatial abilities in their boys, they generally encourage the development of verbal abilities in their girls. Specifically, regardless of infant age, fathers spend much of their time encouraging their boys to explore and play with objects, while they encourage their girls to vocalize. However, the *ways* in which they encourage these behaviors vary with the infant's developmental status. For example, fathers encourage exploratory skills in their boys first by showing them toys (newborn to 3 months), then by encouraging the manipulation of toys (7 to 10 months), and finally by allowing them to explore freely the physical environment on their own (11 to 17 months). Similarly, while fathers of 3-month-old girls encourage vocal behavior through imitating games, fathers of older girls are more likely to encourage vocal behavior through asking questions and labeling objects. (Power, 1981, p. 237)

The only other general pattern identified in the Power (1981) review was a tendency for fathers to show higher levels of stimulation toward their boys and more affection toward their girls.

In his 1987 review, Siegal also concluded that fathers show greater differential treatment of the genders than do mothers. In 20 of the 39 observational studies of mother– and father–child interaction he identified, fathers showed greater differential treatment of the sexes than did mothers. His review included all but two of the published studies of father–infant interaction reviewed by Power (1981), as well as six additional studies published after the 1981 review. Of the 19 studies that did not show greater differential treatment by fathers, only 8 showed no differential patterns of treatment. The remaining studies showed either differential treatment by both mothers and fathers ($n = 10$) or only differential treatment by mothers ($n = 1$). According to Siegal's (1987) conclusions, differential behavior toward boy infants "included more touching of boys, encouragement of more physical play, and exploration in boys, and responding more negatively to boys' doll play" (p. 190). Observational studies of 2- to 12-year-olds showing differential treatment by fathers (many of which were studies not of play but of discipline) showed that "fathers were reported to be firmer, stricter, less affectionate, more physical and directive and to use more positive or negative reinforcement with boys than girls" (p. 190).

Finally, in 1991, Lytton and Romney published a meta-analysis of 172 studies of differential treatment of the genders. Included in the meta-analysis were observational and self-report studies covering eight socialization areas: amount of interaction, encouragement of achievement, warmth/nurturance/responsiveness, encouragement of dependency, restrictiveness, disciplinary strictness, encouragement of gender-typed activities, and clarity of communication/use of reasoning. Using meta-analytic techniques, Lytton and Romney examined the effect sizes for differential treatment of the genders in each of these areas, and explored the effects of several moderating variables, including data collection method, source of study, year of publication, gender of the first author, sample size, and methodological rigor. Separate analyses were conducted for mother and father behavior. With the exception of the encouragement of gender-typed activities, all the remaining effect sizes were nonsignificant, indicating that there were no consistent patterns of differential treatment in any of the other socialization areas listed here. Although the mean effect sizes were in the predicted directions, the variability in effect sizes across studies was so great that only the encouragement of gender-typed activities was significant. The effect sizes for fathers were all greater than those for mothers, but only the effects for the encouragement of gender-typed activities reached statistical significance. Examination of the role of moderating factors showed that effect sizes decreased with child age, and were larger for observational and high-quality studies than for self-report and less methodologically rigorous investigations. However, controlling for these variables did not make any additional effect sizes significant. Surprisingly, given changes in gender roles during the 1970s and 1980s, year of publication showed no moderating effects.

Given the differences in conclusions between the Lytton and Romney (1991) and the Power (1981) and Siegal (1987) reviews, the specific observational studies considered in these reviews were examined. As already mentioned, Siegal (1987) included all but two of Power's (1981) published references, and six newer infant references as well. Surprisingly, however, Lytton and Romney (1991) did not include 8 of Power's 15 observational studies of father–infant interaction; and Lytton and Romney (1991) did not include 16 of Siegal's (1987) 39 observational studies. Of the studies Lytton and Romney (1991) did not include, 8 out of 8 (Power,

1981) and 13 out of 16 (Siegal, 1987) found significant differential treatment of the genders. The studies not included in Lytton and Romney (1991) were Baumrind and Black, (1967); Bell, Johnson, McGillicuddy-DeLisi, and Sigel (1981); Belsky (1979); Bronstein (1984); Field (1979a); Field et al. (1984); Golinkoff and Ames (1979); Mulhern and Passman (1981); Noller (1980); Parke and O'Leary (1975—2 studies); Parke and Sawin (1980); Pedersen, Cain, Zaslow, and Anderson (1982); Power and Parke (1982—2 studies); Stuckey, McGhee, and Bell (1982); Weinraub and Frankel (1977). Therefore, almost half of the studies that led Power (1981) and Siegal (1987) to their convergent conclusions were not included in Lytton and Romney's (1991) meta-analysis.

Besides differences in the pools of studies reviewed, a major difference in method might have led to the discrepancies between these reviews. All a study had to do to qualify for demonstrating differential treatment of the genders in either the Power (1981) or the Siegal (1987) reviews was for one or more of the parent variables to vary significantly as a function of child gender. Because the nature of differential treatment varied considerably across studies, these differences may have been "washed out" in Lytton and Romney's (1991) meta-analysis.

A number of studies of mother– and father–infant interaction in unstructured or play contexts have been published since Lytton and Romney's (1991) review (they only reviewed studies abstracted by 1987). A review of these studies (and also several studies published before 1987 that were not included in the previous reviews) confirms the conclusions offered by Power (1981) and Siegal (1987). Specifically: (a) Mothers show higher levels of verbal stimulation and responsiveness to the verbalizations of girls versus boys (Fagot & Hagan, 1991; Fagot & Kavanagh, 1993; Kaye, 1980; Leaper, Lever, Strasser, & Schwartz, 1995; Masur, 1987; Sagi, Lamb, Shoham, Dvir, & Lewkowicz, 1985), (b) fathers show higher levels of stimulation and overall involvement with boys (Farver & Wimbarti, 1995b; Sagi et al., 1985; Shields & Sparing, 1993), and (c) fathers use more verbal strategies with girls (Farver & Wimbarti, 1995b; Kruper & Uzgiris, 1987; Masur & Gleason, 1980).

It should be noted, however, that the specific behaviors involved in the differential treatment varied from study to study. For example, regarding the conclusion about mothers and the verbal stimulation of girls, Fagot and Hagan (1991) found that parents gave girls more positive responses to their attempts to communicate; Fagot and Kavanagh (1993) noted that mothers gave more instructions/directions to girls trying to communicate; Kaye (1980) observed that mothers directed longer, more varied utterances toward girls; Leaper et al. (1995) determined that mothers were more likely to reciprocate their girls' speech acts; Masur (1987) gleaned that mothers were more likely to imitate their girls' discrete vocalizations and words; and Sagi et al. (1985) noted that mothers vocalized more frequently to their girls than to their boys. Therefore, it is likely that if these studies were put into a meta-analysis such as that employed by Lytton and Romney (1991), these differential patterns of treatment might have been obscured. It is a question for future research whether this variability in findings reflects the impact of situational factors influencing the kinds of differential treatment that is observed, or reflects random variation in results.

Two additional areas of differential treatment have been identified in the more recent studies. The first is that mothers show more positive emotional expressiveness toward their girls than toward their boys (Fish, Stifter, & Belsky, 1993; Malatesta, Culver, Tesman, & Shepard, 1989; Sagi et al., 1985); the second is that, in studies employing sequential analysis, mothers are more likely to match the emotional expression or dyadic states of their boys than their girls (Malatesta & Haviland, 1982; Masur, 1987; Robinson, Little, & Biringen, 1993; Tronick & Cohn, 1989).

Leaper, Anderson, and Sanders (1998) supported two of these conclusions in a recent meta-analysis. In this analysis of observational studies of mothers' verbalizations toward boys versus girls (studies of differential treatment of the genders by fathers were not included due to their low frequency), these authors found that, across studies, mothers talked more to and made more supportive statements toward girls than boys.

In summary, as concluded by Lytton and Romney (1991), the major form of differential treatment of boys and girls concerns the activities in which they engage, with parents (just as with peers) encouraging children's involvement in traditionally gender-typed activities. Other differences in treatment are more subtle and harder to demonstrate. However, despite the conclusions of Lytton and Romney (1991), parents, especially fathers, do appear to engage in differential treatment of the genders during parent–child play. These differences primarily involve higher levels of verbal stimulation and positive emotion and/or affection with girls; and (apparently for fathers only) higher levels of visual, exploratory, and physical stimulation to boys. However, the nature of this differential treatment appears to vary across studies and with the interactional context and age of the child.

Given parental differences in activities, it is possible that some of the differences in parent behavior toward the genders may be a function of the activities they choose, as was the case for play between peers (e.g., Carpenter, 1983; Huston et al., 1986). Only two studies could be found that examined this issue. O'Brien and Nagle (1987) observed mothers and fathers playing with their toddlers in three separate sessions (once with dolls, once with trucks, and once with a shape sorter). Only one main effect of child gender was significant across sessions: Parents used more verbs with their daughters than with their sons. Numerous differences in parent verbal behavior emerged as a function of the toys available. These differences are summarized as follows:

> Each type of toy had a distinctive pattern of language associated with it. With the shape sorters, parents' language was functional; they were helping their child accomplish a specific task. Parent speech included a lot of directives, attentionals and praise, but no playful or imaginative language and little labeling of objects. In this context, they used a lot of active verbs and pronouns. The variety in parents' speech, indexed by number of different nouns, active verbs, and modifiers, was lowest when they played with shape sorters.
>
> With dolls, parents were highly verbal and encouraged verbalizations from their toddlers. Their speech was characterized by more and longer utterances, frequent labeling of objects and a high proportion of questions, particularly yes–no questions. This context also elicited the greatest variety of words, the most nouns, the most different nouns, the most references to objects and the most copulas. Finally, parents' ratio of nouns to pronouns was highest in this context.
>
> With the vehicles, parents used little language. Total utterances and MLU was lowest in this context, as were frequency of pronoun and verb use. The only category of speech that was highest with the vehicles was imaginative sounds; motor noises made up an average of 10% of all utterance types during play with the vehicles. (p. 275)

The results of this study are consistent with the hypothesis that the nature of parental differential treatment of the genders may be a function of the toys they use when playing with their child. Activity choice, in fact, may be a more powerful determinant of parental behavior than gender of the child.

Leaper et al. (1995) also examined this issue. Married and single mothers of preschool children were videotaped while playing with a set of masculine-stereotyped and a set of feminine-stereotyped toys. Analyses of mothers' and children's use of controlling, supportive, and collaborative speech acts showed a number of complex interrelations among gender of the child, the mother's marital status, and the play activity. Unlike the O'Brien and Nagle (1987) study, the gender of the child had an impact on maternal behavior over and above the nature of the activity in which they engaged. Clearly, further research on this issue is needed.

Mother–Father Differences in Play

Not only have investigators examined differential treatment of boys and girls, they have also examined differences in the play styles of mothers versus fathers. With the exception of studies of parental language, these studies have mostly involved parents and infants or toddlers. Studies employing unstructured observations of mothers, fathers, and infants in the home have shown that, in most North American families, mothers display greater overall rates of interaction than do fathers (e.g., Belsky, Gilstrap et al., 1984), and also that the nature of parent involvement varies as well. Several studies have shown that, when after controlling for the overall level of interaction, that mothers devote a greater proportion of their time to caregiving than do fathers, and that fathers devote a greater proportion of their time to play (Belsky, 1979; Belsky, Gilstrap et al., 1984; Clarke-Stewart, 1978; Lamb, 1977; Parke & Sawin, 1980; Russell & Russell, 1987). In Western cultures outside of North America, mothers show higher rates of interaction with infants than do fathers as well. This has been found in a naturalistic observational study of parents and children in playgrounds in France, Germany, and Italy (Best, House, Barnard, & Spiker, 1994), and in two studies of nontraditional fathers: fathers who took paternity leave during the first year of their infant's life in Sweden (Lamb, Frodi, Frodj, & Hwang, 1982), and fathers living in an Israeli kibbutz where primary care was not provided by the parents (Sagi et al., 1985).

Examination of play styles reveals several mother–father differences. Mothers' play, in general, tends to be more verbal, object mediated, and didactic, whereas fathers' play tends to be more physical, active, and unpredictable. Most of these differences have been documented for infants in the first months of life and apparently continue throughout early and middle childhood.

Mothers spend more time vocalizing to their infants during play, during both face-to-face-interactions during the first few months of life (e.g., Papousek et al., 1987) and toy play during the second half year (Kruper & Uzgiris, 1987; Landerholm & Scriven, 1981; Lewis & Gregory, 1987), and during toy play in the second year and later (e.g., Rondal, 1980). In a recent meta-analysis of 35 studies of mother–father differences in parent-to-child verbalizations (mostly during studies of parent–child play—Leaper et al., 1998), mothers were found to talk more frequently to their children and engage in more supportive and more negative speech acts. Fathers, in contrast, engaged in more directive speech acts, informing speech acts, wh-questions, and information questions. The size of the effects varied with several factors, and the greatest differences found for studies involved infants and toddlers versus older children, data collected in the home versus the lab, longer versus shorter observation sessions, unstructured observations versus specific instructions, a choice over which toys to play with versus instructions to play with one specific toy, and where no toys were used. The fact that the extent of mother–father differences in language during play varies considerably with these factors is evidenced by the fact that numerous studies find no differences between the nature of mother- and father–child speech or no more differences than would be expected by chance alone (i.e., either one or two differences in a study of numerous characteristics—e.g., Hummel, 1982; Kavanaugh & Jirkovsky, 1982; Lewis & Gregory, 1987; Lipscomb & Coon, 1983; O'Brien & Nagle, 1987; Pakizegi, 1978; Wilkinson, Hiebert, & Rembold, 1981).

Several other mother–father differences in parental language not examined in the meta-analysis (probably due to the small number of studies that considered these characteristics) suggest that mothers' speech overall is more complex and more responsive than that of fathers. For example, Davidson and Snow (1996) found that mothers' speech was more complex than was fathers' in terms of number of words, verbs to turn ratio, and subordinate clauses to turns ratio; and Hladik and Edwards (1984) noted that a greater percentage of maternal utterances were grammatically correct. Stevenson et al. (1988) observed that mothers engaged in more instructive play with their 12-month-olds than did fathers, whereas fathers engaged in more functional play. Mothers are also more likely to comment on or label objects than are fathers (Rondal, 1980; Vandell, 1979). Masur and Gleason (1980) noted that when fathers *did* label objects they were more likely to label an object that the child had already used, whereas mothers were more likely to reserve their labels for objects about whose names the child was not sure. Fathers also show greater levels of verbal repetition than do mothers (Fash & Madison, 1981; McLaughlin et al., 1983). This may be a consequence of the greater frequency of communication breakdowns that occur for fathers in which requesting clarification becomes necessary (Tomasello, Conti-Ramsden, & Ewert, 1990). Fathers are also more likely to not acknowledge their children's utterances than are mothers (Tomasello et al., 1990).

Mothers engage in more play with objects than do fathers (Clarke-Stewart, 1978; Parke & Sawin, 1980; Russell & Russell, 1987), and are more likely to engage their child in conventional games (Lamb, 1977; MacDonald & Parke, 1984; Yogman, 1981). Mothers' greater experience with toy play is evidenced in microanalyses of their behavior in this context. Mothers are more likely than fathers to follow their infant's looking behavior when presenting a new toy, and to build on the infant's behavior when encouraging a new behavior with a toy (Power, 1985; Power & Parke, 1983). Fathers, in contrast, engage in more physical play by touching their infant with toys (Power & Parke, 1982), and are more likely than mothers to shift to physical play when the infant shows disinterest in a toy (Power & Parke, 1983). Fathers are also less successful than mothers in getting their infants to engage in specific actions with a toy (Power & Parke, 1983), and more likely to interfere with ongoing infant behavior when trying to direct toy play (Power, 1985). However, Hunter and colleagues (1987) found that fathers engaged in more joint object play than did mothers, and that infants were more likely to vocalize on toy presentation when playing with their fathers than with their mothers. The degree to which this may be a function of toy-touching play among the fathers was not examined.

A major difference between mothers and fathers is that fathers engage in higher levels of physical play than do mothers (Clarke-Stewart, 1978; Dickson, Walker, & Fogel, 1997; Field, 1978; Jacklin, Di Pietro, & Maccoby, 1984; Lamb, 1977; MacDonald, 1987; MacDonald & Parke, 1984; Power & Parke, 1982; Ross & Taylor, 1989; Russell & Russell, 1987; Yogman, 1981), and mothers are more likely to engage their child in play involving nonphysical, visual displays (Power & Parke, 1982; Yogman, 1981). Fathers' physical play takes the form of poking the child (Field, 1978) or actively moving the child's limbs (Yogman, 1981) during face-to-face interactions in the early months, and is manifested in a wide range of forms in the second half year (see Table 9.5). In their interview study of mothers and fathers of 1- to 10-year-olds, MacDonald and Parke (1986) noted that fathers engaged in almost all forms of physical play more often than did mothers. In general, the forms that were more common among mothers were those that were either conventional games (i.e., pat-a-cake) or involved lower levels of physical stimulation (i.e., roll on soft surface). Besides these two forms, tumbling was the only other form of physical play found more often in mothers than fathers. Fathers showed higher levels of involvement for all play forms involving high levels of stimulation (i.e., swing child, toss child, and wrestling), as well as forms characterized by more moderate levels (i.e., bounce on knee, horsey, tickle, piggyback).

Besides being more physical, fathers' play is more active and possibly more diverse than that of mothers. Fathers' face-to-face play occurs at a faster tempo than does mother–infant play, with fathers showing a greater number of playful stimuli (Arco, 1983) and utterances (Kruper & Uzgiris, 1987) per minute, and more syllables per second (Papousek et al., 1987). Some researchers have found that fathers' speech to children is more diverse than is mothers' speech (Masur & Gleason, 1980; Rondal, 1980), although others have not (Davidson & Snow, 1996; Ratner, 1988).

Finally, research on toy choices that parents make during play show that there is a general tendency for mothers to spend more time playing with traditionally "feminine " toys and fathers to spend more time with traditionally "masculine" toys during the child's second year of life (Caldera, Huston, & O'Brien, 1989; Idle, Wood, & Desmarais, 1993; Jacklin et al., 1984; Russell & Russell, 1987).

In summary, mothers and fathers appear to provide their children with very different experiences during play. Mothers' play, in general, tends to be more verbal, object mediated, didactic, and responsive to the child's behavior; whereas fathers' play tends to be more physical, idiosyncratic, and unpredictable. Because so many studies of mother–father differences involve observations of parent behavior during play with toys (a context preferred by mothers over fathers), mothers not surprisingly show more competence and responsiveness in this context than do fathers. However, these differences occur within the context of numerous similarities between mothers and fathers, and individual differences within the genders (especially for fathers) are considerable. Unfortunately, with but a couple of exceptions (e.g., Field, 1978), most studies involve comparisons of primary caregiver mothers to secondary caregiver fathers, making it impossible to determine how much these differences reflect actual gender differences versus differences in parenting roles.

Studies of Non-Western Cultures

Studies of parents and children in hunting-and-gathering and agricultural village cultures show that the kind of parent–child play described in the preceding pages is by no means universal. Parent–child play varies considerably across cultures, with play being a major part of the parent's daily activities in some cultures and something that is rarely observed in others. Middle-class parents in Western cultures spend the most time playing. For example, Whiting and Edwards (1988), found that of the 12 cultures they studied, middle-class mothers in the United States showed the highest levels of parent–child play. In most hunting-and-gathering and agriculture village cultures, children tend to be the primary playmates instead of parents, even in the toddler years (e.g., Edwards & Whiting, 1993; Goncu, Mistry, & Mosier, 1991; Lasky, Klein, Yarbrough, Sellars, & Kagan, 1983; Rogoff, 1981). This appears to be a function of sibling caretaking and children's participation in mixed-aged playgroups (see chap. 8). Weisner and Gallimore (1977) summarized data from Barry and Paxson (1971) on caretaking arrangements in 186 traditional societies. They showed that, in *early childhood* (i.e., postinfancy), child care was principally the responsibility of the mother in only about one quarter of the societies for which information was available. In the others, children and/or other adults were responsible for the child's care for half or more of the time; three quarters of the alternate caregivers were children.

Maternal caretaking is more common in infancy. In the Barry and Paxson (1971) data, 3% of the societies were characterized by exclusive maternal care; in about half, care was provided principally by the mother, with others having minor roles; and in about one third, care was provided principally by the mother, with others having important roles. Although mothers usually care for their infants, the amount of mother–infant play or affectionate interaction varies considerably. In some traditional cultures, mother–infant play is rare or nonexistent (e.g., Ainsworth, 1967; Brazelton et al., 1969; Kagan & Klein, 1973; Minturn & Hitchcock, 1963), whereas in others it is observed frequently (e.g., Hogbin, 1943; Maretzki & Maretzki, 1963). Consider the following descriptions:

Uganda (Ainsworth, 1967):

My notes contain only one observation of a mother holding her baby so that there was a face-to-face confrontation—Habatanzi's mother once held her high in the air in a gesture of delight, talking to her and smiling, while the baby responded with delighted laughter. I noted no instance of a mother bending over her baby trying to coax a smile or a coo. Indeed there was only one mother—Nabatanzi's again—who could be described as playing with her baby. Except for her, we noticed no one playing games of peek-a-boo or pat-a-cake. When we inquired about such games we sometimes received the response "when he is older."...I noted no instance of a mother trying to elicit a response from her baby by dangling a plaything in front of him. I saw no instance of a mother attempting to direct a baby's attention to a physical object and away from herself. Their interaction was direct and not mediated by playthings. (pp. 93–94)

Okinawa (Maretzki & Maretzki, 1963):

A happy smiling personality is ideal, and it is this wish for the child which leads grandmothers and mothers to perpetuate the custom of burying the afterbirth of a newborn to the accompaniment of laughter and appeals to the spirits to ensure this characteristic. Adults constantly try to get infants to smile by talking and making funny sounds and faces. Even young children approach infants clapping their hands together in a attempt to elicit smiles...[Mothers] will encourage the little 2-year-olds to imitate older boys and girls, particularly in their baby-tending role. A mother will strap a "doll" on her daughter or son's back. The "dolls" are usually rolled blankets or jackets, for real dolls are expensive and rarely bought in the village. Older children and adults do not fail to comment to a child carrying such a burden. They praise him for being such a good caretaker; they pretend to see the "baby" sleeping and report this to him. If the strapping become loose, they call to him and reassemble the "baby" and the pinning. (pp. 473, 479)

New Guinea (Hogbin, 1943):

She rarely, if ever, plays any sort of games with it, but one sees her crooning it, caressing the fleshy parts of its body with gentle fingers, and implanting breathy kisses over and over again in the region of its genitals. (p. 298)

In many traditional cultures, mothers of infants carry their children while they work, using a sling or other carrying device. These devices, although freeing the mother's hands for various nonchild-care tasks, limit opportunities for face-to-face interaction (e.g., Brazelton et al., 1969; Goldberg, 1977; McSwain, 1981). Moreover, in some cultures, mothers orient their infants away from themselves in an apparent effort to direct the infant's attention toward other people (Ainsworth, 1967; Sostek et al., 1981). These two practices sometimes lead to more playful interactions between infants and children or other adults than with the mothers themselves (Bakeman et al., 1990; Kilbride & Kilbride, 1974; Konner, 1976). For example, in a study in the Marquesas Islands, Martini and Kirkpatrick (1981) found that:

Babies were generally oriented outward. They were directed to attend, and eventually to interact, with the surrounding world. Complex interactions with surrounding others were set up. Marquesans did not engage in pseudo-dialogues in which caregivers act as if infants controlled the detailed turn-taking rules of conversation. Instead, they had infants "recognize" others, even before they seemed capable of distinguishing the many people mentioned. (p. 208)

Rather than play, in many cultures mothers of infants provide high levels of physical contact, vestibular stimulation, and nursing on demand (Brazelton et al., 1969; Kagan & Klein, 1973; Kaplan & Dove, 1987; Kilbride & Kilbride, 1974; Konner, 1976; McSwain, 1981; Richman, Miller, & LeVine, 1992). However, traditional cultures vary in the amount of face-to-face play. Hewlett (1987), in a study of the Akan Pygmies of central Africa, found that between the ages of 1 and 4 months 65% of the father–infant play interactions involved a face-to-face orientation.

Several theorists have offered explanations for the differences between parent–infant interactions in Western and many traditional cultures. In a study of the Rjaputs of Khalapur, India, Minturn and Hitchcock (1963) argued that: "Adult interaction with babies is generally aimed at producing a cessation of response rather than a stimulation of it" (p. 317). LeVine et al. (1994) made a similar point in trying to account for differences in the behavior of Gusii and North American mothers. Gusii mothers, they maintained, act in a way consistent with the pediatric model (see Table 9.7)—that is, the parent's role is to protect the child by modulating excitement and distress through soothing patterns of interaction. As seen in the table, this model predicts decreasing maternal attention over time as children further develop their self-regulatory abilities. In contrast, the behavior of North American mothers is consistent with a pedagogical model, in which the goal is to stimulate the child intellectually and socially through protoconversations, contingent responding, and eliciting infant verbal and emotional responses. This model predicts increasing maternal attention over time as

TABLE 9.7

Models of Early Attention in Early Child Care

	Gusii: Pediatric Model	*American: Pedagogical Model*
Goal	Protection	Active engagement, social exchange
Means	Soothing	Stimulation, protoconversation
Temporal distribution over first 30 months	Decreasing	Increasing
Cultural script for selective responsiveness	Respond to distress	Respond to babble
	Modulate excitement	Elicit excitement
	Commands	Questions, praise

Note. From LeVine et al. (1994), p. 249. Copyright © 1994 by Cambridge University Press. Adapted with permission.

the child becomes increasingly receptive to environmental input. Dixon, Tronick, Keefer, and Brazelton (1981) noted differences in the face-to-face interactions of Gusii and North American mothers that were consistent with these models. When the Gusii mothers played with their infants, their interactions were characterized as even and smooth, with few peaks of affective involvement. Although Gusii mothers responded to their infants and the interactions were cyclical in nature, they showed lower levels of stimulation and joint positive affect than did North American mothers.

Rogoff, Mistry, Goncu, and Mosier (1993) made a similar distinction. They asserted that in Western cultures, where children are segregated from many adult activity contexts, parents take responsibility for child learning by structuring teaching situations for their children, often in the context of parent–child play. In many traditional cultures, where adult–child segregation is less evident, parents expect children to learn through observation and increasing participation in adult tasks. Rather than *actively teach* their children, parents in traditional cultures provide assistance in the context of shared group activities. Thus, Western parents believe that parent–child play is a context for stimulation and learning, whereas parents in many traditional cultures view play as an activity primarily for children. In an observational study of mothers and toddlers from four cultures (United States, Turkey, India, and Guatemala), Rogoff et al. (1993) observed differences in free play and teaching consistent with this view. Goncu et al. (1991) described the kinds of play observed in these sessions (both mothers and other children were present). Children in the two urban, industrialized settings (United States and Turkey) engaged in more pretend play, language play, object play, and games than did children in the traditional village settings (Guatemala and India). Children from the United States showed more physical play than did children in any of the other groups. Children in the urban settings also showed more of the following: animation of inanimate objects, substitution of objects, pretending an imaginary situation exists, sound effects, hiding games, role playing, nonsense word construction, and mothers' mimicking their child's vocalizations. Mothers from India were more likely than those from the other groups to tease their children and mothers from the United States were more likely to label objects.

An issue of considerable debate among language development researchers is whether baby talk and motherese is found in all cultures (e.g., Blount, 1972; Ferguson, 1964; Fernald et al., 1989; Grieser & Kuhl, 1988; Ingram, 1995; Papousek & Papousek, 1991; Ratner & Pye, 1984). A considerable amount of cross-cultural research supports this idea. Parents from a wide range of cultures make similar speech modifications when interacting with infants and young children (for reviews, see Lieven, 1994; Snow, 1995). For example, Papousek and Papousek (1991) and Fernald and colleagues (Fernald et al., 1989; Fernald & O'Neill, 1993) found similar melodic structures and pitch modifications in parents from France, Germany, Italy, England, Japan, Malaysia, Greece, India, Iran, Russia, Brazil, Indonesia, Korea, South Africa, China, and the United States. However, one problem with hypothesizing that such modifications are *necessary* for the development of language is that in some cultures adults apparently rarely talk to prelinguistic children *at all*, and in others the language variations shown toward infants are not of the type that would facilitate language learning (see Lieven, 1994). Cultures also vary in the degree to which adult members actively carry out intentional language tuition. Until more cross-cultural research on the relationship between language environments and children's language development is conducted, this will likely continue to be a source of debate among language acquisition theorists.

Besides the research on hunting-and-gathering and traditional agricultural village cultures already reviewed, the other major source of information on mother–infant play outside of Western settings comes from several studies of mother–infant play in Japan and the United States (e.g., Bornstein, Tal, et al., 1992; Bornstein, Tamis-LeMonda, et al., 1992; Bornstein, Toda, Azuma, Tamis-LeMonda, & Ogino, 1990; Caudill & Weinstein, 1969; Fernald & Morikawa, 1993; Fogel, Toda, & Kawai, 1988; Morikawa, Shand, & Kosawa, 1988; Otaki, Durrett, Richards, Nyquist, & Pennebaker, 1986; Shand & Kosawa, 1985; Tamis-Lemonda, Bornstein, Cyphers, Toda, & Ogino, 1992; Toda, Fogel, & Kawai, 1990). In these studies, mothers and infants were observed interacting either in face-to-face sessions or during unstructured sessions in the home.

In general, Japanese mothers show a more muted style of parent–infant interaction during play, and U.S. mothers exhibit a more stimulating style (Shand & Kosawa, 1985). Summarizing across studies, during parent–infant play mothers from the United States behave in a way consistent with LeVine et al.'s (1994) pedagogical model: making more references to objects and events in the physical environment, labeling objects, asking more questions, and speaking in complete sentences. Japanese mothers, in contrast, are closer to the pediatric mode, not simply reducing infant distress, but gently modulating infant affect through affectionate, moderately stimulating actions. As a group, Japanese mothers are more likely than mothers from the United States to lean close to their infants, call the infant's name, use affect-salient speech, use hand displays, and "lull" versus "chat" with their infants. Japanese mothers seem more concerned with affective engagement than didactic stimulation in that the grammar of their utterances is simpler and they use more onomatopoeic labels, make more interjections and nonsense sounds, and engage in more social routines. They are also more likely to demonstrate and solicit other-directed pretense.

In a study conducted in urban Indonesia, Farver and Wimbarti (1995a) compared mother– and sibling–child play with a set of wooden toys. Families were working class and lived in "a densely populated urbanized version of a rural village with cement brick houses set close together in rows along narrow walkways" (p. 1495). Although participants lived in Yogjakarta, a large urban area, with the exception of the father's occupation (most fathers were employed as shop clerks, mechanics, etc.), the child's lifestyle was closer to that of a rural village than to that of Western urban or suburban neighborhoods. Children lived in extended family households, were often cared for by their siblings, and from the age of 3 or 4 years spent most of their time playing in mixed-age playgroups. Results showed that, in contrast to studies in Western cultures, Indonesian children showed a greater amount of exploratory, combinatorial, pretend-object, and social-pretend play when playing with their siblings than with their mothers. Siblings were more likely than mothers to join in play, make comments, make suggestions for pretend play, and share positive affect. Mothers, in contrast, showed more teaching behaviors. An analysis of play themes showed that play with siblings was more likely to revolve around environmental danger, whereas play with mothers was more likely to center on family issues.

Only a limited number of studies have examined developmental differences in the nature of parent–child play in non-Western settings. Most findings, however, are consistent with the developmental changes identified in Western cultures. In a study on the Marquesas Islands of mothers and infants during the 1st year of life, Martini and Kirkpatrick (1981) observed that, with increasing child age, mothers spent less time holding their infants close, looking at the infant, engaging in mutual gaze, and calling the infant's name. Time that children spent not being held, looking at objects, and interacting with someone besides the mother increased with age.

Studies of infants in the second year, as with studies in Western cultures, show increases in sociability, verbal interactions, and independence. In a study of Embu toddlers, Sigman et al. (1988) found that time spent carrying the infant, time spent in physical care, and the amount of infant crying decreased between 15 and 30 months, whereas infant talking and smiling increased. Similarly, Bakeman et al. (1990) noted that, with increasing child age, Kung mothers were more likely to speak to the infant, and others (children, other adults) were more likely to encourage specific infant activities. The degree that others entertained the infant decreased with age.

In the previously discussed study of Indonesian mothers and children, Farver and Wimbarti (1995a) discerned age-related differences in mother–child play that parallels the results of several Western studies. Between 18 and 36 months, exploratory and combinatorial play decreased and cooperative social-pretend play increased.

Few quantitative studies of hunting-and-gathering or traditional village cultures examine differential treatment of the genders (possibly because of the small sample sizes often involved). However, a good deal of research has been reported on differences between mother and father involvement. Ethnographic studies show that the level of father involvement in child care and childrearing varies considerably by culture (for reviews, see Hewlett, 1992a; Lamb, 1987). Father involvement tends to be highest in cultures where fathers spend considerable time in the home setting. Specific factors associated with high levels of father involvement include monogamy, nuclear family structure, gathering or horticultural subsistence patterns, and low levels of warfare (West & Konner, 1976).

The only consistent mother–father differences demonstrated across cultures are the tendencies of mothers to engage in more caretaking and for fathers to engage in more physical play. The cultural group with the highest levels of father involvement identified to date is the Aka pygmies of central Africa. Hewlett (1992b) observed that Aka fathers hold or are within arm's reach of their infant more than 50% of each 24-hour period. In a study of 1- to 18-month-olds, Hewlett (1987) found that although Aka mothers were more likely to provide the infant with nourishment or to transport the infant, Aka fathers were more likely to hug, kiss, and play with the infant. Fathers engaged in more minor physical play than did mothers (tickling, bounce lap, nibble stomach). Fathers and mothers rarely engaged in major physical play (rough and tumble, chasing, total body movement), and mothers and fathers did not differ in object-mediated or face-to-face play.

Two studies of parents of 12-month-olds living in urban, non-Western cultures yield similar differences. In a study of college-educated parents in India, Roopnarine, Talukder, Jain, Joshi, and Srivastau (1990) noted that mothers were more likely to pick up, feed, comfort, display affection, and play peek-a-boo with their infants than were fathers, whereas fathers were more likely to engage in rough physical play. Mothers and fathers did not differ in the amount of minor physical or object mediated play. In a study of mostly middle-class parents in Taiwan, Sun and Roopnarine (1996) found that mothers were more likely than fathers to hold, feed, smile at, and vocalize to the infant and to engage in object play. Fathers, in contrast, were more likely to engage in rough play. Mothers and fathers did not differ in the amount of affection, comforting, proffering, minor physical play, and peek-a-boo.

FUNCTIONS/EFFECTS

Given the limited attention given to adult–offspring play in the research on animals, there has been much less extensive discussion of its effects or functions than for peer or sibling play. Fagen (1981) posed the issue this way: "Why should a physically fit, fully skilled adult, whose daily feeding and predator avoidance activities give it abundant and perhaps even excessive stimulation, throw itself on its back in front of an infant and wave its legs in the air while the infant flails clumsily, but perhaps painfully, at it?" (p. 442).

Fagen (1981) discussed several possible functions or effects of adult–offspring play, including Zahavi's (1977) hypothesis that young animals engage their parents in play as a way of assessing their willingness to contribute to the offspring's well-being. Fagen (1981) argued that a parent "can facilitate the persistence of its own genes by participating in its offspring's (adaptive) play" (p. 443), and that such play might provide other benefits as well, such as communicating that future interactions will be cooperative instead of competitive. MacDonald (1993a) asserted that parent–offspring play should be strongly selected for because the parent can provide levels of stimulation that are more complex and intense than those offered by peers or siblings. Parent–offspring play, he maintained, might be particularly important for young animals still too immature to engage in play with siblings or peers. Given that parent–offspring play takes many forms (e.g., tickling, play-fighting, play-chasing), it could have many of the same effects or functions hypothesized for peer and sibling play (see discussion in previous chapters).

Data on the functions/effects of parent–offspring play in nonhumans are hard to find. A difference in the level of play shown by adult males toward infants in gorillas versus chimpanzees is consistent with an evolutionary analysis. Bard (1995) pointed out that gorilla fathers frequently play with their infants in the wild, whereas play is uncommon in chimpanzee fathers. She also noted that "gorilla and chimpanzee fathers differ in that chimpanzee paternity is usually not known either by observers or apparently by the chimpanzees…whereas it is certain in gorilla harems" (p. 36). No specific studies of the functions or effects of parent–offspring play in nonhumans could be found.

As was the case for children's social-object play, theorists have argued that parent–child play affects many aspects of children's cognitive, social, and emotional development (see articles in MacDonald, 1993b). However, empirical research on this topic is limited to many of the same categories investigated for children's social-object and social-pretend play (see Table 9.8). Each of these categories is considered in the sections that follow. Within the sections on cognitive and social developmental functions/effects, research from correlational studies is presented before the experimental data.

Correlational Studies: Cognitive Stimulation and Learning

Promoting General Cognitive Development

The results of several correlational studies are consistent with the hypothesis that maternal stimulation during parent–child play promotes general cognitive development in children. In the typical study, parents are observed in either a structured play situation or during unstructured observations of parent–child interaction in the home, and observers collect data on the nature of parental behaviors when interacting with the child. These data are then correlated with standardized assessments of child cognitive functioning, either at the same age at which the observational data were collected or at some future date. Because the maternal behaviors that correlate with children's general cognitive functioning vary somewhat with the age of the child, studies of infants, toddlers, and preschool-age children

TABLE 9.8

Hypothesized Functions/Effects of Parent–Child Play

Cognitive stimulation and learning
Promoting general cognitive development
Promoting linguistic skills
Providing information about the physical environment
Social development
Establishing social relationships
Facilitating the development of social perspective-taking skills
Facilitating self-regulation and control
Facilitating gender role development

are considered separately here. Although some of the most consistent correlational data come from studies using the HOME inventory (e.g., Bradley et al., 1989), these studies are not reviewed here because that semi-structured maternal interview is not a direct assessment of parent–child play. It should be noted, however, that the maternal characteristics identified in the play studies discussed here are much the same as those identified in research using the HOME.

Infants. The best maternal predictor of general cognitive functioning during infancy appears to be the responsiveness of the mother to the infant's behavior. Although studies of the *total amount* of stimulation that mothers direct toward their infants are inconsistent (Coates & Lewis, 1984; Lewis & Coates, 1980; Yarrow, Rubenstein, Pedersen, & Jankowski, 1972—some find positive correlations with general cognitive functioning and some find negative), *responsive* stimulation shows positive correlations (Clarke-Stewart, 1973; Lewis, 1993; Lewis & Coates, 1980). This is particularly true of contingent vocalizations, that is, mothers who vocalize contingently in response to their infant's vocalizations have infants who score higher on standardized assessments of cognitive functioning (Coates & Lewis, 1984; Wachs & Gandour, 1983). In fact, two studies (Coates & Lewis, 1984; Lewis, 1993) revealed that maternal responsiveness at 3 months positively predicted children's performance on standardized cognitive assessments at ages 4 and 6 years. However, Wachs and Gandour (1983) observed that contingent responsiveness to infant vocalizations was only positively correlated with infant cognitive functioning for temperamentally easy infants; the correlation for difficult infants was negative.

Other maternal behaviors positively associated with cognitive functioning during infancy include the variety of stimulation (Yarrow et al., 1972), the total amount of personal-social and object play (Clarke-Stewart, 1973; Wachs & Gandour, 1983; Yarrow, Rubenstein, & Pedersen, 1975), and the amount of eye contact and smiling during unstructured mother–infant observations (Clarke-Stewart, 1973; Crockenberg, 1983). Crockenberg (1983) found that the amount of routine care during unstructured observation sessions was negatively associated with Bayley scores; Clarke-Stewart (1973) noted a negative correlation for restrictiveness.

Toddlers. Some of the same positive correlates have been identified in studies of toddlers (12 to 24 months). These include maternal responsiveness (Clarke-Stewart, 1973; Hunter et al., 1987; O'Connor, Sigman, &

Kasari, 1993; Spangler, 1989), positive emotion (Clarke-Stewart, 1973, 1978; Spangler, 1989), and object play (Clarke-Stewart, 1973, 1978). Based on the work of White and Watts (1973) and Carew (1980), several researchers have examined the relationship between "intellectual experiences" in the home and the child's later cognitive development. For example Spangler (1989) noted that the amount of time that toddler spent between the ages of 14 and 17 months in "high-level" or "verbal-symbolic" experiences with their mothers in their homes was positively correlated with performance on the Bayley at 24 months. High-level experiences included decentered symbolic play, fine-motor activities, and problem solving; verbal-symbolic activities included interactions focused on understanding the relationship between objects or events and language concepts. Interestingly, the degree to which toddlers engaged in solitary high-level experiences did not significantly predict later cognitive development. Using an interview method, Gaiter, Morgan, Jennings, Harmon, and Yarrow (1982) noted that the variety of cognitively oriented caregiver activities in which mothers reported engaging at 1 year positively predicted McCarthy scores at age 3½. Consistent with these results are the findings of Olson, Bates, and Bayles (1984), who observed that maternal teaching behaviors at 13 months positively correlated with Bayley scores.

Teti, Bond, and Gibbs (1988) also examined activities based on the work of Carew (1980), but they found only that maternal activities involving language mastery predicted the infant cognitive assessments. This is consistent with Clarke-Stewart's (1973, 1978) determination that the amount of verbal stimulation was positively correlated with infant cognitive performance.

Preschoolers. As mentioned in the previous section on structure, fewer studies have been conducted on parent–child play with preschoolers than with infants. Most studies examining the correlates of children cognitive functioning have used teaching tasks (e.g., Hess & Shipman, 1965; Laosa, 1978). Of the few studies that have been conducted, several similarities with the research on infants and toddlers arise. Specifically, children's general cognitive functioning is positively correlated with verbal stimulation (Clarke-Stewart, VanderStoep, & Killian, 1979; Olson et al., 1984), praise and affection (Jennings & Connors, 1989; Olson et al., 1984), and positive play (Clarke-Stewart et al., 1979). It is negatively correlated with restrictiveness (Clarke-Stewart et al., 1979; Olson et al., 1984) and maternal directiveness (Jennings & Connors, 1989). In a study of mothers and their 3-year-olds in New Zealand, Podmore (1988) found that children's cognitive development was positively correlated with maternal cooperation, attending, questions, rewards, and teaching behaviors, and was negatively correlated with noncooperation, criticisms, and direct commands.

The Role of Fathers. Little correlational research has demonstrated the role of fathers in promoting children's general cognitive development during play. Of the studies reviewed thus far, only three (Clarke-Stewart, 1978; Hunter et al., 1987; Teti et al., 1988) included fathers, and only one (Clarke-Stewart, 1978) found significant prediction for fathers. Specifically, father encouragement to play during unstructured home observations was the only significant predictor of toddlers' cognitive development. However, causal analyses indicated that the likely direction of influence was from child cognitive development to father play rather than vice versa (in contrast, similar analyses on the mother data suggested that the direction was from mother to child).

Non-Western Cultures. Data on the function/effects of parent–child play in non-Western cultures are more limited. Roberts and Barnes (1992) observed that high and intermediate distancing positively correlated with children's performance on the Weshler Preschool and Primary Scale of Intelligence; low-level distancing showed a negative correlation. In a study of 15- to 30-month-old Embu toddlers in Kenya, Sigman et al. (1988) noted that the frequency of talking to the infant, the amount of social interaction, and maternal responsiveness to infant vocalizations were positively correlated with performance on the Bayley mental scale. In a study of urban Japanese 4-year-olds, Hatano, Miyake, and Tajima (1980) found that children's Binet IQ scores at age 6 were positively associated with maternal sensitivity, praising, and acceptance during mother–child play at age 4.

Summary. In summary, studies of mother–child play reveal a number of consistent predictors of infant's and children's general cognitive functioning. Regardless of child age, the factors of maternal responsiveness, verbal stimulation, positive affect, and object stimulation show positive correlations, whereas negative affect shows a negative association. During the toddler and preschool years, maternal restrictiveness and directiveness emerge as negative predictors, probably because such behaviors interfere with the child's developing autonomy during this period. Insufficient research has been conducted on fathers to draw any firm conclusions.

Promoting Linguistic Skills

With the increased interest during the 1970s and 1980s in documenting the nature of parental speech input to the language-developing child came a large number of studies on the relationship between language input and language development in children. Most of these studies involved collecting language samples of parent-to-child speech in either the home or the laboratory, and then correlating individual differences in the language parents direct toward their children with individual differences in the rate and nature of children's language acquisition.

Because many (if not most) of these studies collected their language samples during parent–child play, this research provides considerable empirical evidence for the possible language-promoting effects/functions of parent–child play. However, a comprehensive review of this literature is beyond the scope of this volume (for reviews, see Pine, 1994; Snow, 1995). However, based on this literature a few general conclusions can be drawn. Because most of these studies have been with language-learning toddlers, researchers have focused primarily on the parental correlates of children's receptive or productive vocabulary versus their syntax. Of the parental correlates that have been identified, the most common are the total amount of parental verbal stimulation (Huttenlocher, Haight, Bryk, Seltzer, & Lyons, 1991; Olson, Bayles, & Bates, 1986; Wachs & Chan, 1986), the amount of descriptive or referential speech (Clarke-Stewart et al., 1979; Hampson & Nelson, 1993; Tamis-LeMonda & Bornstein, 1994; Wachs & Chan, 1986), references to objects during joint attention (Saxon, 1997; Smith, Adamson, & Bakeman, 1988; Tomasello & Farrar, 1986), responsiveness to infant vocalizations (Hardy-Brown, Plomin, & DeFries, 1981; Wachs & Chan, 1986), and repetitions, yes–no questions, wh- questions, verbal reflective questions, expansions, and extensions (Hampson & Nelson, 1993; Hoff-Ginsberg, 1985, 1986; Pine, 1994; Wachs & Chan, 1986). Parent behaviors showing negative correlations with language development include redirecting child attention, directiveness, and negative control (Nelson, 1973; Olson et al., 1986; Podmore, 1988; Tomasello & Farrar, 1986).

Providing Information About the Physical Environment

Along with being a context for the stimulation of language development, parent–child play also appears to be a context in which children may learn about the physical environment. One way that researchers have examined parent–child play as a context for environmental learning has been to look at differences in children's exploratory behavior during parent–child versus solitary play. In the typical study, researchers observe infants playing with a set of objects in two conditions: when their mother is engaged in active play with the infant, and when the infant is engaged in solitary play. Mothers are frequently present in the latter condition, but are often busy in some other task (e.g., questionnaire completion) and are instructed not to direct their infant's behavior. Other studies have compared infant behavior during peer and/or sibling play to parent–child play in the same way.

Infant Object-Directed Behavior During Parent–Child Versus Solitary Play. Results of these studies show that, in general, infants engage in more complex forms of exploration and play when interacting with their mothers than when engaged in solitary play. Specifically, in the mother–child play context, infants or toddlers are more likely to engage in pretend play (Fiese, 1990) and in pretend play that is at a higher developmental level (Fiese, 1990; Slade, 1987b). Children's play in the mother–child context is more likely to exploit the unique characteristics of the objects involved (Cohen & Tomlinson-Keasey, 1980) and to be rated as more "appropriate" (Dale, 1989) or conventional (Fiese, 1990; Turkheimer, Bakeman, & Adamson, 1989). Moreover, the length of play episodes is longer in the mother–child play context (Lawson, Parrinello, & Ruff, 1992; Slade, 1987b). However, Lawson et al. (1992) found that maternal effects on the length of play episodes were restricted to low- and intermediate-attending infants only. In contrast, during solitary play infants and toddlers have been shown to spend more time in unengaged activity (Turkheimer et al., 1989) and sensorimotor functional play (Fiese, 1990), including dropping and throwing objects (Lawson et al., 1992). Play episodes are shorter and more frequent (Slade, 1987b), but are also more statistically unique or "creative" (Cohen & Tomlinson-Keasey, 1980). However, Cohen and Tomlinson-Keasey (1980) found the latter effect only for boys.

Infant Object-Directed Behavior During Parent–Child Versus Peer or Sibling Play. Comparisons of infant and toddler behavior during parent–child versus peer or sibling play reveal similar differences—that is, behavior during play with parents is more complex. For example, compared to behavior during peer play, infant or toddler behavior during mother–child play is more conventional (Turkheimer et al., 1989) and tied to the specific characteristics of objects (Cohen & Tomlinson-Keasey,

1980). Mothers are more likely than peers to offer the child toys (Field, 1979b). This latter tendency was well illustrated in a naturalistic study of children with their parents or their peers in a "touch and see" room in a natural history museum (Henderson, Charlesworth, & Gamradt, 1982). Compared to when they were with their peers, children with their parents spent more time engaged in visual or manipulatory exploration of novel objects than children with their peers. When with peers, there were more object changes, area changes, walking, and running than when with parents. As was the case with solitary play, children with their peers spend more time in unengaged behavior and the "simple" manipulation of objects (Turkheimer et al., 1989), but their play is also more unique or "creative" (Cohen & Tomlinson-Keasey, 1980—boys only). Similar patterns are found with siblings as with peers. Compared to behavior with siblings, behavior with parents is more conventional (Dale, 1989), instructive (Stevenson et al., 1988), and object mediated (Dale, 1989), whereas sibling play involves more sensorimotor or constructive play (Stevenson et al., 1988). One study that compared the amount of fantasy play during parent–child and peer play revealed an interesting developmental difference. At ages 2 to 3 years, children engaged in more fantasy transformations when engaged in play with their mothers or fathers than with their peers; at age 4, this pattern was reversed, with more fantasy transformations occurring during play with peers (Perlmutter & Pellegrini, 1987). In their longitudinal study of pretend play in the home, Haight and Miller (1992) observed that at 12, 24, and 36 months, children engaged in more pretend play with their mothers than with other children, whereas at 4 years the amount of pretend play with mothers and other children was the same.

Parent–Child Correlates of Infant Solitary Exploration and Play. A number of other studies have demonstrated that certain characteristics of parental behavior during parent–child play correlate with individual differences in children's solitary play. Studies reviewed in this section only include those in which independent assessments of parent–child and children's solitary play were collected. In some studies (e.g., Belsky, Goode, & Most, 1980; Endsley, Hutcherson, Garner, & Martin, 1979; Lawson et al., 1992; Rubenstein & Howes, 1979; Tamis-LeMonda & Bornstein, 1989), correlations between child and maternal behavior during the same sessions were examined. These are not considered here, because they may represent only the short-term effects of the behavior of one partner on another during social interaction without demonstrating generalization across contexts.

Three aspects of infant behavior during solitary object exploration or play have been examined in research on parental correlates: visual and manual exploration, developmental level of object play, and behaviors reflecting mastery motivation and persistence. The parental correlates of all three aspects of infant behavior are similar. Positive correlations include the level of maternal "attentiveness" and stimulation (Bornstein, Vibbert, Tal, & O'Donnell, 1992; Gaiter et al., 1982; Rubenstein, 1967; Wenar, 1976; Yarrow et al., 1975, 1984), variety of stimulation (Yarrow et al., 1972, 1975), maternal sensitivity and responsiveness (Frodi, Bridges, & Grolnick, 1985; van Aken & Riksen-Walraven, 1992; Wachs, 1987), object play (Teti et al., 1988; Yarrow et al., 1975), didactic stimulation (Bornstein et al., 1992), maternal demonstrations and affirmations (Stilson & Harding, 1997), the expression of positive affect (Yarrow et al., 1975), and noninterference (Frodi et al., 1985; Wachs, 1987). Frodi et al. (1985) assessed noninterference as part of a more general autonomy-supportive play style.

In a longitudinal study, Schaefer (1989) found that positive mother–infant interactions at 4 and 12 months were positively correlated with teacher ratings of children's curiosity and creativity in kindergarten. In a study of first-grade boys, Saxe and Stollack (1971) observed that high-curiosity boys had mothers who showed greater positive affect when interacting with their child in a laboratory playroom.

Parent–Child Play and Children's Understanding of the Physical Environment. Despite the significant number of studies that have compared the higher level of exploratory and play behavior that infants and young children demonstrate during parent–child versus solitary play, and those showing that parent–child play patterns predict individual differences in children's solitary play and exploration, few studies have examined correlations between parent–child play and infant's understanding of the physical world. Although items assessing such knowledge are part of the standardized assessments of cognitive functioning used in the studies reviewed earlier, few studies report separate analyses for the correlates of verbal and performance measures. In one study that did, Jennings and Connors (1989) noted that, for 3- to 4-year-olds, maternal affective tone (i.e., praise and affection versus criticism) positively predicted verbal performance on the McCarthy scales, whereas maternal directiveness negatively predicted performance scores. Similarly, Hatano et al. (1980) determined that maternal sensitivity, praising, and acceptance during mother–child play at age 4 predicted Japanese children's Binet IQ scores at age 6, but that maternal directiveness (i.e., press, orienting, caregiving, and control) negatively predicted later performance on a standardized conservation task given at the same time as the IQ assessment.

In the previously cited study, Wachs and Gandour (1983) observed that, for temperamentally easy babies, maternal contingent responses to infant vocalizations predicted infant's understanding of causality, and personal-social play predicted the assessment of foresight on the Uzgiriz and Hunt scales. In a study of 9-month-olds Van Aken and Riksen-Walraven (1992) also found that maternal responsiveness was positively correlated with the child's understanding of causality—in this case, a contingency awareness task. Finally, two studies by Rubenstein and colleagues (Rubenstein, 1967; Yarrow et al., 1972) examined correlations between maternal behavior and infant preferences for novelty in a paired comparison task. Positive predictors in these studies included maternal attentiveness, positive affect, and contingent responsiveness to positive vocalizations.

The Role of Fathers. As was the case for research on the correlates of general cognitive functioning, studies of fathers' effects of infant solitary object exploration and play are rare. Of the studies cited earlier, only seven included fathers: three comparing father–child, mother–child, and/or children's solitary play (Henderson et al., 1982; Perlmutter & Pellegrini, 1987; Stevenson et al., 1988); and four examining the correlates of father–infant play (Bornstein, Vibbert et al., 1992; Teti et al., 1988; Wachs, 1987; Yarrow et al., 1984). The studies examining infant behavior across contexts show few differences in the exploratory behavior of infants when playing with their mothers versus their fathers; the studies of parental correlates show no consistent pattern of differences between mothers and fathers.

Summary. In summary, parents directly affect the behavior of their young children when they engage the children in play. When playing with parents, infants' and toddlers' behavior is more complex, more conventional, of longer duration, and more symbolic than when playing with peers, siblings, or alone. During solitary play, infants' and toddlers' behavior is less developmentally advanced and of shorter duration, but more statistically unique. These differences hold up whether the infant is playing with his or her mother or father. It is not clear that these same differences hold for older children. In the one study that included preschoolers (Perlmutter & Pellegrini, 1987), 4-year-olds actually showed more pretend behavior with their peers than with their parents. Parents appear to scaffold the play (especially the pretend and conventional play) of their infants and toddlers but, by the preschool years, such scaffolding may be less necessary (and may actually interfere with the play process).

Research on the relationship between the quality of parent–child play and various aspects of children's solitary play competence reveal similar correlates—whether competence is defined as the total amount of exploration or play, as its developmental level, or as mastery motivation and persistence. Competent children have parents who play frequently with objects, often in a didactic manner; are sensitive and responsive to their child's behavior; and show high levels of positive affect, praise, and social stimulation. In contrast, interfering parents have children who show lower levels of competence. These same parent behaviors are correlated with children's performance on assessments of their knowledge of the physical word. Insufficient research has been conducted to draw conclusions about the effects of mothers versus fathers.

Summary: Correlational Studies

The correlational studies reviewed here suggest several ways in which parents may influence the cognitive development of their child during parent–child play. The parenting correlates of child performance across areas are similar, with positive affect and the amount and responsiveness of parental stimulation being the most consistent correlates. A parenting practice that may become increasingly important as children grow older is support for and encouragement of child autonomy. Based on the research reviewed here, it is reasonable to hypothesize that parents may facilitate their children's cognitive development during play by encouraging them to act on the social and physical world in increasingly independent and complex ways. This can take many forms, such as supporting the child in solving a problem, persisting in obtaining a goal, answering a difficult question, negotiating a complex set of play rules and conventions, and so on. By scaffolding their child's behavior during play, parents may facilitate cognitive development in a variety of ways.

Of course, the problem with the correlational approach is the difficulty in making statements about causality. This is particularly problematic for studies of parent–child play, because parents spend most of their time with their children in nonplay contexts. For example, many researchers who collect data on parent–child play do not do so to collect data on play per se, but instead conduct play observations as a way to get a "snapshot" of the overall quality of the parent–child relationship. This is particularly true in the parental language studies. If parent behavior during play is highly correlated with parent behavior in nonplay contexts, then any correlations between parent–child play and child outcomes tells us little about the effects of play per se. For example, what is more important for language development—the amount of parent verbal stimulation that occurs during *play*, or the total amount of verbal stimulation overall? Does the level of parental responsiveness during play predict something unique about child development, or is it general parental responsiveness that is important? These are questions (along with others concerning the direction of effects, and the effects of other, third variables) that cannot be answered using correlational data.

A third variable that must be considered for studies of parent–child play that is not relevant for studies of peers is the possibility that genetic factors may account for the relationship between the nature of parent–child play patterns and children's cognitive outcomes. Because parents share (on the average) half of their genes with any individual child, it could be, for example, that those parents with genes contributing to the development of high levels of verbal competence not only pass those genes on to their children but also provide a verbally stimulating environment. In this case, there would be a positive correlation between the amount of parental verbal stimulation during play and the child's verbal skills, but it might reflect genetic inheritance rather than environmental learning.

Behavior geneticists have developed various methods to explore this possibility. One of the most powerful is the adoption study, through which one can examine the correlation between parental behavior and child outcomes in individuals who are not biologically related to one another. The Hardy-Brown et al. (1981) study on language development cited earlier was such a study (in that study, maternal responsiveness to infant vocalizations was the only significant predictor of language development in adoptive families). Coon, Fulker, DeFries, and Plomin (1990) provided additional data in a study using the HOME inventory (a structured interview method that provides information about the nature of parent–child interactions, mostly through observation of parent–child interactions in an interview situation). They found, as have others, that in nonadoptive families, HOME scores at age 2 assessing maternal responsiveness, avoidance of restriction, environmental organization, play facilitation, and maternal involvement all positively predicted children's IQ scores at age 7. In adoptive families, however, only the environmental organization score was a significant predictor (all other correlations were less than .02). These findings, along with the results of several causal modeling procedures, are consistent with the interpretation that most environment-development relations were due to indirect genetic mediation.

Experimental Studies: Cognitive Development

The results of numerous intervention studies conducted with at-risk children (usually children growing up in poverty) have shown that increasing the level of parental stimulation and responsiveness toward young children has a positive impact on their cognitive development relative to nonintervention controls (for reviews, see Guralnick, 1997; Ramey & Ramey, 1998). Child outcome variables that have been included in these studies range from general assessments of cognitive functioning (e.g., standardized intelligence or language development assessments) as well as assessments of more specific cognitive skills. Although increasing the amount of time that parents play with their children is often an important part of these programs, many other aspects of parenting are targeted as well (e.g., teaching interactions, verbal stimulation, discipline, the home environment, exposure to stimulation outside the home), making it impossible to isolate the role of parent–child play. Even in those studies that have focused primarily on encouraging parent–child play (e.g., Graul & Zeece, 1990; Madden, Levenstein, & Levenstein, 1976), it is not clear what is responsible for the effects—increases in the amount of play, changes in the nature of play, social support from the experimenter or other participants, information about child development, and so on.

More focused but less intensive interventions provide more specific experimental data on the role of parenting behavior. For example, several experimental studies have examined the effects of parent presence and behavior on the nature of children's exploration of objects in the environment. Although early studies indicated that parental presence alone had positive effects of infant and toddler exploration (especially after attachment behavior has set in; e.g., Ainsworth & Wittig, 1969; Arsenian, 1943; Cox & Campbell, 1968; Gershaw & Schwarz, 1971; Rheingold, 1969), more recent studies have shown that parental presence is not enough and that effects on infant exploration vary as a function of the parent's looking behavior and emotional expression (e.g., Carr, Dabbs, & Carr, 1975; Dickstein & Parke, 1988; Feinman, 1992; Gunnar & Stone, 1984; Source & Emde, 1981). For example, infants were found to be more likely to explore if their mother was attending to them and was displaying positive versus negative emotions.

Experimental studies have also been conducted in which mothers of children were randomly assigned to conditions in which the mother or an adult stranger structured child exploration in a specific way. For example, Parrinello and Ruff (1988) had 10-month-old infants interact with a female adult experimenter who showed low, medium, and high rates of intervention during toy play. The intervention levels varied in the frequency, tempo, and intensity of the adult's motor and vocal behavior during object presentations. The effect of condition varied with the infant's baseline level of attention. For low-attending infants, both medium and high levels of intervention were associated with increased looking, whereas initially high-attending infants were unaffected by the experimental manipulation. In another experimental study, Belsky et al. (1980) reinforced mothers' attention-focusing strategies (e.g., name object, highlight object property, instruct/question) during the course of three 45-minute unstructured observations of mother–infant interactions in the home. At a follow-up assessment of infant solitary free play, infants in the experimental group showed less unfocused behavior and simple exploration and more conventional play than did infants in a control group whose mothers received the same amount of contact with the experimenters.

Several studies of preschoolers have examined the effects of varying parent or adult behavior as well. In these studies, children were found to engage in higher levels of object manipulation or exploration when adult experimenters focused the child's attention on objects (Henderson, 1984; Vliestra, 1980) or if the adult took on a friendly, approving versus an aloof, critical interaction style (Moore & Bulbulian, 1976). Another experiment (Johns & Endsley, 1977) showed that, when children observed their mothers engaged in active exploratory behavior (vs. simply looking at the stimulus objects or engaged in social interaction with another adult), they later imitated their mother's behavior when engaged in exploration on their own.

Two short-term interventions in the Netherlands showed that experimentally increasing maternal responsiveness during the second half-year of life had positive effects on infant exploratory behavior (Riksen-Walraven, 1978; van den Boom, 1994). In both studies, the intervention lasted 3 months. Mothers in the first study (Riksen-Walraven, 1978) received individual training and booklets emphasizing either increasing maternal stimulation, maternal responsiveness, or both; mothers in the second (van den Boom, 1994) received individual coaching during three 2-hour sessions to learn to respond to their infant's idiosyncratic behavior patterns. In the

Riksen-Walraven (1978) study, mothers in the two groups receiving responsiveness training had infants who showed greater manipulation of novel stimuli in a pairs test, greater diversity of exploratory behavior in their manipulation of a set of novel objects, and higher levels of performance on a contingency analysis test. Infants whose mothers received stimulation training only did not differ from controls. In the van den Boom (1994) study, infants of mothers receiving responsiveness training showed more sophisticated exploration and less mouthing in a free-play task. Sophisticated exploration included relational exploration, developmental level of exploration, and specific exploration (see Table 3.3).

Finally, language development researchers have experimentally demonstrated the importance of joint attention in language learning (e.g., Dunham, Dunham, & Curwin, 1993; Tomasello & Farrar, 1986; Tomasello & Kruger, 1992). In these studies, responsive, joint-attention-establishing strategies were more effective in encouraging word learning than were less responsive strategies.

Summary: Cognitive Stimulation and Learning

Together, the correlational and experimental studies suggest that parent–child play may have an impact on the cognitive development of infants and young children. Although the number of experimental studies is small, they are consistent with the much more numerous correlational findings. When parents take an active role in structuring infant exploration and play, the complexity of infant behavior increases, with infants showing longer periods of object involvement and higher developmental levels of object play. Moreover, the effects of parental involvement generalize to subsequent infant solitary play sessions, although only short-term effects have been demonstrated. Experimental manipulations that increase the level of maternal responsiveness, positive affect, and didactic stimulation appear to have positive effects on infant exploration and play as well. Given the promising findings of these initial investigations, further replications and extensions would be useful. In such research, the relation between specific parental practices and specific infant behaviors should be examined in order to understand the processes behind parental influences. The degree to which these relationships vary as a function of child characteristics (e.g., age, gender, temperament) should be examined as well.

Correlational Studies: Social Development

Establishing Social Relationships

Theorists have posited that parent–child play interactions may facilitate the development of children's social relationships with others. Three types of relationships have been examined: attachment relationships with parents, sociability with adults, and relationships with siblings and peers. Each are considered next.

Attachment. In their landmark work on the development of infant-to-parent attachments, Ainsworth and colleagues (for a review, see Ainsworth et al., 1978) noted that the quality of mother–infant interactions during the first year of life—often interactions during play—were predictors of the security of infant attachments at 12 months of age. In their longitudinal study, attachment security was best predicted by global ratings of maternal behavior during the second half-year of life along the following four qualitative rating scales: sensitivity–insensitivity, cooperation–interference, acceptance–rejection, and accessibility–ignoring. They also found that security of attachment was predicted by qualitative ratings of early face-to-face interactions at 6 to 15 weeks of age (Blehar et al., 1977). Mothers whose infants later showed a secure attachment relationship were rated during early face-to-face interactions as higher on contingent pacing, encouraging further interaction, playfulness, and liveliness of stimulation, whereas mothers whose infants later showed an insecure attachment were rated as higher on routine manner and abruptness. These mothers were also more likely to show no response to their infant's initiations, and to engage in silent, unsmiling initiations themselves. Overall, their face-to-face interactions with their infants were of shorter duration than those of mothers of secure infants. Finally, Tracy and Ainsworth (1981) found that mothers of later avoidant infants were proportionally more likely to kiss and less likely to hug or cuddle their infants than were mothers of secure or resistant infants.

During the 1980s and 1990s, numerous studies of the childrearing correlates of attachment were conducted, and for the most part they replicated Ainsworth et al.'s (1978) findings. In three meta-analyses, two of research with mothers (De Wolff & van IJzendoorn, 1997; Goldsmith & Alansky, 1987) and one of research with fathers (van IJzendoorn & De Wolff, 1997), the relationship between parental behavior and the security of infant attachment was confirmed, although the overall effect sizes have been modest. In the most comprehensive review involving 66 studies, De

Wolff and van IJzendoorn (1997) discerned the following effect sizes for the following maternal behaviors (effect sizes were reported as *r*s): sensitivity (.22), contiguity of response (.10), physical contact (.09), cooperation (.13), synchrony (.26), mutuality (.32), support (.16), positive affect (.18), and stimulation (.18). In a meta-analysis of eight studies with fathers, the effect sizes were somewhat smaller, with a mean of .13 (van IJzendoorn & De Wolff, 1997). The reviewers noted that the effect sizes for the original Ainsworth study were considerably larger than for subsequent research, but that studies using the Ainsworth rating scales tend to show the highest effect sizes among the replication studies.

De Wolff and van IJzendoorn (1997) observed that the relationship between the maternal variables and infant attachment were moderated by some other variables, including sample size, socioeconimic status (SES), techniques for measuring maternal behavior, location of observation, child age, and time interval between assessments. However, the effects of the moderators were limited.

Sociability With Adults. Research on the correlations between parent–infant play variables and infant reactions to other adults is much less common. In two studies by Clarke-Stewart and colleagues (Clarke-Stewart et al., 1979; Clarke-Stewart, Umeh, Snow, & Pedersen, 1980), sociability toward adult strangers was positively correlated with the mothers' play and other positive interactions with the child, a questioning versus a commanding language style, and/or language directed toward the child about other people. Spangler (1990) found that maternal responsiveness at 14 to 17 months during unstructured observations in the home were positively correlated with children's cooperation with a stranger in a laboratory play session at 24 months. Finally, Martin (1981) noted that maternal responsiveness at 10 months positively predicted sociability toward strangers at 42 months for boys, but was negatively related for girls.

Peer and Sibling Relationships. A number of recent studies also show that mother–child play variables are positively correlated with the quality of the child's interactions with siblings or peers. Before turning to this research, however, studies comparing children's social behavior during parent–child play to social behavior during peer or sibling play are reviewed, because such studies provide some clues about the effects of parent–child play on children's relationships with others.

Children's Social Behavior During Play With Parents Versus Peers and Siblings. In general, infants and toddlers show higher levels of social competence when interacting with their parents than with their peers or siblings. Two studies comparing parent–infant/toddler to peer play were located (Bakeman & Adamson, 1984; Vandell & Wilson, 1987). In a longitudinal study of infants from 6 to 9 months, Vandell and Wilson (1987) observed that compared to interactions with an unfamiliar peer, infants interacting with their mothers showed more: social acts, nonsocial acts, and social exchanges; extended social interactions; coordinated, responsive social interactions; social initiations; social initiations that were successful; and elicited responses. The only behavior more common in interactions with peers was the frequency of social sequences when there was no clear evidence of social intent by either partner. In a study of 6 to 18-month-olds, Bakeman and Adamson (1984) found that, when interacting with their mothers, infants spent more time in coordinated joint and passive joint activities than they did when interacting with familiar peers. Individual episodes of joint activity were of longer duration in the mother–child sessions as well. In contrast, when infants interacted with peers, they spent a greater proportion of their time in unengaged activities. In a further analysis of these data, Adamson and Bakeman (1985) concluded that infants and toddlers also showed more positive affect when playing with their mothers versus their peers.

Additional differences are found in studies comparing interactions with mothers versus siblings (usually studies of unstructured family interactions in the home). Not only are interactions more complex in parent–child versus sibling play, but sibling interactions are decidedly more negative. Vandell and Wilson (1987) also included sibling interactions in the previously mentioned study, and for the most part the differences between mother–child and peer play outlined earlier characterized the differences between mother–child play and play with older siblings. On several measures (e.g., time in exchanges, proportion of longer exchanges, two of the social responsiveness measures), the means for sibling play were between the means for mother–child and peer play. Whether or not this was due to differences in the familiarity or the developmental level of the play partner could not be determined from the design. In a study of 12-month-olds and their preschool siblings, Stevenson and colleagues (1988) noted that both infants and preschoolers played more complex, rule-governed games with their parents than with their siblings. Finally, a comparison of the nature of 2-year-olds' conversations with their preschool-age siblings and their mothers showed that children were much more likely to request clarification and to show verbal responses (i.e., simple, extended or disruptive) to the utterances of their mothers versus those of their siblings (Mannle, Barton, & Tomasello, 1992).

Data on the negativity of sibling interactions come from two studies (Baskett & Johnson, 1982; Dunn & Munn, 1987). In an observational study of naturalistic interactions in the homes of families of 4- to 8-year-olds, Baskett and Johnson (1982) found that children showed greater levels of high-rate behavior (i.e., repetitive, intense, annoying physical behavior), physical negative behavior (i.e., attacks or attempted attacks), negative commands, and yelling toward their siblings than toward their parents. Children were more likely to display other, less intense, negative behaviors toward their parents, including noncompliance, whining, and demands. Moreover, children exhibited certain positive behaviors more toward parents than toward siblings, including compliance, laughing, talking, and nonaggressive touching. Similarly, Dunn and Munn (1987) observed that toddlers showed more anger and distress in conflicts with their siblings than with their mothers, whereas they were more likely to laugh during conflicts with their mothers. This may have had something to do with differences in the topics of these conflicts. Conflicts with mothers tended to center around caretaking, manners, object destruction, rules of the house, and independence, whereas conflicts with siblings were usually over rights or the use of physical space. In a study of feeling state language, Brown and Dunn (1992) noted that 33-month-olds were more likely to make feeling references to their mothers than to their older siblings, whereas 47-month-olds made more feeling references to their siblings. Overall, children engaged in both more positive and negative affect with their siblings than with their mothers.

A comparison of the differences between mother and sibling behavior in the previous studies shows that mothers and siblings provide very different kinds of social interactional experiences in this context. Not surprisingly, mothers' language toward young children is more complex than that of siblings. In the Mannle et al. (1992) study of language directed toward 2-year-olds, the nature of mothers' conversations with the child were compared to that of the child's 3- to 6-year-old siblings. Mothers' conversations toward children were more frequent and were characterized by greater length, more speaking turns, longer turns and utterances, higher MLU, and more turnabouts, requests for information, simple provisions of information, requests for clarification, requests for confirmation, conversational repairs, and semantic extensions. Sibling conversations had more nonverbal actions, external provision of information, requested action, and nonacknowledgment of the partner's utterance. In Brown and Dunn's (1992) analysis of feeling statements, mothers were more likely than siblings to refer to the child's feelings, control behavior, label or reflect on feelings, and refer to causality. In contrast, siblings' feeling statements were more likely to involve references to the sibling's own feelings, self-interest, and play or humor. These findings are consistent with an earlier study by Dunn, Bretherton, and Munn (1987), who observed that mothers feeling references served to explain or guide child behavior, whereas siblings statements were comments about feeling states.

Parent–Child Play and Peer Relationship Quality. Several studies have examined the relationship between parent–child play and assessments of the quality of children's relationships with their peers (Black & Logan, 1995; Cassidy, Parke, Butkovsky, & Baungart, 1992; Clarke-Stewart et al., 1979; Gottman & Katz, 1989; Johnson, 1978; Kahen, Katz, & Gottman, 1994; MacDonald, 1987; MacDonald & Parke, 1984; Mize & Pettit, 1997; Putallaz, 1987). Comparison of the results across studies is difficult, because these studies employed children of different ages, observed in different settings, using very different methods of coding and analysis. This is especially true given the wide range of peer-outcome measures employed. Quality of peer interactions in these studies was assessed in a wide range of ways, including level of play engagement, amount of positive or negative affect, positive interactions (e.g., ratings of harmoniousness), negative interactions (e.g., ratings of abrasiveness or disagreeableness), dominance, and the degree of self-focus. Several studies included measures of peer popularity as well, using either peer (e.g., Black & Logan, 1995; Cassidy et al., 1992; MacDonald, 1987; Mize & Pettit, 1997; Putallaz, 1987) or teacher (MacDonald & Parke, 1984) ratings.

In addition to the variability in method, the results of these studies are inconsistent with relationships often found for some measures of a construct and not others, or only for some subset of the sample (e.g., boys vs. girls or mothers vs. fathers). Despite this complexity, it is possible to identify some general patterns in the results of these studies by examining, across studies, the types of parent behaviors that showed the most consistent relationships with the quality of peer relationships (broadly defined). When this is done, the following parental play behaviors appear to be consistently related to assessments of positive peer relationships (e.g., positive affect, positive interaction, level of play engagement, peer popularity): play engagement

(MacDonald & Parke, 1984), turn-taking or interactional synchrony (Black & Logan, 1995; Mize & Pettit, 1997), agreeableness or positive affect (Mize & Pettit, 1997; Putallaz, 1987), emotional expressiveness and talk about feelings (Cassidy et al., 1992; Putallaz, 1987), cohesive discourse (Black & Logan, 1995), nondirectiveness (Black & Logan, 1995; MacDonald, 1987), physical play (MacDonald, 1987; MacDonald & Parke, 1984), and verbal interchanges (MacDonald & Parke, 1984). Parental behaviors associated with negative peer relationships (e.g., negative affect, negative interactions, peer rejection) include: directiveness, intrusiveness, dominance, or demandingness (Black & Logan, 1995; Kahen et al., 1994; MacDonald, 1987; MacDonald & Parke, 1984; Putallaz, 1987), overstimulation (MacDonald, 1987), disagreeableness (Putallaz, 1987), unresponsiveness (Black & Logan, 1995; Gottman & Katz, 1989), and criticism and derisive humor (Kahen et al., 1994). In the one study of infants that was located, Vandell and Wilson (1987) noted that turn-taking interactions with parents at 6 months predicted turn taking with peers 3 months later.

Facilitating the Development of Social-Perspective-Taking Skills

Although Howe and Ross (1990) found no relationship between mothers' references to their preschool children about their infant sibling's feelings and social perspective-taking ability, several studies by Dunn and colleagues demonstrate that that maternal feeling-state talk is positively associated with affective perspective-taking abilities and/or performance on a false belief task (Dunn, Brown, & Beardsall, 1991; Dunn, Brown, Slomkowski, Telsa, & Youngblade, 1991). Correlates of perspective-taking skills included the frequency of feeling statements, the frequency of causal statements about emotions, and the diversity of feeling state themes.

Facilitating Self-Regulation and Control

Several studies have uncovered a negative relationship between maternal sensitivity or positive social contact during infancy and the development of internalizing and externalizing behavior problems in the preschool or early school years (e.g., Egeland, Pianta, & O'Brien, 1993; Jacobvitz & Sroufe, 1987; Martin, 1981; McHale & Rasmussen, 1998; Pettit & Bates, 1989; Rothbaum, Rosen, Pott, & Beatty, 1995). Maternal responsiveness to toddlers has also been shown to be associated with their child's later internalization of maternal rules, especially among temperamentally fearless toddlers (Kochanska, 1997a, 1997b). These findings suggest that the parent–child play context may be one setting in which children develop emotional self-regulation and control.

Facilitating Gender Role Development

Many theorists have asserted that one important context for gender-role socialization is the parent–child play context (e.g., Birns, 1976; Block, 1976). The studies usually cited to support this are those demonstrating the differential treatment of the genders (reviewed earlier). However, three studies were located that examined the relationship between parent–child play variables and independent assessments of the child's gender-role development. In these studies (Eisenberg, Wolchik, Hernandez, & Pasternack, 1985; Fagot & Leinbach, 1989; Fagot, Leinbach, & O'Boyle, 1992), parents of children who were early gender labelers reinforced more gender-typed play activities with their child during free play or unstructured home observations than did parents of later gender labelers. In one of these studies (Fagot et al., 1992), mothers of early labelers were more likely to initiate gender-typed play with their children during free play as well.

Experimental Studies: Social Development

Many of the large-scale intervention projects for high-risk children reviewed in Guralnick (1997) and Ramey and Ramey (1998) showed effects on social as well as cognitive development. However, for the same reasons mentioned earlier, it is difficult to determine how many of the effects are due to changes in parent–child play. In a review of intervention studies designed to impact the security of infant attachment, van IJzendoorn, Juffer, and Duyvesteyn (1995) identified 12 studies that provided such data. A meta-analysis showed that although the interventions were successful in increasing parental sensitivity ($d = .58$), the effects on attachment classification were more modest ($d = .17$). Although numerous parent-training studies have documented effectiveness in reducing the level of child behavior problems (for reviews, see Graziano & Diamert, 1992; Todreas & Bunston, 1993), most studies have focused more on changing patterns of discipline than on parent–child play.

In an experimental study designed to examine the effects of maternal responsiveness on preschooler's compliance, Parpal and Maccoby (1985) noted that children whose mothers had been instructed to increase their responsiveness to child behavior in a play session later showed more compliance in a subsequent situation than did children whose mothers had engaged them in free play.

Summary: Social Development

Despite the limited amount of experimental evidence available, the correlational findings suggest that parent–child play may affect children's social development in the same way as earlier described for cognitive development. When parents play with infants and young children, the complexity of children's behavior increases substantially, both in the length of the social interactions, and in the developmental level of children's social behavior. The same individual variations in parent–child play (e.g., engagement, responsiveness, positive affect, nondirectiveness) are correlated with the security of children's attachments to their parents, their developing social perspective-taking and self-regulatory skills, and the quality of children's interactions with other adults and peers. The available experimental evidence suggests that increasing parental responsiveness leads to modest increases in the security of children's attachment to their parents. Given these promising results, further studies should examine the relationship between parent–child play and children's social relationships, with an emphasis on the possible mediating effects of children's social cognitive and self-regulatory skills. In addition, more targeted experimental investigations need to be conducted to discern the directions of cause and effect.

PARENTAL INFLUENCES ON CHILDREN'S PLAY

Besides investigating the ways in which parents play with their children, researchers have also examined the ways in which parents influence the nature of children's solitary, peer, or sibling play. In this section, research on parental effects on the various types of children's play considered in this text is reviewed. The various types of play are considered in the same order as in previous chapters.

Solitary Object Play

Research on parental effects on solitary object play falls into three general categories: the influence of parental presence, the influence of parent–child play, and the influence of parental childrearing practices. Studies of the effects of first two paths of influence were reviewed earlier; it is to the final one that we now turn.

Parents have an impact on their children's solitary play behavior in many more ways than just their presence alone or in the way in which they play with their child. One of the major paths of parental influence (which is not considered in detail in this volume), is through the opportunities they provide for play and exploration via their choice of a house or apartment, the toys or other play objects they provide, or the degree to which they babyproof the home (e.g., Bradley et al., 1989; Parke, 1978; Power & Chapieski, 1986; Wachs, 1987). Also important are the rules that parents have as to which objects can and cannot be touched, which rooms can or cannot be entered with or without parental supervision, and so on. One of the most consistent negative predictors of the nature of infant solitary exploration and play is the level of parental restrictiveness. This factor has been found to correlated negatively with the developmental level of children's play or exploration (Howes & Stewart, 1987; Jennings et al., 1979), their executive competence (Wenar, 1976), and their curiosity (Saxe & Stollak, 1971).

Several studies have investigated the role of parental factors in predicting children's choice of gender-typed play objects (usually assessed during solitary play). Although this research has sometimes demonstrated that children's gender role attitudes and stereotypes are correlated with the parent's attitudes or behavior (e.g., Fagot & Leinbach, 1989; Serbin, Powlishta, & Gulko, 1993; Turner & Gervai, 1995; Weinraub et al., 1984; Weisner & Wilson-Mitchell, 1990), these same studies yield few or no parental correlates of children's gender-typed play preferences.

Play-Fighting

Only two studies could be found that examined parental correlates of play-fighting, and only one of these involved play-fighting with peers. The first, a study of doll play aggression conducted in the 1950s (Levin & Turgeon, 1957), revealed that children engaged in more doll play aggression (i.e., symbolic play-fighting) when their mothers were present than when an adult stranger was present. In the second, using a parent interview methodology, Hart, DeWolf, Wozniak, and Burts (1992) observed that mothers and fathers who reported using more power assertive discipline (vs. induction) had children who were more likely to engage in "rough play."

Locomotor Play

Although no studies of locomotor play could be located, several studies of the parental correlates of children's activity level during play were found.

In these, three positive parental predictors of children's observed free-play activity level were found: parental activity level (Moore et al., 1991), the degree to which the parents were regular exercisers (Poest, Williams, Witt, & Atwood, 1989; Sallis, Patterson, McKenzie, & Nader, 1988), and the degree to which parents were observed prompting their child to be active (Klesges et al., 1984; McKenzie et al., 1991). Negative predictors were parental discouragement of activity (Klesges et al., 1984), and the number of indoor and outdoor play rules (Sallis et al., 1993). Moore et al. (1991) asserted that parent–child similarities in activity level could be due to genetics, role modeling, sharing of activities, and the enhancement and support of child activities. In a study employing behavior genetic analyses of data collected using self-reported activity records, Perusse, Tremblay, LeBlanc, and Bouchard (1989) found evidence for both environmental and genetic influences.

Social Object and Social Pretend Play

Research concerning parental effects on children's social object and social pretend play with peers and siblings takes four forms: studies of the direct parental management of children's behavior during play with peers or siblings; studies of some of the indirect ways in which parents encourage or discourage peer or sibling interaction, such as providing opportunities to play or enrolling children in activities; studies of the relationship between parental childrearing practices and the nature of children's play interactions; and correlational studies of the relationship between parent–child and peer or sibling play. Data on the last of these was already reviewed; data on the first three is considered here. Parental correlates of play with peers are considered before the research with siblings.

Direct Parental Management of Children's Play with Peers

Several investigators observed young children in interaction with their peers in laboratory or home settings in the presence of their parents, in order to study parental management of young children's play (e.g., Bhavnagri & Parke, 1991; Bronson, 1974; Mize & Pettit, 1997; Ross, Telsa, Kenyon, & Lollis, 1990; Rubenstein & Howes, 1979). Studies indicate that unless parents are directly told to intervene, parental interventions are of low frequency in these settings (Bronson, 1974; Rubenstein & Howes, 1979). Studies also show that when mothers intervene it is usually in response to peer conflict. When do mothers intervene, they almost always direct their interventions to their own versus the other mother's child, and usually attempt to get their child to yield to the other child's wishes (Ross et al., 1990). Ross and colleagues (1990) noted that mothers do not follow ownership or entitlement rules in their interventions, but instead appear to focus on restoring harmony between the interacting children.

Bhavnagri and Parke (1991) found that parents were indeed effective in facilitating the interactions of young children. When the nature of peer play interactions was compared between sessions in which the parents had been instructed to play with their child and a same-age peer versus those in which the parents had been told to monitor the peer interactions, those peer interactions were much more complex in the parent play condition: Children showed greater turn taking, more sustained interactions, more simultaneous actions on toys, more complimentary interactions, longer play bouts, more agreement on activity, and overall greater social competency. Interestingly, the differences between the play and monitoring interactions were only significant for younger children (2 to 3½ years); the differences between conditions for 3½ to 5-year-olds were not significant. Mothers and fathers were found to be equally competent in increasing the complexity of peer interactions, although interview data showed that mothers reported playing this role more often in the home.

However, it is important to point out that, among preschool-aged children, parental presence does not always mean more positive peer interactions. In two studies of children in cooperative preschools (Field, 1979b; Smith & Howes, 1994) where parents were present on some days and not on others, children showed more positive and less negative peer-directed behaviors on the days when their parents were absent. The complexity of peer interactions was higher on parent-absent days as well (Smith & Howes, 1994). Similar results were obtained by Gottman (1986), who observed that when best friends played in their mother's presence, 5-year-olds showed fewer information-exchanging sequences and 7-year-olds showed less exploration of similarities than when playing alone.

Finnie and Russell (1988; Russell & Finnie, 1990) examined the relationship between direct parental management of peer interactions and children's peer acceptance. Mothers of preschool children were observed assisting their child in joining a dyad of other children already engaged in play. Mothers of high-social-status children were more active in this situation, using more "socially skillful" strategies, such as encouraging cooperation and communication, making relevant group-oriented suggestions, or joining in the play as an equal. In contrast, mothers of low-status children were more likely to avoid the supervisory role or to use less skillful strategies, such as redirecting the play to a new game to include her child, taking control of the play interaction, or otherwise disrupting the ongoing play interaction.

The observational results of the Finnie and Russell study converged with responses they obtained from a maternal interview, as well as interview studies conducted by other investigators. For example, Ladd and Golter (1988) found that mothers who reported using direct versus indirect peer monitoring strategies had children who were less liked and more likely to be rejected by their peers. Kennedy (1992) determined that rejected children had mothers who were less likely to report teaching their child about group entry strategies, sharing toys, or how to handle hitting. Laird, Pettit, Mize, Brown, and Lindsey (1994) found that peer acceptance was positively correlated with the frequency of mother–child conversations about peers and with the frequency of conversations that included giving advice or having discussions about emotions.

In two studies designed to more directly assess the correlates of maternal coaching styles, Mize and Pettit (1997) had mothers view videotaped vignettes of children experiencing negative peer interactions, and then instructed the mothers to discuss them with their child. Mothers were told to pretend that the child in the tapes was their child and make sure that their child understood "what happened and why it happened and what he/she should do next were her child the child in the story" (p. 316). Results showed that mothers who framed the events in a positive way, elaborated on their meaning, helped the child generate alternative solutions, and offered prosocial strategies for dealing with the events had children with higher levels of peer acceptance and/or teacher ratings of social skills and lower teacher ratings of aggression.

Indirect Parental Influences

The major form of indirect parental influences on children's peer relationships that has been studied is the degree to which parents provide their children with opportunities for peer interaction outside of the school environment. Several studies have shown that the degree to which parents actively provide such experiences is positively correlated with the child's number of frequent playmates (Ladd & Golter, 1988; Ladd & Hart, 1992); children's cooperative, positive interactions with peers in the preschool classroom (Kennedy, 1992; Ladd & Hart, 1992); and children's sociometric popularity among peers (Ladd & Golter, 1988; Ladd & Hart, 1992; Putallaz,

Costanzo, & Smith, 1991). Moreover, it is negatively related to unconstructive, nonsocial behaviors and peer rejection (Ladd & Golter, 1988; Ladd & Hart, 1992). Similarly, Kennedy (1992) noted that mothers of rejected children were less likely than mothers of other children to know their child's best friend. As was the case with direct parental management, mothers play this role more often than fathers (Ladd & Golter, 1988), and parents play this role more with younger than with older children. For example, Bhavnagri and Parke (1991) noted that although mothers initiate most play contacts among younger children, children are more likely to take on the initiating role with increased child age.

Childrearing Practices

Baumrind's (1967, 1971) classic research on parenting styles provides some of the earliest data on the relationship between parental childrearing practices and the quality of interactions between peers during freeplay. In this research, authoritarian parenting was associated with the negative patterns of peer interaction. Several more recent studies employing self-report (Kennedy, 1992; Schwartz, Dodge, Pettit, & Bates, 1997; Strassberg, Dodge, Bates, & Pettit, 1992; Weiss et al., 1992) or observational (Kochanska, 1992; Pettit, Harrist, Bates, & Dodge, 1991) assessments of parenting have found a positive relationship between harsh, coercive discipline and aggressive or conflictual patterns of peer interaction and/or peer rejection. Parental self-reports (Burleson, Delia, & Applegate, 1992; Hart et al., 1992; Roopnarine, 1987) and observations (Dekovic & Janssens, 1992) of the use of reasoning and guidance, in contrast, are correlated with positive peer interactions and peer acceptance.

In an observational study of the relationship between mothers' interventions in sibling conflicts at 33 months and children's play with their best friend at 6 years, Herrera and Dunn (1997) observed that the *type* of maternal justifications appeared to make a difference in how children resolved conflicts that arose during play. Mothers who used other-oriented arguments in addressing earlier sibling conflict (i.e., arguments that "considered the opponent's interests, including incidents of compromise, conciliation, bargaining or agreement, and maternal intervention aimed at resolving conflicts between siblings," p. 872) had children who 3 years later were more likely to compromise and resolve conflicts with their friends than did mothers who used self-oriented arguments (i.e., "giving a reason or justification for the disagreement, solely in defense of one's own position or interest," p. 872). The success of children whose mothers used other-oriented justifications was apparently due to these children's greater use of mitigating and justifying (vs. aggravating) strategies when resolving conflicts with their friends.

Other aspects of parental childrearing related to children's positive interactions with peers and peer acceptance were identified using parent self-reports, including level of maternal involvement (Roopnarine, 1987), involvement in child-centered versus adult-centered activities (Kennedy, 1992), nurturance (Howes & Stewart, 1987), discussions with the child about feelings (i.e., emotion coaching; Hooven, Gottman, & Katz, 1995), and parental discussions about the mechanics of social skills or humanistic values versus defensive values, such as learning to stand up for one's self (Kennedy, 1992). Observational studies identify the following positive correlates of positive peer relationships: parental responsiveness (Pettit et al., 1991); proactive involvement (Pettit et al., 1991); and indirect, polite suggestions (Kochanska, 1992). Observational studies also identified the following correlates of negative peer relationships: negative incentives (Kochanska, 1992), unclear commands (Kochanska, 1992), and intrusiveness (Pettit et al., 1991).

Parental Effects on Sibling Play

Studies of parental management of sibling interactions have largely been restricted to maternal responses to sibling conflict (Dunn & Munn, 1985, 1986b; Kendrick & Dunn, 1983; Ross, 1996; Slomkowski & Dunn, 1992). In general, the research shows that the frequency of parental intervention during sibling conflict correlates positively with the overall frequency of subsequent negative behavior between siblings (Dunn & Munn, 1986b; Kendrick & Dunn, 1983). Also, parents tend to intervene in the most intense conflicts (Perlman & Ross, 1997). This positive correlation has been interpreted in two ways: that parents increase the level of sibling conflicts by responding to them with their attention, and that parents of children who fight a lot are forced to intervene more often. Research examining the specific nature of parental interventions is hard to interpret, because techniques one would expect to be negatively correlated with sibling conflict (e.g., suggest conciliation, discuss rules or feelings) actually show positive relationships (Dunn & Munn, 1986b). This is probably because children who exhibit a great deal of conflict have mothers who attempt to deal with it via a wide range of strategies, including reasoning and explanation. Ideally, such analyses would first control for the level of conflict and then look at the mother's style of conflict resolution. Such a study could not be found.

Parent interventions do, however, appear to have a short-term effect on the nature of sibling conflicts. Perlman and Ross (1997) observed that, after parental intervention, children were less likely to engage in power-based strategies (physical power, verbal power, crying, and oppositional behavior) and more likely to engage in de-escalating moves (ignoring, complying, and other oriented reasoning). Moreover, in a recent experimental study (Vickerman, Reed, & Roberts, 1997), mothers who were trained to verbally intervene in sibling conflict (made a "stop it" statement plus a rationale and redirection) had children who displayed lower levels of coercion during the intervention sessions than did those who were trained to ignore it.

Unlike the research with peer interaction, studies of the effects of parental presence on the nature of sibling interactions are mixed. As was the case for peers, Corter, Abramovitch and Pepler (1983) noted that children displayed more agonistic and less prosocial behavior in their mother's presence than absence, whereas Musun-Miller (1991) found just the opposite. One difference between the Musun-Miller (1991) study and the rest is that the children were somewhat older, although it is not clear whether this was the reason for the inconsistencies.

Parental discussions with their children about feelings outside of the conflict context are positively correlated with positive sibling interactions (Howe & Ross, 1990), as are facilitative and supportive parent–child interaction patterns (Volling & Belsky, 1992). In contrast, parent–child conflict, differential treatment of siblings, parental intrusiveness, power assertion, and overcontrol all are positively correlated with sibling conflict (Erel, Margolin, & John, 1998; Volling & Belsky, 1992).

Mother–Father Differences

Although most of the previously mentioned studies of parent effects on social object play focused only on mothers, 10 included fathers (i.e., Baumrind, 1967, 1971; Bhavnagri & Parke, 1991; Dekovic & Janssens, 1992; Hart et al., 1992; Hooven et al., 1995; Perlman & Ross, 1997; Roopnarine, 1987; Strassberg et al., 1992; Volling & Belsky, 1992). For those papers in which the correlations for father and mother variables were presented separately, there were equally strong relationships between childrearing variables and child outcomes for both mothers and fathers.

Non-Western Cultures

Research on parental management of children's play behavior in non-Western cultures is largely anecdotal. Writers report a number of ways in which parents or other adult caregivers influence children's play behavior, including monitoring the child's whereabouts, intervening in children's conflict, helping the child obtain materials for play, offering suggestions for activities, and directly participating in children's play with peers or siblings (Anderson & Mitchell, 1984; Bloch, 1984; Draper, 1976; Eibl-Eibesfeldt, 1989; Hogbin, 1946; Minturn & Hitchock, 1963; Prosser et al., 1986). As already mentioned in chapter 8 (Feitelson, 1977), two additional ways in which parents indirectly influence their children's play are by assigning tasks or chores that influence the amount of time for play activities, and actively supporting their children's involvement in play by providing a supportive environment in which play canoccur.

Summary

One shortcoming of the literature concerning parental influences on children's play is the lack of experimental studies in the area. Despite the significant number of associations between parenting behaviors and children's play, it is impossible to draw conclusions about the direction of cause and effect. If, however, the correlational findings reflect the operation of causal factors, parents apparently have a significant role in shaping their children's play through the opportunities they provide, the lessons they offer, and the direct monitoring and management of child behavior. General parenting practices appear to be important as well—responsive, child-centered approaches are associated with more positive play outcomes than are harsh, coercive approaches.

Although parents rarely directly structure their children's play with peers, when they do (at least for younger children) the interactions are more complex, and children show more sustained and more developmentally advanced behavior patterns. Although such direct structuring may be counterproductive as children become older, children who engage in more positive play interactions with their peers have parents who provide them with opportunities for interaction, monitor their friendships, and spend time talking to them about their friends.

PARENT–CHILD PLAY: SUMMARY AND CONCLUSIONS

Although mother–offspring play is absent in many species, it is found in almost every mammalian order. The structure of such play is similar to that of younger animals, with mothers (and sometimes fathers or other adults) showing play behaviors typical of their species (play-biting, wrestling, etc.). Adult animals playing with their young often inhibit their own behavior and may self-handicap for a "fairer" fight. The amount of parent–offspring play varies considerably, both within and between species. Even in primates, where play behavior among young animals is ubiquitous, parent–offspring play is rare in many species. However, the level of play within a species is related to ecological factors—if few young play partners are available and the adults and young are healthy, well-nourished, with few competing time demands, then parent–offspring play may be more common. For example, adult male rhesus monkeys, who in natural settings rarely play with young monkeys, engage in considerable play when raised in captivity in a nuclear family setting (e.g., Suomi, 1977).

As found for other types of play, the great apes, particularly the chimpanzees, show the highest inventiveness and diversity in mother–offspring play. Comparisons of early mother–infant interactions for human and chimpanzee mothers show, with the obvious exception of vocal behavior, that, at least superficially, early maternal affiliative and social play behaviors are similar (see Tables 9.1 and 9.3). The major differences are that human mothers engage in few true play-bites (but see Aldis, 1975) and that chimpanzee mothers rarely engage in object-mediated play.

The level and diversity of mother–child play varies greatly across cultures, with the highest levels found in middle-class mothers in industrialized, Western cultures. Middle-class mothers in these cultures (e.g., England, France, Germany, Sweden, United States) show high levels of stimulation, affective involvement, and reciprocal interaction during play with their infants and young children. The nature of mother–infant play varies considerably with the developmental level of the child—mothers encourage and elaborate on the infant's developing social, communicative, cognitive, and motor skills in the play context. In the early months, mothers first engage in "protoconversations" during face-to-face interactions with their infants in response to the infant's developing social responsiveness. During the second half-year, mothers facilitate the infant's manual exploration and play with objects in response to the child's developing manipulatory abilities. During later months, mother–infant play shows similar structural changes in response to the infant's developing locomotor, communicative, and symbolic abilities. Sensitive Western mothers serve as "scaffolds" for their infants' emerging capabilities and show a great deal of responsiveness to the developmental level of the child.

In many traditional, non-Western settings, mother–child play is rare or nonexistent. In fact in some cultures, face-to-face interactions are rare, and adults may rarely speak to preverbal infants. The young child's only playmates may be other children (particularly older siblings) and, beginning in the toddler years, children spend most of their time with other children versus their parents. Both LeVine et al. (1994) and Rogoff et al. (1993) discussed differences between Western, industrialized cultures and traditional cultures that might account for differences in the socialization roles of adults and children in these two settings. MacDonald (1993a) interpreted these cross-cultural variations in parenting in light of parental investment theory. He argued that high levels of parent–child play are part of a high-investment parenting strategy:

> In particular, human societies characterized by monogamy, low fertility (high age of first pregnancy, low number of offspring, high birth-spacing interval), and parent rearing of children (high investment characteristics) are expected to be relatively high in parent–child play, whereas societies characterized by polygyny, high fertility, and sibling rearing and/or fostering of children (low investment characteristics) are expected to be low in parent–child play. (p. 121)

High-investment strategies, MacDonald stated, are more characteristic of industrialized, Western cultures and possibly some hunter-gatherer societies, whereas low-investment strategies are more characteristic of certain contemporary sub-Saharan agricultural societies. Although he cited examples of cultures that fit his theory, a more systematic investigation of this hypothesis across a wider range of cultures is warranted.

The level of father–child play varies considerably across cultures as well. Factors associated with high levels of father involvement (and presumably father–child play) include monogamy, nuclear family structure, gathering or horticultural subsistence patterns, and low levels of warfare. Although studies of Western cultures show a number of differences in the play styles of mothers versus fathers, the only difference that has been found in studies of non-Western cultures is the more physical nature of father–child play.

Researchers have examined a number of possible effects of parent–child play, although discussions of its biological functions in humans are hard to find. Given the variation in parent–child play across cultures, few theorists have made a strong case for the effects of parent–child play on the eventual reproductive success of the young of the human species. Given its limited costs in settings where it is common (e.g., Western, industrialized cultures, some hunting-and-gathering settings), its possible cognitive and social benefits might have an impact on children's ultimate reproductive success, although such effects have not been empirically demonstrated.

Researchers do agree, however, that parent–child play may have beneficial effects on other aspects of child development, at least in some settings. Several parenting correlates of children's general cognitive functioning have been identified in Western, industrialized cultures. Correlates identified in non-Western settings as well include the amount of verbal and social stimulation, and parental responsiveness to children's behavior. (A limited amount of experimental data supports the causal impact of maternal responsiveness on cognitive functioning.)

Parenting correlates of language development and the child's understanding of the physical environment have also been identified, but these have only been investigated in Western, industrialized societies. These studies suggest that parents may provide a significant scaffolding role in the development of linguistic and symbolic abilities, but whether such experiences are necessary for development seems unlikely. Despite a wide range of relevant experiences in these areas across cultures (parents in some cultures rarely talk to or play with their preverbal infants), children around the world develop language and symbolic play. Parent–child play in Western cultures may play a motivational role in the development of language and symbolic abilities (see Fein & Fryer, 1995), or it may encourage the development of more complex or elaborated skills. However, it is unlikely that early scaffolding in these contexts is necessary for development to occur. Also, given the negative correlations sometimes found between the level of parental directiveness and the development of play and language competence, it is possible that too much scaffolding may have a detrimental effect on children's development. Because autonomy is so central to the definition and nature of children's play, too much parental involvement in the toddlerhood and early childhood years may have negative effects on its development and consequences.

Research on the social correlates of parent–child play identifies many of the same correlates as in the cognitive area. Parental sensitivity, responsiveness, and positive affect toward children are associated with the security of infant attachments and/or the infants' sociability toward strangers. Moreover, positive affect and responsiveness during parent–child play are positively associated with the quality of children's peer relationships. Correlational data also suggest that parents may have indirect effects on children's relationships with their peers. Again, given the negative correlation between parental directiveness and peer outcomes, indirect versus direct methods might be most effective in facilitating the development of children's peer relationships (for related research on parental directiveness and children's sports involvement, see Averill & Power, 1995; Power &

Woolger, 1994). Parents of children who have the most positive peer relationships use reasoning and explanation instead of harsh punishment in socializing their children, provide their children with opportunities for play, and talk with their children about their friends and their feelings.

Although these correlational findings provide a good staring point for understanding the possible effects and functions of parent–child play, experimental data are needed to examine the direction of causality. Numerous parent-training programs have shown that changing parenting behavior (including parent–child play) can impact children's cognitive and social development. However, it is not clear which aspects of these large-scale intervention programs account for the results. Without more targeted interventions having carefully chosen controls, the view that parent–child play facilitates children's development will remain a largely untested hypothesis.

10

Conclusions

The first purpose of this book was to provide a state-of-the-art review of the research on play in children and other animals. So what is the state of the art? Let us return to the unanswered questions raised in chapter 1 to see if we are any closer to some answers.

IS PLAY A DISTINCT MOTIVATIONAL/BEHAVIORAL CATEGORY?

Throughout the book, numerous attempts were made to differentiate between play and other forms of behavior. In an attempt to differentiate play, exploration, and tool use in chapter 3, it was necessary to propose a category halfway between "pure" exploration and play that corresponded to mastery behavior: functional exploration/practice play. In chapters 4 and 5, considerable evidence was found for the distinction between play- and serious fighting in young animals and children (e.g., developmental course, signals, contexts, intensity, duration, tactics, targets, consequences), but research suggested that play-fighting often becomes rougher with age and presumably more difficult to distinguish from serious fighting. The distinction between locomotor play and serious flight seems less problematic, but differentiating between play and other affiliative interactions was difficult in social object play, especially for infants and toddlers (see chap. 8). The same difficulty arose in chapter 9, when considering play between parents and infants.

Another question is whether the diverse forms of play considered here (i.e., solitary object play, play-fighting, locomotor play, and the various forms of social object and social pretend play) are under the operation of the same or different motivational systems. That is, when play is activated, are the various play forms interchangeable, with the specific play behaviors a function of the materials or partners present? Or are each separate and under the control of different motivational systems? Data supporting the single-system view are observations, for example, that young animals raised without play partners engage in greater amounts of object play, and that during play "sessions" animals often shift rapidly between various play forms (e.g., play-fighting, play-chasing; Fagen, 1981). According to Eibl-Eibesfeldt (1982), during play "the entire instinct in Tinbergen's model of hierarchy is not activated, but rather the motor patterns of a lower hierarchical order become activated from another source of control" (p. 161). In his view, behaviors during play are detached from their usual emotions (e.g., anger, fear), and this emotional detachment "allows an animal to act without anxiety. An animal is free to experiment with his movement patterns, to combine them in new ways, and even to invent new ones, thus learning to use them as tools in interaction with his natural and social environment" (p. 161). Critics of the single-system view argue that the various forms of play are so diverse that different psychological and physiological processes are likely involved in their control (e.g., Berlyne, 1969).

The research reviewed in this book provides little insight into this second issue. In fact, in our separate consideration of the various forms of play, the interrelationships have been neglected. Few of the articles located have provided data relevant to this question. One interesting exception consistent with the single-system view is Rasa's (1984) study of object play in dwarf mongooses. In this species, young animals engage in a specific, continuous vocalization during play, whether the play is solitary or social. It is unfortunate that such unequivocal play markers are not available in many other species.

Schwartzman (1982) argued that, rather than try to identify specific patterns of play behavior, it is more useful to view play as a behavioral or attitudinal mode:

> In this way play is viewed as an attitude or frame that can be adopted toward anything (an object, an individual, an activity, an event). As a mode, play produces *allusions* to objects, individuals, activities, events, but these allusions are paradoxical (they are and are not what they seem to be). In a play-fight, the bite is both a bite and a non-bite. (p. 168)

Schwartzman's sentiments, along with those of Eibl-Eibesfeldt, were echoed in other commentaries on Smith's (1982) review. For example, Lewis (1982), in a response entitled "Play as whimsy," maintained that "the nature of the play is relatively unimportant" and that "the affective quality of play is its fundamental importance which does not change and is its invariant feature" (p. 166). Similarly, Suomi (1982) stated that we must not overlook the role of motivational factors in the study of play, and that children and other primates play "'because it's fun!' Indeed, when it stops being fun, they stop playing" (p. 170).

WHAT EXACTLY IS PLAY?

Of course, a problem with defining play as a mode or a frame is the difficulty in determining when this mode is in operation. Although play researchers can usually identify play when they see it, play has defied simple definition. With the notable exception of Smith and colleagues' (1985) study of children's conceptions of play, little headway has been made on this issue. Although the literature reviewed for this volume helps describe the various ways in which children and animals play, we are no closer to a definition now than when we began. Play has been defined as a mode, as an observable behavior, and as a context (see Rubin et al., 1983), and the studies reviewed here have used one or more of these definitions.

Given its elusive nature, it is unlikely that researchers will ever come up with a satisfactory definition of play. As Lazarus suggested in his discussion of the difficult concept of "stress," it may be best to define play as a research area, or more precisely as "what play researchers study." From this perspective, the research suggests that play encompasses a wide range of apparently "nonfunctional" solitary or social activities that involve the manipulation or use of objects (real or imagined), language, locomotor activities, or modified agonistic patterns. Given the diversity of activities encompassed by this list, it might be more useful to follow the lead of Schwartzman (1982) and others and conceptualize play as a mode versus a behavior pattern. However, if play is reserved for situations in which children are "having fun," positive affect would probably have to be observed or reported for an activity to be classified as play. Such a procedure would exclude from consideration numerous instances of child behavior currently viewed as play—for example, a toddler "drinks" from an empty cup with neutral affect; an infant engages in repeated variations as she intently and repeatedly drops her spoon to the floor; or a child is so caught up in play-fighting with an action figure that he frequently shows signs of negative, not positive, affect.

A more useful approach than attempting to come up with a one-size-fits-all definition for children's play would be to map out the domain of activities that children and adults (both scientists and laypeople) see as play and then identify homogeneous subcategories of play behavior, such as those proposed for solitary object play in Table 3.10. Once identified, descriptive research could be carried out to see how homogeneous the categories really are. Such research should include molecular descriptions of the behavior patterns, descriptions of the situations in which the behaviors usually occur, and measures of children's emotion and physiological arousal (observational, self-report, and psychophysiological measures). Following this empirical approach, the categories could be revised, and analyses of structural similarities could be used to place the categories into a hierarchical structure (see Allen & Bekoff, 1994, for a discussion of a similar approach to defining behavioral phenotypes). The definition of play itself should best be thought of in prototypical terms, with an understanding that many individual instances of play will not meet many of the criteria usually offered for its definition.

ARE COMPARISONS BETWEEN ANIMAL AND HUMAN PLAY MEANINGFUL?

The research provides numerous examples of the value of the comparative approach. Undoubtedly, some readers will not agree. As stated by Csikszentmihalyi (1982) in his response to Smith's (1982) review, "It reminds this commentator of the truism that, for a psychologist, the only mistake worse than ignoring the evidence of comparative ethology is taking it too seriously" (p. 160). He went on further to assert, "Finally, one might note that at least as far as human play is concerned, what is important and interesting about it does not so much consist of the universal 'design features,' but the variety of forms it develops, and the uses it serves" (p. 160).

Although the first point is a caution well taken, the second is clearly a matter of opinion— interesting findings to one researcher are error variance to another. In the present author's opinion, the most compelling comparative evidence offered are the data for solitary object play and for play-fighting. Cross-species comparisons of the object-directed behavior across a wide range of species in chapter 2 showed that, with increasing phylogenetic proximity to humans, the nature of solitary object manipulation and play (as well as the sequences of sensorimotor development) are increasingly similar to that of human infants. Also, although much less descriptive data were available on this issue, exploration in human infants is similar to that in many other mammalian species. The major differences between humans and other primates concern the role of olfactory investigation, bimanual manipulation, and symbolic play. Smith (1982) stipulated that the capacity for fantasy evolved in hominids and greatly increased the possibilities for the practice of complex skills in the play context. Mitchell (1994b) considered the evolution of pretense in a recent theory of the evolution of iconic representations in primates.

With regard to play-fighting—the most extensively studied type of play in other animals—the parallels with play-fighting in humans were remarkable. This was true for its component behaviors, developmental course, gender differences in occurrence, the play partners chosen, and the numerous ways in which play-fighting differs from serious aggression. The only differences concern the absence of play-biting and play-mounting in humans, differences in structure due to the bipedal nature of human locomotion, and the absence of a verbal and, in some play forms, a pretend component in other mammals.

The comparative data were less compelling in the chapters on locomotor play, social object and social pretend play, and parent–child play. In the case of locomotor play, this was largely due to a lack of descriptive data on humans. It is unfortunate that possibly the most common form of play in preschoolers (Bloch, 1987) has been the focus of so little empirical research. Based on what little data in humans are currently available, however, it appears that this form of play has commonalities with the play of many other species.

The value of the comparative data in chapters 8 and 9 was more limited, although there was still some evidence that the basic play repertoire of preverbal infants and toddlers, as well as the nonverbal behavior of mothers toward their infants and toddlers, differed little from that of the other primates, especially the chimpanzees. In this case, it is the lack of descriptive data on animal play that made detailed comparisons difficult. Moreover, despite differences in the form and content of the play of preschoolers and other primates (e.g., sociodramatic play, games with rules), aspects of children's agonistic and affiliative behaviors in group settings appear to be governed by dominance hierarchies similar to those in other primate groups. It is possible that further studies from an evolutionary perspective will identify other similarities as well (competition vs. cooperation, conflict resolution processes, etc.).

Overall, the rather limited nature of the descriptive data on both humans and other animals made the comparative analyses a frustrating task. Often, the analysis only allowed for generating lists of behaviors observed in various species and looking for cross-species similarities and differences in these lists. The absence of data on the quality, sequencing, and temporal structuring of animal and child behavior made more informative comparisons impossible. Moreover, because most of the studies of animal and human play were conducted in captive, convenience settings (in zoos or research colonies for animals, and in Western preschools for children), the data may have limited generalizability. Clearly, a substantial amount of descriptive research across species, cultures, and settings is needed.

WHAT ARE THE EFFECTS OF PLAY ON DEVELOPMENT?

Despite the frequency with which theorists have argued that play contributes to the development of children and other animals, the research allows for few general conclusions to be drawn. Although children who show the highest levels of play involvement and complexity score high on various assessments of cognitive and social functioning, the experimental research does not allow for many confident inferences about the direction of causality. The same is true for studies of parent–child play and child development. A much more parsimonious explanation is that competent parents and children show their competence in both play and nonplay settings. Research on the effects of play in animals is even more lacking. Although considerable speculations about the evolution and functions of animal play have been offered, the data in support of these hypotheses come mostly from structural analyses of animal play patterns. The danger of arguing for play functions from play structure alone is in committing the logical fallacy of affirming the consequent (Bernstein, 1982; Fassino, 1982): "If play has this function then it will have this structure. If it has this structure, therefore this function. . . . Clearly the. . .evidence for the hypothesis is the hypothesis itself" (Fassino, 1982, p. 163). Very few correlational or experimental studies on animal play have been conducted, and of those that have, the results are for the most part either nonsignificant or inconclusive.

This is not to argue that play is unimportant for development—we simply do not know. Children and the young of many other species spend a considerable time in play, and it is likely that considerable learning occurs in this context. However, it is still not clear what is learned or how this learning occurs. Because the greatest amount of play research in humans was conducted in the 1970s, it is not surprising that researchers focused on its cognitive effects. As argued in chapter 8, researchers need to examine its social and emotional effects as well, and numerous directions for future research have been suggested throughout this text.

IMPLICATIONS FOR PRACTICE

Given our current state of knowledge about the structure, effects, and functions of children's play, it is difficult to make many empirically based recommendations for practice beyond what most competent practitioners who work with young children already know, but some tentative suggestions are offered. Clearly, children should be given opportunities to play—not only because play may be important for early learning, but because it is *fun*, and everybody needs to have fun! Given the differences in experiences offered by different kinds of play, children should be given opportunities to engage in a range of both solitary and social play activities, ideally with partners of different developmental levels (younger, same age, and older). Although children should be given considerable autonomy in how and what they play, it may be valuable to occasionally construct situations in which children engage in play activities or with partners whom they might not choose on their own (e.g., partners of the opposite gender, activities usually engaged in by the opposite gender). In providing such activities, children need to be given sufficient choices about what and how to play so that they benefit from these experiences, rather than resent them.

The best way to facilitate solitary object play would be to provide children with various materials or opportunities: some that elicit functional play and exploration, some designed for constructive play, and some that encourage symbolic play. Children should be given sufficient time with the materials for exploration and play to occur, as well as a supportive emotional environment. Providing opportunities for children to "fool around" with materials for later problem solving may be useful as well.

Given the disruptive nature of play-fighting in many settings, it is difficult to make recommendations about it. Professionals, however, need to be careful not to equate play-fighting with serious fighting, and not to label a child as "aggressive" simply because he or she prefers a particular kind of play. Given many children's interest and enjoyment in active, locomotor play, children should be given numerous opportunities for this type of play as well. Such activities likely contribute to motor development, overall physical fitness, and possibly cognitive development. If certain activities are too disruptive in a given setting, practitioners could consider ways to have children play more structured games that incorporate elements of these activities.

In the realm of social play, children should be given numerous opportunities for play with various partners, and considerable autonomy in working out for themselves the conflicts that arise (up to a point). When a supervising adult needs to intervene, the focus should be on helping the children solve the conflict themselves, rather than solving the problem for them. Overall, adults need to avoid being too controlling of children's solitary or social play, given the potentially negative effects of adult directiveness.

Parents should be encouraged to play with their children as well—again, not only because it is important, but because it is fun. Interventions to consider include increasing parental responsiveness to children's behavior, and helping parents monitor their level of verbal and object stimulation. An optimal level of stimulation is not necessarily a high one, and parents need to learn to provide this. Increasing parental sensitivity to infant cues would facilitate this process. Professionals themselves also need to be sensitive, particularly to the culture of the parents and children with whom they work. If parents rarely play with their children due to cultural practices or beliefs, any potential interventions need to be carefully planned. Children in many cultures around the world develop normally without ever playing with their parents, and professionals need to keep this in mind.

Children's play is a fascinating activity—one that will continue to engage the interest of researchers as long as they realize that children, not researchers, are the experts in this domain. Much has yet to be learned about the normative structure, development, functions, and effects of animals' and children's play, and only through only a combination of descriptive, correlational, and experimental methods can this be accomplished. As the work of Pellis and Gottman shows, detailed, descriptive studies are useful in generating hypotheses, but considerably more research must be conducted before the many claims about the effects of play can be substantiated.

References

Abramovitch, R. (1976). The relation of attention and proximity to dominance in preschool children. In M. R. A. Chance & R. R. Larsen (Eds.), *The social structure of attention* (pp. 153–177). London: Wiley.

Abramovitch, R., Corter, C. M., & Lando, B. (1979). Sibling interaction in the home. *Child Development*, *50*, 997–1003.

Abramovitch, R., Corter, C., & Pepler, D. J. (1980). Observations of mixed-sex sibling dyads. *Child Development*, *51*, 1268–1271.

Abramovitch, R., Corter, C., Pepler, D. J., & Stanhope, L. (1986). Sibling and peer interaction: A final follow-up and comparison. *Child Development*, *57*, 217–229.

Abramovitch, R., & Grusec, J. E. (1978). Peer imitation in a nature setting. *Child Development*, *49*, 60–65.

Abravanel, E. (1968). The development of intersensory patterning with regard to selected spatial dimensions. *Monographs of the Society for Research in Child Development*, *33* (2, Serial No. 118).

Abravanel, E. (1972). How children combine vision and touch when perceiving the shape of objects. *Perception and Psychophysics*, *12*, 171–175.

Adams-Curtis, L. E., & Fragaszy, D. M. (1994). Development of manipulation in capuchin monkeys during the first six months. *Developmental Psychobiology*, *27*, 123–136.

Adamson, L. B., & Bakeman, R. (1985). Affect and attention: Infants observed with mothers and peers. *Child Development*, *56*, 582–593.

Ainsworth, M. D. S. (1967). *Infancy in Uganda: Infant care and the growth of love.* Baltimore: Johns Hopkins Press.

Ainsworth, M. D. S., Blehar, M. C., Waters, E., & Wall, S. (1978). *Patterns of attachment: A psychological study of the strange situation*. Hillsdale, NJ: Lawrence Erlbaum Associates.

Ainsworth, M., & Wittig, B. (1969). Attachment and exploratory behavior of one-year-olds in a strange situation. In B. M. Foss (Ed.), *Determinants of infant behavior.* (Vol. 4, pp. 111–136). New York: Wiley.

Alcock, J. (1972). The evolution of the use of tools by feeding animals. *Evolution*, *26*, 464–474.

Aldis, O. (1975). *Play fighting*. New York: Academic.

Ali, F. (1973). Dimensions of competence in young children. *Genetic Psychology Monographs*, *88*, 305–328.

Allen, C., & Bekoff, M. (1994). Intentionality, social play, and definition. *Biology and Philosophy, 9, 63–74.*

Allen, C., & Bekoff, M. (1995). Function, natural design, and animal behavior: Philosophical and ethological considerations. In N. S. Thompson (Ed.), *Perspectives in ethology.* (Vol. 11, pp. 1–46). New York: Plenum.

Altmann, S. A. (1962). A field study of the sociobiology of rhesus monkeys, *Macaca mulatta. Annuals of the New York Academy of Science*, *102*, 338–435.

Alvarez, F. (1993). Risks of fighting in relation to age and territory holding in fallow deer. *Canadian Journal of Zoology*, *71*, 376–383.

Anderson, B. J., Vietze, P., & Dokecki, P. (1977). Reciprocity in vocal interactions of mothers and infants. *Child Development, 48*, 1676–1681.

Anderson, F. J., & Willis, F. N. (1976). Glancing at others in preschool children in relation to dominance. *The Psychological Record, 26*, 467–472.

Anderson, H. H. (1940). An examination of the concepts of domination and integration in relation to dominance and ascendance. *Psychological Review, 47*, 21–37.

Anderson, R., & Mitchell, E. M. (1984). Children's health and play in rural Nepal. *Social Science Medicine, 19*, 735–740.

Anderson, W. W. (1982). Ecological and sociocultural determinants in Thai children's game-playing events. In J. W. Loy (Ed.), *The paradoxes of play* (pp. 167–175). West Point, NY: Leisure.

Andreas, C. (1969). War toys and the peace movement. *Journal of Social Issues, 25*, 83–99.

Antinucci, F. (1989a). *Cognitive structure and development in nonhuman primates.* Hillsdale, NJ: Lawrence Erlbaum Associates.

Antinucci, F. (1989b). Systematic comparison of early sensorimotor development. In F. Antinucci (Ed.), *Cognitive structure and development in nonhuman primates* (pp. 67–85). Hillsdale, NJ: : Lawrence Erlbaum Associates.

Araba, B. D., & Crowell-Davis, S. L. (1994). Dominance relationships and aggression of foals (*Equus caballus*). *Applied Animal Behavior Science, 41*, 1–25.

Archer, J. (1995). What can ethology offer the psychological study of human aggression? *Aggressive Behavior, 21*, 243–255.

Arco, C. M. B. (1983). Infant reactions to natural and manipulated temporal patterns of paternal communication. *Infant Behavior and Development, 6*, 391–399.

Ariel, S. (1992). Semiotic analysis of children's play: A method for investigating social development. *Merrill-Palmer Quarterly, 38*, 119–138.

Arsenian, J. M. (1943). Young children in an insecure situation. *Journal of Abnormal and Social Psychology, 38*, 225–249.

Asarnow, J. B. (1983). Children with peer adjustment problems: Sequential and nonsequential analyses of school behaviors. *Journal of Consulting and Clinical Psychology, 51*, 709–717.

Asendorphf, J. (1990). Beyond social withdrawal: Shyness, unsociability, and peer avoidance. *Human Development, 33*, 250–259.

Asendorphf, J. (1991). Development of inhibited children's coping with unfamiliarity. *Child Development, 62*, 1460–1474.

Asher, S. R., Markell, R. A., & Hymel, S. (1981). Identifying children at risk in peer relations: A critique of the rate-of-interaction approach to assessment. *Child Development, 52*, 1239–1245.

Atz, J. W. (1970). The application of the idea of homology to behavior. In L. R. Arsonson, E. Tobach, D. S. Lehrman, & J. S. Rosenblatt (Eds.), *Development of the evolution of behavior: Essays in memory of T. C. Schneirla* (pp. 53–74). San Francisco: Freeman.

Aureli, F. (1997). Post-conflict anxiety in nonhuman primates: The mediating role of emotion in conflict resolution. *Aggressive Behavior, 23*, 315–328.

Austin, A. M. B., Salehi, M., & Leffler, A. (1987). Gender and developmental differences in children's conversations. *Sex Roles, 16*, 497–510.

Averill, P. M., & Power, T. G. (1995). Parental attitudes and children's experiences in soccer: Correlates of effort and enjoyment. *International Journal of Behavioral Development, 18*, 263–276.

Axline, V. M. (1947). *Play therapy.* New York: Ballantine.

Bailey, D. B., McWilliam, R. A., Ware, W. B., & Burchinal, M. A. (1993). Social interactions of toddlers and preschoolers in same-age and mixed-age play groups. *Journal of Applied Developmental Psychology, 14*, 261–276.

Bakchine, E., Pham-Delegue, M. H., Kaiser, L., & Masson, C. (1990). Computer analysis of the exploratory behavior of insects and mites in an olfactometer. *Physiology and Behavior, 48*, 183–187.

Bakeman, R., & Adamson, L. B. (1984). Coordinating attention to people and objects in mother–infant and peer–infant interaction. *Child Development, 55*, 1278–1289.

Bakeman, R., Adamson, L. B., Konner, M., & Barr, R. G. (1990). Kung infancy: The social context of object exploration. *Child Development, 61*, 794–809.

Bakeman, R., & Brown, J. V. (1977). Behavioral dialogues: An approach to the assessment of mother–infant interaction. *Child Development, 48*, 195–203.

Bakeman, R., & Brownlee, J. R. (1980). The strategic use of parallel play: A sequential analysis. *Child Development, 51*, 873–878.

Baldwin, D. A., Markman, E. M., & Melartin, R. L. (1993). Infants' ability to draw inferences about nonobvious object properties: Evidence from exploratory play. *Child Development, 64*, 711–728.

Baldwin, J. D., & Baldwin, J. I. (1973). The role of play in social organization: Comparative observation in squirrel monkeys (*Saimiri*). *Primates, 14*, 369–381.

Baldwin, J. D., & Baldwin, J. I. (1974). Exploration and social play in squirrel monkeys (*Saimiri*). *American Zoologist, 14*, 303–315.

Baldwin, J. D., & Baldwin, J. I. (1976). Effects of food ecology on social play: A laboratory simulation. *Zeitschrift fur Tierpsychologie, 40*, 1–14.

Baldwin, J. D., & Baldwin, J. I. (1978a). Exploration and play in howler monkeys (*Aloutta palliata*). *Primates, 19*, 411–422.

Baldwin, J. D., & Baldwin, J. I. (1978b). Reinforcement theories of exploration, play, creativity, and psychological growth. In E. O. Smith (Ed.), *Social play in primates* (pp. 231–258). New York: Academic.

Ball, W. A., & Hagen, J. W. (1978). Visual reinforcement of haptic activity. *Perceptual and Motor Skills, 46*, 1209–1210.

Bandura, A., & Walters, R. H. (1963). *Social learning and personality*. New York: Holt, Rinehart, & Winston.

Baranowski, T., Bouchard, C., Bar-Or, O., Bricker, T., Heath, G., Kimm, S. Y. S., Malina, R., Obarzmek, E., Pate, R., Strong, W. B., Truman, B., & Washington, R. (1992). Assessment, prevalence, and cardiovascular benefits of physical activity and fitness in youth. *Medicine and Science in Sports and Exercise, 24*(Suppl. 1), S237–S247.

Barber, J. T., Ellgaard, E. G., Thien, L. B., & Stack, A. E. (1989). The use of tools for food transportation by the imported fire ant, *Solenopsis invicta*. *Animal Behaviour, 38*, 550–552.

Barber, N. (1991). Play and energy regulation in mammals. *Quarterly Review of Biology, 66*, 129–147.

Bard, K. A. (1994). Evolutionary roots of intuitive parenting: Maternal competence in chimpanzees. *Early Development and Parenting, 3*, 19–28.

Bard, K. A. (1995). Parenting in primates. In M. H. Bornstein (Ed.), *Handbook of parenting: Vol. 2. Biology and ecology of parenting* (pp. 27–58). Mahwah, NJ: Lawrence Erlbaum Associates.

Bard, K. A., & Vauclair, J. (1984). The communicative context of object manipulation in ape and human adult–infant pairs. *Journal of Human Evolution, 13*, 181–190.

Barnes, K. E. (1971). Preschool play norms: A replication. *Developmental Psychology, 5*, 99–103.
Barnett, L. A. (1984). Research note: Young children's resolution of distress through play. *Journal of Child Psychology and Psychiatry and Allied Disciplines, 25*, 477–483.
Barnett, L. A. (1985). Young children's free play and problem-solving ability. *Leisure Sciences, 7*, 25–46.
Barnett, L. A., & Kleiber, D. A. (1984). Playfulness and the early play environment. *Journal of Genetic Psychology, 144*, 153–164.
Barnett, L. A., & Storm, B. (1981). Play, pleasure, and pain: The reduction of anxiety through play. *Leisure Sciences, 4*, 161–175.
Barnett, S. A. (1958a). Experiments on "neophobia" in wild and laboratory rats. *British Journal of Psychology, 49*, 195–201.
Barnett, S. A. (1958b). Exploratory behavior. *British Journal of Psychology, 49*, 289–310.
Barrett, L., Dunbar, R. I. M., & Dunbar, P. (1992). Environmental influences on play behavior in immature gelada baboons. *Animal Behaviour, 44*, 111–115.
Barrett, P., & Bateson, P. P. G. (1978). The development of play in cats. *Behaviour, 66*, 110–120.
Barry, H., III, & Paxson, L. M. (1971). Infancy and early childhood: Cross-cultural codes 2. *Ethology, 10*, 466–508.
Barry, R. J., & Barry, A. (1976). Stereotyping of sex roles in preschool kindergarten children. *Psychological Reports, 38*, 948–950.
Bar-Tal, D., Raviv, A., & Goldberg, M. (1982). Helping behaviors among preschool children: An observational study. *Child Development, 53*, 396–402.
Baskett, L. M., & Johnson, S. M. (1982). The young child's interactions with parents versus siblings: A behavioral analysis. *Child Development, 53*, 643–650.
Bates, E., Camaioni, L., & Volterra, V. (1975). The acquisition of performatives prior to speech. *Merrill-Palmer Quarterly, 21*, 205–226.
Bates, E., Carlson-Luden, E., & Bretherton, I. (1980). Perceptual aspects of tool-using in infancy. *Infant Behavior and Development, 3*, 127–140.
Bateson, M. C. (1975). Mother–infant exchanges: The epigenesis of conversational interaction. *Annals of the New York Academy of Sciences, 263*, 101–113.
Bateson, P. (1976). Specificity and the origins of behavior. In J. Rosenblatt, R. Hinde, E. Shaw, & C. Beer (Eds.), *Advances in the study of behavior* (Vol. 6, pp. 1–20). New York: Academic.
Baumrind, D. (1967). Child care practices anteceding three patterns of preschool behavior. *Genetic Psychology Monographs, 75*, 43–88.
Baumrind, D. (1971). Current patterns of parental authority. *Developmental Psychology Monographs, 4*(1, Part 2), 1–103.
Baumrind, D., & Black, A. E. (1967). Socialization practices associated with dimensions of competence in preschool boys and girls. *Child Development, 38*, 291–327.
Beach, F. A. (1945). Current concepts of play in animals. *American Naturalist, 79*, 523–541.
Beck, B. B. (1972). Tool use in captive Hamadryas baboons. *Primates, 13*, 276–296.
Beck, B. B. (1980). *Animal tool behavior: The use and manufacture of tools by animals.* New York: Garland STPM.
Beck, S., Collins, L., Overholser, J., & Terry, K. (1985). A cross-sectional assessment of the relationship of social competence measures to peer friendship and likability in elementary school age children. *Genetic, Social, and General Psychology Monographs, 111*, 41–63.

Becker, J. M. T. (1977). A learning analysis of the development of peer-oriented behavior in 9-month-old infants. *Developmental Psychology, 13*, 481–491.

Bee, H. L., Van Egeren, L., Streissguth, A. P., Nyman, B. A., & Leckie, M. S. (1969). Social class differences in maternal teaching strategies and speech patterns. *Developmental Psychology, 1*, 726–734.

Beebe, B., Alson, D., Jaffe, J., Feldstein, S., & Crown, C. L. (1988). Vocal congruence in mother–infant play. *Journal of Psycholinguistic Research, 17*, 245–259.

Begin, G. (1986). Sociometric status and social interaction: Are the neglected children less socially adaptive? *Perceptual and Motor Skills, 63*, 823–830.

Bekoff, M. (1974). Social play and play-soliciting by infant canids. *American Zoologist, 14*, 323–340.

Bekoff, M. (1975). The communication of play intention: Are play signals functional? *Semiotica, 15*, 231–240.

Bekoff, M. (1976). The social deprivation paradigm: Who's being deprived of what? *Developmental Psychobiology, 9*, 499–500.

Bekoff, M. (1977). Social communication in canids: Evidence for the evolution of a stereotyped mammalian display. *Science, 197*, 1097–1099.

Bekoff, M. (1989). Social play and physical training: When "not enough" may be plenty. *Ethology, 80*, 330–333.

Bekoff, M. (1995). Play signals as punctuation: The structure of social play in canids. *Behaviour, 132*, 419–429.

Bekoff, M., & Byers, J. (1981). A critical reanalysis of the ontogeny and phylogeny of mammalian social and locomotor play: An ethological hornet's nest. In K. Immelmann, G. W. Barlow, L. Petrinovich, & M. Main (Eds.), *Behavioral development: The Bielefeld Interdisciplinary Project* (pp. 269–337). Cambridge, England: Cambridge University Press.

Bekoff, M., & Byers, J. A. (1992). Time, energy, and play. *Animal Behaviour, 44*, 981–982.

Bekoff, M., & Byers, J. (1998). *Animal play: Evolutionary, comparative, and ecological approaches.* New York: Cambridge University Press.

Bell, C. S., Johnson, S. E., McGillicuddy-DeLisi, A., & Sigel, I.E. (1981). The effects of family constellation and child gender on parental use of evaluative feedback. *Child Development, 52*, 701–704.

Belsky, J. (1979). Mother–father–infant interaction: A naturalistic observational study. *Developmental Psychology, 15*, 601–607.

Belsky, J. (1980). A family analysis of parental influence on infant exploratory competence. In F. Pedersen (Ed.), *The father–infant relationship: Observational studies in a family context* (pp. 87–110). New York: Praeger.

Belsky, J., Gilstrap, B., & Rovine, M. (1984). The Pennsylvania Infant and Family Development Project I: Stability and change in mother–infant and father–infant interaction in a family setting at one, three, and nine months. *Child Development, 55*, 692–705.

Belsky, J., Goode, M. K., & Most, R. K. (1980). Maternal stimulation, and infant exploratory competence: Cross-sectional, correlational, and experimental analyses. *Child Development, 51*, 1163–1178.

Belsky, J., & Most, R. K. (1981). From exploration to play: A cross-sectional study of infant free play behavior. *Developmental Psychology, 17*, 630–639.

Belsky, J., Taylor, D., & Rovine, M. (1984). The Pennsylvania Infant and Family Development Project II: The development of reciprocal interaction in the mother–infant dyad. *Child Development, 55*, 706–717.

Beneson, J. F. (1993). Greater preference among females than males for dyadic interaction in early childhood. *Child Development, 64*, 544–555.

Berenbaum, S. A., & Hines, M. (1992). Early androgens are related to childhood sex-typed toy preferences. *Psychological Science, 3*, 203–206.

Beresin, A. R. (1989). Toy war games and the illusion of two-sided rhetoric. *Play and Culture, 2*, 218–224.

Bergen, D. (1988). (Ed.). *Play as a medium for learning and development: A handbook of theory and practice.* Portsmouth, NH: Heinemann.

Berger, J. (1979). Social ontogeny and behavioral diversity: Consequences for Bighorn sheep, *Ovis canadensis*, inhabiting desert and mountain environments. *Journal of Zoology, 188*, 251–266.

Berger, J. (1980). The ecology, structure and function of social play in bighorn sheep (*Ovis canadensis*). *Journal of Zoology, 192*, 531–542.

Berkowitz, R. I., Argas, W. S., Korner, A. F., Kraemer, H. C., & Zeanah, C. H. (1985). Physical activity and adiposity: A longitudinal study from birth to childhood. *Journal of Pediatrics, 106*, 734–738.

Berlyne, D. E. (1950). Novelty and curiosity as determinants of exploratory behavior. *British Journal of Psychology, 41*, 68–80.

Berlyne, D. E. (1960). *Conflict, arousal, and curiosity.* New York: McGraw-Hill.

Berlyne, D. E. (1966). Curiosity and exploration. *Science, 153*, 25–33.

Berlyne, D. E. (1969). Laughter, humor, and play. In G. Lindzey & E. Aronson (Eds.), *The handbook of social psychology* (Vol. 3, pp. 795–852). Reading, MA: Addison-Wesley.

Bernstein, I. S. (1982). Hypotheses about play. *The Behavioral and Brain Sciences, 5*, 158–159.

Bertrand, M. (1976). Rough-and-tumble play in stumptails. In J. S. Bruner, A. Jolly, & K. Sylva (Eds.), *Play: Its role in development and evolution* (pp. 320–327). New York: Penguin.

Best, D. L., House, A. S., Barnard, A. E., & Spiker, B. S. (1994). Parent–child interactions in France, Germany, and Italy: The effects of gender and culture. *Journal of Cross-Cultural Psychology, 25*, 181–193.

Bhavnagri, N. P., & Parke, R. D. (1991). Parents as direct facilitators of children's peer relationships: Effects of age of child and sex of parent. *Journal of Social and Personal Relationships, 8*, 423–440.

Bianchi, B. D., & Bakeman, R. (1978). Sex-typed affiliation preferences observed in preschoolers: Traditional and open school differences. *Child Development, 49*, 910–912.

Biben, M. (1982a). Object play and social treatment of prey in bush dogs and crab-eating foxes. *Behaviour, 79*, 201–211.

Biben, M. (1982b). Sex differences in the play of young ferrets. *Biology of Behaviour, 7*, 303–308.

Biben, M. (1983). Comparative ontogeny of social behavior in three South American canids: The maned wolf, crab-eating fox, and bush dogs. Implications for sociality. *Animal Behaviour, 31*, 814–826.

Biben, M. (1986). Individual- and sex-related strategies of wrestling play in captive squirrel monkeys. *Ethology, 71*, 229–241.

Biben, M., & Suomi, S. J. (1993). Lessons from primate play. In K. MacDonald (Ed.), *Parent–child play: Descriptions and implications* (pp. 185–196). Albany: State University of New York Press.

Biben, M., & Symmes, D. (1986). Play vocalizations of squirrel monkeys (*Saimiri sciureus*). *Folia Primatologica, 46*, 173–182.

Biben, M., Symmes, D., & Bernard, D. (1989). Vigilance during play in squirrel monkeys. *American Journal of Primatology, 17*, 41–49.

Bickerton, D. (1990). *Language and species*. Chicago: University of Chicago Press.

Bierman, K. L., Smoot, D. L., & Aumiller, K. (1993). Characteristics of aggressive-rejected, aggressive (nonrejected), and rejected (nonaggressive) boys. *Child Development, 64*, 139–151.

Birch, H. G. (1945). The role of motivational factors in insightful problem solving. *Journal of Comparative Psychology, 38*, 295–317.

Biringen, Z., Emde, R. N., Campos, J. J., & Applebaum, M. I. (1995). Affective reorganization in the infant, the mother, and the dyad: The role of upright locomotion and its timing. *Child Development, 66*, 499–514.

Biringen, Z., Robinson, J. L., & Emde, R. N. (1994). Maternal sensitivity in the second year: Gender-based relations in the dyadic balance of control. *American Journal of Orthopsychiatry, 64*, 78–90.

Birke, L. I. A., & Archer, J. (1983). Some issues and problems in the study of animal exploration. In J. Archer & L. Birke (Eds.), *Exploration in animals and humans* (pp. 1–21). Berkshire, England: Van Nostrand Reinhold.

Birns, B. (1976). The emergence and socialization of sex differences in the earliest years. *Merrill-Palmer Quarterly, 22*, 229–254.

Black, B. (1989). Interactive pretense: Social and symbolic skills in preschool play groups. *Merrill-Palmer Quarterly, 35*, 379–397.

Black, B. (1992). Negotiating social pretend play: Communicative differences related to social status and sex. *Merrill-Palmer Quarterly, 38*, 212–232.

Black, B., & Hazen, N. L. (1990). Social status and patterns of communication in acquainted and unacquainted preschool children. *Developmental Psychology, 26*, 379–387.

Black, B., & Logan, A. (1995). Links between communication patterns in mother–child, father–child, and child–peer interactions and children's social status. *Child Development, 66*, 255–271.

Blake, J., O'Rourke, P., & Borzellino, G. (1994). Form and function in the development of pointing and reaching gestures. *Infant Behavior and Development, 17*, 195–203.

Blehar, M., Lieberman, A., & Ainsworth, M. (1977). Early face-to-face interaction and its relation to later infant–mother attachment. *Child Development, 48*, 182–194.

Bloch, D. A., Siber, E., & Perry, S. E. (1956). Some factors in the emotional reactions of children to disaster. *American Journal of Psychiatry, 113*, 416–422.

Bloch, M. N. (1984). Play materials: Consideration from a West African setting. *Childhood Education, 60*, 345–348.

Bloch, M. N. (1987). The development of sex differences in young children's activity at home: The effect of social context. *Sex Roles, 16*, 279–301.

Bloch, M. N. (1989). Young boys' and girls' play at home and in the community: A cultural-ecological framework. In M. N. Bloch & A. D. Pellegrini (Eds.), *The ecological context of children's play* (pp. 120–154). Norwood, NJ: Ablex.

Block, J. H. (1976). Issues, problems, and pitfalls in assessing sex differences: A critical review of *The psychology of sex differences*. *Merrill-Palmer Quarterly, 22,* 283–308.

Bloom, K. (1974). Eye contact as a setting event for infant learning. *Journal of Experimental Child Psychology, 17*, 250–263.

Bloom, K. (1988). Quality of adult vocalizations affects the quality of infant vocalizations. *Journal of Child Language, 15*, 469–480.

Bloom, K., & Esposito, A. (1975). Social conditioning and its proper control procedures. *Journal of Experimental Child Psychology, 19*, 209–222.

Bloom, K., Russell, A., & Wassenberg, K. (1987). Turn taking affects the quality of infant vocalizations. *Journal of Child Language, 14*, 211–227.

Blount, B. (1972). Parental speech and language acquisition: Some Luo and Samoan examples. *Anthropological Linguistics*, *14*, 119–130.

Blount, B., & Padgug, E. (1977). Prosodic, paralinguistic, and interactional features in parent–child speech: English and Spanish. *Journal of Child Language*, *4*, 67–86.

Blurton-Jones, N. G. (1967). An ethological study of some aspects of social behavior of children in nursery school. In D. Morris (Ed.), *Primate ethology* (pp. 347–368). London: Weidenfeld & Nicolson.

Blurton-Jones, N. G., & Konner, M. J. (1973). Sex differences in the behavior of Bushman and London two- to five-year-olds. In J. Crook & R. Michael (Eds.), *Comparative ecology and behavior of primates* (pp. 689–750). New York: Academic.

Boesch, C., & Boesch, H. (1984). Possible causes of sex differences in the use of natural hammers by wild chimpanzees. *Journal of Human Evolution*, *13*, 415–440.

Boesch, C., & Boesch, H. (1990). Tool use and tool making in wild chimpanzees. *Folia Primatologica*, *54*, 86–99.

Bolwig, N. (1964). Observations on the mental and manipulative abilities of a captive baboon (*Papio doguera*). *Behaviour*, *22*, 24–40.

Bonney, M. E. (1955). Social behavior differences between second grade children of high and low sociometric status. *Journal of Educational Research*, *48*, 481–495.

Bonney, M. E., & Powell, J. (1953). Differences in social behavior between sociometrically high and sociometrically low children. *Journal of Educational Research*, *46*, 481–495.

Borman, K. M., & Kurdek, L. A. (1987). Grade and gender differences in and the stability and correlates of the structural complexity of children's playground games. *International Journal of Behavioral Development*, *10*, 241–251.

Bornstein, M. H., Tal, J., Rahn, C., Galperin, C. Z., Pecheux, M. G., Lamour, M., Toda, S., Azuma, H., Ogino, M., & Tamis-LeMonda, C. S. (1992). Functional analysis of the contents of maternal speech to infants of 5 and 13 months in four cultures: Argentina, France, Japan, and the United States. *Developmental Psychology*, *28*, 593–603.

Bornstein, M. H., & Tamis-LeMonda, C. S. (1990). Activities and interactions of mothers and their firstborn infants in the first six months of life: Covariation, stability, continuity, correspondence, and prediction. *Child Development*, *61*, 1206–1217.

Bornstein, M. H., Tamis-LeMonda, C. S., Tal, J., Ludemann, P., Toda, S., Rahn, C. W., Pecheux, M., Azuma, H., & Vardi, D. (1992). Maternal responsiveness to infants in three societies: The United States, France, and Japan. *Child Development*, *63*, 808–821.

Bornstein, M. H., Toda, S., Azuma, H., Tamis-LeMonda, C. S., & Ogino, M. (1990). Mother and infant activity and interaction in Japan and in the United States: II. A comparative microanalysis of naturalistic exchanges focused on the organization of infant attention. *International Journal of Behavioral Development*, *13*, 289–308.

Bornstein, M. H., Vibbert, M., Tal, J., & O'Donnell, K. (1992). Toddler language and play in the second year: Stability, covariation, and influences on parenting. *First Language*, *12*, 323–338.

Bott, H. (1928). Observation of play activities in a nursery school. *Genetic Psychology Monographs*, *4*, 44–88.

Boulton, M. J. (1991a). A comparison of structured and contextual features of middle school children's playful and aggressive fighting. *Ethology and Sociobiology*, *12*, 119–145.

Boulton, M. J. (1991b). Partner preferences in middle school children's playful fighting and chasing. *Ethology and Sociobiology*, *12*, 177–193.

Boulton, M. J. (1993). Children's abilities to distinguish between playful and aggressive fighting: A developmental perspective. *British Journal of Developmental Psychology*, *11*, 249–263.

Boulton, M. J., & Smith, P. K. (1989). Rough and tumble play in children: Research and theory. *Infancia-y-Aprendizaje, 48*, 79–91.

Boyd, R., & Richerson, P. J. (1985). *Culture and the evolutionary process*. Chicago: University of Chicago Press.

Brackbill, Y. (1958). Extinction of the smiling response in infants as a function of reinforcement schedule. *Child Development, 29*, 115–124.

Bradley, R. H., Caldwell, B. M., Rock, S. L., Ramey, C. T., Barnard, K. E., Gray, C., Hammond, M. A., Mitchell, S., Gottfried, A. W., Siegel, L., & Johnson, D. (1989). Home environment and cognitive development in the first three years of life: A collaborative study involving six sites and three ethnic groups in North America. *Developmental Psychology, 25*, 217–235.

Bramblett, C. A. (1978). Sex differences in the acquisition of play among juvenile vervet monkeys. In E. O. Smith (Ed.), *Social play in primates* (pp. 33–48). New York: Academic.

Brazelton, T. B., Koslowski, B., & Main, M. (1974). The origins of reciprocity: The early mother–infant interaction. In M. Lewis & L. A. Rosenblum (Eds.), *The effect of the infant on its caregiver* (pp. 49–76). New York: Wiley.

Brazelton, T. B., Robey, J. S., & Collier, G. A. (1969). Infant development in the Zinacanteco Indians of southern Mexico. *Pediatrics, 44*, 274–290.

Brenner, J., & Mueller, E. (1982). Shared meanings in boy toddlers' peer relations. *Child Development, 53*, 380–391.

Bretherton, I. (1984). Representing the social world in symbolic play: Reality and fantasy. In I. Bretherton (Ed.), *Symbolic play: The development of social understanding* (pp. 3–41). Orlando: Academic.

Bretherton, I. (1989). Pretense: The form and function of make-believe play. *Developmental Review, 9*, 383–401.

Bridges, K. M. B. (1933). A study of social development in early infancy. *Child Development, 4*, 36–49.

Bronson, W. C. (1974). Mother–toddler interaction: A perspective on studying the development of competence. *Merrill-Palmer Quarterly, 20*, 275–301.

Bronson, W. C. (1981). *Toddlers' behaviors with agemates: Issues of interaction, cognition, and affect*. Norwood, NJ: Ablex.

Bronstein, P. (1984). Differences in mothers' and fathers' behavior toward children: A cross-cultural comparison. *Developmental Psychology, 20*, 995–1003.

Brooks-Gunn, J., & Lewis, M. (1979). The effects of age and sex on infant's playroom behavior. *Journal of Genetic Psychology, 134*, 99–105.

Brown, A. L. (1990). Domain-specific principles affect learning and transfer in children. *Cognitive Science, 14*, 107–133.

Brown, J. R., & Dunn, J. (1992). Talk with your mother or your sibling? Developmental changes in early family conversations about feelings. *Child Development, 63*, 336–349.

Brown, K. W., & Gottfried, A. W. (1986). Cross-modal transfer of shape in early infancy: Is there reliable evidence? In L. P. Lipsitt & C. Rovee-Collier (Eds.), *Advances in infancy research* (Vol. 4, pp. 163–170). Norwood, NJ: Ablex.

Brown, S. (1988). Play behavior in lowland gorillas: Age differences, sex differences, and possible functions. *Primates, 29*, 219–228.

Brownell, C. A. (1990). Peer social skills in toddlers: Competencies and constraints illustrated by same-age and mixed-age interaction. *Child Development, 61*, 838–848.

Brownlee, A. (1954). Play in domestic cattle: An analysis of its nature. *British Veterinary Journal, 110*, 48–68.

Brownlee, A. (1984). Animal play. *Applied Animal Behaviour Science, 12*, 307–312.
Bruner, J. S. (1972). Nature and uses of immaturity. *American Psychologist, 27*, 687–708.
Bruner, J. S. (1978). Learning how to do things with words. In J. Bruner & A. Garton (Eds.), *Human growth and development* (pp. 62–84). New York: Oxford University Press.
Burghardt, G. M. (1984). On the origins of play. In P. K. Smith (Ed.), *Play in animals and humans* (pp. 5–41). Oxford, England: Basil Blackwell.
Burghardt, G. M. (1988). Precocity, play, and the ectotherm-endotherm transition: Profound reorganization or superficial adaptation? In E. M. Blass (Ed.), *Handbook of behavioral neurobiology* (Vol. 9, pp. 107–148). New York: Plenum.
Burghardt, G. M. (in press). Play. In G. Greenberg & M. Haraway (Eds.), *Comparative psychology: A handbook*. New York: Garland.
Burghardt, G. M., Ward, B., & Rosscoe, R. (1996). Problem of reptile play: Environmental enrichment and play behavior in a captile Nile soft-shelled turtle, *Trionyx triunguis. Zoo Biology, 15*, 223–238.
Burleson, B. R., Delia, J. G., & Applegate, J. L. (1992). Effects of maternal communication and children's social cognitive and communication skills on children's acceptance by the peer group. *Family Relations, 41,* 264–272.
Burnett, C. N., & Johnson, E. W I. (1971). Development of gait in childhood, Part II. *Developmental Medicine and Child Neurology, 13*, 207–215.
Burns, S. M., & Brainerd, C. J. (1979). Effects of constructive and dramatic play on perspective taking in very young children. *Developmental Psychology, 15*, 512–521.
Bushnell, E. W. (1982). Visual-tactual knowledge in 8-, 9 ½-, and 11-month-old infants. *Infant Behavior and Development, 5*, 63–75.
Buss, D. M., Haselton, M. G., Shackelford, T. K., Bleske, A. L., & Wakefield, J. C. (1998). Adaptations, exaptations, and spandrels. *American Psychologist, 53*, 533–548.
Byers, J. A. (1977). Terrain preferences in the play behavior of Siberian ibex kids (*Capra ibex sibirica*). *Zeitschrift fur Tierpsychologie, 45*, 199–209.
Byers, J. A. (1980). Play partner preferences in Siberian ibex, *Capra ibex sibirica. Zeitschrift fur Tierpsychologie, 53*, 23–40.
Byers, J. A. (1984). Play in ungulates. In P. K. Smith (Ed.), *Play in animals and humans* (pp. 43–65). Oxford, England: Basil Blackwell.
Byers, J. A. (1998). The biology of human play. *Child Development, 69*, 599–600.
Byers, J. A., & Walker, C. (1995). Refining the motor training hypothesis for the evolution of play. *American Naturalist, 146*, 25–40.
Cairns, R. B. (1979). *Social development: The origins and plasticity of behavior*. San Francisco: Freeman.
Cairns, R. B., Cairns, B. D., Neckerman, H. J., Gest, S., & Garieppy, J. L. (1988). Peer networks and aggressive behavior: Peer support or peer rejection? *Developmental Psychology, 24*, 815–823.
Caldera, Y. M., Huston, A. C., & O'Brien, M. (1989). Social interactions and play patterns of parents and toddlers with feminine, masculine, and neutral toys. *Child Development, 60*, 70–76.
Call, J., & Tomasello, M. (1996). The effects of humans on the cognitive development of apes. In A. E. Russon, K. A. Bard, & S. T. Parker (Eds.), *Reaching into thought: The minds of the great apes* (pp. 371–403). Cambridge, England: Cambridge University Press.
Calvin, W. H. (1983). A stone's throw and its launch window: Timing, precision, and implications for language, and hominid brains. *Journal of Theoretical Biology, 104*, 121–135.

Camaioni, L., Baumgartner, E., & Perucchini, P. (1991). Content and structure in toddlers' social competence with peers from 12 to 36 months of age. *Early Child Development and Care, 67*, 17–27.

Camaioni, L., & Laicardi, C. (1985). Early social games and the acquisition of language. *British Journal Developmental Psychology, 3*, 31–39.

Cameron, E., Eisenberg, N., & Tryon, K. (1985). The relations between sex-typed play and preschoolers' social behavior. *Sex Roles, 12*, 601–615.

Campbell, J. D., & Yarrow, M. R. (1961). Perceptual and behavioral correlates of social effectiveness. *Sociometry, 24*, 1–20.

Camras, L. A. (1977). Facial expressions used by children in a conflict situation. *Child Development, 48*, 1431–1435.

Caplan, M., Vespo, J., Pedersen, J., & Hay, D. F. (1991). Conflict and its resolution in small groups of one- and two-year-olds. *Child Development, 62*, 1513–1524.

Carew, J. V. (1980). Experience and the development of intelligence in young children at home and in day care. *Monographs of the Society for Research in Child Development, 45* (6–7, Serial No. 187).

Carlsson-Paige, N., & Levin, D. E. (1987). *War play dilemma: Balancing needs and values in the early childhood classroom.* New York: Teacher's College Press.

Carlsson-Paige, N., & Levin, D. E. (1990). *Who's calling the shots? How to respond effectively to children's fascination with war play and war toys.* Philadelphia: New Society.

Caro, T. M. (1980). Effects of the mother, object play, and adult experience on predation in cats. *Behavioral and Neural Biology, 29*, 29–51.

Caro,T. M. (1981). Predatory behaviour and social play in kittens. *Behaviour, 76*, 1–24.

Caro, T. M. (1988). Adaptive significance of play: Are we getting closer? *Trends in Ecology and Evolution, 3*, 50–54.

Caro, T. M. (1995). Short-term cost and correlates of play in cheetahs. *Animal Behaviour, 49*, 333–345.

Carpenter, C. J. (1983). Activity structure and play: Implications for socialization. In M. B. Liss (Ed.), *Social and cognitive skills: Sex roles and children's play* (pp. 117–145). New York: Academic.

Carpenter, C., Huston, A. C., & Holt, W. (1986). Modification of preschool sex-typed behaviors by participation in adult-structured activities. *Sex Roles, 14*, 603–615.

Carpenter, C., & Huston-Stein, A. (1980). Activity structure and sex-typed behavior in preschool children. *Child Development, 51*, 862–872.

Carr, R. M., & Brown, W. L. (1959). The effect of the introduction of novel stimuli upon manipulation in rhesus monkeys. *Journal of Genetic Psychology, 94*, 107–111.

Carr, S., Dabbs, J., & Carr, T. (1975). Mother–infant attachment: The importance of the mother's visual field. *Child Development, 46*, 331–338.

Caruso, D. A. (1993). Dimensions of quality in infant's exploratory behavior: Relationships to problem-solving ability. *Infant Behavior and Development, 16*, 441–454.

Casby, M. W., & Corte, M. D. (1987). Symbolic play performance and early language development. *Journal of Psycholinguistic Research, 16*, 31–43.

Case, R. (1985). *Intellectual development: Birth to adulthood.* New York: Academic.

Cassidy, J., Parke, R. D., Butkovsky, L., & Baungart, J. (1992). Family–peer connections: The roles of emotional expressiveness within the family and children's understanding of emotions. *Child Development, 63*, 603–618.

Caudill, W., & Weinstein, H. (1969). Maternal care and infant behavior in Japan and America. *Psychiatry, 32*, 12–43.

Chafel, J. A. (1986). A naturalistic investigation of the use of social comparison by young children. *Journal of Research and Development in Education, 19*, 51–61.
Chafel, J. A. (1987). Achieving knowledge about self and others through physical object and social fantasy play. *Early Childhood Research Quarterly, 2*, 27–43.
Chalmers, W. R. (1980). The ontogeny of play in feral olive baboons. *Animal Behaviour, 28*, 570–585.
Chalmers, W. R., & Locke-Haydon, J. (1981). Temporal patterns of play bouts in captive common marmosets (*Callitrhrix jacchus*). *Animal Behaviour, 29*, 1229–1238.
Chalmers, W. R., & Locke-Haydon, J. (1984). Correlations among measures of playfulness and skillfulness in captive common marmosets (*Callitrhrix jacchus jacchus*). *Developmental Psychobiology, 17*, 191–208.
Chambers, J. H., & Ascione, F. R. (1987). The effects of prosocial and aggressive videogames on children's donating and helping. *Journal of Genetic Psychology, 148*, 409–505.
Chance, M. R. A. (1967). Attention structure as the basis of primate rank orders. *Man, 2*, 503–518.
Charlesworth, R., & Hartup, W. W. (1967). Positive reinforcement in the nursery school peer group. *Child Development, 38*, 993–1002.
Cheney, D. L. (1978). The play patterns of immature baboons. *Animal Behaviour, 26*, 1038–1050.
Chepko, B. D. (1971). A preliminary study of the effect of play deprivation on very young goats. *Zeitschrift fur Tierpsychologie, 28*, 517–526.
Chevalier-Skolnikoff, S. (1974). The primate play face: A possible key to the determinants and evolution of play. *Rice University Studies, 60*, 9–29.
Chevalier-Skolnikoff, S. (1977). A Piagetian model for describing and comparing socialization in monkey, ape, and human infants. In S. Chevalier-Skolnikoff & F. E. Poirer (Eds.), *Primate biosocial development: Biological, social, and evolutionary determinants* (pp. 159–187). New York: Garland.
Chevalier-Skolnikoff, S. (1983). Sensorimotor development in orangutans and other primates. *Journal of Human Evolution, 12*, 545–561.
Chevalier-Skolnikoff, S. (1989). Spontaneous tool use and sensorimotor intelligence in Cebus compared with other monkeys and apes. *Behavioral and Brain Sciences, 12*, 561–627.
Chevalier-Skolnikoff, S., & Liska, J. (1993). Tool use by wild and captive elephants. *Animal Behaviour, 46*, 209–219.
Cheyne, J. A. (1982). Object play and problem-solving: Methodological problems and conceptual promise. In D. J. Pepler & K. H. Rubin (Eds.), *The play of children: Current theory and research* (pp. 79–96). Basel: S. Karger.
Cheyne, J. A., & Rubin, K. H. (1983). Playful precursors of problem solving in preschoolers. *Developmental Psychology, 19*, 577–584.
Chiszar, D. A., Scudder, K., Knight, L., & Smith, H. M. (1978). Exploratory behavior in prairie rattlesnakes (*Crotalus viridis*) and water moccasins (*Agkistrodon piscivorus*). *Psychological Record, 28*, 363–368.
Christie, J. F. (1983). The effects of play tutoring on young children's cognitive performance. *Journal of Educational Research, 76*, 326–330.
Christie, J. F., & Johnsen, E. P. (1985). Questioning the results of play training research. *Educational Psychology, 20*, 7–11.
Christie, J. F., & Johnsen, E. P. (1987). Reconceptualizing constructive play: A review of the empirical literature. *Merrill-Palmer Quarterly, 33*, 439–452.

Clark, A., Wyon, S., & Richards, M. (1969). Freeplay in nursery children. *Journal of Child Psychology and Psychiatry, 10*, 205–216.

Clarke-Stewart, K. A. (1973). Interactions between mothers and their young children: Characteristics and consequences. *Monographs of the Society for Research in Child Development, 38* (6–7, Serial No. 153).

Clarke-Stewart, K. A. (1978). And daddy makes three: The father's impact on mother and young child. *Child Development, 44*, 466–478.

Clarke-Stewart, K. A., & Hevey, C. M. (1981). Longitudinal reported observations of mother– child interactions from 1 to 2½ years. *Developmental Psychology, 17*, 127–145.

Clarke-Stewart, K. A., Umeh, B. J., Snow, M. E., & Pedersen, J. A. (1980). Development and prediction of children's sociability from 1 to 2½ years. *Developmental Psychology, 16*, 290–302.

Clarke-Stewart, K. A., VanderStoep, L. P., & Killian, G. A. (1979). Analysis and replication of mother–child relations at two years of age. *Child Development, 50*, 777–793.

Coates, D. L., & Lewis, M. (1984). Early mother–infant interaction and infant cognitive status as predictors of school performance and cognitive behavior in six-year-olds. *Child Development, 55*, 1219–1230.

Coates, S., Lord, M., & Jakabovics, E. (1975). Field dependence–independence, social–non-social play, and sex differences in preschool children. *Perceptual and Motor Skills, 40*, 195–202.

Cohen, N. L., & Tomlinson-Keasey, C. (1980). The effects of peers and mothers on toddlers' play. *Child Development, 51*, 921–924.

Cohen, S. (1974). Exploratory task behavior and creativity in young children. *Home Economics Research Journal, 2*, 262–267.

Cohn, J. F., & Elmore, M. (1988). Effect of contingent changes in mothers' affective expression on the organization of behavior in 3-month-old infants. *Infant Behavior and Development, 11*, 493–505.

Cohn, J. F., & Tronick, E. Z. (1987). Mother–infant interaction: The sequence of dyadic states at 3, 6, and 9 months. *Developmental Psychology, 23*, 68–77.

Cohn, J. F., & Tronick, E. Z. (1988). Mother–infant interaction: Influence is bidirectional and unrelated to periodic cycles in either partners' behavior. *Developmental Psychology, 24*, 386–392.

Coie, J. D. (1974). An evaluation of the cross-situational stability of children's curiosity. *Journal of Personality, 42*, 93–116.

Coie, J. D., Belding, M., & Underwood, M. (1988). Aggression and peer rejection in childhood. In B. B. Lahey & A. E. Kazdin (Eds.), *Advances in clinical child psychology* (Vol. 11, pp. 125–158). New York: Plenum.

Coie, J. D., & Dodge, K. A. (1983). Continuities and changes in children's social status: A four-year longitudinal study. *Merrill-Palmer Quarterly, 29*, 261–282.

Coie, J. D., & Dodge, K. A. (1988). Multiple sources of data on social behavior and social status in the school: A cross-age comparison. *Child Development, 59*, 815–829.

Coie, J. D., Dodge, K. A., & Kupersmidt, J. (1990). Peer group behavior and social status. In S. R. Asher & J. D. Coie (Eds.), *Peer rejection in childhood* (pp. 17–59). New York: Cambridge University Press.

Coie, J. D., Dodge, K. A., Terry, R., & Wright, V. (1991). The role of aggression in peer relations: An analysis of aggression episodes in boys' play groups. *Child Development, 62*, 812–826.

Coie, J. D., & Kupersmidt, J. B. (1983). A behavioral analysis of emerging social status in boys' groups. *Child Development, 54*, 1400–1416.

Cole, D., & La Voie, J. C. (1985). Fantasy play and related cognitive development in 2- and 6-year-olds. *Developmental Psychology, 21*, 233–240.

Collard, R. R. (1971). Exploration and play behaviors of infants reared in an institution and lower and middle class homes. *Child Development, 42*, 1003–1015.

Conner, J. M., & Serbin, L. A. (1977). Behaviorally based masculine- and feminine-activity preference scales for preschoolers. Correlates with other classroom behaviors and cognitive tests. *Child Development, 48*, 1411–1416.

Connolly, J. A., & Doyle, A. (1984). Relation of social fantasy play to social competence in preschoolers. *Developmental Psychology, 20*, 797–806.

Connolly, J. A., Doyle, A., & Ceschin, F. (1983). Forms and functions of social fantasy play in preschoolers. In M. B. Liss (Ed.), *Social and cognitive skills: Sex roles and children's play* (pp. 71–92). New York: Academic.

Connolly, J. A., Doyle, A. B., & Reznick, E. (1988). Social pretend play and social interaction in preschoolers. *Journal of Applied Developmental Psychology, 9*, 301–313.

Connolly, K., & Dalgleish, M. (1989). The emergence of a tool-using skill in infancy. *Developmental Psychology, 6*, 894–912.

Connolly, K., & Elliott, J. (1972). The evolution and ontogeny of hand function. In N. B. Jones (Eds.), *Ethological studies of child behavior* (pp. 329–383). London: Cambridge University Press.

Cook, A. S., Fritz, J. J., McCornack, B. L., & Visperas, C. (1985). Early gender differences in the functional use of language. *Sex Roles, 12*, 909–915.

Coon, H., Fulker, D. W., DeFries, J. C., & Plomin, R. (1990). Home environment and cognitive ability of 7-year-old children in the Colorado Adoption Project: Genetic and environmental etiologies. *Deveopmental Psychology, 26*, 459–468.

Cooper, J., & Mackie, D. (1986). Video games and aggression in children. *Journal of Applied Social Psychology, 16*, 726–744.

Coplan, R. J., Rubin, K. H., Fox, N. A., Calkins, S. D., & Stewart, S. L. (1994). Being alone, playing alone, and acting alone: Distinguishing among reticence and passive and active solitude in young children. *Child Development, 65*, 129–137.

Copple, C. E., Cocking, R. R., & Matthews, W. S. (1984). Objects, symbols, and substitutes: The nature of cognitive activity during symbolic play. In T. D. Yawkey & A. D. Pellegrini (Eds.), *Child's play: Developmental and applied* (pp. 105–123). Hillsdale, NJ: Lawrence Erlbaum Associates.

Corrigan, R. (1982). The control of animate and inanimate components in pretend play and language. *Child Development, 53*, 1343–1353.

Corrigan, R. (1987). A developmental sequence of actor-object pretend play in young children. *Merrill-Palmer Quarterly, 33*, 87–106.

Corter, C., Abramovitch, R., & Pepler, D. J. (1983). The role of the mother in sibling interaction. *Child Development, 54*, 1599–1605.

Corter, C., Pepler, D. J., & Abramovitch, R. (1982). The effects of situation and sibling status on sibling interaction. *Canadian Journal of Behavioral Science, 14*, 380–392.

Cossette, L., Malcuit, G., & Pomerleau, A. (1991). Sex differences in motor activity during early infancy. *Infant Behavior and Development, 14*, 175–186.

Cossette, L., Pomerleau, A., Malcuit, G., & Kaczorowski, J. (1996). Emotional expressions of female and male infants in a social and a nonsocial context. *Sex Roles, 35*, 693–709.

Costabile, A., Genta, M. L., Zucchini, E., Smith, P. K., & Harker, R. (1992). Attitudes of parents toward war play in young children. *Early Education and Development, 3*, 356–369.

Costabile, A., Smith, P. K., Matheson, L., Aston, J., Hunter, T., & Boulton, M. (1991). Cross-national comparison of how children distinguish serious and playful fighting. *Developmental Psychology*, *27*, 881–887.

Costello, M., & Fragaszy, D. M. (1988). Comparison of prehension in squirrel monkeys (*Saimiri sciureus*) and capuchins (*Cebus apella*) I: Grip type and hand preference. *American Journal of Primatology*, *15*, 235–245.

Cowan, P. E. (1983). Exploration in small mammals: Ethology and ecology. In J. Archer & L. Birke (Eds.), *Exploration in animals and humans* (pp. 147–175). Berkshire, England: Van Nostrand Reinhold.

Cowey, A., & Weikrantz, L. (1975). Demonstration of cross-modal matching in Rhesus monkeys (*Macca mulatta*). *Neuropsychologia*, *13*, 117–121.

Cox, F. N., & Campbell, D. (1968). Young children in a new situation without their mothers. *Child Development*, *39*, 123–131.

Crawley, S. B., Rogers, P. P., Friedman, S., Iacobbo, M., Criticos, A., Richardson, L., & Thompson, M. A. (1978). Developmental changes in the structure of mother–infant play. *Developmental Psychology*, *14*, 30–36.

Crawley, S. B., & Sherrod, K. B. (1984). Parent–infant play during the first year of life. *Infant Behavior and Development*, *7*, 65–75.

Creasey, G. L., & Myers, B. J. (1986). Video games and children: Effects on leisure activities, schoolwork, and peer involvement. *Merrill-Palmer Quaterly*, *32*, 251–262.

Crick, N. R., Casas, J. F., & Mosher, M. (1997). Relational and overt aggression in preschool. *Developmental Psychology*, *33*, 579–588.

Crockenberg, S. (1983). Early mother and infant antecedents of Bayley Scale performance at 21 months. *Developmental Psychology*, *19*, 727–730.

Croft, D. B., & Snaith, F. (1991). Boxing in red kangaroos, *Macropus rufus*: Aggression or play? *International Journal of Comparative Psychology*, *4*, 221–236.

Crowell-Davis, S. L., Houpt, K. A., & Kane, L. (1987). Play development in Welsh pony (*Equus cabellus*) foals. *Applied Animal Behavioral Science*, *18*, 119–131.

Crum, R. A., Thornburg, K., Benninga, J., & Bridge, C. (1983). Preschool children's object substitutions during symbolic play. *Perceptual and Motor Skills*, *56*, 947–955.

Csikszentmihalyi, M. (1982). Does being human matter? On some interpretive problems of comparative ludology. *The Behavioral and Brain Sciences*, *5*, 160.

Czerniecki, J. G., Deitz, J. C., Crowe, T. K., & Booth, C. L. (1993). Attending behavior: A descriptive study of children aged 18 through 23 months. *The American Journal of Occupational Therapy*, *47*, 708–716.

Daldry, A. D., & Russell, P. A. (1982). Sex differences in the behavior of preschool children with novel and familiar toys. *The Journal of Genetic Psychology*, *141*, 3–6.

Dale, N. (1989). Pretend play with mother and siblings: Relations between early performance and partners. *Journal of Child Psychology and Psychiatry and Allied Disciplines*, *30*, 751–759.

Damast, A. M., Tamis-LeMonda, C. S., & Bornstein, M. H. (1996). Mother–child play: Sequential interactions and the relation between maternal beliefs and behaviors. *Child Development*, *67*, 1752–1766.

Damon, W. (1998). *Handbook of child psychology* (5th ed.). New York: Wiley.

Dansen, P. R. (1973). Preliminary study of sensorimotor development in Baoule children. *Early Child Development and Care*, *2*, 345–354.

Dansky, J. L. (1980a). Cognitive consequences of sociodramatic play and exploration training for disadvantaged preschoolers. *Journal of Child Psychology and Psychiatry*, *20*, 47–58.

Dansky, J. L. (1980b). Make believe: A mediation of the relationship between play and associative frequency. *Child Development*, *51*, 576–579.

Dansky, J. L. (1985). Questioning "a paradigm questioned": A commentary on Simon and Smith. *Merrill-Palmer Quarterly*, *31*, 279–284.

Dansky, J. L., & Silverman, I. W. (1973). Effects of play on associative fluency in preschool-aged children. *Developmental Psychology*, *9*, 38–43.

Dansky, J. L., & Silverman, I. W. (1975). Play: A general facilitation of associative fluency. *Developmental Psychology*, *11*, 104.

Darwin, C. (1874). *The descent of man and selection in relation to sex*. Appleton: New York.

Darwin, C. (1898). *The expression of emotions in man and animals*. New York: Appleton.

Dashiell, J. F. (1925). A quantitative demonstration of animal drive. *Journal of Comparative Psychology*, *5*, 205–208.

Davenport, R. K., & Rogers, C. M. (1970). Intermodal equivalence of stimuli in apes. *Science*, *168*, 279–280.

Davenport, R. K., Rogers, C. M., & Russel, I. S. (1973). Cross-modal perception in apes. *Neuropsychologia*, *11*, 21–28.

Davenport, R. K., Rogers, C. M., & Russel, I. S. (1975). Cross-modal perception in apes: Altered visual cues and delay. *Neuropsychologia*, *13*, 229–236.

Davidson, R. G., & Snow, C. E. (1996). Five-year-olds' interactions with fathers versus mothers. *First Language*, *16*, 223–242.

Davies, V. A., & Kemble, E. D. (1983). Social play behaviors and insect predation in northern grasshopper mice (*Onychomys levcogaster*). *Behavioural Processes*, *8*, 197–204.

Dawe, H. C. (1934). An analysis of zoo quarrels of preschool children. *Child Development*, *5*, 139–157.

Deaux, K., & Major, B. (1987). Putting gender into context: An interactive model of gender-related behavior. *Psychological Review*, *94*, 369–389.

Deitz, W., & Gortmaker, S. L. (1985). Do we fatten our children at the television set? Obesity and television viewing in children and adolescents. *Pediatrics*, *75*, 807–812.

Dekovic, M., & Janssens, J. M. A. M. (1992). Parents' child-rearing style children's sociometric status. *Developmental Psychology*, *28*, 925–932.

DeLoache, J. S., Sugarman, S., & Brown, A. L. (1985). The development of error correction strategies in young children's manipulative play. *Child Development*, *56*, 928–939.

de Lorimier, S., Doyle, A., & Tessier, O. (1995). Social coordination during pretend play: Comparisons with nonpretend play and effects on expressive content. *Merrill-Palmer Quarterly*, *41*, 497–516.

Denham, S. A., McKinley, M., Couchoud, E. A., & Holt, R. (1990). Emotional and behavioral predictors of preschool peer ratings. *Child Development*, *61*, 1145–1152.

D'Entremont, B., & Muir, D. W. (1997). Five-month-olds' attention and affective responses to still-faced emotional expressions. *Infant Behavior and Development*, *20*, 563–568.

DePaulo, B. M., & Bonvillian, J. D. (1978). The effect on language development of the special characteristics of speech addressed to children. *Journal of Psycholinguistic Research*, *7*, 189–211.

Derevensky, J. L. (1979). Relative contributions of active and passive touch to child's knowledge of physical objects. *Perceptual and Motor Skills*, *48*, 1331–1346.

Deriagina, M. (1982). Note on the manipulatory activity of apes. *Journal of Human Evolution*, *11*, 171–172.

De Wolff, M. S., & van IJzendoorn, M. H. (1997). Sensitivity and attachment: A meta-analysis on parental antecedents of infant attachment. *Child Development*, *68*, 571–591.

Dickson, N. L., Walker, H., & Fogel, A. (1997). The relationship between smile type and play type during parent–infant play. *Developmental Psychology*, *33*, 925–933.
Dickstein, S., & Parke, R. D. (1988). Social referencing in infancy: A glance from fathers and marriage. *Child Development*, *59*, 509–511.
Di Pietro, J. A. (1981). Rough and tumble play: A function of gender. *Developmental Psychology*, *17*, 50–58.
Dixon, S., Tronick, E., Keefer, C., & Brazelton, T. B. (1981). Mother–infant interaction among the Gusii of Kenya. In T. M. Field, A. M. Sotek, P. Vietze, & P. H. Leiderman (Eds.), *Culture and early interactions* (pp. 149–168). Hillsdale, NJ: Lawrence Erlbaum Associates.
Dixon, W. E., & Shore, C. (1993). What shall we play? Play content as a vehicle for parent-infant interactions. *Infant Behavior and Development*, *16*, 121–125.
Dodge, K. A. (1983). Behavioral antecedents of peer social status. *Child Development*, *54*, 1386–1399.
Dodge, K. A., Coie, J. D., & Brakke, W. P. (1982). Behavior patterns of socially rejected and neglected preadolescents: The roles of social approach and aggression. *Journal of Abnormal Child Psychology*, *10*, 389–409.
Dodge, K. A., Coie, J. D., Pettit, G. S., & Price, J. M. (1990). Peer status and aggression in boys' groups: Developmental and contextual analyses. *Child Development*, *61*, 1289–1309.
Dodge, K. A., & Frame, C. L. (1982). Social cognitive biases and deficits in aggressive boys. *Child Development*, *53*, 620–635.
Dodge, K. A., Pettit, G. S., McCluskey, C. L., & Brown, M. M. (1987). Social competence in children. *Monographs of the Society for Research in Child Development*, *51* (2, Serial No. 213).
Dodge, K. A., Schlundt, D. C., Schocken, I., & Delugach, J. D. (1983). Social competence and children's sociometric status: The role of peer group entry strategies. *Merrill-Palmer Quaterly*, *29*, 309–336.
Dominick, J. R. (1984). Videogames, television violence, and aggression in teenagers. *Journal of Communication*, *34*, 136–147.
Dore, F. Y., & Duman, C. (1987). Psychology of animal cognition: Piagetian studies. *Psychological Bulletin*, *102*, 219–233.
Doyle, A. B., Bowker, A., Serbin, L., Gold, D., & Sherman, L. (1989). Developmental changes in pretend play during middle childhood. *Canadian Journal of Research in Early Childhood Education*, *3*, 43–51.
Doyle, A., Ceschin, F., Tessier, O., & Doehring, P. (1991). The relation of age and social class factors in children's social pretend play to cognitive and symbolic ability. *International Journal of Behavioral Development*, *14*, 395–410.
Doyle, A. B., & Connolly, J. (1989). Negotiation and enactment in social pretend play: Relations to social acceptance and social cognition. *Early Childhood Research Quaterly*, *4*, 289–302.
Doyle, A., Connolly, J., & Rivest, L. (1980). The effects of playmate familiarity on the social interactions of young children. *Child Development*, *51*, 217–223.
Doyle, A., Doehring, P., Tessier, O., de Lorimier, S., & Shapiro, S. (1992). Transitions in children's play: A sequential analysis of states preceding and following social pretense. *Developmental Psychology*, *28*, 137–144.
Draper, P. (1976). Social and economic constraints on child life among the Kung. In R. B. Lee & I. De Vore (Eds.), *Kalahari hunter-gatherers: Studies of the Kung San and their neighbors* (pp. 200–217). Cambridge, MA: Harvard University Press.

Drea, C. M., Hawk, J. E., & Glickman, S. E. (1996). Aggression decreases as play emerges in infant spotted hyaenas: Preparation for joining the clan. *Animal Behaviour, 51*, 1323–1336.

Dunbar, R. I. M. (1993). Coevolution of neocortical size, group size, and language in humans. *Behavioral and Brain Sciences, 16*, 661–736.

Dunham, P. J., Dunham, F., & Curwin, A. (1993). Joint-attentional states and lexical acquisition at 18 months. *Developmental Psychology, 29*, 827–831.

Dunn, J., Bretherton, I., & Munn, P. (1987). Conversations about feelings between mothers and their young children. *Developmental Psychology, 23*, 132–139.

Dunn, J., Brown, J., & Beardsall, L. (1991). Family talk about feeling states and children's later understanding of others' emotions. *Developmental Psychology, 27*, 448–455.

Dunn, J., Brown, J., Slomkowski, C., Telsa, C., & Youngblade, L. (1991). Young children's understanding of other people's feelings and beliefs: Individual differences and their antecedents. *Child Development, 62*, 1352–1366.

Dunn, J., & Dale, N. (1984). I a daddy: 2-year-olds' collaboration in joint-pretend with sibling and with mother. In I. Bretherton (Ed.), *Symbolic play: The development of social understanding* (pp. 131–158). New York: Academic.

Dunn, J., & Kendrick, C. (1981). Social behavior of young siblings in the family context: Differences between same-sex and different-sex dyads. *Child Development, 52*, 1265–1273.

Dunn, J., & Munn, P. (1985). Becoming a family member: Family conflict and the development of social understanding in the second year. *Child Development, 56*, 480–492.

Dunn, J., & Munn, P. (1986a). Siblings and the development of prosocial behavior. *International Journal of Behavioral Development, 9*, 265–284.

Dunn, J., & Munn, P. (1986b). Sibling quarrels and maternal intervention: Individual differences in understanding and aggression. *Journal of Child Psychology and Psychiatry and Allied Disciplines, 27*, 583–595.

Dunn, J., & Munn, P. (1987). Development of justification in disputes with mother and sibling. *Developmental Psychology, 23*, 791–798.

Dunn, J., & Richards, M. (1977). Observations on the developing relationship between mother and baby in the neonatal period. In H. R. Schaffer (Ed.), *Studies in mother–infant interaction* (pp. 427–455). London: Academic.

Dunn, J., Slomkowski, C., & Beardsall, L. (1994). Sibling relationshps from the preschool period through middle childhood and early adolescence. *Development Psychology, 30*, 315–324.

Dunn, J., & Wooding, C. (1977). Play in the home and its implications for learning. In B. Tizard & D. Harvey (Eds.), *Biology of play* (pp. 45–58). London: Heinemann.

Dunn, L., & Herwig, J. E. (1992). Play behaviors and convergent and divergent thinking skills of young children attending full-day preschool. *Child Study Journal, 22*, 23–38.

Dunnington, M. J. (1957). Behavioral differences of sociometric status groups in a nursery school. *Child Development, 28*, 103–111.

Durrett, E., & Huffman, W. (1968). Playfulness and divergent thinking among Mexican-American children. *Journal of Home Economics, 60*, 355–358.

Duveen, G., & Lloyd, B. (1988). Gender as an influence on the development of scripted pretend play. *British Journal of Developmental Psychology, 6*, 89–95.

Duveen, G., Lloyd, B., & Smith, C. (1988). A note on the effects of age and gender on children's social behaviour. *British Journal of Social Psychology, 27*, 275–278.

Eaton, W., & Enns, L. (1986). Sex differences in human motor activity. *Psychological Bulletin, 100*, 19–28.

Eaton, W., & Keats, J. G. (1982). Peer presence, stress, and sex differences in the motor activity levels of preschoolers. *Developmental Psychology, 18*, 534–540.

Eaton, W., & Yu, A. (1989). Are sex differences in child motor activity level a function of sex differences in motivational status? *Child Development, 60*, 1005–1011.

Eckerman, C. O., Davis, C. C., & Didow, S. M. (1989). Toddlers' emerging ways of achieving social coordinations with a peer. *Child Development, 60*, 440–453.

Eckerman, C. O., & Didow, S. M. (1996). Nonverbal imitation and toddlers' mastery of verbal means of achieving coordinated action. *Developmental Psychology, 32*, 141–152.

Eckerman, C. O., & Rheingold, H. L. (1974). Infants' exploratory responses to toys and people. *Developmental Psychology, 10*, 255–259.

Eckerman, C. O., & Whatley, J. (1977). Toys and social interaction between infant peers. *Child Development, 48*, 1645–1656.

Eckerman, C. O., Whatley, J., & Kutz, S. (1975). Growth of social play with peers during the second year of life. *Developmental Psychology, 11*, 42–49.

Eckler, J. A., & Weininger, O. (1989). Structural parallels between pretend play and narratives. *Developmental Psychology, 25*, 736–743.

Edwards, C. P., & Whiting, B. B. (1993). "Mother, older sibling, and me": The overlapping roles of caregivers and companions in the social world of two- and three-year-olds in Ngeca, Kenya. In K. MacDonald (Ed.), *Parent-child play: Descriptions and implications* (pp. 305–329). Albany: State University of New York Press.

Egan, J. (1976). Object play in cats. In J. S. Bruner, A. Jolly, & K. Sylva (Eds.), *Play: Its role in development and evolution* (pp. 161–165). Harmondsworth, England: Penguin.

Egeland, B., Pianta, R., & O'Brien, M. A. (1993). Maternal intrusiveness in infancy and child maladaptation in early school years. *Development and Psychopathology, 5*, 359–370.

Egli, E. A., & Meyers, L. S. (1984). The role of video game playing in adolescent life: Is there reason to be concerned? *Bulletin of the Psychonomic Society, 22*, 309–312.

Eibl-Eibesfedlt, I. (1970). *Ethology: The biology of behavior*. New York: Holt, Rinehart, & Winston.

Eibl-Eibesfedlt, I. (1982). The flexibility and affective autonomy of play. *The Behavioral and Brain Sciences, 5*, 160–162.

Eibl-Eibesfeldt, I. (1989). *Human ethology*. New York: de Gruyter.

Eifermann, R. R. (1971). Social play in childhood. In R. E. Herron & B. Sutton-Smith (Eds.), *Child's play* (pp. 270–297). New York: Wiley.

Einon, D., Morgan, M., & Kibbler, C. C. (1978). Brief periods of socialization and later behaviour in the rat. *Developmental Psychobiology, 11*, 213–225.

Einon, D., & Potegal, M. (1991). Enhanced defense in adult rats deprived of playfighting experience as juveniles. *Aggressive Behavior, 17*, 27–46.

Eisenberg, N. (1983). Sex-typed toy choices: What do they signify? In M. B. Liss (Ed.), *Social and cognitive skills: Sex roles and children's play* (pp. 45–70). New York: Academic.

Eisenberg, N., Fabes, R. A., Karbon, M., Murphy, B. C., Wosinski, M., Polazzi, L., Carlo, G., & Juhnke, C. (1996). The relations of children's dispositional prosocial behavior to emotionality, regulation, and social functioning. *Child Development, 67*, 974–992.

Eisenberg, N., Fabes, R. A., Minore, D., Mathy, R., Hanish, L., & Brown, T. (1994). Children's enacted interpersonal strategies: Their relations to social behavior and negative emotionality. *Merrill-Palmer Quarterly, 40*, 212–232.

Eisenberg, N., Fabes, R. A., Nyman, M., Bernzweig, J., & Pinuelas, A. (1994). The relations of emotionality and regulation to children's anger-related reactions. *Child Development, 65*, 109–128.

Eisenberg, N., Tyron, K., & Cameron, E. (1984). The relation of preschoolers' peer interaction to their sex-typed toy choices. *Child Development, 55*, 1044–1050.

Eisenberg, N., Wolchik, S. A., Hernandez, R., & Pasternack, J. F. (1985). Parental socialization of young children's play: A short-term longitudinal study. *Child Development, 56*, 1506–1513.

Eisenberg-Berg, N., & Hand, M. (1979). The relationship of preschooler's reasoning about prosocial moral conflicts to prosocial behavior. *Child Development, 50*, 356–363.

Eisenberg-Berg, N., & Lennon, R. (1980). Altruism and the assessment of empathy in the preschool years. *Child Development, 51*, 552–557.

Elder, J. L., & Pederson, D. R. (1978). Preschool children's use of objects in symbolic play. *Child Development, 49*, 500–504.

Ellis, S., Rogoff, B., & Cromer, C. C. (1981). Age segregation in children's social interactions. *Developmental Psychology, 17*, 399–407.

Ellsworth, C. P., Muir, D. W., & Hains, S. M. (1993). Social competence and person-object differentiation: An analysis of the still-face effect. *Developmental Psychology, 29*, 63–73.

Emde, R. N., Gaensbauer, T. J., & Harmon, R. J. (1976). Emotional expression in infancy: A biobehavioral study [monograph]. *Psychological Issues, 10*(37). New York: International Universities Press.

Endsley, R. C., Hutcherson, M. A., Garner, A. P., & Martin, M. J. (1979). Interrelationships among selected maternal behaviors, authoritarianism, and preschool children's verbal and nonverbal curiosity. *Child Development, 50*, 331–339.

Enomoto, T. (1990). Social play and sexual behavior of the bonobo (*Pan paniscus*) with special reference to flexibility. *Primates, 31*, 469–480.

Eppler, M. A. (1995). Development of manipulatory skills and the development of attention. *Infant Behavior and Development, 18*, 391–405.

Erel, O., Margolin, G., & John, R. S. (1998). Observed sibling interaction: Links with the marital and mother–child relationship. *Developmental Psychology, 34*, 288–298.

Erikson, E. H. (1950). *Childhood and society*. New York: Norton.

Erikson, E. H. (1977). *Toys and reasons*. New York: Norton.

Etaugh, C., Collins, G., & Gerson, A. (1975). Reinforcement of sex-typed behaviors of two-year-old children in a nursery school setting. *Developmental Psychology, 11*, 255.

Etaugh, C., & Happach, A. (1979). Effects of aggressive play on children's subsequent aggressive behavior. *Psychological Reports, 45*, 656–568.

Fabes, R. A., & Eisenberg, N. (1992). Young children's coping with interpersonal anger. *Child Development, 63*, 116–129.

Fady, J. C. (1976). Social play: The choice of playmates observed in the young of the crab-eating macaque. In J. S. Bruner, A. Jolly, & K. Sylva (Eds.), *Play: Its role in development and evolution* (pp. 328–335). New York: Penguin.

Fagen, R. (1981). *Animal play behavior*. New York: Oxford University Press.

Fagen, R. (1992). Play, fun, and communication of well-being. *Play and Culture, 5*, 40–58.

Fagen, R. (1993). Primate juveniles and primate play. In M. E. Pereira & L. A. Fairbanks (Eds.), *Juvenile primates: Life history, development, and behavior* (pp. 183–196). New York: Oxford University Press.

Fagen, R. (1995). Animal play, games of angels, biology, and brain. In A. D. Pellegrini (Ed.), *The future of play theory: A multidisciplinary inquiry into the contributions of Brian Sutton-Smith* (pp. 23–44). Albany: State University of New York Press.

Fagen, R., & George, T. K. (1977). Play behavior and exercise in young ponies (*Equus caballus L.*). *Behavior, Ecology, and Sociobiology*, *2*, 267–269.

Fagot, B. I. (1977). Consequences of moderate cross-gender behavior in preschool children. *Child Development*, *48*, 902–907.

Fagot, B. I. (1984). Teacher and peer reaction to boys' and girls' play styles. *Sex Roles*, *11*, 691–702.

Fagot, B. I. (1985). Beyond the reinforcement principle: Another step toward understanding sex role development. *Developmental Psychology*, *21*, 1097–1104.

Fagot, B. I., & Hagan, R. (1991). Observations of parent reactions to sex-stereotyped behaviors: Age and sex effects. *Child Development*, *62*, 617–628.

Fagot, B. I., & Kavanagh, K. (1993). Parenting during the second year: Effects of children's age, sex, and attachment classification. *Child Development*, *64*, 258–271.

Fagot, B. I., & Leinbach, M. D. (1989). The young child's gender schema: Environment input, internal organization. *Child Development*, *60*, 663–672.

Fagot, B. I., Leinbach, M. D., & Hagan, R. (1986). Gender labeling and the adoption of sex-typed behaviors. *Developmental Psychology*, *22*, 440–443.

Fagot, B. I., Leinbach, M. D., & O'Boyle, C. (1992). Gender labeling, gender stereotyping, and parenting behaviors. *Developmental Psychology*, *28*, 225–230.

Fagot, B. I., & Littman, I.(1976). Relation of preschool sex-typing to intellectual performance in elementary school. *Psychological Reports*, *39*, 699–704.

Fagot, B. I., & O'Brien, M. (1994). Activity level in young children: Cross-age stability, situational influences, correlates with temperament, and the perception of problem behaviors. *Merrill-Palmer Quarterly*, *40*, 378–398.

Fagot, B. L., & Patterson, G. R. (1969). An *in vivo* analysis of reinforcing contingencies for sex-role behaviors in the preschool child. *Developmental Psychology*, *1*, 563–568.

Farentinos, R. C. (1971). Some observations on the play behavior of the Stellar sea lion (*Eumetopias jubata*). *Zeitschrift fur Tierpsychologie*, *28*, 428–438.

Farran, D. C., & Kasari, C. (1990). A longitudinal analysis of the development of synchrony in mutual gaze in mother–child dyads. *Journal of Applied Developmental Psychology*, *11*, 419–430.

Farver, J. (1992). Communicating shared meaning in social pretend play. *Early Childhood Research Quarterly*, *7*, 501–516.

Farver, J., & Howes, C. (1988). Cross-cultural differences in social interaction: A comparison of American and Indonesian children. *Journal of Cross-Cultural Psychology*, *19*, 203–215.

Farver, J., & Wimbarti, S. (1995a). Indonesian children's play with their mothers and their older siblings. *Child Development*, *66*, 1493–1503.

Farver, J., & Wimbarti, S. (1995b). Parental participation in toddlers' pretend play. *Social Development*, *4*, 17–31.

Fash, D. S., & Madison, C. L. (1981). Parents' language interaction with young children: A comparative study of mothers' and fathers.' *Child Study Journal*, *11*, 137–153.

Fasold, R. (1990). *Sociolinguistics of language*. Oxford: Basil Blackwell.

Fassino, M. (1982). Play: Structure and function. *The Behavioral and Brain Sciences*, *5*, 162–163.

Fay, J. M., & Carroll, R. W. (1994). Chimpanzee tool use for honey and termite extraction in central Africa. *American Journal of Primatology*, *34*, 309–317.

Feddersen-Petersen, D. (1991). The ontogeny of social play and agonistic behaviour in selected canid species. *Bonner Zoologische Beitrage*, *42*, 97–114.

Fedigen, L. (1972). Social and solitary play in a colony of vervet monkeys (*Cercoptihecus aethiops*). *Primates*, *13*, 347–364.
Fein, G. G. (1975). A transformational analysis of pretending. *Developmental Psychology*, *11*, 291–296.
Fein, G. G., & Apfel, N. (1979). The development of play: Style, structure, and situation. *Genetic Psychology Monographs*, *99*, 231–250.
Fein, G. G., & Fryer, M. G. (1995). Maternal contributions to early symbolic play competence. *Developmental Review*, *15*, 367–381.
Feinman, S. (1992). *Social referencing and the social construction of reality in infancy*. New York: Plenum.
Feitelson, D. (1959). Aspects of the social life of Kurdish Jews. *Jewish Journal of Sociology*, *1*, 201–216.
Feitelson, D. (1977). Cross-cultural studies of representational play. In B. Tizard & D. Harvey (Eds.), *Biology of play* (pp. 6–14). London: Heinemann.
Fenson, L. (1984). Developmental trends for action and speech in pretend play. In I. Bretherton (Ed.), *Symbolic play: The development of social understanding* (pp. 249–270). New York: Academic.
Fenson, L., Kagan, J., Kearsley, R. B., & Zelazo, P. (1976). The developmental progression of manipulative play in the first two years. *Child Development*, *47*, 232–236.
Fenson, L., & Ramsay, D. S. (1980). Decentration and integration of the child's play in the second year. *Child Development*, *51*, 171–178.
Fenson, L., Sapper, V., & Minner, D. G. (1974). Attention and manipulative play in the 1-year-old child. *Child Development*, *45*, 757–764.
Ferguson, C. A. (1964). Baby talk in six languages. *American Anthropologist*, *66*, 103–144.
Fernald, A., & Kuhl, P. K. (1987). Acoustic determinants of infant preference for motherese speech. *Infant Behavior and Development*, *10*, 279–293.
Fernald, A., & Morikawa, H. (1993). Common themes and cultural variations in Japanese and American mothers' speech to infants. *Child Development*, *64*, 637–656.
Fernald, A., & O'Neill, D. K. (1993). Peek-a-boo across cultures: How mothers and infants play with voices, faces, and expectations. In K. MacDonald (Ed.), *Parent-child play: Descriptions and implications* (p. 259–285). Albany: State Univeristy of New York Press.
Fernald, A., & Simon, T. (1984). Expanded intonation contours in mothers' speech to newborns. *Developmental Psychology*, *20*, 104–113.
Fernald, A., Taeschner, T., Dunn, J., Papousek, M., Boysson-Bardies, B., & Fukui, I. (1989). A cross-language study of prosodic modifications in mothers' and fathers' speech to preverbal infants. *Journal of Child Language*, *16*, 477–501.
Fernandes, M. E. (1991). Tool use and predation of oysters (*Crassostrea rhizophorae*) by the tufted capuchin, *Cebus apella apella*, in brackish water mangrove swamps. *Primates*, *32*, 529–531.
Ferron, J. (1975). Solitary play of the red squirrel *(Tamiasciurus hudsonicus*). *Canadian Journal of Zoology*, *53*, 1495–1499.
Feshbach, S. (1956). The catharsis hypothesis and some consequences of interaction with neutral play objects. *Journal of Personality*, *24*, 449–462.
Field, T. (1977). Effects of early separation, interactive deficits, and experimental manipulations on infant–mother face-to-face interaction. *Child Development*, *48*, 763–771.
Field, T. (1978). Interactive behaviors of primary versus secondary caretaker fathers. *Developmental Psychology*, *14*, 183–184.

Field, T. (1979a). Games parents play with normal and high-risk infants. *Child Psychiatry and Human Development, 10*, 41–48.
Field, T. (1979b). Infant behaviors toward peers and adults in the presence and absence of mother. *Infant Behavior and Development, 2*, 47–54.
Field, T. (1982). Same-sex preferences of preschool children: An artifact of same-age grouping? *Child Study Journal, 12*, 151–159.
Field, T., De Stefano, L., & Koewler, J. H. (1982). Fantasy play of toddlers and preschoolers. *Developmental Psychology, 18*, 503–508.
Field, T., Gewirtz, J. L., Cohen, D., Garcia, R., Greenberg, R., & Collins, K. (1984). Leave-taking and reunions of infants, toddlers, preschoolers, and their parents. *Child Development, 55*, 628–635.
Field, T., Vega-Lahr, N., Goldstein, S., & Scafidi, F. (1987). Face-to-face interaction behavior across early infancy. *Infant Behavior and Development, 10*, 111–116.
Field, T. M., Vega-Lahr, N., Scafidi, F., & Goldstein, S. (1986). Effects of maternal unavailability on mother–infant interactions. *Infant Behavior and Development, 9*, 473–478.
Fiese, B. H. (1990). Playful relationships: A contextual analysis of mother–toddler interaction and symbolic play. *Child Development, 61*, 1648–1656.
Fifer, F. C. (1987). The adoption of bipedalism by the hominids: A new hypothesis. *Human Evolution, 2*, 135–147.
Findji, F. (1993). Attentional abilities and maternal scaffolding in the first year of life. *International Journal of Psychology, 28*, 681–692.
Fine, G. A. (1988). Good children and dirty play. *Play and Culture, 1*, 43–56.
Fink, R. S. (1976). Role of imaginative play in cognitive development. *Psychological Reports, 39*, 895–906.
Finkelstein, N., Dent, C., Gallagher, K., & Ramey, C. (1977). Social behavior of infants and toddlers in a daycare environment. *Developmental Psychology, 14*, 257–262.
Finley, G. E., & Layne, O. (1971). Play behavior in young children: A cross-cultural study. *Journal of Genetic Psychology, 119*, 203–210.
Finnie, V., & Russell, A. (1988). Preschool children's social status and their mother's behavior and knowledge in the supervisory role. *Developmental Psychology, 24*, 789–801.
Fischer, R. B., & Nadler, R. D. (1978). Affiliative, playful, and homosexual interactions of adult female lowland gorillas. *Primates, 19*, 657–664.
Fish, M., Stifter, C. A., & Belsky, J. (1993). Early patterns of mother–infant dyadic interaction: Infant, mother, and family demographic antecedents. *Infant Behavior and Development, 16*, 1–18.
Fishbein, H. D., & Imai, S. (1993). Preschoolers select playmates on the basis of gender and race. *Journal of Applied Developmental Psychology, 14*, 303–316.
Fisher, E. D. (1992). The impact of play on development: A meta-analysis. *Play and Culture, 5*, 159–181.
Flannery, K. A., & Watson, M. W. (1993). Are individual differences in fantasy play related to peer acceptance levels? *Journal of Genetic Psychology, 154*, 407–416.
Fling, S., Smith, L., Rodriguez, T., Thorton, D., Atkins, E., & Nixon, K. (1992). Videogames, aggression, and self-esteem: A survey. *Social Behavior and Personality, 20*, 39–46.
Fogel, A. (1977). Temporal organization in mother–infant face-to-face interaction. In H. R. Schaffer (Ed.), *Studies in mother–infant interaction* (pp. 119–151). London: Academic.
Fogel, A. (1982). Social play, positive affect, and coping skills in the first six months of life. *Topics in Early Childhood Special Education, 2*, 53–65.

Fogel, A., & Hannan, T. E. (1985). Manual actions of 2- and 3-month-old human infants during social interaction. *Chilc Development*, *56*, 1271–1279.

Fogel, A., Toda, S., & Kawai, M. (1988). Mother–infant face-to-face interaction in Japan and the United States: A laboratory comparison using 3-month-old infants. *Child Development*, *24*, 398–406.

Fontaine, R. P. (1994). Play as physical flexibility training in five ceboid primates. *Journal of Comparative Psychology*, *108*, 203–212.

Fonzi, A., Schneider, B. H., Tani, F., & Tomada, G. (1997). Predicting children's friendship status from their dyadic interaction in structured situations of potential conflict. *Child Development*, *68*, 496–506.

Forbes, D., & Yablick, G. (1984). The organization of dramatic content in children's fantasy play. In F. Kessel & A. Goncu (Eds.), *Analyzing children's play dialogues* (pp. 23–36). San Francisco: Jossey-Bass.

Fox, M. W. (1969). Ontogeny of prey-killing behavior in Canidae. *Behaviour*, *35*, 259–272.

Fragaszy, D. M., & Adams-Curtis, L. E. (1991). Generative aspects of manipulation in tufted capuchin monkeys (*Cebus apella*). *Journal of Comparative Psychology*, *105*, 387–397.

Fragaszy, D. M., Baer, J., & Adams-Curtis, L. (1991). Behavioral development and maternal care in tufted capuchins (*Cebus apella*) and squirrel monkeys (*Saimiri sciureus*) from birth through seven months. *Developmental Psychobiology*, *24*, 375–393.

Fragaszy, D. M., Vitale, A. F., & Richie, B. (1994). Variation among juvenile capuchins in social influences on exploration. *American Journal of Primatology*, *32*, 249–260.,

Frankel, D. G., & Arbel, T. (1980a). Group formation by two-year-olds. *International Journal of Behavioral Development*, *3*, 287–298.

Frankel, D. G., & Arbel, T. (1980b). Social organization of young children's groups: A comparison of two models of dominance relations. *Psychological Reports*, *46*, 911–915.

Frankel, M., & Rollins, H. A. (1983). Does mother know best? Mothers and fathers interacting with preschool sons and daughters. *Developmental Psychology*, *19*, 694–702.

French, D. C., & Waas, G. A. (1987). Social cognitive and behavioral characteristics of peer-rejected boys. *Professional School Psychology*, *2*, 103–112.

Freud, S. (1959). Creative writers and daydreaming. In J. Strachey (Ed.), *The standard edition of the complete works of Sigmund Freud* (Vol. IX, pp. 143–153). London: Hogarth.

Frey, C., & Hoppe-Graff, S. (1994). Serious and playful aggression in Brazilian girls and boys. *Sex Roles*, *30*, 249–268.

Freyberg, J. (1973). Increasing the imaginative play of urban disadvantaged kindergarten children through systematic training. In J. L. Singer (Ed.), *The child's world of make-believe: Experimental studies of imaginative play* (pp. 129–154). New York: Academic.

Friedman, S., Thompson, M. A., Crawley, S., Critilos, A., Drake, D., Iacobbo, M., Rogers, P. P., & Richardson, L. (1976). Mutual visual regard during mother–infant play. *Perceptual and Motor Skills*, *42*, 427–431.

Friedrich, L. K., & Stein, A. H. (1973). Aggressive and prosocial television programs and the natural behavior of preschool children. *Monographs of the Society for Research in Child Development*, *38* (4, Serial No. 151).

Frodi, A., Bridges, L., & Grolnick, W. (1985). Correlates of mastery-related behavior: A short-term longitudinal study of infants in their second year. *Child Development*, *56*, 1291–1298.

Fry, D. P. (1987). Differences between playfighting and serious fighting among Zapotec children. *Ethology and Sociobiology*, *8*, 285–306.

Fry, D. P. (1990). Play aggression among Zapotec children: Implications for the practice hypothesis. *Aggressive Behavior, 16*, 321–340.

Fu, V. R., Kelso, G. B., & Moran, J. D. (1984). The effects of stimulus dimension and mode of exploration on original thinking in preschool children. *Educational and Psychological Measurement, 44*, 431–440.

Furman,W., Rahe, D., & Hartup, W. W. (1979). Rehabilitation of socially-withdrawn preschool children through mixed-age and same-age socialization. *Child Development, 50*, 915–922.

Gaiter, J. L., Morgan, G. A., Jennings, K. D., Harmon, R. J., & Yarrow, L. J. (1982). Variety of cognitively oriented caregiver activities: Relationship to cognitive and motivational functioning at 1 and 3½ years of age. *Journal of Genetic Psychology, 141*, 49–56.

Galda, L., Pellegrini, A. D., & Cox, S. (1989). Preschoolers' emergent literacy: A short term longitudinal study. *Research in the Teaching of English, 23*, 292–309.

Gard, G., & Meier, G. (1977). Social and contextual factors of play behavior of subadult rhesus monkeys. *Primates, 18*, 367–377.

Garvey, C. (1974). Some properties of social play. *Merrill-Palmer Quarterly, 20*, 163–180.

Gaskins, S. (1994). Integrating interpretive and qualitative methods in socialization research. *Merrill-Palmer Quarterly, 40*, 313–333.

Gaskins, S., & Goncu, A. (1992). Cultural variation in play: A challenge to Piaget and Vygotsky. *The Quarterly Newletter of the Laboratory of Comparative Human Cognition, 14*, 31–35.

Geist, V. (1971). *Mountain sheep*. Chicago: University of Chicago Press.

Gentry, R. L. (1974). The development of social behavior through play in the Stellar sea lion. *American Zoologist, 14*, 391–403.

Gergen, M. (1990). Beyond the evil empire: Horseplay and aggression. *Aggressive Behavior, 16*, 381–398.

Gershaw, N. J., & Schwarz, J. C. (1971). The effects of a familiar toy and mothers' presence on exploratory and attachment behaviors in young children. *Child Development, 42*, 1662–1666.

Gesell, A. (1940). *The first five years of life: A guide to the study of the preschool child*. New York: Harper & Brothers.

Ghiaci, G., & Richardson, J. T. (1980). The effects of dramatic play upon cognitive structure and development. *Journal of Genetic Psychology, 136*, 77–83.

Gibson, E. J. (1988). Exploratory behavior in the development of perceiving, acting, and the acquiring of knowledge. *Annual Review of Psychology, 39*, 1–41.

Gibson, E. J., & Walker, A. S. (1984). Development of knowledge of visual-tactual affordances of substance. *Child Development, 55*, 453–461.

Gibson, J. J. (1962). Observations of active touch. *Psychological Review, 69*, 477–491.

Gibson, K. R., & Ingold, T. (Eds.). (1993). *Tools, language, and cognition in human evolution*. Cambridge, England: Cambridge University Press.

Giddings, M., & Halverson, C. F. (1981). Young children's use of toys in home environments. *Family Relations, 30*, 69–74.

Giffin, H. (1984). The coordination of meaning in the creation of a shared make-believe reality. In I. Bretherton (Ed.), *Symbolic play: The development of social understanding* (pp. 73–100). New York: Academic.

Gilmore, J. B. (1966). The role of anxiety and cognitive factors in children's play behavior. *Child Development, 37*, 397–416.

Ginsburg, G. P., & Kilbourne, B. K. (1988). Emergence of vocal alternation in mother–infant interchanges. *Journal of Child Language, 15*, 221–235.

Ginsburg, H. J., Pollman, V. A., & Wauson, M. S. (1977). An ethological analysis of nonverbal inhibitors of aggressive behavior in male elementary school children. *Developmental Psychology, 13,* 417–418.

Girdner, J. B. (1953). An experimental analysis of the behavioral effects of a perceptual consequence unrelated to organic drive states. *American Psychologist, 8,* 354–355.

Glick, B. B., Eaton, G. G., Johnson, D. F., & Worlein, J. (1986). Development of partner preferences in Japanese macaques (*Macaca fuscata*): Effects of gender and kinship during the second year of life. *International Journal of Primatology, 7,* 467–479.

Glickman, S. E., & Sroges, R. W. (1966). Curiosity in zoo animals. *Behaviour, 27,* 151–187.

Goedeking, P., & Immelmann, K. (1986). Vocal cues in cotton-top tamarin play vocalizations. *Ethology, 73,* 219–224.

Gold, K. C. (1992). Nonsocial behavior of captive infant gorillas. *American Journal of Primatology, 26,* 65–72.

Goldberg, S. (1972). Infant care and growth in urban Zambia. *Human Development, 15,* 77–89.

Goldberg, S. (1977). Infant development and mother–infant interaction in urban Zambia. In P. H. Leiderman, S. R. Tulkin, & A. Rosenfeld (Eds.), *Culture and infancy: Variations in the human experience* (pp. 211–243). New York: Academic.

Goldberg, S., & Lewis, M. (1969). Play behavior in the year-old infant: Early sex differences. *Child Development, 40,* 21–32.

Goldman, B. D., & Ross, H. S. (1978). Social skills in action: An analysis of early peer games. In J. Glick & K. A. Clarke-Stewart (Eds.), *The development of social understanding* (pp. 177–212). New York: Gardner.

Goldman, J. (1981). The social participation of preschool children in same-age versus mixed-age groupings. *Child Development, 52,* 644–650.

Goldsmith, H. H., & Alansky, J. A. (1987). Maternal and infant temperamental predictors of attachment: A meta-analytic review. *Journal of Consulting and Clinical Psychology, 55,* 805–816.

Goldstein, J. (1995). Aggressive toy play. In A. D. Pellegrini (Ed.), *The future of play theory: A multidisciplinary inquiry into the contributions of Brian Sutton-Smith* (pp. 127–147). Albany: State University of New York Press.

Goldstein, S., Field, T., & Healy, B. (1989). Concordance of play behavior and physiology in preschool friends. *Journal of Applied Developmental Psychology, 10,* 337–351.

Golinkoff, R. M., & Ames, G. F. (1979). A comparison of fathers' and mothers' speech with their young children. *Child Development, 50,* 28–40.

Golomb, C., & Bonen, S. (1981). Playing games of make-believe: The effectiveness of symbolic play training with children who failed to benefit from each conservation training. *Genetic Psychology Monographs, 104,* 137–159.

Golomb, C., & Cornelius, C. B. (1977). Symbolic play and its cognitive significance. *Developmental Psychology, 13,* 246–252.

Golomb, C., Gowing, E. D. G., & Friedman, L. (1982). Play and cognition: Studies of pretense play and conservation of quantity. *Journal of Experimental Child Psychology, 33,* 257–279.

Gomendio, M. (1988). The development of different kinds of play in gazelles: Implications for the nature and functions of play. *Animal Behaviour, 36,* 825–836.

Gomez, J. C. (1996). Ostensive behavior in great apes: The role of eye contact. In A. E. Russon, K. A. Bard, & S. T. Parker (Eds.), *Reaching into thought: The minds of the great apes* (pp. 131–151). Cambridge, England: Cambridge University Press.

Goncu, A. (1993). Development of intersubjectivity in the dyadic play of preschoolers. *Early Childhood Research Quarterly, 8*, 99–116.
Goncu, A., & Kessel, F. (1984). Children's play: A contextual-functional perspective. *New Directions for Child Development, 25*, 5–22.
Goncu, A., & Kessel, F. (1988). Preschoolers' collaborative construction in planning and maintaining imaginative play. *International Journal of Behavioral Development, 11*, 327–344.
Goncu, A., Mistry, J., & Mosier, C. (1991, April). *Cultural variations in the play of toddlers*. Paper presented at the Society for Research in Child Development, Seattle.
Gonzalez-Kirchner, J. P., & Sainz-de-la-Maza, M. (1992). Sticks used by wild chimpanzees: A new locality in Rio Muni. *Folia Primatologica, 58*, 99–102.
Goodfader, R. A. (1982). Sex differences in the play constructions of preschool children. *Smith College Studies in Social Work, 52*, 129–144.
Goodnow, J. J. (1969). Effects of active handling, illustrated by uses for objects. *Child Development, 40*, 201–212.
Gottfried, A. W., Rose, S. A., & Bridges, W. H. (1978). Effects of visual, haptic, and manipulatory experiences on infants' visual recognition memory of objects. *Developmental Psychology, 14*, 305–312.
Gottman, J. M. (1977). Toward a definition of social isolation in children. *Child Development, 48*, 513–517.
Gottman, J. M. (1983). How children become friends. *Monographs of the Society for Research in Child Development, 48* (3, Serial No. 201).
Gottman, J. M. (1986). The world of coordinated play: Same- and cross-sex friendship in young children. In J. Gottman & J. G. Parker (Eds.), *Conversations of friends: Speculation on affective development* (pp. 139–191). Cambridge, England: Cambridge University Press.
Gottman, J. M., Gonso, J., & Rasmussen, B. (1975). Social interaction, social competence, and friendship in children. *Child Development, 46*, 709–718.
Gottman, J. M., & Katz, L. F. (1989). Effects of marital discord on young children's peer interactions and health. *Developmental Psychology, 25*, 373–381.
Gottman, J. M., & Ringland, J. T. (1981). The analysis of dominance and bidirectionality in social development. *Child Development, 52*, 393–412.
Gove, D., & Burghardt, G. M. (1983). Context-correlated parameters of snake and lizard tongue-flicking. *Animal Behaviour, 31*, 718–723.
Govindarajulu, P., Hunte, W., Vermeer, L. A., & Horrocks, J. A. (1993). The ontogeny of social play in a feral troop of vervet monkeys (*Cercopithecus aethiops sabaeus*): The function of early play. *International Journal of Primatology, 14*, 701–719.
Grant, E. C. (1969). Human facial expression. *Man, 4*, 525–536.
Graul, S. K., & Zeece, P. D. (1990). Effects of play training of adults on the cognitive development and play behavior of preschool children. *Early Child Development and Care, 57*, 15–22.
Graves, P. L. (1978). Infant behavior and maternal attitudes: Early sex differences in West Bengal, India. *Journal of Cross-Cultural Psychology, 9*, 45–60.
Gray, P., & Feldman, J. (1997). Patterns of age mixing and gender mixing among children and adolescents at an ungraded democratic school, *Merrill-Palmer Quarterly, 43*, 67–86.
Grayhill, D., Strawniak, M., Hunter, T., & O'Leary, M. (1987). Effects of playing versus observing violent versus nonviolent video games on children's aggression. *Psychology: A Quarterly Journal of Human Behavior, 24*, 1–7.

Graziano, A. M., & Diamert, D. M. (1992). Parent behavioral training: An examination of the paradigm. *Behavior Modification, 16*, 3–38.

Green, E. H. (1933). Friendships and quarrels among preschool children. *Child Development, 4*, 237–252.

Green, J. A., Gustafson, G. E., & West, M. J. (1980). Effects of infant development on mother– infant interactions. *Child Development, 51*, 199–207.

Greenberg, R. (1990). Ecological plasticity, neophobia, and resource use in birds. *Studies in Avian Biology, 13*, 431–437.

Greenfield, P. M. (1991). Language, tools, and brain: The ontogeny and phylogeny of hierarchically organized sequential behavior. *Behavioral and Brain Science, 14*, 531–595.

Greenfield, P. M., Nelson, K., & Saltzman, E. (1972). The development of rulebound strategies for manipulating seriated cups. A parallel between action and grammar. *Cognitive Psychology, 3*, 291–310.

Greenfield, P. M., Yut, E., Chung, M., Land, D., Kreider, H., Pantoja, M., & Horsby, K. (1990). The program-length commercial: A study of the effects of television/toy tie-ins on imaginative play. *Psychology and Marketing, 7*, 237–255.

Greenwood, C. R., Walker, H. M., Todd, N. M., & Hops, H. (1981). Normative and descriptive analysis of preschool free play social interaction rates. *Journal of Pediatric Psychology, 6*, 343–367.

Gresham, F. M. (1982). Social interactions as predictors of children's likability and friendship patterns: A multiple regression analysis. *Journal of Behavioral Assessment, 4*, 39–53.

Grieser, D. L., & Kuhl, K. (1988). Maternal speech to infants in a tonal language: Support for universal prosodic features in motherese. *Developmental Psychology, 24*, 14–20.

Griffing, P., Stewart, L. W., McKendry, M. A., & Anderson, R. M. (1983). Sociodramatic play: A follow-up study of imagination, self-concept, and school achievement among Black school children representing two social class groups. *Genetic Psychology Monographs, 107*, 249–301.

Groos, K. (1898). *The play of animals*. New York: Appleton.

Gunderson, V. M., Rose, S. A., & Grant-Webster, K. S. (1990). Cross-modal transfer in high- and low-risk pigtail macaque monkeys. *Developmental Psychology, 26*, 576–581.

Gunnar, M. R., Senior, K., & Hartup, W. W. (1984). Peer presence and the exploratory behavior of eighteen- and thirty-month-old children. *Child Development, 55*, 1103–1109.

Gunnar, M. R., & Stone, C. (1984). The effects of positive maternal affect on infant responses to pleasant, ambiguous, and fear-provoking toys. *Child Development, 55*, 1231–1236.

Guralnick, M. J. (1997). *The effectiveness of early intervention*. Baltimore: Brookes.

Gusella, J. L., Muir, D. W., & Tronick, E. Z. (1988). The effect of manipulating maternal behavior during an interaction on 3- and 6-month-olds' affect and attention. *Child Development, 59*, 1111–1124.

Gustafson, G. E., Green, J. A., & West, M. J. (1979). The infant's changing role in mother–infant games: The growth of social skills. *Infant Behavior and Development, 2*, 301–308.

Guthrie, K., & Hudson, L. M. (1979). Training conservation through symbolic play. *Child Development, 50*, 1269–1271.

Haas, A. (1979). The acquisition of genderlect. In J. Orasanu, M. K. Slater, & L. L. Adler (Eds.), *Annals of the New York Academy of Sciences: Language, sex, and gender, 327*, 101–113.

Haas, A. (1981). Partner influence of sex-associated spoken language of children. *Sex Roles*, *7*, 925–935.

Haemig, P. D. (1989). A comparative study of exploratory behavior in Santa Cruz Island and mainland California scrub jays, *Apelocoma coerulescens*. *Bird Behavior*, *8*, 38–42.

Haight, W. L., Masiello, T., Dickson, K. L., Huckeby, E., & Black, J. E. (1994). The everyday contexts and social contexts of spontaneous mother–child pretend play in the home. *Merrill-Palmer Quarterly*, *40*, 509–522.

Haight, W. L., & Miller, P. J. (1992). The development of everyday pretend play: A longitudinal study of mothers' participation. *Merrill-Palmer Quarterly*, *38*, 331–349.

Haight, W. L., & Miller, P. J. (1993). *Pretending at home: Early development in a sociocultural context*. Albany: State University of New York Press.

Hains, S. M., & Muir, D. W. (1996a). Effects of stimulus contingency in infant–adult interactions. *Infant Behavior and Development*, *19*, 49–61.

Hains, S. M., & Muir, D. W. (1996b). Infant sensitivity to adult eye direction. *Child Development*, *67*, 1940–1951.

Hall, G. S. (1904). *Adolescence: Its psychology and its relation to physiology, anthropology, sex, crime, religion, and education* (Vol. 1). New York: Appleton.

Halliday, S., & Leslie, J. C. (1986). A longitudinal semi-cross-sectional study of the development of mother–child interaction. *British Journal of Developmental Psychology*, *4*, 211–222.

Halliday, J., & McNaughton, S. (1982). Sex differences in play at kindergarten. *New Zealand Journal of Educational Studies*, *17*, 161–170.

Halliday-Scher, K., Urberg, K. A., & Kaplan-Estrin, M. (1995). Learning to pretend: Preschoolers' use of meta-communication in sociodramatic play. *International Journal of Behavioral Development*, *18*, 451–461.

Hallock, M. B., & Worobey, J. (1984). Cognitive development in chimpanzee infants (*Pan troglodytes*). *Journal of Human Evolution*, *13*, 441–447.

Halverson, C., & Waldrop, M. (1973). The relations of mechanically recorded activity level to varieties of preschool play behavior. *Child Development*, *44*, 678–681.

Hamilton, W. J. III., Buskirk, R. E., & Buskirk, W. H. (1975). Defensive stoning by baboons. *Nature*, *256*, 488–489.

Hampson, J., & Nelson, K. (1993). The relation of maternal language to variation in the rate and style of language acquisition. *Journal of Child Language*, *20*, 313–342.

Hanfmann, E. (1935). Social structure of a group of kindergarten children. *American Journal of Orthopsychiatry*, *5*, 407–410.

Hannan, T. E. (1987). A cross-sectional assessment of the occurrences of pointing in three- to twelve-month-old infants. *Infant Behavior and Development*, *10*, 11–22.

Harcourt, R. (1991). The development of play in the South American fur seal. *Ethology*, *88*, 191–202.

Harding, C. G., & Golinkoff, R. M. (1979). The origins of intentional vocalizations in prelinguistic infants. *Child Development*, *50*, 33–40.

Hardy-Brown, K., Plomin, R., & De Fries, J. C. (1981). Genetic and environmental influences on rate of communicative development in the first year of life. *Developmental Psychology*, *17*, 704–717.

Harkness, S., & Super, C. M. (1985). The cultural context of gender segregation in children's peer groups. *Child Development*, *56*, 219–224.

Harlow, H. F. (1950). Learning and satiation of response in intrinsically motivated complex puzzle performance by monkeys. *Journal of Comparative and Physiological Psychology*, *43*, 289–294.

Harlow, H. F. (1962). The heterosexual affection system in monkeys. *American Psychologist, 17*, 1–9.

Harlow, H. F. (1965). Sexual behavior in rhesus monkeys. In F. A. Beach (Ed.), *Sex and behavior* (pp. 234–265). New York: Wiley.

Harlow, H. F., Blazek, N. C., & McClearn, G. E. (1956). Manipulatory motivation in the infant rhesus monkey. *Journal of Comparative and Physiological Psychology, 49*, 444–448.

Harlow, H. F., Harlow, M. K., & Meyer, D. R. (1950). Learning motivated by a manipulation drive. *Journal of Experimental Psychology, 40*, 228–234.

Harper, L., & Huie, K. (1985). The effects of prior group experience, age, and familiarity on the quality and organization of preschoolers' social relationships. *Child Development, 56*, 704–717.

Harris, M. R. (1985). Peer interactional abilities of children from four age-groups. *Folia Phoniatrica, 37*, 119–128.

Harris, M., Jones, D., & Grant, J. (1983). The nonverbal context of mothers' speech to infants. *First Language, 4*, 21–30.

Harris, P. L., & Kavanaugh, R. D. (1993). Young children's understanding of pretense. *Monographs of the Society for Research in Child Development, 58* (1, Serial No. 231).

Harrist, A. W., Zaia, A. F., Bates, J. E., Dodge, K. A., & Pettit, G. S. (1997). Subtypes of social withdrawal in early childhood: Sociometric status and social-cognitive differences across four years. *Child Development, 68*, 278–294.

Hart, B. L., & Hart, L. A. (1994). Fly switching by Asian elephants: Tool use to control parasites. *Animal Behaviour, 48*, 35–45.

Hart, C. H., DeWolf, D. M., Wozniak, P., & Burts, D. C. (1992). Maternal and paternal disciplinary styles: Relations with preschoolers' playground behavioral orientations and peer status. *Child Development, 63*, 879–892.

Hartup, W. (1974). Aggression in childhood: Developmental perspectives. *American Psychologist, 29*, 336–341.

Hartup, W. W. (1983). Peer relations. In E. M. Hetherington (Ed.), *Socialization, personality, and social development* (pp. 103–196). New York: Wiley.

Hartup, W. W., Glazer, J. A., & Charlesworth, R. (1967). Peer reinforcement and sociometric status. *Child Development, 38*, 1017–1024.

Hartup, W. W., & Keller, E. D. (1960). Nurturance and its relation to dependency. *Child Development, 31*, 681–690.

Hartup, W., Laursen, B., Stewart, H., & Eastenson, A. (1988). Conflict and friendship relations of young children. *Child Development, 59*, 1590–1610.

Haslett, B. J. (1983). Communicative functions and strategies in children's conversations. *Human Communciation Research, 9*, 114–129.

Hass, C. C., & Jenni, D. A. (1993). Social play among juvenile bighorn sheep. Structure, development, and relationship to adult behavior. *Ethology, 93*, 105–116.

Hatano, G., Miyake, K., & Tajima, N. (1980). Mother behavior in an unstructured situation and child's acquisition of number conservation. *Child Development, 51*, 379–385.

Hattwick, L. A. (1937). Sex differences in behavior of nursery school children. *Child Development, 8*, 343–355.

Havkin, Z., & Fentress, J. C. (1985). The form of combative strategy in interactions among wolf pups (*Canis lupus*). *Zeitschrift fur Tierpsychologie, 68*, 177–200.

Hay, D. F. (1985). Learning to form relationships in infancy: Parallel attainments with parents and peers. *Developmental Review, 5*, 122–161.

Hay, D. F., Nash, A., & Pedersen, J. (1983). Interaction between six-month-old peers. *Child Development*, *54*, 557–562.
Hay, D. F., & Ross, H. S. (1982). The social nature of early conflict. *Child Development*, *53*, 105–113.
Hayaki, H. (1983). The social interactions of juvenile Japanese monkeys on Koshima Islet. *Primates*, *24*, 139–153.
Hayaki, H. (1985). Social play of juvenile and adolescent chimpanzees in the Mahale Mountanis National Park, Tanzania. *Primates*, *26*, 343–360.
Haywood, K. M. (1986). *Life span motor development*. Champaign, IL: Human Kinetics Publishers.
Hazen, N. L., & Black, B. (1989). Preschool peer communication skills: The role of social status and interaction context. *Child Development, 60*, 867–876.
Heidecker, J. L., & Leuthold, R. H. (1984). The organization of collective foraging in the harvester termite *Hodotermes mossambicus (Isoptera)*. *Behavioral Ecology and Sociobiology*, *14*, 195–202.
Henderson, B. B. (1984). Social support and exploration. *Child Development*, *55*, 1246–1251.
Henderson, B. B. (1988). Individual differences in exploration: A replication and extension. *Journal of Genetic Psychology*, *149*, 555–557.
Henderson, B. B. (1991). Describing parent–child interaction during exploration: Situation definitions and negotiations. *Genetic, Social, and General Psychology Monographs*, *117*, 77–89.
Henderson, B. B., Charlesworth, W. R., & Gamradt, J. (1982). Children's exploratory behavior in a novel field setting. *Ethology and Sociobiology*, *3*, 93–99.
Henderson, B. B., & Moore, S. G. (1979). Measuring exploratory behavior in young children: A factor analytic study. Developmental Psychology, *15*, 113–119.
Henderson, B. B., & Wilson, S. E. (1991). Intelligence and curiosity in preschool children. *Journal of School Psychology*, *29*, 167–175.
Henry, J. D., & Herrero, S. M. (1974). Social play in the American black bear: Its similarity to Canid social play and an examination of its identifying characteristics. *American Zoologist*, *14*, 371–389.
Henschel, J. R. (1995). Tool use by spiders: Stone selection and placement by corolla spiders *Ariadna* (Segestrilidae) of the Namib desert. *Ethology*, *101*, 187–199.
Herrera, C., & Dunn, J. (1997). Early experiences with family conflict: Implications for arguments with a close friend. *Developmental Psychology, 33,* 869–881.
Hertz-Lazarowitz, R., Feitelson, D., Zahavi, S., & Hartup, W. (1981). Social interaction and social organization of Israel's five- to seven-year-olds. *International Journal of Behavioral Development*, *4*, 143–155.
Hess, R. D., & Shipman, V. C. (1965). Early experience and the socialization of cognitive modes in children. *Child Development*, *36*, 869–886.
Hewlett, B. S. (1987). Intimate fathers: Patterns of paternal holding among Aka pygmies. In M. E. Lamb (Ed.), *The father's role: Cross-cultural perspective* (pp. 295–330). Hillsdale, NJ: Lawrence Erlbaum Associates.
Hewlett, B. (1992a). *Father–child relations: Cultural and biosocial contexts*. New York: De Gruyter.
Hewlett, B. S. (1992b). Husband–wife reciprocity and the father–infant relationship among Aka pygmies. In B. S. Hewlett (Ed.), *Father–child relations: Cultural and biosocial contexts* (pp. 153–176). New York: De Gruyter.

Higley, J. D., & Suomi, S. J. (1986). Parental behavior in non-human primates. In W. Sluckin & M. Herbert (Eds.), *Parental behavior* (pp. 152–207). Oxford, England: Basil Blackwell.

Hill, H. L., & Bekoff, M. (1977). The variability of some motor components of social play and agonistic behavior in infant eastern coyotes, *Canis latrans var. Animal Behaivour*, *25*, 907–909.

Hinde, R. A., & Spencer-Booth, Y. (1967). The behavior of socially-living rhesus monkeys in the first two and a half years. *Animal Behaviour*, *15*, 169–196.

Hinde, R. A., Tamplin, A., & Barrett, J. (1993). A comparative study of relationship structure. *British Journal of Social Psychology*, *32*, 191–207.

Hinde, R. A., Titmus, G., Easton, D., & Tamplin, A. (1985). Incidence of "friendship" and behavior toward strong associates versus nonassociates in preschoolers. *Child Development*, *56*, 234–245.

Hirsch, J. (1976). Multiple review of Wilson's *Sociobiology*. *Animal Behaviour*, *24*, 707–709.

Hladik, E. G., & Edwards, H. T. (1984). A comparative analysis of mother–father speech in the naturalistic home environment. *Journal of Psycholinguistic Research*, *13*, 321–332.

Hobson, R. P. (1990). On acquiring knowledge about people and the capacity to pretend: Response to Leslie (1987). *Psychological Review*, *97*, 114–121.

Hodapp, R. M., Goldfield, E. C., & Boyatzis, C. J. (1984). The use and effectiveness of maternal of scaffolding in mother–infant games. *Child Development*, *55*, 772–781.

Hoff, M. P., & Nadler, R. D., & Maple, T. L. (1981). The development of infant play in a captive group of lowland gorillas (*Gorilla gorilla gorilla*). *American Journal of Primatology*, *1*, 65–72.

Hoff-Ginsberg, E. (1985). Relations between discourse properties of mothers' speech and their children's syntactic growth. *Journal of Child Language*, *12*, 367–385.

Hoff-Ginsberg, E. (1986). Function and structure in maternal speech: Their relation to the child's development of syntax. *Developmental Psychology*, *22*, 155–163.

Hogbin, H. I. (1943). A New Guinea infancy. *Oceania*, *13*, 285–309.

Hogbin, H. I. (1946). A New Guinea childhood: From weaning till the eighth year in Wogeo. *Oceania*, *16*, 275–354.

Hold, B. C. L. (1976). Attention structure and rank specific behavior in preschool children. In M. R. A. Chance & R. R. Larsen (Eds.), *The social structure of attention* (pp. 177–201). London: Wiley.

Hold, B. C. (1980). Attention structure and behavior G/wi San children. *Ethology & Sociobiology*, *1*, 275–290.

Hold-Cavell, B. C. L. (1985). Showing-off and aggression in young children. *Aggressive Behavior, 11*, 303–314.

Hold-Cavell, B. C., & Attili, G., & Schleidt, M. (1986). A cross-cultural comparison of children's behaviour during their first year in a preschool. *International Journal of Behavioral Development*, *9*, 471–483.

Hole, G. (1988). Temporal features of social play in the laboratory rat. *Ethology*, *78*, 1–20.

Hole, G. (1991). Proximity measures of social play in the laboratory rat. *Developmental Psychobiology*, *24*, 117–133.

Holmberg, A. R. (1969). *Nomads of the long bow: The Siriono of Eastern Bolivia*. Garden City, NY: Natural History Press.

Holmberg, M. C. (1980). The development of social interaction patterns from 12 to 42 months. *Child Development*, *51*, 448–456.

Hooven, C., Gottman, J. M., & Katz, L. F. (1995). Parental meta-emotion structure predicts family and child outcomes. *Cognition and Emotion, 9*, 229–264.

Hopkins, W. D., Bard, K. A., Jones, A., & Bales, S. L. (1993). Chimpanzee hand preferences in throwing and infant cradling: Implications for the origin of human handedness. *Current Anthropology, 34*, 786–790.

Hovell, M., Bursick, J., Sharkey, R., & McClure, J. (1978). An evaluation of elementary students' voluntary physical activity during recess. *Research Quarterly for Exercise and Sport, 49*, 460–474.

Howe, N., & Ross, H. S. (1990). Socialization, perspective-taking, and the sibling relationship. *Developmental Psychology, 26*, 160–165.

Howes, C. (1985). Sharing fantasy: Social pretend play in toddlers. *Child Development, 56*, 1253–1258.

Howes, C. (1987). Social competence with peers in young children: Developmental consequences. *Developmental Review, 7*, 252–272.

Howes, C. (1988a). Peer interaction of young children. *Monographs of the Society for Research in Child Development, 53* (1, Serial No. 217).

Howes, C. (1988b). Same- and cross-sex friends: Implications for interaction and social skills. *Early Childhood Research Quarterly, 3*, 21–37.

Howes, C., Droege, K., & Matheson, C. C. (1994). Play and communicative processes within long- and short-term friendship dyads. *Journal of Social and Personal Relationships, 11*, 401–410.

Howes, C., & Farver, J. (1987). Social pretend play in 2-year-olds: Effects of age of partner. *Early Childhood Research Quarterly, 2*, 305–314.

Howes, C., & Matheson, C. C. (1992). Sequences in the development of competent play with peers. Social and social pretend play. *Developmental Psychology, 28*, 961–974.

Howes, C., & Stewart, P. (1987). Children's play with adults, toys, and peers: An examination of family and child care influences. *Developmental Psychology, 23*, 423–430.

Howes, C., Unger, O., Seidner, L. B. (1989). Social pretend play in toddlers: Parallels with social play and solitary pretend. *Child Development, 60*, 77–84.

Huffman, M. A. (1984). Stone play of *Macaca fuscata* in Arashiyama B troop: Transmission of non-adaptive behavior. *Journal of Human Evolution, 13*, 725–735.

Hughes, J. N., & Sullivan, K. A. (1988). Outcome assessment in social skills training with children. *Journal of School Psychology, 26*, 167–183.

Hughes, M. (1978). Sequential analysis of exploration and play. *International Journal of Behavioral Development, 1*, 83–97.

Hughes, M. (1979). Exploration and play revisited: A hierarchical analysis. *International Journal of Behavioral Development, 2*, 215–224.

Hughes, M. (1987). The relationship between symbolic and manipulative (object) play. In D. Gurlitz & J. F. Wohwill (Eds.), *Curiosity, imagination, and play: On the development of spontaneous cognitive and motivational processes* (pp. 247–257). Hillsadle, NJ: Lawrence Erlbaum Associates.

Hughes, M., & Hutt, C. (1979). Heart-rate correlates of childhood activities: Play, exploration, problem-solving, and day-dreaming. *Biological Psychology, 8*, 253–263.

Hull, J. W. (1985). Videogames: Transitional phenomena in adolescence. *Child and Adolescent Social Work*, 106–113.

Hummel, D. D. (1982). Syntactic and conversational characteristics of fathers' speech. *Journal of Psycholinguistic Research, 11*, 465–483.

Humphreys, A. P., & Smith, P. K. (1984). Rough-and-tumble in preschool and playground. In P. K. Smith (Ed.), *Play in animals and humans* (pp. 241–266). Oxford, England: Basil Blackwell.

Humphreys, A. P., & Smith, P. K. (1987). Rough and tumble, friendship, and dominance in schoolchildren: Evidence for continuity and change with age. *Child Development*, *58*, 201–212.

Hunt, J. M. (1965). Intrinsic motivation and its role in psychological development. In D. Levine (Ed.), *Nebraska symposium on motivation* (pp. 189–282). Lincoln: University of Nebraska Press.

Hunter, F. T., McCarthy, M. E., MacTurk, R. H., & Vietze, P. M. (1987). Infants' social-constructive interactions with mothers and fathers. *Developmental Psychology*, *23*, 249–254.

Huston, A. C. (1983). Sex-typing. In E. M. Hetherington (Ed.), *Socialization, personality and social development* (pp. 387–467). New York: Wiley.

Huston, A. C., Carpenter, J. C., Atwater, J. B., & Johnson, L. M. (1986). Gender, adult structuring of activities, and social behavior in middle childhood. *Child Development*, *57*, 1200–1209.

Hutt, C. (1966). Exploration and play in children. *Symposia of the Zoological Society of London*, *18*, 61–81.

Hutt, C. (1967). Effects of stimulus novelty on manipulatory exploration in an infant. *Journal of Child Psychology and Psychiatry*, *8*, 241–247.

Hutt, C. (1970a). Curiosity in young children. *Science Journal*, *6*, 68–72.

Hutt, C. (1970b). Specific and diversive exploration. In H. W. Reese & L.P. Lipsitt (Eds.), *Advances in child development and behavior* (Vol. 5, pp. 119–180). New York: Academic.

Hutt, C., & Bhavnani, R. (1972). Predictions from play. *Nature*, *237*, 171–172.

Huttenlocher, J., Haight, W., Bryk, A., Seltzer, M., & Lyons, T. (1991). Early vocabulary growth: Relation to language input and gender. *Developmental Psychology*, *27*, 236–248.

Idle, T., Wood, E., & Desmarais, S. (1993). Gender role socialization in toy play situations: Mothers and fathers with their sons and daughters. *Sex Roles*, *28*, 679–691.

Iijima, F. (1986). An analysis of preschool children's interpersonal behavior by a longitudinal method. *Japanese Psychological Research*, *28*, 30–40.

Imakawa, S. (1990). Playmate relationships of free-ranging Japanese monkeys at Katsuyama. *Primates*, *31*, 509–521.

Inagaki, K. (1978). Effects of object curiosity on exploratory learning in young children. *Psychological Reports*, *42*, 899–908.

Inglis, I. R. (1983). Towards a cognitive theory of exploratory behavior. In J. Archer & L. Birke (Eds.), *Exploration in animals and humans* (pp. 72–115). Berkshire, England: Van Nostrand Reinhold.

Inglis, I. R., & Shepard, D. S. (1994). Rats work for food they then reject: Support for the information-primary approach to learned industriousness. *Ethology*, *98*, 154–164.

Ingram, D. (1995). The cultural basis of prosodic modifications to infants and children: A response to Fernald's universalist theory. *Journal of Child Language*, *22*, 223–233.

Ironsmith, S., & Poteat, G. M. (1990). Behavioral correlates of preschool sociometric status and the prediction of teacher ratings of behavior in kindergarten. *Journal of Clilnical Child Psychology*, *19*, 17–25.

Islam, M. A., & Husain, K. Z. (1982). A preliminary study on the ecology of the capped langur. *Folia Primatologica*, *39*, 145–159.

Jacklin, C. N., Di Pietro, J. A., & Maccoby, E. E. (1984). Sex-typing behavior and sex-typing pressure in child/parent interaction. *Archives of Sexual Behavior, 13*, 413–425.
Jacklin, C. N., & Maccoby, E. E. (1978). Social behavior at thirty three months in same-sex and mixed-sex dyads. *Child Development, 49*, 557–569.
Jacobvitz, D., & Sroufe, L. A. (1987). The early caregiver–child relationship and attention-deficit disorder with hyperactivity in kindergarten: A prospective study. *Child Development, 58*, 1496–1504.
Jackowitz, E. R., & Watson, M. W. (1980). Development of object transformations in early pretend play. *Developmental Psychology, 16*, 543–549.
Jackson, E. (1964). Native toys of the Guarayu Indians. *American Anthropologist, 66*, 1153–1155.
Jacobson, J. L. (1981). The role of inanimate objects in early peer interaction. *Child Development, 52*, 618–626.
Jamieson, S. H., & Armitage, K. B. (1987). Sex differences in the play behavior of yearling yellow-bellied marmots. *Ethology, 74*, 237–253.
Janus, M. (1989). Reciprocity in play, grooming, and proximity in sibling and nonsibling young rhesus monkeys. *International Journal of Primatology, 10*, 243–261.
Jennings, K. D. (1975). People versus object orientation, social behavior, and intellectual abilities in preschool children. *Developmental Psychology, 11*, 511–519.
Jennings, K. D., & Connors, R. E. (1989). Mothers' interactional style and children's competence at 3 years. *International Journal of Behavioral Development, 12*, 155–175.
Jennings, K. D., Harmon, R. J., Morgan, G. A., Gaiter, J. L., & Yarrow, L. J. (1979). Exploratory play as an index of mastery motivation: Relationships to persistence, cognitive functioning, and environmental measures. *Developmental Psychology, 15*, 386–394.
Johns, C., & Endsley, R. C. (1977). The effects of a maternal model on young children's tactual curiosity. *Journal of Genetic Psychology, 131*, 21–28.
Johnson, D., & Brody, N. (1977). Visual habituation, sensorimotor development, and tempo of play in one-year-old infants. *Child Development, 48*, 315–319.
Johnson, H. M. (1974). The art of block building. In E. S. Hirsch (Ed.), *The block book* (pp. 9–24). Washington, DC: National Association for the Education of Young Children.
Johnson, J. E. (1976). Relations of divergent thinking and intelligence to test scores with social and nonsocial make-believe play of preschool children. *Child Development, 47*, 1200–1203.
Johnson, J. E. (1978). Mother–child interaction and imaginative behavior of preschool children. *Journal of Psychology, 100*, 123–129.
Johnson, J. E., & Ershler, J. (1981). Developmental trends in preschool play as a function of classroom program and child gender. *Child Development, 52*, 995–1004.
Johnson, J. E., Ershler, J., & Lawton, J. T. (1982). Intellective correlates of preschoolers' spontaneous play. *Journal of General Psychology, 106*, 115–122.
Jones, A., & Glenn, S. M. (1991). Gender differences in pretend play in a primary school group. *Early Child Development and Care, 72*, 61–67.
Jones, B. (1972). Facilitation of visual perception through voluntary movement in elementary school children. *Journal of Experimental Child Psychology, 14*, 408–415.
Jones, D. C. (1984). Dominance and affiliation as factors in the social organization of same-sex groups of elementary school children. *Ethology and Sociobiology, 5*, 193–202.
Joseph, R. M. (1998). Intention and knowledge in preschoolers' conception of pretend. *Child Development, 69*, 966–980.

Kagan, J., Kearsley, R. B., & Zelazo, P. R. (1978). *Infancy: Its place in human development*. Cambridge, MA: Harvard University Press.

Kagan, J., & Klein, R. E. (1973). Cross-cultural perspectives on early development. *American Psychologist*, *28*, 947–961.

Kagan, J., Lapidus, D. R., & Moore, M. (1978). Infant antecedents of cognitive functioning: A longitudinal study. *Child Development*, *49*, 1005–1023.

Kahen, V., Katz, L. F., & Gottman, J. M. (1994). Linkages between parent–child interaction and conversations of friends. *Social Development*, *3*, 238–254.

Kaiser, J., Snyder, T. D., & Rogers, C. S. (1995). Adult choice of toys affects children's prosocial and antisocial behavior. *Early Child Development and Care*, *111*, 181–193.

Kaitz, M., Zvi, H., Levy, M., Berger, A., & Eidelman, A. I. (1995). The uniqueness of mother–own infant interactions. *Infant Behavior and Development*, *18*, 247–252.

Kaplan, H., & Dove, H. (1987). Infant development among the Ache of Eastern Paraguay. *Developmental Psychology*, *23*, 190–198.

Karniol, R. (1989). The role of manual manipulative stages in the infant's acquisition of perceived control over objects. *Developmental Review*, *9*, 205–233.

Karpoe, K. P., & Olney, R. L. (1983). The effect of boys' and girls' toys on sex-typed play in preadolescents. *Sex Roles*, *9*, 507–518.

Katz, G. S., Cohn, J. F., & Moore, C. A. (1996). A combination of vocal f_0 dynamic and summary features discriminates between three pragmatic categories of infant-directed speech. *Child Development*, *67*, 205–217.

Kavanaugh, R. D., & Jirkovsky, A. M. (1982). Parental speech to young children: A longitudinal analysis. *Merrill-Palmer Quarterly*, *28*, 297–311.

Kavanaugh, R. D., Whittington, S., & Cerbone, M. J. (1983). Mothers' use of fantasy in speech to young children. *Journal of Child Language*, *10*, 45–55.

Kaye, K. (1977). Toward the origin of dialogue. In H. R. Schaffer (Ed.), *Studies in mother–infant interaction* (pp. 89–117). London: Academic.

Kaye, K. (1980). Why we don't talk "baby talk" to babies. *Journal of Child Langauge*, *7*, 489–507.

Kaye, K., & Fogel, A. (1980). The temporal structure of face-to-face communication between mothers and infants. *Developmental Psychology*, *16*, 454–464.

Keating, C. F., & Bai, D. L. (1986). Children's attribution of social dominance from facial cues. *Child Development*, *57*, 1269–1276.

Keating, C. F., & Heltman, K. R. (1994). Dominance and deception in children and adults: Are leaders the best misleaders? *Personality and Social Psychology Bulletin*, *20*, 312–321.

Keller, H., Scholmerich, A., Miranda, D., & Gauda, G. (1987). The development of exploratory behavior in the first four years of life. In D. Gorlitz & J. F. Wohwill (Eds.), *Curiosity, imagination, and play: On the development of spontaneous cognitive and motivational processes* (pp. 127–150). Hillsdale, NJ: Lawrence Erlbaum Associates.

Kelly, C. A., & Dale, P. S. (1989). Cognitive skills associated with the onset of multiword utterances. *Journal of Speech and Hearing Research*, *32*, 645–656.

Kemple, K., Speranza, H., & Hazen, N. (1992). Cohesive discourse and peer acceptance: Longitudinal relations in the preschool years. *Merrill-Palmer Quarterly*, *38*, 364–381.

Kendrick, C., & Dunn, J. (1983). Sibling quarrels and maternal responses. *Developmental Psychology*, *19*, 62–70.

Kennedy, J. H. (1992). Relationship of maternal beliefs and childrearing strategies to social competence in preschool children. *Child Study Journal*, *22*, 39–60.

Kenny, D. A., & La Voie, L. (1984). The social relations model. In L. Berkowitz (Ed.), *Advances in experimental social psychology* (Vol. 19, pp. 141–182). New York: Academic.

Kesterbaum, G. I., & Weinstein, L. (1985). Personality, psychopathology, and developmental issues in male adolescent video game use. *Journal of the American Academy of Child Psychiatry*, *24*, 329–337.

Kilbride, P. & Kilbride, J. (1974). Sociocultural factors and early manifestation of sociability behavior among Baganda infants. *Ethos*, *2*, 296–314.

Kiley-Worthington, M. (1976). The tail movements of ungulates, canids, and felids with particular reference to their causation and function as displays. *Behaviour*, *56*, 69–115.

Kimmerle, M., Mick, L. A., & Michel, G. F. (1995). Bimanual role-differentiated toy play during infancy. *Infant Behavior and Development*, *18*, 299–307.

Kinnaman, A. J. (1902). Mental life of two Macacus rhesus monkeys in captivity. I. *American Journal of Psychology*, *13*, 98–148.

Kleiman, D. (1977). Monogamy in mammals. *Quarterly Review of Biology*, *52*, 39–69.

Klein, M. (1955). The psychoanalytic play technique. *American Journal of Orthopsychiatry*, *25*, 223–237.

Kleinman, J. M. (1979). Developmental changes in haptic exploration and matching accuracy. *Developmental Psychology*, *15*, 480–481.

Klesges, R. C., Coates, T. J., Moldenhauer-Klesges, L. M., Holzer, B., Gustavson, J., & Barnes, J. (1984). The FATS: An observational system for assessing physical activity in children and associated parent behavior. *Behavioral Assessment*, *6*, 333–345.

Klesges, R. C., Eck, L., Hanson, C., Haddock, C. K., & Klesges, L. M. (1990). Effects of obesity, social interactions, and physical environment in preschoolers. *Health Psychology*, *9*, 435–449.

Klesges, R. C., Malott, J. M., Boschee, P. F., & Weber, J. M. (1986). The effects of parental influences on children's food intake, physical activity, and relative weight. *International Journal of Eating Disorders*, *5*, 335–346.

Knobloch, H., & Pasamanick, B. (1959). The development of adaptive behavior in an infant gorilla. *Journal of Comparative and Physiological Psychology*, *52*, 699–703.

Knusel, C. J. (1992). The throwing hypothesis and hominid origins. *Human Evolution*, *7*, 1–7.

Koch, H. L. (1933). Popularity in preschool children: Some related factors and a technique for its measurement. *Child Development*, *4*, 164–175.

Kochanska, G. (1992). Children's interpersonal influence with mothers and peers. *Developmental Psychology*, *28*, 491–499.

Kochanska, G. (1997a). Multiple pathways to conscience for children with different temperaments: From toddlerhood to age 5. *Developmental Psychology*, *33*, 228–240.

Kochanska, G. (1997b). Mutually responsive orientation between mothers and their young children: Implications for early socialization. *Child Development*, *68*, 94–112.

Kogan, N., & Carter, A. S. (1996). Mother–infant reengagement following the still-face: The role of maternal emotional availability in infant affect recognition. *Infant Behavior and Development*, *19*, 359–370.

Kohler, W. (1925). *The mentality of apes*. New York: Harcourt & Brace.

Kolominskii, I., & Zhiznevskii, B. P. (1992). A sociopsychological analysis of conflicts among children during play. *Journal of Russian and East European Psychology*, *30*, 72–86.

Konner, M. J. (1972). Aspects of the developmental ethology of a foraging people. In N. Blurton-Jones (Eds.), *Ethological studies of child behavior* (pp. 285–304). Cambridge, England: Cambridge University Press.

Konner, M. J. (1976). Maternal care, infant behavior, and development among the Kung. In R. B. Lee, & I. DeVore (Eds.), *Kalahari hunter–gatherers. Studies of the Kung San and their neighbors* (pp. 218–245). Cambridge, MA: Harvard University Press.

Konner, M. (1977). Infancy among the Kalahari desert San. In P. H. Leiderman, S. R. Tulkin, & A. Rosenfeld (Eds.), *Culture and infancy: Variations in the human experience* (pp. 287–328). New York: Academic.

Kopp, C. B. (1974). Fine motor abilities of infants. *Developmental Medicine and Clinical Neurology, 16*, 629–636.

Kopp, C. B., Khoka, E., & Sigman, M. (1977). A comparison of sensorimotor development among infants in India and the United States. *Journal of Cross-Cultural Psychology, 8*, 435–452.

Kopp, C. B., & Vaughn, B. E. (1982). Sustained attention during exploratory manipulation as a predictor of cognitive competence in preterm infants. *Child Development, 53*, 174–182.

Kortland, A., & Kooij, M. (1963). Protohominid behaviour in primates. In J. Napier & N. A. Barnicot (Eds.), *The primates* (pp. 61–88). London: Symposia of the Zoological Society.

Koslowski, B., & Bruner, J. S. (1972). Learning to use a lever. *Child Development, 43*, 790–799.

Koyama, N. (1985). Playmate relationships among individuals of the Japanese monkey troop on Arashiyama. *Primates, 26*, 390–406.

Krahenbuhl, G. S., Skinner, J. S., & Kohrt, W. M. (1985). Developmental aspects of maximal aerobic power in children. In R. L. Terjung (Ed.), *Exercise science and sports research* (pp. 503–538). New York: Macmillan.

Kramer, L. (1996). What's real in children's fantasy play? Fantasy play across the transition to becoming a sibling. *Journal of Child Psychology and Psychiatry and Allied Disciplines, 37*, 329–337.

Kramer, L., & Gottman, J. M. (1992). Becoming a sibling: "With a little help from my friends." *Developmental Psychology, 28*, 685–699.

Kramer, L., & Schaefer-Hernan, P. (1994). Patterns of fantasy play engagement across the transition to becoming a sibling. *Journal of Child Psychology and Psychiatry and Allied Disciplines, 35*, 749–767.

Kramer, M., & Burghardt, G. M. (1998). Precocious courtship and play in emydid turtles. *Ethology, 104*, 38–56.

Krasnor, L. R., & Rubin, K. H. (1983). Preschool social problem solving: Attempts and outcomes in naturalistic interaction. *Child Development, 54*, 1545–1558.

Kreitler, S., Zigler, E., & Kreitler, H. (1975). The nature of curiosity in children. *Journal of School Psychology, 13*, 185–200.

Kreitler, S., Zigler, E., & Kreitler, H. (1984). Curiosity and demographic factors as determinants of children's probability learning strategies. *The Journal of Genetic Psychology, 145*, 61–73.

Kruper, J. C., & Uzgiris, I. C. (1987). Fathers' and mothers' speech to young infants. *Journal of Psycholinguistic Research, 16*, 597–614.

Kubey, R., & Larson, R. (1990). The use and experience of the new video media among children and young adolescents. *Communication Research, 17*, 107–130.

Kummer, H. (1968). *Social organization of hamadryas baboons: A field study*. Chicago: University of Chicago Press.

Lacey, G. N. (1972). Observations on Aberfan. *Journal of Psychosomatic Research, 16*, 257–260.

Ladd, G. (1983). Social networks of popular, average, and rejected children in school settings. *Merrill-Palmer Quarterly, 29*, 283–307.

Ladd, G. W., & Golter, B. S. (1988). Parents' management of preschoolers' peer relations: Is it related to children's social competence? *Developmental Psychology, 24*, 109–117.

Ladd, G. W., & Hart, C. H. (1992). Creating informal play opportunities: Are parents' and preschoolers' initiations related to children's competence with peers? *Developmental Psychology, 28*, 1179–1187.

Ladd, G. W., & Price, J. M. (1987). Predicting children's social and school adjustment following the transition from preschool to kindergarten. *Child Development, 58*, 1168–1189.

Ladd, G. W., Price, J. M., & Hart, C. H. (1988). Predicting preschoolers' status from their playground behaviors. *Child Development, 59*, 986–992.

La Freniere, P., & Charlesworth, W. R. (1983). Dominance, attention, and affiliation in a preschool group: A nine-month longitudinal study. *Ethology and Sociobiology, 4*, 55–67.

La Freniere, P., Strayer, F. F., & Gauthier, R. (1984). The emergence of same-sex preferences among preschool peers. *Child Development, 55*, 1958–1966.

La Greca, A. M. (1993). Social skills training with children: Where do we go from here? *Journal of Clinical Child Psychology, 22*, 288–298.

Laird, R. D., Pettit, G. S., Mize, J., Brown, E. G., & Lindsey, E. (1994). Mother–child conversations about peers: Contributions to competence. *Family Relations, 43*, 425–432.

Lamb, M. E. (1977). Father–infant and mother–infant interaction in the first year of life. *Child Development, 48*, 167–181.

Lamb, M. E. (1978a). Interactions between eighteen-month-olds and their preschool-aged sibling. *Child Development, 49*, 51–59.

Lamb, M. E. (1978b). The development of sibling relationships in infancy: A short-term longitudinal study. *Child Development, 49*, 1189–1196.

Lamb, M. E. (1987). *The father's role: Cross-cultural perspectives*. Hillsdale, NJ: Lawrence Erlbaum Associates.

Lamb, M. E., Easterbrooks, M. A., & Holden, G. W. (1980). Reinforcement and punishment among preschoolers: Characteristics, effects, and correlates. *Child Development, 51*, 1230–1236.

Lamb, M. E., Frodi, A. M., Frodj, M., & Hwang, C. D. (1982). Characteristics of maternal and paternal behavior in traditional and nontraditional Swedish families. *International Journal of Behavioral Development, 5*, 131–141.

Lamb, M. E., Morrison, D. C., & Malkin, C. M. (1987). The development of infant social expectations in face-to-face interaction: A longitudinal study. *Merrill-Palmer Quarterly, 33*, 241–254.

Lamb, M. E., & Roopnarine, J. L. (1979). Peer influences on sex-role development in preschoolers. *Child Development, 50*, 1219–1222.

Lancaster, J. B. (1971). Play-mothering: The relations between juvenile females and young infants among free-ranging vervet monkeys (*Cercopithecus aethiops*). *Folia Primatologica, 15*, 161–182.

Lancy, D. F. (1977). The play behavior of Kpelle children during rapid cultural change. In D. F. Lancy & B. A. Tindall (Eds.), *The study of play: Problems and prospects* (pp. 84–91). West Point, NY: Leisure.

Lancy, D. F. (1980). Play in species adaptation. *Annual Review of Anthropology, 9*, 471–495.
Landau, R. (1981). Object grasping versus self-stimulation: Findings for a cross-cultural study. *Infant Behavior and Development, 4*, 373–385.
Landerholm, E. J., & Scriven, G. (1981). A comparison of mother and father interaction with their six-month-old male and female infants. *Early Child Development and Care, 7*, 317–328.
Langevin, R. (1971). Is curiosity a unitary construct? *Canadian Journal of Psychology, 25*, 360–374.
Langlois, J., H., Gottfried, N. W., Barnes, B., & Hendricks, D. (1978). Effects of peer age on social behavior of preschool children. *Journal of Genetic Psychology, 132*, 11–19.
Langlois, J. H., Gottfried, N. W., & Seay, B. (1973). The influence of sex of peer on the social behavior of preschool children. *Developmental Psychology, 8*, 93–98.
Lansink, J. M., & Richards, J. E. (1997). Heart rate and behavioral measures of attention in six-, nine-, and twelve-month-old infants during object exploration. *Child Development, 68*, 610–620.
Laosa, L. M. (1978). Maternal teaching strategies in Chicano families of varied educational and socioeconomic levels. *Child Development, 49*, 1129–1135.
Largo, R. H., & Howard, J. A. (1979). Developmental progression of play behavior of children between nine and thirty months: Part 1. Spontaneous play and imitation. *Developmental Medicine and Child Neurology, 21*, 299–310.
Lasky, R. E., Klein, R. E., Yarbrough, L., Sellars, M., & Kagan, J. (1983). Social interactions of Guatemalan infants: The importance of different caregivers. *Journal of Cross-Cultural Psychology, 14*, 17–28.
Latour, B., & Strum, S. C. (1986). Human social origins: Oh please, tell us another story. *Journal of Social and Biological Structures, 9*, 169–187.
Laursen, B., & Hartup, W. W. (1989). The dynamics of preschool children's conflicts. *Merrill-Palmer Quarterly, 35*, 281–297.
Laursen, B., Hartup, W. W., & Koplas, A. L. (1996). Towards understanding peer conflict. *Merrill-Palmer Quarterly, 42*, 76–102.
Lawson, K. R., Parrinello, R., & Ruff, H. A. (1992). Maternal behavior and infant attention. *Infant Behavior and Development, 15*, 209–229.
Lazar, J. W., & Beckhorn, G. D. (1974). Social play or the development of social behavior in ferrets (*Mustela putorius*). *American Zoologist, 14*, 405–414.
Leacock, E. (1971, December). At play in African villages. *Natural History*, pp. 60–65.
Leaper, C. (1991). Influence and involvement in children's discourse: Age, gender, and partner effects. *Child Development, 62*, 797–811.
Leaper, C., Anderson, K. J., & Sanders, P. (1998). Moderators of gender effects on parents' talk to their children: A meta-analysis. *Developmental Psychology, 34*, 3–27.
Leaper, C., Lever, L., Strasser, T., & Schwartz, R. (1995). Mother–child communication sequences: Play activity, child gender, and marital status effects. *Merrill-Palmer Quarterly, 41*, 307–327.
Leavens, D. A., & Hopkins, W. D. (1998). Intentional communication by chimpanzees: A cross-sectional study of the use of referential gestures. *Developmental Psychology, 34*, 813–822.
Lederman, S. J., & Klatsky, R. L. (1987). Hand movements: A window into haptic object recognition. *Cognitive Psychology, 19*, 342–368.
Lee, P. C. (1984). Ecological constraints on the social development of vervet monkeys. *Behaviour, 91*, 245–261.

Legerstee, M. (1991). Changes in the quality of infant sounds as a function of social and nonsocial stimulation. *First Language, 11*, 327–343.

Le Normand, M. T. (1986). A developmental exploration of language used to accompany symbolic play in young, normal children (2-4 years old). *Child: Care, Health, and Development, 12*, 121–134.

Le Resche, L. A. (1976). Dyadic play in hamadryas baboons. *Behaviour, 57*, 190–205.

Lerner, R. M., & von Eye, A. (1992). Sociobiology and human development: Arguments and evidence. *Human Development, 35*, 12–33.

Leslie, A. M. (1987). Pretense and representation: The origins of "theory of mind." *Psychological Review, 94*, 412–426.

Lester, B., Hoffman, J., & Brazelton, T. (1985). The rhythmic structure of mother–infant interaction in term and preterm infants. *Child Development, 56*, 15–27.

Lever, J. (1976). Sex differences in the games children play. *Social Problems, 23*, 479–487.

Lever, J. (1978). Sex differences in the complexity of children's play and games. *American Sociological Review, 43*, 471–483.

Levin, H., & Turgeon, V. F. (1957). The influence of the mother's presence on children's doll play aggression. *Journal of Abnormal and Social Psychology, 55*, 304–308.

Levin, H., & Wardwell, E. (1962). The research use of doll play. *Psychological Bulletin, 59*, 27–56.

LeVine, R. A., Dixon, S., LeVine, S., Richman, A., Leiderman, P. H., Keefer, C. H., & Brazelton, T. B. (1994). *Child care and culture: Lessons from Africa*. Cambridge, England: Cambridge University Press.

LeVine, R. A., & LeVine, B. B. (1963). Nyansongo: A Gusii community in Kenya. In B. B. Whiting (Ed.), *Six cultures: Studies of childrearing* (pp. 15–202). New York: Wiley.

Lewis, C., & Gregory, S. (1987). Parents' talk to their infants: The importance of context. *First Language, 7*, 201–216.

Lewis, M. (1982). Play as whimsy. *The Behavioral and Brain Sciences, 5*, 166.

Lewis, M., & Coates, D. (1980). Mother–infant interaction and cognitive development in twelve-week-old infants. *Infant Behavior and Development, 3*, 95–105.

Lewis, M. D. (1993). Early socio-emotional predictors of cognitive competency at 4 years. *Developmental Psychology, 29*, 1036–1049.

Liberman, P. (1991). *Uniquely human: The evolution of speech, thought, and selfless behavior*. Cambridge, MA: Harvard University Press.

Liddell, C., Kvalsvig, J., Strydom, N., Qotyana, P., & Shabalala, A. (1993). An observational study of 5-year-old South African children in the year before school. *International Journal of Behavioral Development, 16*, 537–561.

Lieberman, J. N. (1965). Playfulness and divergent thinking. An investigation of their relationship at the kindergarten level. *Journal of Genetic Psychology, 107*, 219–224.

Lieven, E. V. M. (1994). Cross-linguistic and cross-cultural aspects of language addressed to children. In C. Galloway & B. Richards (Eds.), *Input and interaction in language acquisition* (pp. 56–73). London: Cambridge University Press.

Lillard, A. S. (1993). Pretend play skills and the child's theory of mind. *Child Development, 64*, 348–371.

Lillard, A. S. (1998). Wanting to be it: Children's understanding of intentions underlying pretense. *Child Development, 69*, 981–993.

Lin, S., & Lepper, M. R. (1987). Correlates of children's usage of videogames and computers. *Journal of Applied Social Psychology, 17*, 72–93.

Lipscomb, T. J., & Coon, R. C. (1983). Parental speech modification to young children. *Journal of Genetic Psychology, 143*, 181–187.

Liss, M. B. (1981). Patterns of toy play: An analysis of sex differences. *Sex Roles, 7,* 1143–1150.

Liss, M. B. (1983). Learning gender-related skills through play. In M. B. Liss (Ed.), *Social and cognitive skills: Sex roles and children's play* (pp. 147–166). New York: Academic.

Lloyd, B., & Goodwin, R. (1993). Girls' and boys' use of directives in pretend play. *Social Development, 2,* 122–130.

Lloyd, B., & Smith, C. (1985). The social representation of gender and young children's play. *British Journal of Developmental Psychology, 3,* 65–73.

Lockheed, M. E. (1986). Reshaping the social order: The case of gender segregation. *Sex Roles, 14,* 617–628.

Loizos, C. (1969). Play behavior in higher primates: A review. In D. Morris (Ed.), *Primate ethology* (pp. 176–218). New York: Doubleday.

Longhurst, T. M., & Stepanich, L. (1975). Mothers' speech addressed to one-, two-, and three-year-old normal children. *Child Study Journal, 5,* 3–11.

Lorenz, K. Z. (1956). Plays and vacuum activities. In M. Autuori (Ed.), *L'instinct dans le comportement des animaux et de l'homme* (pp. 633–645). Paris: Masson et Cie.

Lougee, M. D., Grueneich, R., & Hartup, W. W. (1977). Social interaction in same- and mixed-age dyads of preschool children. *Child Development, 48,* 1353–1361.

Lovinger, S. L. (1974). Sociodramatic play and language development in preschool disadvantaged children. *Psychology in the Schools, 11,* 313–320.

Lowe, M. (1975). Trends in the development of representational play in infants from one to three years: An observational study. *Journal of Child Psychology and Psychiatry, 16,* 33–47.

Lowenstein, G. (1994). The psychology of curiosity: A review and reinterpretation. *Psychological Bulletin, 116,* 75–98.

Luria, Z., & Herzog, E. W. (1991). Sorting gender out in a children's museum. *Gender and Society, 5,* 224–232.

Lytton, H., & Romney, D. M. (1991). Parents' differential socialization of boys and girls: A meta-analysis. *Psychological Bulletin, 109,* 267–296.

Lyytinen, P. (1991a). Developmental trends in children's pretend play. *Child: Care, Health, and Development, 17,* 9–25.

Lyytinen, P. (1991b). Peer interaction in children's dyadic play. *Early Child Development and Care, 71,* 105–115.

Lyytinen, P., Poikkeus, A. M., & Laakso, M. L. (1997). Language and symbolic play in toddlers. *International Journal of Behavioral Development, 21,* 289–302.

Maccoby, E. E. (1966). *The development of sex differences.* Stanford, CA: Stanford University Press.

Maccoby, E. E. (1988). Gender as a social category. *Developmental Psychology, 24,* 755–765.

Maccoby, E. E. (1990). Gender and relationships: A developmental account. *American Psychologist, 45,* 513–520.

Maccoby, E. E. (1998). *The two sexes: Growing up apart, coming together.* Cambridge, MA: Belknap.

Maccoby, E. E., & Jacklin, C. N. (1974). *The psychology of sex differences.* Stanford, CA: Stanford University Press.

Maccoby, E. E., & Jacklin, C. N. (1987). Gender segregation in childhood. In E. H. Reese (Ed.), *Advances in child development and behavior* (Vol. 20, pp. 239–287). New York: Academic.

Maccoby, M., Modiano, N., & Lander, P. (1964). Games and social character in a Mexican village. *Psychiatry, 27*, 150–162.
MacDonald, K. (1987). Parent-child physical play with rejected, neglected, and popular boys. *Developmental Psychology, 23,* 705–711.
MacDonald, K. (1993a). Parent–child play: An evolutionary perspective. In K. MacDonald (Ed.), *Parent–child play: Descriptions and implications* (pp. 113–143). Albany: State University of New York Press.
MacDonald, K. (Ed.). (1993b). *Parent–child play: Descriptions and implications*. Albany: State University of New York Press.
MacDonald, K., & Parke, R. D. (1984). Bridging the gap: Parent–child play and peer interactive competence. *Child Development, 55*, 1265–1277.
MacDonald, K., & Parke, R. D. (1986). Parent–child physical play: The effects of sex and age of children and parents. *Sex Roles, 15*, 367–378.
MacDonald, S. E., & Pinel, J. P. (1991). Information gathering: A component of the defensive behavior of rats and Old-World monkeys. *Psychological Record, 41*, 207–215.
Machida, S. (1990). Standing and climbing a pole by members of a captive group of Japanese monkeys. *Primates, 31*, 291–298.
Madden, J., Levenstein, P., & Levenstein, S. (1976). Longitudinal IQ outcomes of the mother–child home program. *Child Development, 47*, 1015–1025.
Magee, M. A. (1989). Social play as performance. *Play and Culture, 2*, 193–196.
Maguire, M. C., & Dunn, J. (1997). Friendships in early childhood and social understanding. *International Journal of Behavioral Development, 21*, 669–686.
Malatesta, C. Z., Culver, C., Tesman, J. R., & Shepard, B. (1989). The development of emotion expression during the first year of life. *Monographs of the Society for Research in Child Development, 54* (1–2, Serial No. 219).
Malatesta, C. Z., Grigoryev, P., Lamb, C., Albin, M., & Culver, C. (1986). Emotion socialization and expressive development in preterm and full-term infants. *Child Development, 57*, 316–330.
Malatesta, C. Z., & Haviland, J. M. (1982). Learning display rules: The socialization of emotion expression in infancy. *Child Development, 53*, 991–1003.
Mallay, H. (1935). A study of some of the techniques underlying the establishment of successful social contacts at the preschool level. *Journal of Genetic Psychology, 47*, 431–457.
Mannle, S., Barton, M., & Tomasello, M. (1992). Two-year-olds' conversations with their mothers and preschool-aged siblings. *First Language, 12*, 57–71.
Manwell, E. M., & Mengert, I. G. (1934). A study of the development of two- and three-year-old children with respect to play activities. *University of Iowa Studies in Child Welfare, 9*, 1–111.
Marbach, E. S., & Yawkey, T. D. (1980). The effect of imaginative play actions on language development in five-year-old children. *Psychology in the Schools, 17*, 257–263.
Marcus, R. F., & Jenny, B. (1977). A naturalistic study of reciprocity in the helping behavior of young children. *Alberta Journal of Educational Research, 23*, 195–206.
Maretzki, T. W., & Maretzki, H. (1963). Taira: An Okinawan village. In B. B. Whiting (Ed.), *Six cultures: Studies of childrearing* (pp. 363–539). New York: Wiley.
Marshack, A. (1984). The ecology and brain of two-handed bipedalism: An analytic, cognitive, and evolutionary assessment. In H. L. Roitblat, T. G. Bever, & H. S. Terrance (Eds.), *Animal cognition* (pp. 491–511). Hillsdale, NJ: Lawrence Erlbaum Associates.

Martin, J. A. (1981). A longitudinal study of the consequences of early mother–infant interaction: A microanalytic approach. *Monographs of the Society for Research in Child Development, 46* (3, Serial No. 190).

Martin, P. (1982). The energy cost of play: Definition and estimation. *Animal Behavior, 30,* 294–295.

Martin, P. (1984a). The (four) whys and wherefores of play in cats: A review of functional, evolutionary, developmental, and causal issues. In P. K. Smith (Ed.), *Play in animals and humans* (pp. 71–94). Oxford, England: Basil Blackwell.

Martin, P. (1984b). The time and energy costs of play behaviour in the cat. *Zeitschrift fur Tierpsychologie, 64,* 298–312.

Martin, P., & Caro, T. M. (1985). On the functions of play and its role in behavioral development. *Advances in the Study of Behavior, 15,* 59–103.

Martin, R. D. (1990). *Primate origins and evolution: A phylogenetic reconstruction.* Princeton, NJ: Princeton University Press.

Martini, M. (1994). Peer interactions in Polynesia: A view from the Marquesas. In J. L. Roopnarine, J. E. Johnson, & F. H. Hooper (Eds.), *Children's play in diverse cultures* (pp. 73–103). Albany: State University of New York Press.

Martini, M., & Kirkpatrick, J. (1981). Early interactions in the Marquesas Islands. In T. M. Field, A. M. Sostek, P. Vietze, & P. H. Leiderman (Eds.), *Culture and early interactions* (pp. 189–213). Hillsdale, NJ: Lawrence Erlbaum Associates.

Masataka, N., & Kohda, M. (1988). Primate play vocalizations and their functional significance. *Folia Primatologica, 50,* 152–156.

Mason, W. A. (1961). Effects of age and stimulus characteristics on manipulatory responsiveness of monkeys raised in a restricted environment. *Journal of Genetic Psychology, 99,* 301–308.

Mason, W. A., Harlow, H. F., & Rueping, R. R. (1959). The development of manipulatory responses in the infant rhesus monkey. *Journal of Comparative and Physiological Psychology, 52,* 555–558.

Masters, J. C., & Furman, W. C. (1981). Popularity, individual friendship selection, and specific peer interaction among children. *Developmental Psychology, 17,* 344–350.

Masur, E. F. (1982). Mothers' responses to infants' object-related gestures: Influences on lexical development. *Journal of Child Language, 9,* 23–30.

Masur, E. F. (1983). Gestural development, dual directional signaling, and the transition to words. *Journal of Psycholinguistic Research, 12,* 93–109.

Masur, E. F. (1987). Imitative interchanges in a social context: Mother–infant matching behavior at the beginning of the second year. *Merrill-Palmer Quarterly, 33,* 453–472.

Masur, E. F. (1997). Maternal labeling of novel and familiar objects: Implications for children's development of lexical constraints. *Journal of Child Language, 24,* 427–439.

Masur, E. F., & Gleason, J. B. (1980). Parent–child interaction and the acquisition of lexical information during play. *Developmental Psychology, 16,* 404–409.

Mathieu, M., & Bergeron, G. (1981). Piagetian assessment of cognitive development in chimpanzee (*Pan troglodytes*). In B. Chiarelli & R. S. Corruccini (Eds.), *Primate behavior and sociobiology* (pp. 142–147). Berlin: Springer.

Mathieu, M., Daudelin, N., Dagennis, Y., & Decarie, T. (1980). Piagetian causality in two house-reared chimpanzees (*Pan troglodytes*). *Canadian Journal of Psychology, 34,* 179–185.

Matthews, W. S. (1977). Modes of transformation in the initiation of fantasy play. *Developmental Psychology, 13,* 212–216.

Maudry, M., & Nekula, M. (1939). Social relations between children of the same age during the first two years of life. *Journal of Genetic Psychology, 54*, 193–215.

Mayes, L. C., & Carter, A. S. (1990). Emerging social regulatory capabilities as seen in the still-face situation. *Child Development, 61*, 754–763.

McCall, R. B. (1974). Exploratory manipulation and play in the human infant. *Monographs of the Society for Research in Child Development, 39* (2, Serial No. 155).

McCandless, B. R., & Hoyt, J. M. (1961). Sex, ethnicity, and play preferences of preschool children. *Journal of Abnormal and Social Psychology, 62*, 683–685.

McCune, L. (1995). A normative study of representational play in the transition to language. *Developmental Psychology, 31*, 198–206.

McCune-Nicolich, L., & Fenson, L. (1984). Methodological issues in studying pretend play. In T. D. Yawkey & A. D. Pellegrini (Eds.), *Child's play: Developmental and applied* (pp. 81–124). Hillsdale, NJ: Lawrence Erlbaum Associates.

McDonald, D. (1977). Play and exercise in the Columbian ground squirrel (*Spermophilus beecheyi*). *Animal Behaviour, 25*, 782–784.

McFarlane, A. C. (1987). Posttraumatic phenomena in a longitudinal study of children following a natural disaster. *Journal of the American Academy of Child and Adolescent Psychiatry, 26*, 764–769.

McGrew, W. C. (1972). Introduction into the group and social development. In W. C. McGrew (Ed.), *An ethological study of children's behavior* (pp. 149–173). London: Academic.

McGrew, W. C. (1977). Socialization and object manipulation of wild chimpanzees. In S. Chevalier-Skolnikoff & F. E. Poirier (Eds.), *Primate bio-social development: Biological, social, and ecological determinants* (pp. 261–288). New York: Garland.

McGrew, W. C., & Marchant, L. F. (1992). Chimpanzees, tools, and termites: Hand preference or handedness? *Current Anthropology, 33*, 114–119.

McHale, J. P., & Rasmussen, J. L. (1998). Coparental and family group level dynamics during infancy: Early family precursors of child and family functioning during preschool. *Development and Psychopathology, 10*, 39–59.

McKenzie, T. L., Sallis, J. F., Nader, P. R., Broyles, S. L., & Neslon, J. A. (1992). Anglo- and Mexican-American preschoolers at home and at recess: Activity patterns and environmental influences. *Developmental and Behavioral Pediatrics, 13*, 173–180.

McKenzie, T. L., Sallis, J. F., Nader, P. R., Patterson, T. L, Elder, J. P., Berry, C. C., Rupp, J. W., Atkins, C. J., Buono, M. J., & Nelson, J. A. (1991). BEACHES: An observational system for assessing children's eating and physical activity behaviors and associated events. *Journal of Applied Behavior Analysis, 24*, 141–151.

McLaughlin, B., White, D., McDevitt, T., & Raskin, R. (1983). Mothers' and fathers' speech to their young children: Similar or different? *Journal of Child Language, 10*, 245–252.

McLoyd, V. C. (1980). Verbally expressed modes of transformation in the fantasy play of Black preschool children. *Child Development, 51*, 1133–1139.

McLoyd, V. (1982). Social class differences in sociodramatic play: A critical review. *Developmental Review, 2*, 1–30.

McLoyd, V. C., & Ratner, H. H. (1983). The effects of sex and toy characteristics on exploration in preschool children. *Journal of Genetic Psychology, 142*, 213–224.

McLoyd, V. C., Warren, D., & Thomas, E. A. (1984). Anticipatory fantastic role enactment in preschool triads. *Developmental Psychology, 20*, 807–814.

McReynolds, P., Acker, M., & Pietila, C. (1961). Relation of object curiosity to psychological adjustment in children. *Child Development, 32*, 393–400.

McSwain, R. (1981). Care and conflict in infant development. *Infant Behavior and Development, 4*, 225–246.

Meaney, M. J., & Stewart, J. (1981). A descriptive study of social development in the rat. *Animal Behaviour, 29*, 34–45.

Meaney, M. J., Stewart, J., & Beatty, W. W. (1985). Sex differences in social play: The socialization of sex roles. *Advances in the Study of Behavior, 15*, 1–58.

Meckley, A. (1994). Shared knowledge of play events in young children's social play construction. *Communication and Cognition, 27*, 287–300.

Melson, G. F., & Dyan, D. (1987). Dominance and visual attention rank in preschool groups. *Perceptual and Motor Skills, 65*, 570.

Meltzoff, A. N., & Borton, R. W. (1979). Intermodal matching by human neonates. *Nature, 282*, 403–404.

Mendoza, D. L., & Ramirez, J. M. (1987). Play in kittens (*Felis domesticus*) and its association with cohesion and aggression. *Bulletin of the Psychonomic Society, 25*, 27–30.

Mendoza-Granados, D., & Sommer, V. (1995). Play in chimpanzees of the Anaheim zoo: Self-serving compromises. *Primates, 36*, 57–68.

Mengert, I. G. (1931). A preliminary study of the reactions of two-year-old children to each other when paired in a semi-controlled situation. *Journal of Genetic Psychology, 39*, 393–398.

Menzel, E. W. (1962). The effects of stimulus size and proximity upon avoidance of complex objects in rhesus monkeys. *Journal of Comparative and Physiological Psychology, 55*, 1044–1046.

Menzel, E. W. (1966). Responsiveness to objects in free-ranging Japanese monkeys. *Behaviour, 26*, 130–150.

Menzel, E. W., Davenport, R. K., & Rogers, C. M. (1961). Some aspects of behavior toward novelty in young chimpanzees. *Journal of Comparative and Physiological Psychology, 54*, 16–19.

Menzel, E. V., & Menzel, C. R. (1979). Cognitive, developmental, and social aspects of responsiveness to novel objects in a family group of marmosets (*Saguinus fuscicollis*). *Behaviour, 70*, 251–279.

Merrick, N. (1977). Social grooming and play behavior of a captive group of chimpanzees. *Primates, 18*, 215–224.

Messer, D. J. (1978). The integration of mothers' referential speech with joint play. *Child Development, 49*, 781–787.

Messer, D. J., McCarthy, M. E., McQuiston, S., MacTurk, R. H., Yarrow, L. J., & Vietze, P. M. (1986). Relation between mastery behavior in infancy and competence in early childhood. *Developmental Psychology, 22*, 366–372.

Messer, D. J., Rachford, D., McCarthy, M. E., & Yarrow, L. J. (1987). Assessment of mastery behavior at 30 months: Analysis of task-directed activities. *Developmental Psychology, 23*, 771–781.

Messer, D. J., & Vietze, P. M. (1984). Timing and transitions in mother–infant gaze. *Infant Behavior and Development, 7*, 167–181.

Mignault, C. (1985). Transition between sensorimotor and symbolic activities in nursery-reared chimpanzees (*Pan troglodytes*). *Journal of Human Evolution, 14*, 747–758.

Milich, R. (1984). Cross-sectional and longitudinal observations of activity level and sustained attention in a normative sample. *Journal of Abnormal Child Psychology, 12*, 261–276.

Miller, F. L. (1975). Play activities of black-tailed deer in northwestern Oregon. *Canadian Field Naturalist, 89*, 149–155.

Miller, M. N., & Byers, J. A. (1991). Energetic cost of locomotor play in pronghorn fawns. *Animal Behaviour, 41*, 1007–1013.

Miller, N. E., & Dollard, J. (1941). *Social learning and imitation*. New Haven, CT: Yale University Press.

Miller, P., & Garvey, C. (1984). Mother–baby role play: Its origins in social support. In I. Bretherton (Ed.), *Symbolic play: The development of social understanding* (pp. 101–130). Orlando, FL: Academic.

Miller, P. M., Danaher, D. L., & Forbes, D. (1986). Sex-related strategies for coping with interpersonal conflict in children aged five to seven. *Developmental Psychology, 22*, 543–548.

Miller, S. N. (1973). Ends, means, and gallumphing: Some leitmotifs of play. *American Anthropologist, 75*, 87–98.

Minturn, L., & Hitchcock, J. T. (1963). The Rajputs of Khalapur, India. In B. B. Whiting (Ed.), *Six cultures: Studies of child rearing* (pp. 203–361). New York: Wiley.

Minuchin, P. (1971). Correlates of curiosity and exploratory behavior in preschool disadvantaged children. *Child Development, 42*, 939–950.

Miracle, A. W. (1977). Some functions of Aymara games and play. In P. Stevens (Ed.), *Studies in the anthropology of play: Papers in memory of B. Allan Tindall* (pp. 98–105). West Point, NY: Leisure.

Mitchell, R. W. (1994a). Multiplicities of self. In S. T. Parker, R. W. Mitchell, & M. L. Boccia (Eds.), *Self-awareness in animals and humans: Developmental perspectives* (pp. 81–107). Cambridge, England: Cambridge University Press.

Mitchell, R. W. (1994b). The evolution of primate cognition: Simulation, self-knowledge, and knowledge of other minds. In D. Quiatt & J. Itani (Eds.), *Hominid culture in primate perspective* (pp. 177–232). Niwot: University of Colorado Press.

Mize, J., & Cox, R. A. (1990). Social knowledge and social competence: Number and quality of strategies as predictors of peer behavior. *Journal of Genetic Psychology, 151*, 117–127.

Mize, J., & Ladd, G. W. (1988). Predicting preschoolers' peer behavior and status from their interpersonal strategies: A comparison of verbal and enactive responses to hypothetical social dilemmas. *Developmental Psychology, 24*, 782–788.

Mize, J., & Pettit, G. S. (1997). Mothers' social coaching, mother–child relationship style, and children's peer competence: Is the medium the message? *Child Development, 68*, 312–332.

Moerk, E. L. (1974). Changes in verbal mother–child interactions with increasing language skills of the child. *Journal of Psycholinguistic Research, 3*, 101–116.

Moerk, E. L. (1975). Verbal interaction between children and their mothers during the preschool years. *Developmental Psychology, 11*, 788–794.

Moller, L. C., Hymel, S., & Rubin, K. H. (1992). Sex-typing in play and popularity in middle childhood. *Sex Roles, 26*, 331–353.

Moller, L. C., & Serbin, L. A. (1996). Antecedents of toddler gender segregation: Cognitive consonance, gender-typed toy preferences, and behavioral compatibility. *Sex Roles, 35*, 445–460.

Montgomery, K. C. (1953). Exploratory behavior as a function of "similarity" of stimulus situations. *Journal of Comparative and Physiological Psychology, 46*, 129–133.

Montgomery, K. C. (1954). The role of the exploratory drive in learning. *Journal of Comparative and Physiological Psychology, 47*, 60–64.

Montgomery, K. C. (1955). The relation between fear induced novel stimulation and exploratory behavior. *Journal of Comparative and Physiological Psychology, 48,* 254–260.

Montgomery, K. C., & Segall, M. (1955). Discriminative learning based upon the exploratory drive. *Journal of Comparative and Physiological Psychology, 48,* 225–228.

Moore, L. L., Lombardi, D. A., White, M. J., Cambell, J. L., Oliveria, S. A., & Ellison, R. C. (1991). Influence of parents' physical activity levels on activity levels of young children. *Journal of Pediatrics, 118,* 215–219.

Moore, S. G., & Bulbulian, K. N. (1976). The effects of contrasting styles of adult-child interaction on children's curiosity. *Developmental Psychology, 12,* 171–172.

Moran, G., Krupka, A., Tutton, A., & Symons, D. (1987). Patterns of maternal and infant imitation during play. *Infant Behavior and Development, 10,* 477–491.

Moran, J. D., Sawyers, J. K., Fu, V. R., & Milgram, R. M. (1984). Predicting imaginative play in preschool children. *Gifted Child Quarterly, 28,* 92–94.

Morgan, G. A., & Ricciuti, H. (1969). Infants' responses to strangers during the first year. In B. M. Foss (Ed.), *Determinants of infant behavior* (Vol. 4, pp. 253–272). London: Methuen.

Morikawa, H., Shand, N., & Kosawa, Y. (1988). Maternal speech to prelingual infants in Japan and the United States: Relationships among functions, forms, and referents. *Journal of Child Language, 15,* 237–256.

Morris, D. (1964). The response of animals to a restricted environment. *Symposia of the Zoological of London, 13,* 99–118.

Mosatche, H., & Bragonier, P. (1981). An observational study of social comparison in preschoolers. *Child Development, 52,* 376–378.

Mosier, C. E., & Rogoff, B. (1994). Infants' instrumental use of mothers to achieve their goals. *Child Development, 65,* 70–79.

Moss, H. A. (1967). Sex, age, and state as determinants of mother–infant interaction. *Merrill-Palmer Quaterly, 13,* 19–36.

Mounts, N. S., & Roopnarine, J. L. (1987). Social-cognitive play patterns in same-age and mixed-age preschool classrooms. *American Educational Research Journal, 24,* 463–476.

Mueller, E. (1972). The maintenance of verbal exchanges between young children. *Child Development, 43,* 930–938.

Mueller, E., & Brenner, J. (1977). The origins of social skills and interaction among playgroup toddlers. *Child Development, 48,* 854–861.

Mueller, E., & Lucas, T. (1975). A developmental analysis of peer interaction among toddlers. In M. Lewis & L. Rosenblum (Eds.), *Friendship and peer relations* (pp. 223–257). New York: Wiley.

Mulhern, R. K., Jr., & Passman, R. H. (1981). Parental discipline as affected by sex of the parent, the sex of the child, and the child's apparent responsiveness to discipline. *Developmental Psychology, 17,* 604–613.

Mummery, D. V. (1947). An analytical study of ascendant behavior of preschool children. *Child Development, 18,* 40–81.

Munn, P., & Dunn, J. (1989). Temperament and the developing relationship between siblings. *International Journal of Behavioral Development, 12,* 433–451.

Murphy, C. M. (1978). Pointing in the context of a shared activity. *Child Development, 49,* 371–380.

Murphy, L. B. (1978). The practical problems of recognizing and measuring fear and exploration in the domestic fowl. *Animal Behaviour, 26,* 422–471.

Murray, L., & Trevarthen, C. (1985). Emotion regulation of interactions between two-month-olds and their mothers. In T. Field & N. A. Fox (Eds.), *Social perception in infants* (pp. 177–197). Norwood, NJ: Ablex.
Mussen, P. H. (1983). *Handbook of child psychology* (4th ed.). New York:Wiley.
Musun-Miller, L. (1991). Effects of maternal presence on sibling behavior. *Journal of Applied Development Psychology*, *12*, 145–157.
Myers, A. K., & Miller, N. E. (1954). Failure to find a learned drive based on hunger; evidence for learning motivated by "exploration." *Journal of Comparative and Physiological Psychology*, *47*, 428–436.
Nadler, R. D., & Braggio, J. T. (1974). Sex and species differences in captive-reared juvenile chimpanzees and orangutans. *Journal of Human Evolution*, *3*, 541–550.
Napier, J. R. (1962). The evolution of the hand. *Scientific American*, *207*, 56–62.
Napier, J. R., & Napier, P. H. (1985). *The natural history of primates*. Cambridge, MA: MIT Press.
Nash, W. R. (1975). The effects of warm-up activities on small group divergent problem-solving with young children. *The Journal of Psychology*, *89*, 237–241.
Natale, F. (1989). Patterns of object manipulation. In F. Antinucci (Ed.), *Cognitive structure and development in nonhuman primates* (pp. 145–161). Hillsdale, NJ: Lawrence Erlbaum Associates.
Navon, R., & Ramsey, P. G. (1989). Possession and exchange of materials in Chinese and American preschools. *Journal of Research in Childhood Education*, *4*, 18–29.
Negro, J. J., Bustamante, J., Milward, J., & Bird, D. M. (1996). Captive fledging American kestrels prefer to play with objects resembling natural prey. *Animal Behaviour*, *52*, 707–714.
Neill, S. R. St. J. (1976). Aggressive and non-aggressive fighting in 12-13 year old preadolescent boys. *Journal of Child Psychology and Psychiatry*, *17*, 213–220.
Nelson, K. (1973). Structure and strategy in learning to talk. *Monographs of the Society for Research in Child Development*, *38* (1–2, Serial No. 149).
Nelson, K. (1981). Social cognition in a script framework. In J. H. Flavell & L. Ross (Eds.), *Social cognitive development* (pp. 97–118). Cambridge, England: Cambridge University Press.
Nelson, K., & Seidman, S. (1984). Playing with scripts. In I. Bretherton (Ed.), *Symbolic play: The development of social understanding* (pp. 45–71). Orlando, FL: Academic.
Neppl, T. K., & Murray, A. D. (1997). Social dominance and play patterns among preschoolers: Gender comparisons. *Sex Roles*, *36*, 381–393.
Newberry, R. C., Wood-Gush, D. G. M., & Hall, J. W. (1988). Playful behavior of piglets. *Behavioural Processes*, *17*, 205–216.
Newcomb, A., & Bagwell, C. (1995). Children's friendship relations: A meta-analytic review. *Psychological Review*, *117*, 306–347.
Newcomb, A. F., Bukowski, W. M., & Pattee, L. (1993). Children's peer relations: A meta-analytic review of popular, rejected, neglected, controversial, and average sociometric status. *Psychological Bulletin*, *113*, 99–128.
Nicolich, L. M. (1977). Beyond sensorimotor intelligence: Assessment of symbolic maturity through analysis of pretend play. *Merrill-Palmer Quarterly*, *23*, 89–99.
Nilsson, N. (1995). Children and war toys: Who had the idea to introduce children to violence? *International Play Theory*, *3*, 20–33.
Nishida, T., & Hiraiwa, M. (1982). Natural history of a tool-using behavior by wild chimpanzees feeding upon wood-boring ants. *Journal of Human Evolution*, *11*, 73–99.

Nisson, H. W. (1930). A study of exploratory behavior in the white rat by means of the obstruction method. *Journal of Genetic Psychology, 37*, 361–376.
Noller, P. (1980). Cross-gender effect in two-child families. *Developmental Psychology, 16*, 159–160.
Novak, M. A., Musante, A., Munroe, H., O'Neill, P. L., Price, C., & Suomi, S. J. (1993). Old, socially housed rhesus monkeys manipulate objects. *Zoo Biology, 12*, 285–298.
Nowicki, S., & Armitage, K. (1979). Behavior of juvenile yellow-bellied marmots: Play and social integration. *Zeitschrift fur Tierpsychologie, 51*, 85–105.
Nunnally, J. C., & Lemond, J. C. (1973). Exploratory behavior and human development. In H. W. Reese (Ed.), *Advances in child development and behavior* (Vol. 8, pp. 59–109). San Diego: Academic.
Nydegger, W. F., & Nydegger, C. (1963). Tarong: An Ilocos barrio in the Philippines. In B. B. Whiting (Ed.), *Six cultures: Studies of childrearing* (pp. 693–867). New York: Wiley.
Oakes, L. M., Madole, K. L., & Cohen, L. B. (1991). Infants' object examining: Habituation and categorization. *Child Development, 6*, 377–392.
Oakes, L. M., & Tellinghuisen, D. J. (1994). Examining in infancy: Does it reflect active processing? *Developmental Psychology, 30*, 748–756.
O'Brien, M., & Huston, A. C. (1985). Development of sex-typed play behavior in toddlers. *Developmental Psychology, 21*, 866–871.
O'Brien, M., Huston, A. C., & Risley, T. R. (1983). Sex-typed play of toddlers in a day care center. *Journal of Applied Developmental Psychology, 4*, 1–9.
O'Brien, M., & Nagle, K. J. (1987). Parents' speech to toddlers: The effect of play context. *Journal of Child Language, 14*, 269–279.
O'Connell, B., & Bretherton, I. (1984). Toddlers' play alone and with mother: The role of maternal guidance. In I. Bretherton (Ed.), *Symbolic play* (pp. 337–369). Orlando, FL: Academic.
O'Connor, M. J., Sigman, M., & Kasari, C. (1993). Interactional model for the association among maternal alcohol use, mother–infant interaction, and infant cognitive development. *Infant Behavior and Development, 16*, 177–192.
Ogbu, J. V. (1981). Origins of human competence: A cultural ecological perspective. *Child Development, 52*, 413–429.
Ogilvy, C. M. (1994). Social skills training with children and adolescents: A review of the evidence on effectiveness. *Educational Psychology, 14*, 73–83.
Olioff, M., & Stewart, J. (1978). Sex differences in the play behavior of prepubescent rats. *Physiology and Behavior, 20*, 113–115.
Olson, S. L., Bates, J. E., & Bayles, K. (1984). Mother–infant interaction and the development of individual differences in children's cognitive competence. *Developmental Psychology, 20*, 166–179.
Olson, S. L., Bayles, K., & Bates, J. E. (1986). Mother–child interaction and children's speech progress: A longitudinal study of the first two years. *Merrill-Palmer Quarterly, 32*, 1–20.
O'Neill, P. L., & Price, C. S. (1991). Customizing an enrichment program: Rhesus monkeys. *Lab Animal, 20*, 29–41.
O'Neill-Wagner, P. L., Bolig, R., & Price, C. S. (1994). Do play activity levels tell us something about psychosocial welfare in captive monkey groups? *Communication and Cognition, 27*, 261–272.
Opie, I., & Opie, P. (1969). *Children's games in streets and playgrounds*. London: Clarendon.

Ortega, J. C., & Bekoff, M. (1987). Avian play: Comparative, evolutionary, and developmental trends. *Auk, 104*, 338–341.

Otaki, M., Durrett, M., Richards, P., Nyquist, L., & Pennebaker, J. (1986). Maternal and infant behavior in Japan and America: A partial replication. *Journal of Cross-Cultural Psychology, 17*, 251–268.

Overton, W. F., & Jackson, J. P. (1973). The representation of imagined objects in action sequences: A developmental study. *Child Development, 44*, 309–314.

Owens, N. W. (1975a). A comparison of aggressive play and aggression in free-living baboons, *Papio anubis*. *Animal Behaviour, 23*, 757–765.

Owens, N. W. (1975b). Social play behaviour in free-living baboons, *Papio anubis*. *Animal Behaviour, 23*, 387–408.

Pakizegi, B. (1978). The interaction of mothers and fathers with their sons. *Child Development, 49*, 479–482.

Palmer, C. F. (1989). The discriminating nature of infants' exploratory actions. *Developmental Psychology, 25*, 885–893.

Pan, B. A., Imbens-Bailey, A., Winner, K., & Snow, C. (1996). Communicative intents expressed by parents in interaction with young children. *Merrill-Palmer Quarterly, 42*, 248–267.

Pan, H. W. (1994). Children's play in Taiwan. In J. Roopnarine, J. E. Johnson, & F. H. Hooper (Eds.), *Children's play in diverse cultures* (pp. 31–50). Albany: State University of New York Press.

Panksepp, J. (1981). The ontogeny of play in rats. *Developmental Psychobiology, 14*, 327–332.

Panksepp, J., Siviy, S., & Normansell, L. (1984). The psychobiology of play: Theoretical and methodological perspectives. *Neuroscience and Biobehavioral Reviews, 8*, 465–492.

Papousek, H., & Papousek, M. (1991). Innate and cultural guidance of infants' integrative competencies: China, the United States, and Germany. In M. H. Bornstein (Ed.), *Cultural approaches to parenting* (pp. 23–44). Hillsdale, NJ: Lawrence Erlbaum Associates.

Papousek, M., & Papousek, H. (1989). Forms and functions of vocal matching in interactions between mothers and their precanonical infants. *First Language, 9*, 137–158.

Papousek, M., Papousek, H., & Haekel, M. (1987). Didactic adjustments in fathers' and mothers' speech to their three-month-old infants. *Journal of Psycholinguistic Research, 16*, 491–516.

Parke, R. D. (1978). Children's home environments. In I. Altman & J. F. Wohlwill (Eds.), *Children and the environment* (pp. 33–81). New York: Plenum.

Parke, R. D., & O'Leary, S. (1975). Father–mother–infant interaction in the newborn period: Some feelings, observations, and some unresolved issues. In K. Riegel & J. Meacham (Eds.), *The developing individual in a changing world. Vol. 2. Social and environmental issues* (pp. 653–663). The Hague: Mouton.

Parke, R. D., & Sawin, D. B. (1980). The family in early infancy: Social interactional and attitudinal analyses. In F. Pedersen (Ed.), *The father–infant relationship in a family context* (pp. 44–179). New York: Prager.

Parker, C. E. (1974a). Behavioral diversity in ten species of nonhuman primates. *Journal of Comparative and Physiological Psychology, 87*, 930–937.

Parker, C. E. (1974b). The antecedents of man the manipulator. *Journal of Human Evolution, 3*, 493–500.

Parker, C. E. (1978). Opportunism and the rise of intelligence. *Journal of Human Evolution, 7*, 597–608.

Parker, J. (1986). Becoming friends: Conversational skills for friendship formation in young children. In J. M. Gottman & J. G. Parker (Eds.), *Conversations of friends: Speculations on affective development* (pp. 103–138). Cambridge, England: Cambridge University Press.
Parker, S. T. (1977). Piaget's sensorimotor series in an infant macaque: A model for comparing unstereotyped behavior and intelligence in human and nonhuman primates. In S. Chevalier-Skolnikoff & F. E. Poirier (Eds.), *Primate bio-social development: Biological, social, and ecological determinants* (pp. 43–112). New York: Garland.
Parker, S. T., & Gibson, K. (1977). Object manipulation, tool-use, and sensorimotor intelligence as feeding adaptations in cebus monkeys and great apes. *Journal of Human Evolution, 6*, 623–641.
Parker, S. T., & Milbrath, C. (1994). Contributions of imitation and role-playing games to the construction of self in primates. In S. T. Parker, R. W. Mitchell, & M. L. Boccia (Eds.), *Self-awareness in animals and humans* (pp. 108–128). Cambridge, England: Cambridge University Press.
Parks, K. A., & Novak, M. A. (1993). Observations of increased activity and tool use in captive rhesus monkeys exposed to troughs of water. *American Journal of Primatology, 29*, 13–25.
Parpal, M., & Maccoby, E. E. (1985). Maternal responsiveness and subsequent child compliance. *Child Development, 56*, 1326–1334.
Parquette, D. (1994). Fighting and play fighting in captive adolescent chimpanzees. *Aggressive Behavior, 20*, 49–65.
Parquette. D., & Prescott, J. (1989). Use of novel objects to enhance environments of captive chimpanzees. *Zoo Biology, 7*, 15–24.
Parrinello, R. M., & Ruff, H. A. (1988). The influence of adult intervention on infants' level of attention. *Child Development, 59*, 1125–1135.
Parten, M. B. (1932). Social participation among preschool children. *Journal of Abnormal and Social Psychology, 27*, 243–269.
Parten, M. B. (1933). Social play among preschool children. *Journal of Abnormal and Social Psychology, 28*, 136–147.
Pawlby, S. J. (1977). Imitative interaction. In H. R. Schaffer (Ed.), *Studies in mother–infant interaction* (pp. 203–224). London: Academic.
Pedersen, F. A., Cain, R. L., Jr., Zaslow, M. J., & Anderson, B. J. (1982). Variation in infant experience associated with alternative family roles. In L. M. Laosa & I. E. Sigel (Eds.), *Families as learning environments for children* (pp. 203–221). New York: Plenum.
Pedersen, F. A., & Wender, P. H. (1968). Early social correlates of cognitive functioning in six-year-old boys. *Child Development, 39*, 185–194.
Pedersen, J. M., Glickman, S. E., Frank, L. G., & Beach, F. A. (1990). Sex differences in the play behavior of immature spotted hyenas, *Crocuta crocuta. Hormones and Behavior, 24*, 403–420.
Pederson, D. R., Rook-Green, A., & Elder, J. L. (1981). The role of action in the development of pretend play in young children. *Developmental Psychology, 17*, 756–759.
Pelaez-Nogueros, M., Field, T., Gewirtz, J. L., Cigales, M., Gonzalez, A., Sanchez, A., & Richardson, S. C. (1997). The effects of systematic stroking versus tickling and poking on infant behavior. *Journal of Applied Developmental Psychology, 18*, 169–178.
Pelaez-Nogueros, M., Gewirtz, J. L., Field, T., Cigales, M. Malphurs, J., Clasky, S., & Sanchez, A. (1996). Infants' preference for touch stimulation in face-to-face interactions. *Journal of Applied Developmental Psychology, 17*, 199–213.

Pellegrini, A. D. (1980). The relationship between kindergartners' play and achievement in prereading, language, and writing. *Psychology in the Schools*, *17*, 530–535.

Pellegrini, A. D. (1982). Development of preschoolers' social cognitive play behaviors. *Perceptual and Motor Skills*, *55*, 1109–1110.

Pellegrini, A. D. (1984a). The effects of dramatic play on children's generation of cohesive text. *Discourse Processes*, *7*, 57–67.

Pellegrini, A. D. (1984b). Identifying causal elements in the thematic-fantasy play paradigm. *American Educational Research Journal*, *21*, 691–701.

Pellegrini, A. D. (1984–1985). The effects of exploration and play on young children's associative fluency. *Imagination, Cognition, and Personality*, *4*, 29–40.

Pellegrini, A. D. (1985). The narrative organization of children's fantasy play: The effects of age and play context. *Educational Psychology*, *5*, 17–25.

Pellegrini, A. D. (1988). Elementary school children's rough-and-tumble play and social competence. *Developmental Psychology*, *24*, 802–806.

Pellegrini, A. D. (1989a). Categorizing children's rough-and-tumble play. *Play and Culture*, *2*, 48–51.

Pellegrini, A. D. (1989b). Elementary school children's rough-and-tumble play. *Early Childhood Research Quarterly*, *4*, 245–260.

Pellegrini, A. D. (1989c). What is a category? The case of rough-and-tumble play. *Ethology and Sociobiology*, *10*, 331–341.

Pellegrini, A. D. (1993). Boys' rough-and-tumble play, social competence, and group composition. *British Journal of Developmental Psychology*, *11*, 237–248.

Pellegrini, A. D. (1994). The rough play of adolescent boys of differing sociometric status. *International Journal of Behavioral Development*, *17*, 525–540.

Pellegrini, A. D. (1995). A longitudinal study of boys' rough-and-tumble play and dominance during early adolescence. *Journal of Applied Developmental Psychology*, *16*, 77–93.

Pellegrini, A. D., & Davis, P. D. (1993). Relations between children's playground and classroom behaviour. *British Journal of Educational Psychology*, *63*, 88–95.

Pellegrini, A. D., & Galda, L. (1982). The effects of thematic-fantasy play training on the development of children's story comprehension. *American Educational Research Journal*, *19*, 443–452.

Pellegrini, A. D., & Perlmutter, J. C. (1987). A re-examination of the Smilansky-Parten matrix of play behavior. *Journal of Research in Childhood Education*, *2*, 89–96.

Pellegrini, A. D., & Perlmutter, J. C. (1989). Classroom contextual effects on children's play. *Developmental Psychology*, *25*, 289–296.

Pellegrini, A. D., & Smith, P. K. (1998). Physical activity play: The nature and function of a neglected aspect of play. *Child Development*, *69*, 577–598.

Pellis, S. M. (1981). A description of social play by the Australian magpie *Gymnorhina tibicen* based on Eshkul-Wachman notation. *Bird Behaviour*, *3*, 61–79.

Pellis, S. M. (1983). Development of head and foot coordination in the Australian magpie *Gymnorhina tibicen*, and the function of play. *Bird Behaviour*, *4*, 57–62.

Pellis, S. M. (1984). Two aspects of play fighting in a captive group of Oriental small-clawed otters, *Amblonyx cinerea*. *Zeitschrift fur Tierpsychologie*, *65*, 77–83.

Pellis, S. M. (1988). Agonistic versus amicable targets of attack and defense: Consequences for the origins, function, and descriptive classification of play-fighting. *Aggressive Behavior*, *14*, 85–104.

Pellis, S. M. (1991). How motivationally distinct is play? A preliminary case study. *Animal Behaviour*, *42*, 851–853.

Pellis, S. M. (1993). Sex and the evolution of play fighting: A review and model based on the behavior of muroid rodents. *Play Theory and Research, 1*, 55–75.
Pellis, S. M., & McKenna, M. M. (1992). Intrinsic and extrinsic influences on play fighting in rats: Effects of dominance, partner's playfulness, temperament, and neonatal exposure to testosterone propionate. *Behavioral Brain Research, 50*, 135–145.
Pellis, S. M., & McKenna, M. (1995). What do rats find rewarding in play fighting? An analysis using drug-induced non-playful partners. *Behavioural Brain Research, 68*, 65–73.
Pellis, S. M., McKenna, M. M., Field, E. F., Pellis, V. C., Prusky, G. T., & Whishaw, I. Q. (1996). Uses of vision by rats in play fighting and other close-quarter social interactions. *Physiology and Behavior, 59*, 905–913.
Pellis, S. M., & Pellis, V. C. (1983). Locomotor-rotational movements in the ontogeny and play of the laboratory rat *Rattus novegicus*. *Developmental Psychobiology, 16*, 269–286.
Pellis. S. M., & Pellis,V. C. (1987). Play-fighting differs from serious fighting in both target of attack and tactics of fighting in the laboratory rat *Rattus norvegicus*. *Aggressive Behavior, 13*, 227–242.
Pellis, S. M., & Pellis, V. C. (1988a). Identification of the possible origin of the body targets that differentiate play fighting from serious fighting in Syrian golden hamsters (*Mesocricetus auratus*). *Aggressive Behavior, 14*, 437–449.
Pellis, S. M., & Pellis, V. C. (1988b). Play-fighting in the Syrian Golden hamster *Mesocricetus auratus* waterhouse and its relationship to serious fighting during post-weaning development. *Developmental Psychobiology, 21*, 323–337.
Pellis, S. M., & Pellis, V. C. (1989). Targets of attack and defense in play-fighting of the Djungarian hamster *Phodopus campbelli*: Links to fighting and sex. *Aggressive Behavior, 15*, 217–234.
Pellis, S. M., & Pellis, V. C. (1990). Differential rates of attack, defense, and counter-attack during the developmental decreases in play fighting by male and female rats. *Developmental Psychobiology, 23*, 215–231.
Pellis, S. M., & Pellis, V. C. (1991a). Attack and defense during play fighting appear to be motivationally independent behaviors in muroid rodents. *The Psychological Record, 41*, 175–184.
Pellis, S. M., & Pellis, V. C. (1991b). Role reversal changes during the ontogeny of play fighting in male rats: Attack vs. defense. *Aggressive Behavior, 17*, 179–189.
Pellis, S. M., & Pellis, V. C. (1992). Juvenilized play fighting in subordinate male rats. *Aggressive Behavior, 18*, 449–457.
Pellis, S. M., & Pellis, V. C. (1993). Influence of dominance on the development of play fighting in pairs of male Syrian golden hamsters. *Aggressive Behavior, 19*, 293–302.
Pellis, S. M., & Pellis, V. C. (1996). On knowing it's only play: The role of play signals in play fighting. *Aggression and Violent Behavior, 1*, 249–268.
Pellis, S. M., & Pellis, V. C. (1997a). The pre-juvenile onset of play fighting in laboratory rats *Rattus norvegicus*. *Developmental Psychobiology, 31*, 193–205.
Pellis, S. M., & Pellis, V. C. (1997b). Targets, tactics, and the open mouth face during play fighting in three species of primates. *Aggressive Behavior, 23*, 41–57.
Pellis, S. M., & Pellis, V. C. (1998). The structure–function interface in the analysis of play fighting. In M. Bekoff & J. A. Byers (Eds.), *Animal play: Evolutionary, comparative, and ecological aspects* (pp. 115–140). Cambridge, England: Cambridge University Press.

Pellis, S. M., Pellis, V. C., & Dewsbury, D. A. (1989). Different levels of complexity in the play fighting by muroid rodents appear to result from different levels of intensity of attack and defense. *Aggressive Behavior, 15*, 297–310.

Pellis, S. M., Pellis, V. C., Manning, C. J., & Dewsbury, D. A. (1991). The paucity of social play in juvenile *Mus domesticus*: What is missing from the behavioral repertoire? *Animal Behaviour, 42*, 686–687.

Pellis, S. M., Pellis, V. C., & Whishaw, I. Q. (1992). The role of the cortex on play fighting in rats: Developmental and evolutionary implications. *Brain, Behavior, and Evolution, 39*, 270–284.

Penman, R., Cross, T., Milgram-Friedman, J., & Meares, R. (1983). Mothers' speech to prelingual infants: A pragmatic analysis. *Journal of Child Language, 10*, 17–34.

Pepler, D. J. (1982). Play and divergent thinking. In D. J. Pepler & K. H. Rubin (Eds.), *The play of children: Research and theory* (pp. 64–78). Basel, Switzerland: Karger.

Pepler, D. J., Abramovitch, R., & Corter, C. (1981). Sibling interaction in the home: A longitudinal study. *Child Development, 52*, 1344–1347.

Pepler, D. J., & Ross, H. S. (1981). The effects of play on convergent and divergent problem solving. *Child Development, 52*, 1202–1210.

Perlman, M., & Ross, H. S. (1997). The benefits of parent intervention in children's disputes: An examination of concurrent changes in children's fighting styles. *Child Develpoment, 64*, 690–700.

Perlmutter, J. C., & Pellegrini, A. D. (1987). Children's verbal fantasy with parents and peers. *Educational Psychology, 7*, 269–281.

Perusse, L., Tremblay, A., LeBlanc, C., & Bouchard, C. (1989). Genetic and environmental influences on the level of physical activity and exercise participation. *American Journal of Epidemiology, 129*, 1012–1022.

Pettit, G. S., Bakshi, A., Dodge, K. A., & Coie, J. D. (1990). The emergence of social dominance in young boys' play groups: Developmental differences and behavioral correlates. *Developmental Psychology, 26*, 1017–1025.

Pettit, G. S., & Bates, J. E. (1989). Family interaction patterns and children's behavior problems from infancy to 4 years. *Developmental Psychology, 25*, 413–420.

Pettit, G. S., Harrist, A. W., Bates, J. E., & Dodge, K. A. (1991). Family interaction, social cognition, and children's subsequent relationships with peers at kindergarten. *Journal of Social and Personal Relationships, 8*, 383–402.

Pfeiffer, S. (1985). Sex differences in social play of scimitar-horned oryx calves (*Oryx dammah*). *Zeitschrift fur Tierpsychologie, 69*, 281–292.

Phillips, J. (1973). Syntax and vocabulary of mothers' speech to young children: Age and sex comparisons. *Child Development, 44*, 182–185.

Phinney, J. S. (1986). The structure of 5-year-olds' verbal quarrels with peers and siblings. *Journal of Genetic Psychology, 147*, 47–60.

Phinney, J. S., & Rotherman, M. (1982). Sex differences between same-sex and cross-sex preschool pairs. *Child Study Journal, 12*, 259–269.

Piaget, J. (1952). *The origins of intelligence in children.* New York: International Universities Press.

Piaget, J. (1962). *Play, dreams, and imagination in childhood.* New York: Norton.

Piaget, J., & Inhelder, B. (1956). *The child's conception of space.* London: Routledge & Kegan Paul.

Pieraut-Le-Bonniec, G. (1985). From visual–motor anticipation to conceptualization: Reaction to solid and hollow objects and knowledge of the function of containment. *Infant Behavior and Development, 8*, 413–424.

Pierce, D. E. (1991). Early object rule acquisition. *The American Journal of Occupational Therapy*, *45*, 438–449.
Pierce, J. D., Pellis, V. C., Dewsbury, D. A., & Pellis, S. M. (1991). Targets and tactics of agonistic and precopulatory behavior in Montane and prairie voles: Their relationship to juvenile play fighting. *Aggressive Behavior*, *17*, 337–349.
Pine, J. M. (1994). The language of primary caregivers. In C. Gallaway & B. J. Richards (Eds.), *Input and interaction in language acquisition* (pp. 15–37). Cambridge, England: Cambridge University Press.
Plooij, F. X. (1979). How wild chimpanzee babies trigger the onset of mother–infant play and what the mother makes of it. In S. T. Parker & K. R. Gibson (Eds.), *Before speech: The beginning of interpersonal communication* (pp. 223–243). Cambridge, England: Cambridge University Press.
Podmore, V. N. (1988). Mothers' interactive behaviors and their children's preschool assessments: Relationships and issues. *New Zealand Journal of Educational Studies*, *23*, 165–174.
Poest, C., Williams, J., Witt, D., & Atwood, M.E. (1989). Physical activity patterns of preschool children. *Early Childhood Research Quarterly*, *4*, 367–376.
Poirier, F. E., & Smith, E. O. (1974). Socializing functions of primate play. *American Zoologist*, *14*, 275–287.
Pomerleau, A., Malcuit, G., & Seguin, R. (1992). Five-month-old girls' and boys' exploratory behaviors in the presence of familiar and unfamiliar toys. *Journal of Genetic Psychology*, *153*, 47–61.
Poole, T. B. (1966). Aggressive play in polecats. *Symposia of the Zoological Society of London*, *18*, 23–44.
Poole, T. B. (1978). An analysis of social play in polecats (Mustelidae) with comments on the form and evolutionary history of the open mouth play face. *Animal Behaviour*, *26*, 36–49.
Poole, T. B. (1985). *Social behavior in mammals*. Glasgow, Scotland: Blackie & Son, LTD.
Poole, T. B. (1987). Social behavior of a group of orangutans (*Pongo pygmaeus*) on an artificial island in Singapore Zoological Gardens. *Zoo Biology*, *6*, 315–330.
Poole, T. B., & Fish, J. (1975). An investigation of playful behavior in *Rattus norvegicus* and *Mus musculus* (Mammalia). *Journal of Zoology*, *175*, 61–71.
Poole, T. B., & Fish, J. (1976). An investigation of individual, age, and sexual differences in the play of *Rattus norvegicus* (Mammalia: Rodentia). *Journal of Zoology*, *179*, 249–260.
Potegal, M., & Einon, D. (1989). Aggressive behaviors in adult rats deprived of play fighting experience as juveniles. *Developmental Psychobiology*, *22*, 159–172.
Poti, P. (1989). Early sensorimotor development in macaques (*M fuscata*, *M fascicularis*). In F. Antinucci (Ed.), *Cognitive structure and development in nonhuman primates* (pp. 39–53). Hillsdale, NJ: Lawrence Erlbaum Associates.
Poti, P., & Spinozzi, G. (1994). Early sensorimotor development in chimpanzees (*Pan troglodytes*). *Journal of Comparative Psychology*, *108*, 93–103.
Power, T. G. (1981). Sex-typing in infancy: The role of the father. *Infant Mental Health Journal*, *2*, 226–240.
Power, T. G. (1985). Mother– and father–infant play: A developmental analysis. *Child Development*, *56*, 1514–1524.
Power, T. G., & Chapieski, M. L. (1986). Childrearing and impulse control in toddlers: A naturalistic investigation. *Developmental Psychology*, *22*, 271–275.

Power, T. G., Chapieski, M. L., & McGrath, M. P. (1985). Assessment of individual differences in infant exploration and play. *Developmental Psychology, 21*, 974–981.

Power, T. G., & Parke, R. D. (1982). Play as a context for early learning: Lab and home analyses. In L. M. Laosa & I. E. Sigel (Eds.), *Families as learning environments for children* (pp. 147–178). New York: Plenum.

Power, T. G., & Parke, R. D. (1983). Patterns of mother and father play with their 8-month-olds: A multiple analyses approach. *Infant Behavior and Development, 6*, 453–459.

Power, T. G., & Woolger, C. (1994). Parenting practices and age-group swimming: A correlational study. *Research Quarterly for Exercise and Sport, 65*, 59–66.

Powlishta, K. K., Serbin, L. A., & Moller, L. C. (1993). The stability of individual differences in gender typing: Implications for understanding gender segregation. *Sex Roles, 29*, 723–737.

Prosser, G. V., Hutt, C., Hutt, S. J., Mahindadasa, K. J., & Goonetilleke, M. D. J. (1986). Children's play in Sri Lanka: A cross-cultural study. *British Journal of Developmental Psychology, 4*, 179–186.

Provost, M. A., & La Freniere, P. J. (1991). Social participation and peer competence in preschool children: Evidence for discriminant and convergent validity. *Child Study Journal, 21*, 57–72.

Putallaz, M. (1983). Predicting children's sociometric status from their behavior. *Child Development, 54*, 1417–1426.

Putallaz, M. (1987). Maternal behavior and children's sociometric status. *Child Development, 58*, 324–340.

Putallaz, M., Costanzo, P. R., & Smith, R. B. (1991). Maternal recollections of childhood peer relationships: Implications for their children's social competence. *Journal of Social and Personal Relationships, 8*, 403–422.

Putallaz, M., & Sheppard, B. H. (1990). Social status and children's orientation to limited resources. *Child Development, 61*, 2022–2027.

Putallaz, M., & Wasserman, A. (1989). Children's naturalistic entry behavior and sociometric status: A developmental perspective. *Developmental Psychology, 25*, 297–305.

Ramey, C. T., & Ramey, S. L. (1998). Early intervention and early experience. *American Psychologist, 53*, 109–120.

Ramsey, P. G., & Lasquade, C. (1996). Preschool children's entry attempts. *Journal of Applied Developmental Psychology, 17*, 135–150.

Rasa, A. E. (1984). A motivational analysis of object play in juvenile dwarf mongooses (*Helogale undulata rufula*). *Animal Behaviour, 32*, 579–589.

Rasa, O. A. E. (1971). Social interaction and object manipulation in weaned pups of the Northern Elephant Seal *Mirounga angustirogtris. Zeitschrift fur Tierpsychologie, 29*, 82–102.

Ratner, N. B. (1988). Patterns of parental vocabulary selection in speech to very young children. *Journal of Child Language, 15*, 481–492.

Ratner, N., & Bruner, J. (1978). Games, social exchange, and the acquisition of language. *Journal of Child Language, 5*, 391–401.

Ratner, N. B., & Pye, C. (1984). Higher pitch in BT is not universal. Acoustic evidence from Quiche Mayan. *Journal of Child Language, 11*, 515–522.

Redican, W. K., & Mitchell, G. (1974). Play between adult male and infant rhesus monkeys. *American Zoologist, 14*, 295–302.

Redshaw, M. (1978). Cognitive development in human and gorilla infants. *Journal of Human Evolution, 7*, 133–141.

Reichle, J. E., Longhurst, T. M., & Stepanich, L. (1976). Verbal interaction in mother-child dyads. *Developmental Psychology, 12*, 273–277.

Reifel, S. (1984). Block construction: Children's developmental landmarks in representation of space. *Young Children, 40*, 61–67.

Reinhart, V., & Reinhart, A. (1982). Mock fighting in cattle. *Behaviour, 81,* 1–13.

Renner, M. J. (1990). Neglected aspects of exploratory and investigatory behavior. *Psychology, 18*, 16–22.

Renner, M. J., Bennett, A. J., Ford, M. L., & Pierre, P. J. (1992). Investigation of objects by the greater bushbaby (*Otolemur garnettii*). *Primates, 33*, 315–327.

Renner, M. J., Bennett, A. J., & White, J. C. (1992). Age and sex as factors influencing spontaneous exploration and object investigation by preadult rats (*Rattus norvegicus*). *Journal of Genetic Psychology, 106*, 217–227.

Renner, M. J., & Seltzer, C. P. (1994). Sequential structure in behavioral components of object investigation by Long-Evans rats (*Rattus norvegicus*). *Journal of Comparative Psychology, 108*, 335–343.

Renouf, D., & Lawson, J. (1986). Play in harbour seals (*Phonca vitulina*). *Journal of the Zoological Society of London, 208*, 73–82.

Reppucci, D. N. (1970). Individual differences in the consideration of information among two-year-old children. *Developmental Psychology, 2*, 240–246.

Reuter, J., & Yunik, G. (1973). Social interaction in nursery schools. *Developmental Psychology, 9*, 319–325.

Reynolds, P. C. (1993). The complementation theory of language and tool use. In K. R. Gibson & T. Ingold (Eds.), *Tools, language, and cognition in human evolution* (pp. 407–428). Cambridge, England: Cambridge University Press.

Rheingold, H. L. (1969). The effect of a strange environment on behavior in infants. In B. M. Foss (Ed.), *Determinants of infant behavior* (Vol. 4, pp. 137–166). New York: Barnes and Noble.

Rheingold, H. L., Gewirtz, J. L., & Ross, B. G. (1959). Social conditioning of infant vocalizations in the infant. *Journal of Comparative and Physiological Psychology, 52*, 68–73.

Rheingold, H. L., Hay, D. F., & West, M. J. (1976). Sharing in the second year of life. *Child Development, 47*, 1148–1158.

Richman, A. L., Miller, P. M., & LeVine, R. A. (1992). Cultural and educational variations in maternal responsiveness. *Developmental Psychology, 28*, 614–621.

Riksen-Walraven, J. (1978). Effects of caregiver behavior on habituation rate and self-efficacy in infants. *International Journal of Behavioral Development, 1*, 105–130.

Roberts, R. N., & Barnes, M. L. (1992). "Let momma show you how": Maternal-child interactions and their effects on children's cognitive performance. *Journal of Applied Developmental Psychology, 13*, 363–376.

Robinson, C. C., & Jackson, R. (1987). The effects of varying toy detail within a prototypical play object on the solitary object pretend play of preschool children. *Journal of Applied Developmental Psychology, 8*, 209–220.

Robinson, J., Little, C., & Biringen, Z. (1993). Emotional communication in mother–toddler relationships: Evidence for early gender differentiation. *Merrill-Palmer Quarterly, 39*, 496–517.

Rochat, P. (1989). Object manipulation and exploration in 2- to 5-month-old infants. *Developmental Psychology, 25*, 871–884.

Rogers, L. J., & Kaplan, G. (1994). A new form of tool use by orangutans in Sabah, East Malaysia. *Folia Primatologica, 63*, 50–52.

Rogoff, B. (1981). Adults and peers as agents of socialization: A highland Guatemalan profile. *Ethos, 9,* 18–36.

Rogoff, B., Mistry, J., Goncu, A., & Mosier, C. (1993). Guided participation in cultural activity by toddlers and caregivers. *Monographs of the Society for Research in Child Development, 58* (7, Serial No. 236).

Romanes, G. L. (1892). *Animal intelligence.* New York: Appleton.

Rome-Flanders, T., Cossette, L., Richard, M., & Decarie, T. G. (1995). Comprehension of rules and structures in mother–infant games: A longitudinal study of the first two years of life. *International Journal of Behavioral Development, 18,* 83–103.

Rome-Flanders, T., Cronk, C., & Gourde, C. (1995). Maternal scaffolding in mother–infant games and its relationship to language development: A longitudinal study. *First Language, 15,* 339–355.

Rome-Flanders, T., & Richard, M. (1992). Infant timing of vocalizations in two mother–infant games: A longitudinal study. *First Language, 12,* 285–297.

Romney, K., & Romney, R. (1963). The Mixtecans of Juxtlahuaca, Mexico. In B. B. Whiting (Ed.), *Six cultures: Studies of childrearing* (pp. 541–691). New York: Wiley.

Rondal, J. A. (1980). Fathers' and mothers' speech in early language development. *Journal of Child Language, 7,* 353–369.

Roopnarine, J. L. (1981). Peer play interaction in a mixed-age preschool setting. *Journal of General Psychology, 104,* 161–166.

Roopnarine, J. L. (1984). Sex-typed socialization in mixed-age preschool classrooms. *Child Development, 55,* 1078–1084.

Roopnarine, J. L. (1987). Social interaction in the peer group: Relationship to perceptions of parenting and to children's interpersonal awareness and problem solving ability. *Journal of Applied Developmental Psychology, 8,* 351–362.

Roopnarine, J. L., & Adams, G. R. (1987). Social interaction patterns of high, average, and low sociometric status children. *Journal of the American Academy of Child and Adolescent Psychiatry, 26,* 28–32.

Roopnarine, J. L., Hossain, Z., Gill, P., & Brophy, H. (1994). Play in the East Indian context. In J. L. Roopnarine, J. E. Johnson, & F. H. Hooper (Eds.), *Children's play in diverse cultures* (pp. 9–30). Albany: State University of New York Press.

Roopnarine, J. L, & Johnson, J. E. (1983). Kindergartner's play with preschool- and school-aged children within a mixed age classroom. *Elementary School Journal, 85,* 578–586.

Roopnarine, J. L., Talukder, E., Jain, D., Joshi, P., & Srivastau, P. (1990). Characteristics of holding, patterns of play, and social behaviors between parents and infants in New Dehli, India. *Developmental Psychology, 26,* 667–673.

Rose, S. A., Gottfried, A. W., & Bridges, W. H. (1981). Cross-modal transfer in 6-month-old infants. *Developmental Psychology, 17,* 661–669.

Rose, S. A., & Ruff, H. A. (1987). Cross-modal abilities in human infants. In J. D. Osofsky (Ed.), *Handbook of infant development* (pp. 318–362). New York: Wiley.

Rosen, C. E. (1974). The effects of sociodramatic play on problem-solving behavior among culturally disadvantaged preschool children. *Child Development, 45,* 920–927.

Rosenberg, B. G., & Sutton-Smith, B. (1960). A revised conception of masculine-feminine differences in play activities. *Journal of Genetic Psychology, 96,* 165–170.

Rosenberg, R. A., & Butler, R. A. (1982). Sound as a reinforcer for infants' manipulation of toys. *Journal of Speech and Hearing Disorders, 47,* 292–295.

Rosenblatt, D. (1977). Developmental trends in infant play. In B. Tizard & D. Harvey (Eds.), *Biology of play* (pp. 33–44). London: Lippincott.

Rosenstiel, A. (1977). The role of traditional games in the process of socialization among the Motu of Papua New Guinea. In D. F. Lancy & B. A. Tindall (Eds.), *The study of play: Problems and prospects* (pp. 64–70). West Point, NY: Leisure.

Roskos, K. (1990). A taxonomic view of pretend play activity among 4- and 5-year-old children. *Early Childhood Research Quarterly, 5*, 495–512.

Ross, H. S. (1974). The influence of novelty and complexity on exploratory behavior in 12-month-old infants. *Journal of Experimental Child Psychology, 17*, 436–451.

Ross, H. S. (1982). Toddler peer relations: Differentiation of games and conflicts. *Canadian Journal of Behavioural Science, 14*, 364–379.

Ross, H. S. (1996). Negotiating principles of entitlement in sibling property disputes. *Developmental Psychology, 32*, 90–101.

Ross, H. S., & Lollis, S. P. (1989). A social relations analysis of toddler peer relationships. *Child Development, 60*, 1082–1091.

Ross, H. S., Rheingold, H. L., & Eckerman, C. O. (1972). Approach and exploration of a novel alternative by 12-month-old infants. *Journal of Experimental Child Psychology, 13*, 85–93.

Ross, H., & Taylor, H. (1989). Do boys prefer daddy or his physical style of play? *Sex Roles, 20*, 23–33.

Ross, H., Tesla, C., Kenyon, B., & Lollis, S. (1990). Maternal intervention and toddler peer conflict: The socialization of principles of justice. *Developmental Psychology, 26*, 994–1003.

Rothbaum, F., Rosen, K. S., Pott, M., & Beatty, M. (1995). Early parent–child relationships and later problem behavior: A longitudinal study. *Merrill-Palmer Quarterly, 41*, 133–151.

Rothstein-Fisch, C., & Howes, C. (1988). Toddler peer interaction in mixed-age groups. *Journal of Applied Developmental Psychology, 9*, 211–218.

Routh, D. K., Schroeder, C. S., & O'Tuama, L. A. (1974). Development of activity level in children. *Developmental Psychology, 10*, 163–168.

Routh, D. K., Walton, M. D., & Paden-Belkin, E. (1978). Development of activity level in children revisited: Effects of mother presence. *Developmental Psychology, 14*, 571–581.

Rowley, J. (1929). Life history of the sea lions on the California Coast. *Journal of Mammology, 10*, 1–36.

Rubenstein, J. (1967). Maternal attentiveness and subsequent exploratory behavior in the infant. *Child Development, 38*, 1089–1100.

Rubenstein, J. (1974). Concordance of visual and manipulative responsiveness to novel and familiar stimuli in six-month-old infants. *Child Development, 45*, 194–195.

Rubenstein, J., & Howes, C. (1979). Caregiving and infant behavior in day care and in homes. *Developmental Psychology, 15*, 1–24.

Rubin, K. H. (1976). Relation between social participation and role-taking in preschool children. *Psychological Reports, 39*, 823–826.

Rubin, K. H. (1977). The social and cognitive value of preschool toys and activities. *Canadian Journal of Behavioral Science, 9*, 382–385.

Rubin, K. H. (1982). Nonsocial play in preschoolers: Necessarily evil? *Child Development, 53*, 651–657.

Rubin, K. H., Coplan, R. J., Fox, N. A., & Calkins, S. D. (1995). Emotionality, emotion regulation, and preschoolers' social adaptation. *Development and Psychopathology, 7*, 49–62.

Rubin, K. H., & Daniels-Beirness, T. (1983). Concurrent and predictive correlates of sociometric status in kindergarten and grade 1 children. *Merrill-Palmer Quarterly, 29*, 337–351.

Rubin, K. H., Daniels-Beirness, T., & Hayvren, M. (1982). Social and social-cognitive correlates of sociometric status in preschool and kindergarten children. *Canadian Journal of Behavioral Science, 14*, 338–349.

Rubin, K. H., Fein, G. G., & Vandenberg, B. (1983). Play. In P. H. Mussen & E. M. Hetherington (Eds.), *Handbook of child psychology* (Vol. 4, pp. 693–774). New York: Wiley.

Rubin, K. H., Hymel, S., & Mills, R. S. L. (1989). Sociability and social withdrawal in childhood: Stability and outcomes. *Journal of Personality, 57*, 238–255.

Rubin, K. H., & Krasnor, L. R. (1980). Changes in the play behaviors of preschoolers: A short-term longitudinal investigation. *Canadian Journal of Behavioral Science, 12*, 278–282.

Rubin, K. H., & Maioni, L. (1975). Play preference and its relationship to egocentrism, popularity, and classification skills in preschoolers. *Merrill-Palmer Quarterly, 21*, 171–179.

Rubin, K. H., Maioni, T. L., & Hornung, M. (1976). Free play behaviors in middle- and lower-class preschoolers: Piaget revisited. *Child Development, 47*, 414–419.

Rubin, K. H., Watson, K. S., & Jambor, T. W. (1978). Free-play behaviors in preschool and kindergarten children. *Child Development, 49*, 534–536.

Ruff, H. A. (1981). Effect of context on infants' responses to novel objects. *Developmental Psychology, 17*, 87–89.

Ruff, H. A. (1984). Infants' manipulative exploration of objects: Effects of age and object characteristics. *Developmental Psychology, 20*, 9–20.

Ruff, H. A. (1986). Components of attention during infants' manipulative exploration. *Child Development, 57*, 105–114.

Ruff, H. A. (1989). The infants' use of visual and haptic information in the perception and recognition of objects. *Canadian Journal of Psychology, 43*, 302–319.

Ruff, H. A., Capozzoli, M., & Saltarelli, L. M. (1996). Focused visual attention and distractibility in 10-month-old infants. *Infant Behavior and Development, 19*, 281–293.

Ruff, H. A., & Dubiner, K. (1987). Stability of individual differences in infants' manipulation and exploration of objects. *Perceptual and Motor Skills, 64*, 1095–1101.

Ruff, H., & Lawson, K. R. (1990). Development of sustained, focused attention in young children during free play. *Developmental Psychology, 26*, 85–93.

Ruff, H. A., Lawson, K. R., Parrinello, R., & Weissberg, R. (1990). Long-term stability of individual differences in sustained attention in the early years. *Child Development, 61*, 60–75.

Ruff, H. A., Saltarelli, L. M., Capozzoli, M., & Dubiner, K. (1992). The differentiation of activity in infants' exploration of objects. *Developmental Psychology, 28*, 851–861.

Russell, A., & Finnie, V. (1990). Preschool children's social status and maternal instructions to assist group entry. *Developmental Psychology, 26*, 603–611.

Russell, C. L., & Russnaik, R. N. (1981). Language and symbolic play in infancy: Independent or related abilities? *Canadian Journal of Behavioral Science, 13*, 95–104.

Russell, G., & Russell, A. (1987). Mother–child and father–child relationships in middle childhood. *Child Development, 58*, 1573–1585.

Russon, A. E., Bard, K. A., & Parker, S. T. (1996). *Reaching into thought: The minds of the great apes*. Cambridge, England: Cambridge University Press.

Rutter, D. R., & Durkin, K. (1987). Turn-taking in mother–infant interaction: An examination of vocalizations and gaze. *Developmental Psychology, 23*, 54–61.
Sachs, B. D., & Harris, V. S. (1978). Sex differences and developmental changes in selected juvenile activities (play) of domestic lambs. *Animal Behaviour, 26*, 678–684.
Sackin, S., & Thelen, E. (1984). An ethological study of peaceful associative outcomes to conflict in preschool children. *Child Development, 55*, 1098–1102.
Sagi, A., Lamb, M. E., Shoham, R., Dvir, R., & Lewkowicz, K. (1985). Parent–infant interaction in families on Israeli Kibbutzism. *International Journal of Behavioral Development, 8*, 273–284.
Sakamoto, A. (1994). Video game use and the development of sociocognitive abilities in children: Three surveys of elementary school students. *Journal of Applied Social Psychology, 24*, 21–42.
Sallis, J. F., Nader, P. R., Broyles, S. L., Berry, C. C., Elder, J. P., McKenzie, T. L., & Nelson, J. A. (1993). Correlates of physical activity at home in Mexican-American and Anglo-American preschool children. *Health Psychology, 12*, 390–398.
Sallis, J. F., Patterson, T. L., McKenzie, T. L., & Nader, P. R. (1988). Family variables and physical activity in preschool children. *Developmental and Behavioral Pediatrics, 9*, 57–61.
Saltz, E., Dixon, D., & Johnson, J. (1977). Training disadvantaged preschoolers on various fantasy activities: Effects on cognitive functioning and impulse control. *Child Development, 48*, 367–380.
Saltz, E., & Johnson, J. (1974). Training for thematic fantasy play in culturally disadvantaged children: Preliminary results. *Journal of Educational Psychology, 66*, 623–630.
Sander, L. W. (1962). Issues in early mother–child interaction. *Journal of the American Academy of Child Psychiatry, 1*, 141–166.
Sanders, K. M., & Harper, L. V. (1976). Free-play fantasy behavior in preschool children: Relations among gender, age, season, and location. *Child Development, 47*, 1182–1185.
Saracho, O. N. (1984). Construction and validation of the play rating scale. *Early Child Development and Care, 17*, 199–200.
Saracho, O. N. (1987). Cognitive style in the play of young children. *Early Child Development and Care, 29*, 163–179.
Saracho, O. N. (1993). A factor analysis of young children's play. *Early Child Development and Care, 84*, 91–102.
Sarid, M., & Breznitz, Z. (1997). Developmental aspects of sustained attention among 2- to 6-year-old children. *International Journal of Behavioral Development, 21*, 303–312.
Savage, E. S., & Malick, C. (1977). Play and socio-sexual behavior in a captive chimpanzee (*Pan troglodytes*) group. *Behaviour, 60*, 179–194.
Saxe, R. M., & Stollack, G. R. (1971). Curioisty and the parent–child relationship. *Child Development, 42*, 375–384.
Saxon, T. F. (1997). A longitudinal study of early mother–infant interaction and later competence. *First Language, 17*, 271–281.
Saylor, C. F., Swenson, C. C., & Powell, P. (1992). Hurricane Hugo blows down the broccoli: Preschoolers' post-disaster play and adjustment. *Child Psychiatry and Human Development, 22*, 139–149.
Schaefer, E. S. (1989). Dimensions of mother–infant interaction: Measurement, stability, and predictive validity. *Infant Behavior and Development, 12*, 379–393.
Schaffer, H. R, & Parry, M. R. (1970). The effects of short-term familiarization on infants' perceptual motor co-ordination in a simultaneous discrimination situation. *British Journal of Psychology, 61*, 559–569.

Schaller, G. B. (1963). *The mountain gorilla.* Chicago: University of Chicago Press.
Schaller, G. B. (1972). *The Serengeti lion.* Chicago: University of Chicago Press.
Schenkel, R. (1966). Play, exploration, and territoriality in the wild lion. *Symposia of the Zoological Society of London, 18,* 11–22.
Schiller, P. H. (1952). Innate constituents of complex responses in primates. *Psychological Review, 59,* 177–191.
Schlosberg, H. (1947). The concept of play. *Psychological Review, 54,* 229–231.
Schneider, K., Moch, M., Sandfort, R., Averswald, M., & Walther-Weckman, K. (1983). Exploring a novel object by preschool children: A sequential analysis of perceptual, manipulatory, and verbal exploration. *International Journal of Behavioral Development, 6,* 477–496.
Schultz, A. H. (1956). Postembryonic age changes. In H. Hofer, A. H. Schultz, & D. Starck (Eds.), *Primatologia* (Vol. 1, pp. 887–964). Basel: S. Karger.
Schutte, N. S., Malouff, J. M., Post-Gorden, J. C., & Rodasta, A. L. (1988). Effects of playing videogames on children's aggression and other behaviors. *Journal of Applied Social Psychology, 18,* 454–460.
Schwartz, D., Dodge, K. A., & Coie, J. D. (1993). The emergence of chronic peer victimization in boys' play groups. *Child Development, 64,* 1755–1772.
Schwartz, D., Dodge, K. A., Pettit, G. S., & Bates, J. E. (1997). The early socialization of aggressive victims of bullying. *Child Development, 68,* 665–675.
Schwartzman, H. B. (1976). The anthropological study of children's play. *Annual Review of Anthropology, 5,* 289–328.
Schwartzman, H. B. (1982). Play as a mode. *The Behavioral and Brain Sciences, 5,* 168–169.
Schwartzman, H. B. (1986). A cross-cultural perspective on child-structured play activities and materials. In A. W. Gottfried & C. C. Brown (Eds.), *Play interactions: The contribution of play materials and parental involvement to children's development* (pp. 13–29). Lexington, MA: Lexington.
Scudder, K. M., & Chiszar, D. (1977). Effects of six visual stimulus conditions on defensive and exploratory behavior in two species of rattlesnakes. *Psychological Record, 27,* 519–526.
Seitz, S., & Stewart, C. (1975). Imitations and expansions: Some developmental aspects of mother-child communication. *Developmental Psychology, 11,* 763–768.
Selman, R. L. (1980). *The growth of interpersonal understanding.* Orlando, FL: Academic.
Selnow, G. W. (1984). Playing videogames: The electronic friend. *Journal of Communication, 34,* 148–156.
Serbin, L. A., & Connor, J. M. (1979). Sex-typing children's play preferences and patterns of cognitive performance. *The Journal of Genetic Psychology, 134,* 315–316.
Serbin, L. A., Connor, J. M., Burchardt, C. J., & Citron, C. C. (1979). Effects of peer presence on sex-typing of children's play behavior. *Journal of Experimental Child Psychology, 23,* 303–309.
Serbin, L. A., Powlishta, K. K., & Gulko, J. (1993). The development of sex typing in middle childhood. *Monographs of the Society for Research in Child Development, 58*(2, Serial No. 232).
Serbin, L., Sprafkin, C., Elman, M., & Doyle, A. B. (1982). The early development of sex differentiation patterns and social influence. *Canadian Journal of Social Sciences, 14,* 350–368.
Serbin, L. A., Tonick, I. J., & Sternglanz, S. (1977). Shaping cooperative cross-sex play. *Child Development, 48,* 924–929.

Shand, N., & Kosawa, Y. (1985). Japanese and American behavior types at three months: Infants and infant–mother dyads. *Infant Behavior and Development*, *8*, 225–240.
Shantz, C. U. (1987). Conflicts between children. *Child Development*, *58*, 283–305.
Shantz, D. (1986). Conflict, aggression, and peer status: An observational study. *Child Development*, *57*, 1322–1332.
Sheldon, A. (1990). Pickle fights: Gendered talk in preschool disputes. *Discourse Processes*, *13*, 5–31.
Sheldon, A. (1992). Conflict talk: Sociolinguistic challenges to self-assertion and how young girls meet them. *Merrill-Palmer Quarterly*, *38*, 95–117.
Sherrod, K. B., Crawley, S., Petersen, G., & Bennett, P. (1978). Maternal language to prelinguistic infants: Semantic aspects. *Infant Behavior and Development*, *1*, 335–345.
Shields, M. J., & Sparing, J. W. (1993). Fathers' play and touch behaviors with their three-month-old infants. *Physical and Occupational Therapy in Pediatrics*, *13*, 39–59.
Shimai, S., Masuda, K., & Kishimoto, Y. (1990). Influences of TV games on physical and psychological development of Japanese kindergarten children. *Perceptual and Motor Development*, *70*, 771–776.
Shimai, S., Yamada, F., Masuda, K., & Tada, M. (1993). TV game play and obesity in Japanese school children. *Perceptual and Motor Skills*, *76*, 1121–1122.
Shmukler, D. (1982–83). Preschool imaginative play predisposition and its relationship to subsequent third grade assessment. *Imagination, Cognitive, and Personality*, *2*, 231–240.
Shmukler, D., & Naveh, I. (1984-85). Structured vs. unstructured play training with economically disadvantaged preschoolers. *Imagination, Cognition, and Personality*, *4*, 293–304.
Shoemaker, N. (1964). Toys of Chama (Eseejja) Indian children. *American Anthropologist*, *66*, 1151–1153.
Shore, C. (1986). Combinatorial play, conceptual development, and early multiword speech. *Developmental Psychology*, *22*, 184–190.
Siegal, M. (1987). Are sons and daughters treated more differently by fathers than mothers? *Developmental Review*, *7*, 183–209.
Siegler, R. S., & Jenkins, E. (1989). *How children discover new strategies*. Hillsdale, NJ: Lawrence Erlbaum Associates.
Sigman, M., Newmann, C., Carter, E., Cattle, D., D'Souza, S., & Bwido, N. (1988). Home interactions and the development of Embu toddlers in Kenya. *Child Development*, *59*, 1251–1261.
Silvern, S., & Williamson, P. (1987). The effects of video game play on young children's aggression, fantasy, and prosocial behavior. *Journal of Applied Developmental Psychology*, *8*, 453–462.
Silvern, S., Williamson, P., & Waters, B. (1982). Play as a mediator of comprehension. *Early Education Research Quarterly*, *7*, 16–21.
Simon, T., & Smith, P. K. (1983). The study of play and problem solving in preschool children: Have experimenter effects been responsible for previous results? *British Journal of Developmental Psychology*, *1*, 289–297.
Simon, T., & Smith, P. K. (1986). Problems with a play paradigm: A reply to Dansky. *Merrill-Palmer Quarterly*, *32*, 205–209.
Simons-Morton, B. G., O'Hara, N. M., Simons-Morton, D. G., & Parcel, G. S. (1987). Children and fitness: A public health perspective. *Research Quarterly for Exercise and Sport*, *58*, 295–302.
Singer, D. G., & Rummo, J. (1973). Ideational creativity and behavioral style in kindergarten aged children. *Developmental Psychology*, *8*, 154–161.

Singer, J. L. (1973). *The child's world of make-believe: Experimental studies of imaginative play*. New York: Academic.

Siviy, S. M., & Atrens, D. M. (1992). The energetic costs of rough-and-tumble play in the juvenile rat. *Developmental Psychobiology, 25*, 137–148.

Skeate, S. T. (1985). Social play behavior in captive white-fronted Amazon parrots *Amazona albifrons*. *Bird Behaviour, 6*, 46–48.

Slade, A. (1987a). A longitudinal study of maternal involvement and symbolic play during the toddler period. *Child Development, 58*, 367–375.

Slade, A. (1987b). Quality of attachment and early symbolic play. *Developmental Psychology, 23*, 78–85.

Slomkowski, C. L., & Dunn, J. (1992). Arguments and relationships within the family: Differences in young children's disputes with mother and sibling. *Developmental Psychology, 28*, 919–924.

Slomkowski, C. L., & Dunn, J. (1996). Young children's understanding of other people's beliefs and feelings and their connected communication with their friends. *Developmental Psychology, 32*, 442–447.

Slonaker, J. R. (1912). The normal activity of the albino rat from birth to natural death, its rate of growth, and the duration of life. *Animal Behaviour, 2*, 20–43.

Sluckin, A. M., & Smith, P. K. (1977). Two approaches to the concept of dominance in preschool children. *Child Development, 48*, 917–923.

Small, W. S. (1899). Notes on the psychic development of the young white rat. *American Journal of Psychology, 11*, 80–100.

Smetana, J. G., & Letourneau, K. J. (1984). Development of gender constancy and children's sex-typed free play behavior. *Developmental Psychology, 20*, 691–696.

Smilansky, S. (1968). *The effects of sociodramatic play on disadvantaged preschool children*. New York: Wiley.

Smith, A. B., & Inder, P. M. (1993). Social interaction in same and cross gender preschool peer groups: A participant observational study. *Educational Psychology, 13*, 29–42.

Smith, C. B., Adamson, L. B., & Bakeman, R. (1988). Interactional predictors of early language. *First Language, 8*, 143–156.

Smith, E. O. (1978). Social play in rhesus macaques (*Macaca mulatta*): A cluster analysis. In E. O. Smith (Ed.), *Social play in primates* (pp. 79–112). New York: Academic.

Smith, E. W., & Howes, C. (1994). The effect of parents' presence on children's social interaction in preschool. *Early Childhood Research Quarterly, 9*, 45–59.

Smith, L. K., Field, E. F., Forgie, M. L., & Pellis, S. M. (1996). Domiance and age-related changes in the play fighting of intact and post-weaning castrated male rats (*Rattus norvegicus*). *Aggressive Behavior, 22*, 215–226.

Smith, P. K. (1978). A longitudinal study of social participation in preschool children: Solitary and parallel play reexamined. *Developmental Psychology, 14*, 517–523.

Smith, P. K. (1982). Does play matter? Functional and evolutionary aspects of animal and human play. *Behavioral and Brain Sciences, 5*, 139–184.

Smith, P. K. (1994). The war play debate. In J. H. Goldstein (Ed.), *Toys, play, and child development* (pp. 67–84). Cambridge, England: Cambridge University Press.

Smith, P. K., & Boulton, M. (1990). Rough-and-tumble play, aggression, and dominance: Perception and behavior in children's encounters. *Human Development, 33*, 271–282.

Smith, P. K., & Connolly, K. (1972). Patterns of play and social interaction in preschool children. In N. Blurton Jones (Ed.), *Ethological studies of child behaviour* (pp. 65–95). Cambridge, England: Cambridge University Press.

Smith, P. K., & Connolly, K. (1980). *The ecology of preschool behavior.* New York: Cambridge University Press.

Smith, P. K., Dalgleish, M., & Herzmark, G. (1981). A comparison of the effects of fantasy play tutoring and skills tutoring in nursery classes. *International Journal of Behavioral Development, 4,* 421–441.

Smith, P. K., & Dutton, S. (1979). Play training in direct and innovative problem solving. *Child Development, 50,* 830–836.

Smith, P. K., & Lewis, K. (1985). Rough-and-tumble play, fighting, and chasing in nursery school children. *Ethology and Sociobiology, 6,* 175–181.

Smith, P. K., & Simon, T. (1984). Object play, problem-solving, and creativity in children. In P. K. Smith (Ed.), *Play in animals and humans* (pp. 199–216). Oxford, England: Basil Blackwell.

Smith, P. K., Simon, T., & Emberton, R. (1985). Play, problem-solving, and experimenter effects: A replication of Simon & Smith (1983). *British Journal of Developmental Psychology, 3,* 105–107.

Smith, P. K., & Syddall, S. (1978). Play and non-play tutoring in preschool children: Is it play or tutoring which matters? *British Journal of Educational Psychology, 48,* 315–325.

Smith, P. K., Takhvar, M., Gore, N., & Vollstedt, R. (1985). Play in young children: Problems of definition, categorization, and measurement. *Early Child Development and Care, 19,* 25–41.

Smith, P. K., & Whitney, S. (1987). Play and associative fluency: Experimenter effects may be responsible for previous positive findings. *Developmental Psychology, 23,* 49–53.

Smith, P.M. (1985). *Language, the sexes, and society.* Oxford, England: Basil Blackwell.

Snow, C. E. (1995). Issues in the study of input: Finetuning, universality, individual, and developmental differences, and necessary causes. In P. Fletcher & B. MacWhinney (Eds.), *The handbook of child language* (pp. 180–193). Oxford, England: Basil Blackwell.

Snyder, J., West, L., Stockemer, V., Gibbons, S., & Almquist-Parks, L. (1996). A social learning model of peer choice in the natural environment. *Journal of Applied Developmental Psychology, 17,* 215–237.

Sommer, V., & Mendoza-Granados, D. (1995). Play as an indicator of habitat quality: A field study of langur monkeys (*Presbytis entellus*). *Ethology, 99,* 177–192.

Soper, W. B., & Miller, M. J. (1983). Junk-time junkies: An emerging addiction among students. *The School Counselor, 31,* 40–43.

Sostek, A. M., Vietze, P., Zaslow, M., Kreiss, L., van der Waals, F., & Rubinstein, D. (1981). Social context in caregiver–infant interaction: A film study of Fais and the United States. In T. M. Field, A. M. Sostek, P. Vietze, & P. H. Leiderman (Eds.). *Culture and early interactions* (pp. 21–37). Hillsdale, NJ: Lawrence Erlbaum Associates.

Source, J. F., & Emde, R. N. (1981). Mothers' presence is not enough: Effects of emotional availability on infant exploration. *Developmental Psychology, 17,* 737–745.

Southern, H. M. (1954). *Control of rats and mice: Vol. 3. Mice.* Oxford, England: Clarendon Press.

Spangler, G. (1989). Toddlers' everyday experiences as related to preceding mental and emotional disposition and their relationship to subsequent mental and motivational development: A short-term longitudinal study. *International Journal of Behavioral Development, 12,* 285–303.

Spangler, G. (1990). Mother, child, and situational correlates of toddlers' social competence. *Infant Behavior and Development, 13,* 405–419.

Spencer, C. C. (1943). Notes on the life history of Rocky Mountain bighorn sheep in the Tanyall Mountains of Colorado. *Journal of Mammalogy*, *24*, 1–11.
Spijkerman, R. P., Dienske, H., van Hooff, J. A. R. A. M., & Jens, W. (1996). Differences in variability, interactivity, and skills in social play of young chimpanzees living in peer groups and in a large family zoo group. *Behaviour*, *133*, 717–739.
Spinozzi, G. (1989). Early sensorimotor development in Cebus (*Cebus apella*). In F. Antinucci (Ed.), *Cognitive structure and development in nonhuman primates* (pp. 55–66). Hillsdale, NJ: Lawrence Erlbaum Associates.
Spinozzi, G., & Natale, F. (1989). Early sensorimotor development in Gorilla. In F. Antinucci (Ed.), *Cognitive structure and development in nonhuman primates* (pp. 21–38). Hillsdale, NJ: Lawrence Erlbaum Associates.
Spivak, G., & Shure, M. (1974). *Social adjustment of young children*. San Francisco: Jossey-Bass.
Sroufe, L. A. (1977). Wariness of strangers and the study of infant development. *Child Development*, *48*, 731–746.
Sroufe, L. A., & Waters, E. (1976). The ontogenesis of smiling and laughter: A perspective on the organization of development in infancy. *Psychological Review*, *83*, 173–189.
Sroufe, L. A., Waters, E., & Matas, L. (1974). Contextual determinants of infant affectional response. In M. Lewis & L. Rosenblum (Eds.), *Origins of fear* (pp. 49–72). New York: Wiley.
Stack, D. M., & Muir, D. W. (1992). Adult tactile stimulation during face-to-face interactions modulates five-month-olds' affect and attention. *Child Development*, *63*, 1509–1525.
Stamps, J. (1995). Motor learning and the value of familiar space. *American Naturalist*, *146*, 41–58.
Starin, E. D. (1990). Object manipulation by wild red colobus monkeys living in the Abuko Nature Reserve, The Gambia. *The Primates*, *31*, 385–391.
Steele, D., & Pederson, D. R. (1977). Stimulus variables which affect the concordance of visual and manipulative exploration in six-month-old infants. *Child Development*, *48*, 104–111.
Steiner, A. L. (1971). Play activity in Columbian ground squirrels. *Zeitschrift fur Tierpsychologie*, *28*, 247–261.
Steinkamp, M. W. (1989). Factors mediating the relationship between preschool children's play patterns and peer ratings: Verbal communication styles. *Journal of Applied Developmental Psychology*, *10*, 505–525.
Stern, D. N. (1971). A micro-analysis of mother–infant interaction: Behavior regulating social contact between a mother and her 3½-month-old twins. *Journal of the American Academy of Child Psychiatry*, *10*, 501–517.
Stern, D. N. (1974). Mother and infant at play: The dyadic interaction involving facial, vocal, and gaze behaviors. In M. Lewis & L. A. Rosenblum (Eds.), *The effect of the infant on its caregiver* (pp. 187–213). New York: Wiley.
Stern, D. N. (1977). *The first relationship*. Cambridge, MA: Harvard University Press.
Stern, D. N. (1985). *The interpersonal world of the infant*. New York: Basic.
Stern, D. N., Beebe, B., Jaffe, J., & Bennett, S. L (1977). The infant's stimulus world during social interaction: A study of caregiver behaviors with particular reference to repetition and timing. In H. R. Schaffer (Ed.), *Studies in mother–infant interaction* (pp. 177–202). London: Academic.
Stern, D. N., Spieker, S., Barnett, R. K., & MacKain, K. (1983). The prosody of maternal speech: Infant age and context related changes. *Journal of Child Language*, *10*, 1–15.

Stern, D. N., Spieker, S., & MacKain, K. (1982). Intonation contours as signals in maternal speech to prelinguistic infants. *Developmental Psychology*, *18*, 727–735.

Sternberg, R. J. (1984). Mechanisms of cognitive development: A componential approach. In R. J. Sternberg (Ed.), *Mechanisms of cognitive development* (pp. 163–186). New York: Freeman.

Stevenson, M. B., Leavitt, L. A., Thompson, R. H., & Roach, M. A. (1988). A social relations model analysis of parent and child play. *Developmental Psychology*, *24*, 101–108.

Stevenson, M. B., Ver Hoeve, J. N., Roach, M. A., & Leavitt, L. A. (1986). The beginning of conversation: Early patterns of mother–infant vocal responsiveness. *Infant Behavior and Development*, *9*, 423–440.

Stevenson, M. F. (1983). The captive environment: Its effect on exploratory and related behavioural responses in wild animals. In J. Archer & L. I. A. Birke (Eds.), *Exploration in animals and humans* (pp. 176–197). Bershire, England: Van Nostrand Reinhold.

Stevenson, M. F., & Poole, T. B. (1982). Playful interactions in family groups of the common marmoset (*Callithrix jacchus jacchus*). *Animal Behaviour*, *30*, 886–900.

Stifter, C. A., & Moyer, D. (1991). The regulation of positive affect: Gaze aversion activity during mother–infant interaction. *Infant Behavior and Development*, *14*, 111–123.

Stiles-Davis, J. (1988). Developmental change in young children's spatial grouping activity. *Developmental Psychology*, *24*, 522–531.

Stilson, S. R., & Harding, C. G. (1997). Early social context as it relates to symbolic play: A longitudinal investigation. *Merrill-Palmer Quarterly*, *43*, 682–693.

Stoneman, Z., Brody, G. H., & MacKinnon, C. (1984). Naturalistic observations of children's activities and roles while playing with their siblings and friends. *Child Development*, 55, 617–627.

Stoneman, Z., Brody, G. H., & MacKinnon, C. E. (1986). Same-sex and cross-sex siblings: Activity choices, roles, behavior, and gender stereotypes. *Sex Roles*, *15*, 495–511.

Storey, K. S. (1977). Field study: Children's play in Bali. In D. F. Lancy & B. A. Tindall (Eds.), *The study of play: Problems and prospects* (pp. 78–84). West Point, NY: Leisure.

Stott, L. H., & Ball, R. S. (1957). Consistency and change in ascendance-submission in the social interaction of children. *Child Development*, *28*, 259–272.

Strassberg, Z., Dodge, K. A., Bates, J. E., & Pettit, G. S. (1992). The longitudinal relations between parental conflict strategies and children's sociometric standing in kindergarten. *Merrill-Palmer Quarterly*, *38*, 477–493.

Strayer, F. (1980). Social ecology of the preschool peer group. In W. A. Collins (Ed.), *Minnesota Symposia on Child Psychology* (Vol. 13, pp. 165–196). Hillsdale, NJ: Lawrence Erlbaum Associates.

Strayer, F., & Strayer, J. (1976). An ethological analysis of social agonism and dominance among preschool children. *Child Development*, *47*, 980–989.

Strayer, F., & Trudel, M. (1984). Developmental changes in the nature and function of social dominance among young children. *Ethology and Sociobiology*, *5*, 279–295.

Strayer, J., & Strayer, F. (1978). Social aggression and power relations among preschool children. *Aggressive Behavior*, *4*, 175–182.

Streri, A., & Pecheux, M. G. (1986). Tactual habituation and discrimination of form in infancy: A comparison with vision. *Child Development*, *57*, 100–104.

Stuckey, M. F., McGhee, P. E., & Bell, N. J. (1982). Parent–child interaction: The influence of maternal employment. *Developmental Psychology*, *18*, 635–644.

Sugar, M. (1989). Children in a disaster: An overview. *Child Psychiatry and Human Development*, *19*, 163–179.

Sun, L., & Roopnarine, J. L. (1996). Mother–infant, father–infant interaction and involvement in childcare and household labor among Taiwanese families. *Infant Behavior and Development, 19*, 121–129.

Suomi, S. J. (1977). Adult male–infant interactions among monkeys living in nuclear families. *Child Development, 48*, 1255–1270.

Suomi, S. J. (1982). Why does play matter? *The Behavioral and Brain Sciences, 5*, 169–170.

Super, C. M. (1981). Behavioral development in infancy. In R. H. Munroe, R. L. Munroe, & B. B. Whiting (Eds.), *Handbook of cross-cultural human development* (pp. 181–270). New York: Garland STPM.

Sutton-Smith, B. (1973). *The folk-games of children*. Austin: University of Texas Press.

Sutton-Smith, B. (1977). Towards an anthropology of play. In P. Stevens (Ed.), *Studies in the anthropology of play* (pp. 222–232). West Point, NY: Leisure.

Sutton-Smith, B. (1988). War toys and childhood aggression. *Play and Culture, 1*, 57–69.

Sutton-Smith, B., Gerstmyer, J., & Meckley, A. (1988). Playfighting as folk play amongst preschool children. *Western Folklore, 47*, 161–176.

Sutton-Smith, B., & Kelly-Byrne, D. (1984). The phenomenon of bipolarity in play theories. In T. D. Yawkey & A. D. Pellegrini (Eds.), *Child's play: Developmental and applied* (pp. 29–47). Hillsdale, NJ: Lawrence Erlbaum Associates.

Sutton-Smith, B., Rosenberg, B. G., & Morgan, E. F. (1963). Development of sex differences in play choices during preadolescence. *Child Development, 34*, 119–126.

Suzuki, S., Kuroda, S., & Nishihara, T. (1995). Tool-set for termite fishing by chimpanzees in the Ndoki forest, Congo. *Behaviour, 132*, 219–235.

Switzky, H. N., Haywood, H. C., & Isett, R. (1974). Exploration, curiosity, and play in young children: Effects of stimulus complexity. *Developmental Psychology, 10*, 321–329.

Sylva, K. (1977). Play and learning. In B. Tizard & D. Harvey (Eds.), *Biology of play* (pp. 59– 73). London: Heinemann.

Sylva, K., Bruner, J. S., & Genova, P. (1976). The role of play in the problem-solving of children 3-5 years old. In J. Bruner, A. Jolly, & K. Sylva (Eds.), *Play: Its role in development and evolution* (pp. 244–257). New York: Penguin.

Symons, D. (1978). *Play and aggression: A study of rhesus monkeys*. New York: Columbia University Press.

Symons, D. K., & Moran, G. (1987). The behavioral dynamics of mutual responsiveness in early face-to-face mother–infant interactions. *Child Development, 58*, 1488–1495.

Symons, D., & Moran, G. (1994). Responsiveness and dependency are different aspects of social contingency: An example from mother and infant smiles. *Infant Behavior and Development, 17*, 209–214.

Taharally, L. C. (1991). Fantasy play, language, and cognitive ability of four-year-old children in Guyana, South America. *Child Study Journal, 21*, 37–56.

Takahashi, L. K., & Lore, R. K. (1983). Play fighting in juvenile rats and the development of agonistic behavior in male and female rats. *Aggressive Behavior, 9*, 217–227.

Takeuchi, M. (1994). Children's play in Japan. In J. L. Roopnarine, J. E. Johnson, & F. H. Hooper (Eds.), *Children's play in diverse cultures* (pp. 51–72). Albany: State University of New York Press.

Takhvar, M., & Smith, P. K. (1990). A review and critique of Smilansky's classification scheme and the "nested hierarchy" of play categories. *Journal of Research in Childhood Education, 4*, 112–122.

Tamis-LeMonda, C. S., & Bornstein, M. H. (1989). Habituation and maternal encouragement of attention in infancy as predictors of toddler language, play, and representational competence. *Child Development, 60*, 738–751.

Tamis-LeMonda, C. S., & Bornstein, M. H. (1991). Individual variation, correspondence, stability, and change in mother and toddler play. *Infant Behavior and Development, 14*, 143–162.

Tamis-LeMonda, C. S., & Bornstein, M. H. (1994). Specificity in mother–toddler language-play relations across the second year. *Developmental Psychology, 30*, 283–292.

Tamis-LeMonda, C. S., Bornstein, M. H., Cyphers, L., Toda, S., & Ogino, M. (1992). Language and play at one year: A comparison of toddlers and mothers in the United States and Japan. *International Journal of Behavioral Development, 15*, 19–42.

Tartabini, A., & Dienske, H. (1979). Social play and rank order in rhesus monkeys *Macaca Mulatta*. *Behavioural Processes, 4*, 375–383.

Tellinghuisen, D. J., & Oakes, L. M. (1997). Distrability in infancy: The effects of distracter characteristics and type of attention. *Journal of Experimental Child Psychology, 64*, 232–254.

Terr, L. C. (1981). "Forbidden games": Post-traumatic children's play. *Journal of the American Academy of Child Psychiatry, 20*, 741–760.

Teti, D. M., Bond, L. A., & Gibbs, E. D. (1988). Mothers, fathers, and siblings: A comparison of play styles and their influence upon infant cognitive level. *International Journal of Behavioral Development, 11*, 415–432.

Thompson, K. V. (1996a). Behavioral development and play. In D. Kleiman, M. E. Allen, K. V. Thompson, & S. Lumpkin (Eds.), *Wild mammals in captivity: Principles and techniques* (pp. 352–371). Chicago: University of Chicago Press.

Thompson, K. V. (1996b). Play-partner preferences and the function of social play in infant sable antelope, *Hippotragus niger*. *Animal Behaviour, 52*, 1143–1155.

Thor, D. H., & Holloway, H. R. (1983). Play solicitation behavior in juvenile male and female rats. *Animal Learning and Behavior, 11*, 173–178.

Tizard, B., Philips, J., & Plewis, I. (1976). Play in pre-school centres: I. Play measures and their relation to age, sex, and I.Q. *Journal of Child Psychology and Psychiatry, 17*, 251–264.

Toates, F. M. (1983). Exploration as a motivational and learning system: A cognitive incentive view. In J. Archer & L. Birke (Eds.), *Exploration in animals and humans* (pp. 55–71). Berkshire, England: Van Nostrand Reinhold.

Toda, S., & Fogel, A. (1993). Infant response to the still-face situation at 3 and 6 months. *Developmental Psychology, 29*, 532–538.

Toda, S., Fogel, A., & Kawai, M. (1990). Maternal speech to three-month-old infants in the United States and Japan. *Journal of Child Language, 17*, 279–294.

Todreas, R., & Bunston, T. (1993). Parent education program evaluation: A review of the literature. *Canadian Journal of Community Mental Health, 12*, 225–257.

Tokida, E., Tanaka, I., Takefushi, H., & Hagiwara,T. (1994). Tool-using in Japanese macaques: Use of stones to obtain fruit from a pipe. *Animal Behaviour, 47*, 1023–1030.

Tolan, J. C., Rogers, C. M., & Malone, D. R. (1981). Cross-modal matching in monkeys: Altered visual cues and delay. *Neuropsycholgia, 19*, 289–300.

Tomasello, M., Conti-Ramsden, G., & Ewert, B. (1990). Young children's conversations with their mothers and fathers: Differences in breakdown repair. *Journal Child Language, 17*, 115–130.

Tomasello, M., & Farrar, M. J. (1986). Joint attention and early language. *Child Development, 57*, 1454–1463.

Tomasello, M., & Kruger, A. C. (1992). Joint attention on actions: Acquiring verbs in ostensive and non-ostensive contexts. *Journal of Child Langage, 19*, 311–333.

Tonkinson, R. (1978). *The Mardujara aborigines*. New York: Holt, Rinehart, & Winston.

Torigoe, T. (1985). Comparison of object manipulation among 74 species of nonhuman primates. *Primates, 26*, 182–194.

Torigoe, T. (1986). Development of object manipulation in infants of Japanese monkeys. *Japanese Psychological Research, 28,* 149–154.

Torigoe, T. (1987). Object manipulation in a captive troop of Japanese monkeys (*Macaca fuscata*): A developmental analysis. *Primates, 28*, 497–506.

Tracy, R. L., & Ainsworth, M. D. S. (1981). Maternal affectionate behavior and infant-mother attachment patterns. *Child Development, 52*, 1341–1343.

Tronick, E. Z., Als, H., Adamson, L., Wise, L., & Brazelton, T. B. (1978). The infant's response to entrapment between contradictory messages in face-to-face interaction. *Journal of the American Academy of Child Psychiatry, 17*, 1–13.

Tronick, E., Als, H., & Brazelton, T. (1980). Monadic phases: A structural descriptive analysis of infant–mother face-to-face interaction. *Merrill-Palmer Quaterly, 26*, 3–24.

Tronick, E. Z., & Cohn, J. F. (1989). Infant–mother face-to-face interaction: Age and gender differences in coordination and the occurrence of mis-coordination. *Child Development, 60*, 85–92.

Truhon, S. A. (1983). Playfulness, play, and creativity: A path analytic model. *Journal of Genetic Psychology, 143*, 19–28.

Tsumori, A. (1966). Delayed response of wild Japanese monkeys by the sand digging method (IT). *Primates, 7*, 363–380.

Turkheimer, M., Bakeman, R., & Adamson, L. B. (1989). Do mothers support and peers inhibit skilled object play in infancy? *Infant Behavior and Development, 12*, 37–44.

Turnbull, C. M. (1962). *The forest people*. New York: Simon & Schuster.

Turner, P. J., & Gervai, J. (1995). A multidimensional study of gender typing in preschool children and their parents: Personality, attitudes, preferences, behavior, and cultural differences. *Developmental Psychology, 31*, 759–779.

Turner, P. J., Gervai, J., & Hinde, R. A. (1993). Gender-typing in young children: Preferences, behaviour, and cultural differences. *British Journal of Developmental Psychology, 11*, 323–342.

Turvey, M. T. (1996). Dynamic touch. *American Psychologist, 51*, 1134–1152.

Udwin, O. (1983). Imaginative play training as an intervention method with institutionalized preschool children. *British Journal of Educational Psychology, 53*, 32–39.

Ungerer, J. A., & Sigman, M. (1984). The relation of play and sensorimotor behavior to language in the second year. *Child Development, 55*, 1448–1455.

Urberg, K. A., & Kaplan, M. G. (1989). An observational study of race-, age-, and sex-heterogeneous interaction in preschoolers. *Journal of Applied Developmental Psychology, 10*, 299–311.

Uzgiris, I. C. (1967). Ordinality in the development of schemas for relating to objects. In J. Hellmuth (Ed.), *Exceptional infant* (Vol. 1, pp. 317–334). Seattle: Special Child Publications.

Uzgiris, I. C., & Hunt, J. McV. (1975). *Assessment in infancy: Ordinal scales for psychological development*. Urbana: University of Illinois Press.

Valantin, S. (1973). Problems raised by observation of children in various cultural environments. *Early Child Development and Care, 2*, 276–289.

van Aken, M. A. G., & Riksen-Walraven, J. M. (1992). Parental support and the development of competence in children. *International Journal of Behavioral Development, 15*, 101–123.

Vandell, D. L. (1979). A microanalysis of toddlers' social inteaction with mothers and fathers. *Journal of Genetic Psychology, 134*, 299–312.

Vandell, D. L., & Wilson, K. S. (1987). Infant's interactions with mother, sibling, and peer: Contrasts and relations between interaction systems. *Child Development*, *58*, 176–186.

Vandell, D. L., Wilson, K. S., & Buchanan, N. R. (1980). Peer interaction in the first year of life: An examination of its structure, content, and sensitivity to toys. *Child Development*, *51*, 481–488.

Vandenberg, B. (1978). Play and development from an ethological perspective. *American Psychologist*, *33*, 724–738.

Vandenberg, B. (1981). The role of play in the development of insightful tool-using strategies. *Merrill-Palmer Quarterly*, *27*, 97–109.

Vandenberg, B. (1984). Developmental features of exploration. *Developmental Psychology*, *20*, 3–8.

Vandeberg, B. (1985). Beyond the ethology of play. In A. W. Gottfried & C.C. Brown (Eds.), *Play interactions: The contribution of play material and parental involvement to children's development* (pp. 3–11). Lexington, MA: Lexington.

Vandenberg, B. (1990). Play and problem solving: An elusive connection. *Merrill-Palmer Quarterly*, *36*, 261–272.

van den Boom, D. C. (1994). The influence of temperament and mothering on attachment and exploration: An experimental manipulation of sensitive responsiveness among lower class mothers with irritable infants. *Child Development*, *65*, 1457–1477.

van de Rijt-Plooij, H., & Plooij, F. X. (1987). Growing independence, conflict, and learning in mother–infant relations in free-ranging chimpanzees. *Behaviour*, *101*, 1–86.

van Hooff, J. A. R. A. M. (1978). The facial displays of the Catarrhine monkeys and apes. In D. Morris (Ed.), *Primate ethology* (pp. 7–68). London: Weidenfeld & Nicolson.

van IJzendoorn, M. H., & De Wolff, M. S. (1997). In search of the absent father—meta-analyses of infant–father attachment: A rejoinder to our discussants. *Child Development*, *68*, 604–609.

van IJzendoorn, M. H., Juffer, F., & Duyvesteyn, M. G. C. (1995). Breaking the intergenerational cycle of insecure attachment: A review of the effects of attachment-based interventions on maternal sensitivity and infant security. *Journal of Child Psychology and Psychiatry*, *36*, 225–248.

van Lawick-Goodall, J. (1968). The behavior of free-living chimpanzees in the Gombe Stream Reserve. *Animal Behavior Monographs*, *1*, 161–311.

van Lawick-Goodall, J. (1970). Tool-using in primates and other vertebrates. *Advances in the Study of Behavior*, *3*, 195–249.

van Leeuwen, L., Smitsman, A., & van Leeuwen, C. (1994). Affordances, perceptual complexity, and the development of tool use. *Journal of Experimental Psychology: Human Perception and Performance*, *20*, 174–191.

Vauclair, J. (1984). Phylogenetic approach to object manipulation in human and ape infants. *Human Development*, *27*, 321–328.

Vauclair, J. (1993). Tool use, hand cooperation, and the development of object in human and non-human primates. In A. F. Kalverboer, B. Hopkins, & R. Geuze (Eds.), *Motor deveopment in early and later childhood: Longitudinal approaches* (pp. 205–216). Cambridge, England: Cambridge University Press.

Vauclair, J., & Bard, K A. (1983). Development of manipulation with objects in apes and human infants. *Journal of Human Evolution*, *12*, 631–645.

Vaughn, B. E., & Langlois, J. H. (1983). Physical attractiveness as a correlate of peer status and social competence in preschool children. *Developmental Psychology*, *19*, 561–567.

Vaughn, B. E., & Martino, D. G. (1988). Q-sort correlates of visual regard in groups of young preschool children. *Developmental Psychology, 24*, 589–594.

Vaughn, B. E., & Waters, E. (1981). Attention structure, sociometric status, and dominance: Interrelations, behavioral correlates, and relationships to social competence. *Developmental Psychology, 17*, 275–288.

Verba, M. (1993). Cooperative formats in pretend play among children. *Cognition and Instruction, 11*, 265–280.

Verbeek, M. E. M., Drent, P. J., & Wiepkema, P. R. (1994). Consistent individual differences in early exploratory behavior of male great tits. *Animal Behaviour, 48*, 1113–1121.

Vespo, J. (1991). Features of preschoolers' relationships. *Early Child Development, 68*, 19–26.

Vespo, J., Pedersen, J., & Hay, D. F. (1995). Young children's conflicts with peers and siblings: Gender effects. *Child Study Journal, 25*, 189–212.

Vickerman, R. C., Reed, M. D., & Roberts, M. W. (1997). Maternal intervention in sub-clinical sibling coercion. *Journal of Applied Developmental Psychology, 18*, 23–35.

Vincent, L. E., & Bekoff, M. (1978). Quantitative analyses of the ontogeny of predatory behavior in coyotes (*Canis latrans*). *Animal Behaviour, 26*, 225–231.

Visalberghi, E. (1988). Responsiveness to objects in two social groups of tufted capuchin monkeys (*Cebus apella*). *American Journal of Primatology, 15*, 349–360.

Visalberghi, E. (1990). Tool use in cebus. *Folia Primatologica, 54*, 146–154.

Vlietstra, A. G. (1978). Exploration and play in preschool children and young adults. *Child Development, 49*, 235–238.

Vliestra, A. G. (1980). Effects of adult-directed activity, number of toys, and sex of child on social and exploratory behavior in young children. *Merrill-Palmer Quarterly, 26*, 231–238.

Voland, E. (1977). Social play behavior of the common marmoset (*Callithrix jacchus Erxl.*, 1777) in captivity. *Primates, 18*, 883–902.

Volling, B. L., & Belsky, J. (1992). The contribution of mother–child and father–child relationship to the quality of sibling interaction: A longitudinal study. *Child Development, 63*, 1209–1222.

Volling, B. L., Youngblade, L. M., & Belsky, J. (1997). Young children's social relationships with sibling and friends. *American Journal of Orthopsychiatry, 67*, 102–111.

Voss, H., & Keller, H. (1983). *Curiosity and exploration: Theories and results.* New York: Academic.

Vygotsky, L. S. (1967). Play and its role in the mental development of the child. *Soviet Psychology, 12*, 62–76.

Wachs, T. D. (1987). Specificity of environmental action as manifest in environmental correlates of infant's mastery motivation. *Developmental Psychology, 23*, 782–790.

Wachs, T. D., & Chan, A. (1986). Specificity of environmental action, as seen in environmental correlates of infants' communication performance. *Child Development, 57*, 1464–1474.

Wachs, T. D., & Gandour, M. (1983). Temperament, environment, and six months cognitive-intellectual development. *International Journal of Behavioral Development, 6*, 135–152.

Waddington, C. H. (1962). *Principles of development and differentiation.* New York: MacMillan.

Waelder, R. (1933). The psychoanalytic theory of play. *Psychoanalytic Quarterly, 2*, 208–224.

Wanska, S. K., & Bedrosian, J. L. (1986). Topic and communicative intent in mother-child discourse. *Journal of Child Language, 13*, 523–535.

Wasser, S. K. (1978). Structure and function of play in the tiger. *Carnivore, 1*, 27–40.

Waterman, J. M. (1988). Social play in free-ranging Columbian ground squirrels. *Ethology, 77*, 225–236.

Waters, E., Garber, J., Gornal, M., & Vaughn, B. E. (1983). Q-sort correlates of visual regard among preschool peers: Validation of a behavioral index of social competence. *Child Development, 19*, 550–560.

Watson, D. M. (1993). The play associations of red-necked wallabies (*Macropus rufogriseus banksianus*) and relations to other social contexts. *Ethology, 94*, 1–20.

Watson, D. M., & Croft, D. (1993). Play fighting in captive red-necked wallabies *Macropus rufogriseus banksianus*. *Behaviour, 125*, 219–245.

Watson, D. M., & Croft, D. (1996). Age-related differences in play fighting strategies of captive male red-necked wallabies (*Macropus rufogriseus banksianus*). *Ethology, 102*, 336–346.

Watson, M. W. (1994). The relation between anxiety and pretend play. In A. Slade, & D. P. Wolf (Eds.), *Children at play: Clinical and developmental approaches to meaning and representation* (pp. 33–47). New York: Oxford University Press.

Watson, M., & Fisher, K. (1977). A developmental sequence of agent use in late infancy. *Child Development, 48*, 828–836.

Watson, M. W., & Jackowitz, E. R. (1984). Agents and recipient objects in the development of early symbolic play. *Child Development, 55*, 1091–1097.

Watson, M. W., & Peng, Y. (1992). The relation between toy gun play and children's aggressive behavior. *Early Education and Development, 3*, 370–389.

Waxman, M., & Stunkard, A. J. (1980). Caloric intake and expenditure of obese boys. *The Journal of Pediatrics, 96*, 187–193.

Wegener-Spohring, G. (1989). War toys and aggressive games. *Play and Culture, 2*, 35–47.

Weinberg, M. K., & Tronick, E. Z. (1996). Infant affective reactions to the resumption of maternal interaction after the still face. *Child Development, 67*, 905–914.

Weinraub, M., Clemens, L. P., Sockloff, A., Ethridge, T., Gracely, E., & Meyers, B. (1984). The development of sex role stereotypes in the third year: Relationship to gender labeling, identity, sex-typed toy preferences, and family characteristics. *Child Development, 55*, 1493–1503.

Weinraub, M., & Frankel, J. (1977). Sex differences in parent–infant interaction during free play, departure, and separation. *Child Development, 48*, 1240–1249.

Weisler, A., & McCall, R. B. (1976). Exploration and play: Resume and redirection. *American Psychologist, 31*, 492–508.

Weisner, T., & Gallimore, R. (1977). My brother's keeper: Child and sibling caretaking. *Current Anthropology, 18*, 169–190.

Weisner, T. S., & Wilson-Mitchell, J. E. (1990). Nonconventional family lifestyles and sex-typing in six-year-olds. *Child Development, 61*, 1915–1933.

Weiss, B., Dodge, K. A., Bates, J. E., & Pettit, G. S. (1992). Some consequences of early harsh discipline: Child aggression and a maladaptive social information processing style. *Child Development, 63*, 1321–1335.

Welker, W. I. (1956a). Effects of age and experience on play and exploration in chimpanzees. *Journal of Comparative Physiological Psychology, 49*, 223–226.

Welker, W. I. (1956b). Some determinants of play and exploration in chimpanzees. *Journal of Comparative and Physiological Psychology, 49*, 84–89.

Wellman, H., & Lempers, J. (1977). The naturalistic communicative abilities of two-year-olds. *Child Development, 48*, 1052–1057.
Wenar, C. (1976). Executive competence in toddlers: A prospective, observational study. *Genetic Psychology Monographs, 93*, 189–285.
Wenckstern, S., Weizmann, F., & Leenaars, A. A. (1984). Temperament and tempo of play in eight-month-old infants. *Child Development, 55*, 1195–1199.
Werebe, M. J. G., & Baudonniere, P. M. (1991). Social pretend play among friends and familiar preschoolers. *International Journal of Behavioral Development, 14*, 411–428.
Werker, J. F., & McLeod, P. J. (1989). Infant preference for both male and female infant-directed talk: A developmental study of attentional and affective responsiveness. *Canadian Journal of Psychology, 43*, 230–246.
West, M. (1974). Social play in the domestic cat. *American Zoologist, 14*, 427–436.
West, M. J. (1977). Exploration and play with objects in domestic kittens. *Developmental Psychobiology, 10*, 53–57.
West, M. M., & Konner, M. J. (1976). The role of the father: An anthropological perspective. In M. E. Lamb (Ed.), *The role of the father in child development* (pp. 185–217). New York: Wiley.
Westergaard, G. C. (1992). Object manipulation and the use of tools by infant baboons (*Papio cynocephalus anubis*). *Journal of Comparative Psychology, 106*, 398–403.
Westergaard, G. C. (1993). Development of combinatorial manipulation in infant baboons (*Papio cynocephalus anubis*). *Journal of Comparative Psychology, 107*, 34–38.
Westergaard, G. C. (1995). The stone-tool technology of capuchin monkeys: Possible implications for the evolution of symbolic communication in hominids. *World Archaeology, 27*, 1–9.
Westergaard, G. C., Greene, J. A., Menuhin-Hauser, C., & Suomi, S. J. (1996). The use of naturally-occurring copper and iron tools by monkeys: Possible implications for the emergence of metal-tool technology in hominids. *Human Evolution, 11*, 17–25.
Westergaard, G. C., & Suomi, S. J. (1993). Use of a tool-set by Capuchin monkeys (*Cebus apella*). *Primates, 34*, 459–462.
Westergaard, G. C., & Suomi, S. J. (1994a). Aimed throwing of stones by tufted capuchin monkeys (*Cebus apella). Human Evolution, 9*, 323–329.
Westergaard, G. C., & Suomi, S. J. (1994b). Asymmetrical manipulation in the use of tools by tufted capuchin monkeys (*Cebus apella*). *Folia Primatologica, 63*, 96–98.
Westergaard, G. C., & Suomi, S. J. (1995a). The manufacture and use of bamboo tools by monkeys: Possible implications for the development of material culture among East Asian hominids. *Journal of Archaeological Science, 22*, 677–681.
Westergaard, G. C., & Suomi, S. J. (1995b). The production and use of digging tools by monkeys: A nonhuman primate model of a hominid subsistence activity. *Journal of Anthropological Research, 51*, 1–8.
White, B. I., & Watts, J. C. (1973). *Experience and environment*. Englewood Cliffs, NJ: Prentice-Hall.
White, R. W. (1959). Motivation reconsidered: The concept of competence. *Psychological Review, 66*, 297–333.
Whiting, B. B. (1963). *Six cultures—studies of childrearing*. New York: Wiley.
Whiting, B., & Edwards, C. (1973). A cross-cultural analysis of sex-differences in the behavior of children age three through 11. *Journal of School Psychology, 91*, 171–188.

Whiting, B. B., & Edwards, C. P. (1988). *The company they keep: The effect of age, gender, and culture on social behavior of children aged two to ten.* Cambridge, MA: Harvard University Press.

Whiting, J. W. M., & Child, I. L. (1953). *Child training and personality.* New Haven, CT: Yale University Press.

Wilcoxon, H. C., Meier, G. W., Orlando, R., & Paulson, D. G. (1968). Visual self-stimulation in socially living rhesus monkeys. *Proceedings of the Second International Congress of Primatology, 1,* 261–266.

Wilkinson, L. C., Hiebert, E., & Renhold, K. (1981). Parents' and peers' communication to toddlers. *Journal of Speech and Hearing Research, 24,* 383–388.

Willatts, P. (1983). Effects of object novelty on the visual and manual exploration of infants. *Infant Behavior and Development, 6,* 145–149.

Wilson, E. O. (1975). *Sociobiology.* Cambridge, MA: Harvard University Press.

Wilson, S. (1974). Juvenile play of the common seal *Phoca vitulina vitulina* with comparative notes on the grey seal *Halichoerus grypus. Behaviour, 48,* 37–60.

Wilson, S., & Kleiman, D. (1974). Eliciting play: A comparative study. *American Zoologist, 14,* 341–370.

Winkel, M., Novak, D. M., & Hopson, H. (1987). Personality factors, subject gender, and the effects of aggressive videogames on aggression in adolescents. *Journal of Research in Personality, 21,* 211–223.

Wohwill, J. F. (1975). Children's voluntary exploration and preference for tactually presented nonsense shapes differing in complexity. *Journal of Experimental Child Psychology, 20,* 159–167.

Wohwill, J. F. (1984). Relationship between exploration and play. In T. D. Yawkey & A. D. Pellegrini (Eds.), *Child's play: Developmental and applied* (pp. 143–170). Hillsdale, NJ: Lawrence Erlbaum Associates.

Wolff, P. (1972). The role of stimulus-correlated activity in children's recognition of nonsense forms. *Journal of Experimental Child Psychology, 14,* 427–441.

Wolff, P. H. (1963). Observations on the early development of smiling. In B. M. Foss (Ed.), *Determinants of infant behavior* (Vol. 2, pp. 113–138). London: Methuen.

Wolff, R. J. (1981). Solitary and social play in wild *Mus musculus* (Mammalia). *Journal of Zoology, 195,* 405–412.

Wood-Gush, D. G., & Westergaard, K. (1991). The seeking of novelty and its relation to play. *Animal Behaviour, 42,* 599–606.

Woolley, J. D. (1997). Thinking about fantasy: Are children fundamentally different thinkers and believers from adults? *Child Development, 68,* 991–1011.

Yarrow, L. J., MacTurk, R. H., Vietze, P. M., McCarthy, M. E., Klein, R. P., & McQuiston, S. (1984). Developmental course of parental stimulation and its relationship to mastery motivation during infancy. *Developmental Psychology, 20,* 492–503.

Yarrow, L. J., McQuiston, S., MacTurk, R. H., McCarthy, M. E., Klein, R. P., & Vietze, P. M. (1983). Assessment of mastery motivation during the first year of life: Contemporaneous and cross-age relationships. *Developmental Psychology, 19,* 159–171.

Yarrow, L. J., Rubenstein, J. L., & Pedersen, F. A. (1975). *Infant and environment: Early cognitive and motivational development.* New York: Wiley.

Yarrow, L. J., Rubenstein, J. L., Pedersen, F. A., & Jankowski, J. J. (1972). Dimensions of early stimulation and their differential effects on infant development. *Merrill-Palmer Quarterly, 18,* 205–218.

Yarrow, M., & Waxler, C. (1976). Dimensions and correlates of prosocial behavior in young children. *Child Development, 47,* 118–125.

Yerkes, R. M., & Tomlin, M. I. (1935). Mother–infant relations in chimpanzees. *Journal of Comparative Psychology*, *20*, 321–359.

Yogman, M. W. (1981). Games fathers and mothers play with their infants. *Infant Mental Health Journal*, *2*, 241–248.

Young, G. H., Coelho, A. M., & Bramblett, C. A. (1982). The development of grooming, sociosexual behavior, play, and aggression in captive baboons in their first two years. *Primates*, *23*, 511–519.

Zahavi, A. (1977). The testing of a bond. *Animal Behaviour*, *25*, 246–247.

Zaporozhets, A. V. (1969). Some of the physiological problems of sensory training in early childhood and the preschool period. In M. Cole & I. Maltzman (Eds.), *A handbook of contemporary Soviet psychology* (pp. 86–120). New York: Basic.

Zelazo, P. R., & Kearsley, R. B. (1980). The emergence of functional play in infants: Evidence for a major cognitive transition. *Journal of Applied Developmental Psychology*, *1*, 95–117.

Zervigon-Hakes, A. (1984). Materials, mastery, and symbolic development in construction play: Stages of development. *Early Child Development and Care*, *17*, 37–48.

Zivin, G. (1977a). Facial gestures predict preschoolers' encounter outcomes. *Social Science Information*, *17*, 715–730.

Zivin, G. (1977b). On becoming subtle: Age and social rank changes in the use of a facial gesture. *Child Development*, *48*, 1314–1321.

Zucker, E. L., & Clarke, M. R. (1992). Developmental and comparative aspects of social play of mantled howling monkeys in Costa Rica. *Behaviour*, *123*, 144–170.

Zucker, E. L., Dennon, M. B., Puleo, S. G., & Maple, T. L. (1986). Play profiles of captive adult orangutans: A developmental perspective. *Developmental Psychobiology*, *19*, 315–326.

Zucker, E. L, Mitchell, G., & Maple, T. (1978). Adult-male offspring play inter-actions within a captive group of orangutans (*Pongo pygmaeus*). *Primates*, *19*, 379–384.

Author Index

A

B

C

D

E

F

G

I

J

K

L

N

O

P

Q

R

S

T

U

V

W

Y

Z

Subject Index

T

U

V

W